STATE-SELECTED AND STATE-TO-STATE
ION–MOLECULE REACTION DYNAMICS
Part 2. Theory

ADVANCES IN CHEMICAL PHYSICS

VOLUME LXXXII

STATE-SELECTED AND STATE-TO-STATE ION–MOLECULE REACTION DYNAMICS
Part 2. Theory

Edited by

MICHAEL BAER
Department of Physics and
Applied Mathematics
Soreq Nuclear Research Center
Yavne, Israel

CHEUK-YIU NG
Ames Laboratory
U.S. Department of Energy and
Department of Chemistry
Iowa State University
Ames, Iowa

ADVANCES IN CHEMICAL PHYSICS
VOLUME LXXXII

Series Editors

ILYA PRIGOGINE
University of Brussels
Brussels, Belgium
and
University of Texas
Austin, Texas

STUART A. RICE
Department of Chemistry
and
The James Frank Institute
University of Chicago
Chicago, Illinois

AN INTERSCIENCE® PUBLICATION
JOHN WILEY & SONS, INC.
NEW YORK • CHICHESTER • BRISBANE • TORONTO • SINGAPORE

An Interscience® Publication

Library of Congress Catalog Number: 58-9935

ISBN 0-471-53263-0

Printed in the United States of America

10 9 8 7 6 5 4 3 2 1

CONTRIBUTORS TO VOLUME LXXXII
Part 2

MICHAEL BAER, Department of Physics and Applied Mathematics, Soreq Nuclear Research Center, Yavne, Israel

S. CHAPMAN, Department of Chemistry, Barnard College, Columbia University, New York, NY

F. A. GIANTURCO, Department of Chemistry, The University of Rome, Città Universitaria, Rome, Italy

ERIC A. GISLASON, Department of Chemistry, University of Illinois at Chicago, Chicago, Illinois

BYRON H. LENGSFIELD III, Theoretical Atomic and Molecular Physics Group, Lawrence Livermore National Laboratory, Livermore CA

HIROKI NAKAMURA, Division of Theoretical Studies, Institute for Molecular Science, Myodaiji, Okazaki, Japan

GERARD PARLANT,* Department of Chemistry, The Johns Hopkins University, Baltimore, MD

F. SCHNEIDER, Central Institute for Physical Chemistry, Akademie der Wissenschaften der DDR, Berlin, Germany

V. SIDIS, Laboratoire des Collisions Atomiques et Moléculaires, Université de Paris-Sud, Orsay Cedex, France

MURIEL SIZUN, Laboratoire des Collisions Atomiques et Moléculaires,[†] Université de Paris-Sud, Orsay Cedex, France

JÜRGEN TROE, Institut für Physikalishe Chemie, Universität Göttingen, Göttingen, Germany

DAVID R. YARKONY, Department of Chemistry, The Johns Hopkins University, Baltimore, MD

*Permanent address: Physico-Chimie des Rayonnements (UA 75, Associé au CNRS) Bâtiment 350, Université de Paris-Sud, Orsay Cedex, France
†Unité associé au CNRS-No. 281

INTRODUCTION

Few of us can any longer keep up with the flood of scientific literature, even in specialized subfields. Any attempt to do more and be broadly educated with respect to a large domain of science has the appearance of tilting at windmills. Yet the synthesis of ideas drawn from different subjects into new, powerful, general concepts is as valuable as ever, and the desire to remain educated persists in all scientists. This series, *Advances in Chemical Physics*, is devoted to helping the reader obtain general information about a wide variety of topics in chemical physics, a field which we interpret very broadly. Our intent is to have experts present comprehensive analyses of subjects of interest and to encourage the expression of individual points of view. We hope that this approach to the presentation of an overview of a subject will both stimulate new research and serve as a personalized learing text for beginners in a field.

ILYA PRIGOGINE
STUART A. RICE

PREFACE

This volume is composed of eight chapters all devoted to the theory of charge-transfer processes during an atom (ion)–molecule (ion) collision. In three chapters are presented different attitudes for treating the potential energy surfaces (and the corresponding coupling terms), which govern the motion of the interacting atoms and ions, and in five chapters is treated the dynamics of these particles. The range of the topics on this subject covers the pure quantum-mechanical approach, various semiclassical approaches, and several statistical approaches. The order of the chapters are as follows.

The first chapter, by Lengsfield and Yarkony, deals with the adiabatic representation of the potential energy surfaces. Special emphasis is given to the importance and the derivation of the radial nonadiabatic coupling terms. These terms represent *what* must be determined in order to characterize an electronically nonadiabatic process. Equally significant is the efficient determination of *where* it is most important to determine these quantities. Nonadiabatic effects are most important in regions of nuclear coordinate space for which the potential energy surfaces in question are in close proximity or actually touch or cross. Thus, this chapter will also discuss efficient techniques for locating actual and avoided surface crossings.

The second chapter, by Sidis, deals with the diabatic representation of the potential energy surfaces. Here are discussed the problems associated with the strict formal definitions of diabatic states and with the adiabatic–diabatic transformations when applied for reduced electronic basis sets. Next is presented in great detail a method that yields from first principles potential energy surfaces and potential coupling terms such that the corresponding nonadiabatic coupling terms are relatively small. Results for several systems are discussed.

The third chapter by Gianturco and Schneider describes a semiempirical method for the calculations of several potential energy surfaces that are usually coupled together in the case of ion–molecule interactions. The method, which is based on the diatomics-in-molecules (DIM) wave expansion, is briefly reviewed and the physical meaning of the terms included in the expansion is discussed. The rigorous handling of the nonadiabatic coupling terms between surfaces for specific geometries of the relative orientation is reduced, within their treatment, to a more simplified approach that lends itself to rather rapid evaluation via the full DIM matrix. Examples are discussed for several simple ion (atom)–diatom systems for which either scattering experiments are available or dynamical computations have been attempted. It is shown in conclusions that the present model approach is one of the most useful and physically realistic pathways for treating ion–molecule collisions based on the general topology of the corresponding potential energy surfaces coupled with the dynamics.

The fourth chapter by Baer deals with the quantum-mechanical approach of low-energy charge-transfer processes between ions and neutrals. First is discussed the formal theory based on the Born–Oppenheimer treatment (with the emphasis on the adiabatic and the diabatic representations and the strict and approximate relations between them), which is then followed by deriving the Schrödinger equation for studying the dynamics. Special attention is given to those instances when exchange (chemical reaction) processes are competing with the charge-transfer process. Integral and differential state-to-state cross sections were calculated for several systems and compared with experiment.

The fifth chapter, by Nakamura, deals with the semiclassical approach to charge-transfer processes. Various transitions such as the Landau–Zener and the Rosen–Zener type of nonadiabatic transitions, transitions induced by Coriolis coupling, spin–orbit interaction, and the coupling due to the electron momentum transfer or the electron translation factor in charge transfer are discussed. It is shown that the semiclassical theory can be utilized to analyze all these transitions uniformly by introducing the new (dynamical-state) representation. Qualitative discussions are also presented for reactive transition or particle rearrangement by emphasizing the role of the potential ridge. Particular emphasis is given to the two-state case, which is then extended to a multistate system. Also, a simple semiclassical generalization of the trajectory-surface-hopping method is discussed.

The sixth chapter, by Gislason, Parlant, and Sizum, is devoted to classical path calculations of charge transfer for ion–molecule collisions. Inelastic processes that occur at the same time, such as fine-structure transitions and vibrational excitation, are also discussed. After a brief introduction the classical path technique is discussed in detail. In particular, the choice of basis sets, the coupled equations, and the classical trajectory are described, and the best numerical techniques to be used are summarized. The calculations of both total and differential cross sections are covered. Next are reviewed previous classical path calculations of total state-to-state cross sections and differential cross sections, and representative results are shown. Certain general features of charge-transfer collisions such as the Franck–Condon principle, the adiabatic vibronic potential energy surfaces, and the general energy dependence of charge-transfer cross sections are also discussed.

The seventh chapter, by Chapman, deals with the classical trajectory-surface-hopping method. Here is not only described in detail the method itself and various extensions to it but also a number of procedures for locating the position for the potential hopping. The main advantage of the trajectory-surface-hopping method is that it enables one to describe relatively simply and self-consistently the wide range of processes that may occur in molecular collisions, particularly in ion–molecule reactions. These include charge transfer, chemical reaction, collision-induced dissociation, and collision-induced predissociation. With the addition of some simplifying assumptions, it is possible to include electron-detachment channels as well. Results for several ion–molecule systems are described in some detail, including $(H + H_2)^+, (Ar + H_2)^+, (He + H_2)^+, (Ne + He_2)^+, (Cl + H_2)^+$. Electron jump reactions, including $M + X_2$ (M = alkali, X = halogen) and $M + O_2$ are also discussed.

The eighth chapter, by Troe, discusses the application of statistical rate theories to unimolecular and bimolecular ion–molecule reactions. Rovibrationally adiabatic channel potential curves and threshold energies are analyzed for various reactive systems. Consequently, the corresponding rate constants and cross sections for capture processes, bimolecular reactions involving predissociation of the collision complexes and unimolecular reactions are derived. Finally, also nonadiabatic corrections to these rate expressions are considered.

MICHAEL BAER
CHEUK-YIU NG

Ames Iowa
November 1991

CONTENTS

NONADIABATIC INTERACTIONS BETWEEN POTENTIAL ENERGY SURFACES: THEORY AND APPLICATIONS

BYRON H. LENGSFIELD III*

Theoretical Atomic and Molecular Physics Group
Lawrence Livermore National Laboratory
Livermore, CA

and

DAVID R. YARKONY†

Department of Chemistry
The Johns Hopkins University
Baltimore, MD

CONTENTS

*Work at Lawrence Livermore National Laboratory was performed under the auspices of the U.S. Department of Energy under Contract # W-7405-Eng-48 and by the Air Force Astronautics Laboratory under Contract #6761-01.
†Supported in part by NSF Grant CHE-87-23020 and AFOSR Grant 90-0051.

State-Selected and State-to-State Ion–Molecule Reaction Dynamics, Part 2: Theory, Edited by Michael Baer and Cheuk-Yiu Ng. Advances in Chemical Physics Series, Vol. LXXXII. ISBN 0-471-53263-0 © 1992 John Wiley & Sons, Inc.

I. INTRODUCTION

The introduction of analytic gradient and higher-derivative techniques has had an enormous impact on the progress of *ab initio* electronic-structure theory. Analytic derivative methods were first introduced as a practical computational tool for use with single configuration wavefunctions[1,2] in the late 1960s as an extension of the previous work of Bratoz[3] and Bishop.[4] Progress in both the theory and application of these techniques has been quite rapid, so that the use of analytic first and higher derivatives at various post-Hartree–Fock levels of theory is now routine.[5-13] Testimony to the importance of this area of research is given by the large number of reviews[14-20] of the field. In this chapter we discuss the theory and application of analytic gradient techniques to the study of electronically nonadiabatic processes. These are processes in which the description of the nuclear motion involves more than one Born–Oppenheimer potential energy surface and include such familiar areas as electronic quenching and charge-transfer reactions, predissociation and photofragmentation, and vibronic spectroscopy. Studies of electronic quenching are frequently motivated by practical considerations such as the design of gas-phase chemical lasers and the interpretation of laser-induced-fluorescence-excitation experiments, which probe intermediate species in combustion reactions. Molecular-photodissociation processes often proceed via excitation to a bound excited state, which is predissociated by nonadiabatic coupling with a (set of) dissociative states. Alternatively, this process may involve excitation to a set of nonadiabatically coupled dissociative states corresponding to distinct exit-channel arrangements. These situations arise in the photofragmentation of H_3^+, where the reactions

$$H_3^+ + hv \rightarrow H_2^+ + H$$

and

$$H_3^+ + hv \rightarrow H^+ + H + H$$

compete.[21] Vibronic absorption spectra involving states of mixed valence-Rydberg character are strongly influenced by nonadiabatic interactions. This is clearly seen in the recent analysis[22] of the $^1\Sigma_u^+$ and $^1\Pi_u$ states of N_2 and the kinetic energy release studies[23,24] of $d\,^1\Pi_g$ state of O_2. Systems exhibiting Jahn–Teller and Renner–Teller[25–27] intersections also require a knowledge nonadiabatic interactions for theoretical study. Theoretical and experimental studies of these phenomena have profitted from the work of Berry[28] and Simon[29] on the adiabatic evolution of the wavefunction in a region of degenerate surfaces.

Our description of these processes employs an adiabatic states approach.[30–33] In this approach it is necessary to determine both the adiabatic state potential energy surfaces $E^I(\mathbf{R})$ as well as the first[34–37]

$$f_\alpha^{IJ}(\mathbf{R}) \equiv \left\langle \Psi_I(\mathbf{r};\mathbf{R}) \left| \frac{\partial}{\partial R_\alpha} \right| \Psi_J(\mathbf{r};\mathbf{R}) \right\rangle_{\mathbf{r}}$$

and second[34,38]

$$h_{\alpha\beta}^{IJ}(\mathbf{R}) \equiv \left\langle \Psi_I(\mathbf{r};\mathbf{R}) \left| \frac{\partial^2}{\partial R_\alpha \partial R_\beta} \right| \Psi_J(\mathbf{r};\mathbf{R}) \right\rangle_{\mathbf{r}},$$

$$k_{\alpha\beta}^{IJ}(\mathbf{R}) \equiv \left\langle \frac{\partial}{\partial R_\alpha} \Psi_I(\mathbf{r};\mathbf{R}) \left| \frac{\partial}{\partial R_\beta} \Psi_J(\mathbf{r};\mathbf{R}) \right\rangle_{\mathbf{r}} \right.,$$

derivative couplings, which result in the breakdown of the single potential energy surface approximation. In this chapter it is shown how a series of recent advances[36–40] in electronic-structure theory have facilitated the computation of derivative coupling matrix elements in the case where the adiabatic electronic states are represented as large-scale multireference configuration-interaction (CI) wavefunctions.[41,42] The analytic derivative procedures that are discussed in this work represent a significant improvement over the finite-difference techniques[43–47] that had been introduced previously to determine these quantities. In Section II these analytic-gradient-based techniques for the determination of the first derivative (Sections II A and II B)[36,37] and second derivative (Section II C)[38,40] nonadiabatic couplings are reviewed. In Section II A a structural isomorphism between the energy gradient and the first-derivative nonadiabatic coupling matrix elements will be developed. This isomorphism is the key to the efficient evaluation of these quantities using analytic gradient methods.

Section II D discusses the additional efficiencies that can be achieved by incorporating body-fixed-frame symmetry into the evaluation of derivative couplings. In Section II F the interrelation between second-derivative couplings, $k_{\alpha\beta}^{IJ}(\mathbf{R})$, and the absence of *rigorous* diabatic bases in polyatomic systems is discussed.

Derivative couplings, nonadiabatic effects, are most important in regions of coordinate space for which two or more potential energy surfaces are in close proximity or actually touch or cross. Such regions of nuclear coordinate space will frequently be characterized by *actual or avoided-crossing seams* of the potential energy surfaces in question. A recently introduced methodology for locating actual/avoided crossings[48,49] will be discussed in Section II E. In that subsection the isomorphism established between the energy gradient and the first-derivative nonadiabatic coupling matrix elements will be extended to include the energy-difference gradient. The method that is developed for the efficient determination of the energy-difference gradient will be essential for the determination of actual/avoided surface crossings.

Section III considers applications of the methodology discussed in Section II. As discussed in Section II the first-derivative nonadiabatic coupling matrix elements $f_{\alpha}^{IJ}(\mathbf{R})$ are responsible for transitions between potential energy surfaces, while the second-derivative nonadiabatic coupling matrix elements $h_{\alpha\beta}^{IJ}(\mathbf{R})$ both couple and modify the potential energy surfaces. While the Born–Oppenheimer potential energy surfaces themselves are mass independent, the modifications resulting from $h_{\alpha\beta}^{II}(\mathbf{R})$, referred to as the adiabatic correction or the Born–Oppenheimer diagonal correction, are mass dependent. Several groups have been interested in the Born–Oppenheimer diagonal correction.[39,50–60] *Ab initio* studies include those of Bishop and Cheung[50–52] who considered H_2, HeH^+, and LiH using specialized CI wavefunctions[50–52] as well as SCF level[58–60] or SCF/limited-CI treatments[53–57] of more complex molecules. In this work the second-derivative methodology is used to discuss the adiabatic correction to the $X\,^1\Sigma^+$ state potential energy curve of LiH. This subtle mass-dependent feature of the $X\,^1\Sigma^+$ potential energy curve had been the object of some controversy, since experimental and theoretical determinations of this quantity had yielded qualitatively different results. We then turn to a discussion of the electronic structure aspects of the radiationless electronic quenching process

$$He + H_2(B\ ^1\Sigma_u^+) \rightarrow He + H_2(X\ ^1\Sigma_g^+).$$

This reaction is of particular interest since it is facilitated by *a seam of actual surface crossings* permitted by the multidimensional breakdown[61-64] of the von Neumann–Wigner[65] noncrossing rule. *Ab initio* characterizations of an actual crossing of states of the same symmetry are rare. The characterization of an actual crossing seam using multiconfiguration self-consistent field (MCSCF)[66-68]/CI wavefunctions discussed in this chapter is unique.[69] The only previous characterization of an actual crossing seam is that of Katriel and Davidson[63] who employed a Frost model[70] in approximate treatment of CH_4^+.

The methodology outlined in Section II E is also capable of determining *seams of avoided crossings*.[48,49] Seams of avoided crossings represents the parametric solution of the equations defining an avoided crossing in a space of reduced dimensionality. Since an actual crossing of two states of the same symmetry is permitted, but not required, by the noncrossing rule, an actual crossing seam may merge into an avoided crossing seam when the solution permitted by the noncrossing rule ceases to exist. Thus, it is appropriate to determine avoided-crossing seams in the same parameter space used to describe an actual crossing 'seam.' In this regard the fact that both classes of solutions can be obtained from the same set of equations is particularly convenient. Although not discussed in this presentation, it is relevant to note that the role of an avoided-crossing seam, unrelated to an actual surface crossing, in the nonadiabatic reactive quenching process

$$Na(^2P) + HCl \rightarrow NaCl + H(^2S)$$

has recently been studied[49,71] using both the *ab initio* techniques discussed in this work as well as the more empirical optimized[72] diatomics-in-molecules (DIM)[73] approach.

II. THEORY

In an electronically nonadiabatic process the rovibronic wavefunction can be expanded in a basis of Born–Oppenheimer electronic states, that is the total wavefunction for the system has the form[30,31]

$$\Psi_L^T(\mathbf{r}, \mathbf{R}) = \sum_{I,K} \Psi_I^e(\mathbf{r}; \mathbf{R})\chi_K^I(\mathbf{R})\beta_{KI}^L \tag{2.1a}$$

$$= \sum_I \Psi_I^e(\mathbf{r}; \mathbf{R})\chi^{I,L}(\mathbf{R}), \tag{2.1b}$$

where $\mathbf{R}$ denotes the nuclear degrees of freedom and $\mathbf{r}$ denotes the electronic degrees of freedom. Here $\Psi_I^e(\mathbf{r};\mathbf{R})$ is the electronic wavefunction and $\chi_K^I(\mathbf{R})$ is the Kth rovibronic wavefunction corresponding to the Ith electronic potential energy surface. Equation 2.1 is valid for any electronic wavefunction that depends parametrically on nuclear coordinates. As a practical matter it is necessary to make a particular choice of $\Psi_I^e(\mathbf{r};\mathbf{R})$ in order to limit the size of the expansion in electronic states. In this work $\Psi_I^e(\mathbf{r};\mathbf{R})$ will be taken to be an *adiabatic* electronic state and denoted $\Psi_I(\mathbf{r};\mathbf{R})$. Thus, the electronic wavefunctions are eigenfunctions of nonrelativistic Born–Oppenheimer electronic hamiltonian,

$$H^e(\mathbf{r};\mathbf{R})\Psi_I(\mathbf{r};\mathbf{R}) = E_I^e(\mathbf{R})\Psi_I(\mathbf{r};\mathbf{R}) \qquad (2.2)$$

and the total hamiltonian in the space-fixed coordinate frame is

$$H^T = \sum_\alpha \frac{-1}{2M_\alpha} \nabla_{R_\alpha}^2 + H^e \equiv T^N + H^e \qquad (2.3)$$

Inserting Eq. (2.1) into the time-independent Schrödinger $H^T\Psi^T = E\Psi^T$ gives the following system of coupled equations for the rovibronic functions $\chi^{I,L}(\mathbf{R})$:

$$[T^N + E_I^e(\mathbf{R}) - K^{II}(\mathbf{R}) - E]\chi^I(\mathbf{R}) = \sum_{J(\neq I)} \left[-H^{IJ}(\mathbf{R}) + \sum_\alpha \left(\frac{1}{M_\alpha} f_\alpha^{IJ}(\mathbf{R}) \frac{\partial}{\partial R_\alpha} \right) \right] \chi^J(\mathbf{R}), \qquad (2.4)$$

where the state label L on $\chi^{I,L}(\mathbf{R})$ has been suppressed and

$$K^{JI}(\mathbf{R}) = \sum_\alpha \frac{-1}{2M_\alpha} k_{\alpha\alpha}^{JI}(\mathbf{R}), \qquad (2.5a)$$

$$H^{JI}(\mathbf{R}) = \sum_\alpha \frac{-1}{2M_\alpha} h_{\alpha\alpha}^{JI}(\mathbf{R}), \qquad (2.5b)$$

$$f_\alpha^{JI}(\mathbf{R}) = \left\langle \Psi_J(\mathbf{r};\mathbf{R}) \left| \frac{\partial}{\partial R_\alpha} \Psi_I(\mathbf{r};\mathbf{R}) \right. \right\rangle_{\mathbf{r}}, \qquad (2.6)$$

$$h_{\alpha\beta}^{JI}(\mathbf{R}) = \left\langle \Psi_J(\mathbf{r};\mathbf{R}) \left| \frac{\partial^2}{\partial R_\alpha \partial R_\beta} \Psi_I(\mathbf{r};\mathbf{R}) \right. \right\rangle_{\mathbf{r}}, \qquad (2.7a)$$

$$k_{\alpha\beta}^{JI}(\mathbf{R}) = \left\langle \frac{\partial}{\partial R_\alpha} \Psi_J(\mathbf{r};\mathbf{R}) \left| \frac{\partial}{\partial R_\beta} \Psi_I(\mathbf{r};\mathbf{R}) \right. \right\rangle_{\mathbf{r}}. \qquad (2.7b)$$

and the subscript $\mathbf{r}$ on the matrix elements in Eqs. (2.5)–(2.7) denotes integration over all electronic coordinates. Thus, the basic matrix elements that couple adiabatic electronic states are $f_\alpha^{JI}(\mathbf{R})$ and $h_{\alpha\beta}^{JI}(\mathbf{R})$. The term $-K^{II}(\mathbf{R})$ is referred to as the adiabatic correction or the Born–Oppenheimer diagonal correction. From Eqs. (2.4) and (2.7) it is seen that $K^{II}(\mathbf{R})$, which is a linear combination of the $k_{\alpha\beta}^{II}(\mathbf{R})$, gives rise to *mass-dependent* corrections to the Born–Oppenheimer potential energy surface. Note that the $k_{\alpha\beta}^{JI}(\mathbf{R})$ and $h_{\alpha\beta}^{JI}(\mathbf{R})$ are related as follows:

$$\frac{\partial}{\partial R_\alpha} f_\beta^{JI}(\mathbf{R}) = k_{\alpha\beta}^{JI}(\mathbf{R}) + h_{\alpha\beta}^{JI}(\mathbf{R}). \tag{2.8}$$

This equation will provide the basis for a procedure discussed in Section II C for the evaluation of $h_{\alpha\beta}^{JI}(\mathbf{R})$ using only analytic first-derivative techniques. Note, too, that in the diatomic case Romelt[74] has discussed a reformulation of the dynamical problem [Eq. (2.4)] in which $k^{JI}(\mathbf{R})$ completely replaces $h^{JI}(\mathbf{R})$ in the coupled electronic state equations.

The adiabatic wavefunctions considered in this chapter are multireference configuration-interaction wavefunctions, $\Psi_I(\mathbf{r};\mathbf{R})$, which are given as finite (but large, on the order of 10^4–10^6 terms) expansion in terms of configuration-state functions (CSFs):[41]

$$\Psi_I(\mathbf{r};\mathbf{R}) = \sum_\lambda C_\lambda^I(\mathbf{R})\psi_\lambda(\mathbf{r};\mathbf{R}), \tag{2.9}$$

where the CSF, $\psi_\lambda(\mathbf{r};\mathbf{R})$, is a linear combination of Slater determinants with the appropriate spin and spatial symmetry. Inserting Eq. (2.9) into the electronic Schrödinger equation, Eq. (2.2), gives the matrix eigenvalue equation, the CI problem

$$\mathbf{H}C^I(\mathbf{R}) = E^I(\mathbf{R})C^I(\mathbf{R}). \tag{2.10}$$

Here the adiabatic electronic energy $E_I^e(\mathbf{R})$ in Eq. (2.2) determined from CI wavefunctions is denoted $E^I(\mathbf{R})$.

The molecular orbitals $\phi(\mathbf{r};\mathbf{R})$ used to build $\psi_\lambda(\mathbf{r};\mathbf{R})$ are in turn constructed from a linear combination of atomic orbitals $\chi(r;\mathbf{R})$:

$$\phi_i(r;\mathbf{R}) = \sum_p \tau_{pi}(\mathbf{R})\chi_p(r;\mathbf{R}). \tag{2.11}$$

Here and throughout this chapter the subscripts i, j, k, l, m and n will be used to denote molecular orbitals, and p, q, r, s and t will be used to label atomic orbitals.

To facilitate the evaluation of the nonadiabatic coupling matrix elements, we employ a common set of orthonormal molecular orbitals in the CSF expansion for each state. Since the use of a common set of orthonormal orbitals permits us to assure state orthogonality at the CI level, it is also key to the determination of the actual crossing seam between two states of the SAME symmetry discussed in Sections II E and III B. The common set of orthonormal molecular orbitals is determined from a quadratically convergent[45,67,75] state-averaged MCSCF[45,66,67,75] (SA-MCSCF) procedure. Although a discussion of the details of the SA-MCSCF procedure is deferred to later in the development, it is significant to note that in a well-designed calculation the quantities of interest should not be sensitive to the details of the state averaging scheme, or the reference (MCSCF) space, used in developing the CI wavefunctions. The weighting factors for the states [see Eq. (2.45)] and reference space used in the SA-MCSCF procedure should yield a multireference CI wavefunction of equivalent quality to that which would be obtained from a reliable multireference CI wavefunction based on orbitals optimized for an individual state. When it is not computationally tractable to verify or guarantee that this situation obtains, the viability of the wavefunction description will be considered by reference to the available experimental data.

A.   Evaluation of the First-Derivative Nonadiabatic Coupling Matrix Elements and Energy Gradients: A Unified Approach

Differentiation of the CI wavefunction defined in Eq. (2.9) gives

$$\frac{\partial}{\partial R_\alpha} \Psi_I(\mathbf{r}; \mathbf{R}) = \sum_\lambda \left[\left(\frac{\partial}{\partial R_\alpha} C_\lambda^I(\mathbf{R}) \right) \psi_\lambda(\mathbf{r}; \mathbf{R}) + C_\lambda^I(\mathbf{R}) \left(\frac{\partial}{\partial R_\alpha} \psi_\lambda(\mathbf{r}; \mathbf{R}) \right) \right]. \quad (2.12)$$

Thus, the first derivative nonadiabatic coupling matrix element consists of two terms

$$f_\alpha^{JI}(\mathbf{R}) = {}^{CI}f_\alpha^{JI}(\mathbf{R}) + {}^{CSF}f_\alpha^{JI}(\mathbf{R}) \quad (2.13)$$

where the CI contribution is given by

$${}^{CI}f_\alpha^{JI}(\mathbf{R}) = \sum_\lambda C_\lambda^J(\mathbf{R}) \left(\frac{\partial}{\partial R_\alpha} C_\lambda^I(\mathbf{R}) \right) \quad (2.14)$$

and the CSF contribution has the form

$${}^{CSF}f_\alpha^{JI}(\mathbf{R}) = \sum_{\lambda,\mu} C_\lambda^J(\mathbf{R}) \left\langle \psi_\lambda(\mathbf{r}; \mathbf{R}) \left| \frac{\partial}{\partial R_\alpha} \psi_\mu(\mathbf{r}; \mathbf{R}) \right. \right\rangle_\mathbf{r} C_\mu^I(\mathbf{R}). \quad (2.15)$$

From Eq. (2.14) it would appear that the derivative of the CI coefficients $\partial C^I(\mathbf{R})/\partial R_\alpha \equiv V_\alpha^I(\mathbf{R})$ would be required to evaluate the CI contribution to $f_\alpha^{JI}(\mathbf{R})$, $^{CI}f_\alpha^{JI}(\mathbf{R})$. However, this is in fact not the case since only the projection onto the state $\Psi_J(\mathbf{R})$ is actually required. Equation (2.14) for $^{CI}f_\alpha^{JI}(\mathbf{R})$ can be recast in a form similar to that of an energy gradient $E_\alpha^I(\mathbf{R}) \equiv \partial E^I(\mathbf{R})/\partial R_\alpha$. In the following discussion the first derivative of the CI energy with respect to a nuclear displacement R_α will be referred to as the CI gradient. The transformation of Eq. (2.14) to a form that obviates the explicit determination of $V_\alpha^I(\mathbf{R})$ is key to the efficient use of analytic gradient techniques in the evaluation of $f^{JI}(\mathbf{R})$ and provides the basis for the unified approach to the evaluation of the energy gradients, nonadiabatic coupling matrix elements, and energy difference gradients noted in the Introduction.

The manipulation of Eq. (2.14) begins with consideration of the coupled-perturbed CI (CP-CI) equations,[76,77] which provide a formal expression for the derivative of the CI coefficients $V_\alpha^I(\mathbf{R})$. This system of linear equations is obtained by differentiating Eq. (2.10) to give

$$\frac{\partial}{\partial R_\alpha}[\mathbf{H}C^I(\mathbf{R}) - E^I(\mathbf{R})\mathbf{C}^I(\mathbf{R})] = 0 \tag{2.16}$$

so that

$$[\mathbf{H} - E^I(\mathbf{R})]\frac{\partial}{\partial R_\alpha}\mathbf{C}^I(\mathbf{R}) = -\left[\frac{\partial}{\partial R_\alpha}\{\mathbf{H} - E^I(\mathbf{R})\}\right]\mathbf{C}^I(\mathbf{R}). \tag{2.17}$$

Taking the dot product of Eq. (2.17) with $\mathbf{C}^J(\mathbf{R})$ gives

$$^{CI}f_\alpha^{JI}(\mathbf{R}) \equiv \mathbf{C}^J(\mathbf{R})^\dagger \frac{\partial}{\partial R_\alpha}\mathbf{C}^I(\mathbf{R}) \tag{2.18}$$

$$= [E^J(\mathbf{R}) - E^I(\mathbf{R})]^{-1}\mathbf{C}^J(\mathbf{R})^\dagger \frac{\partial \mathbf{H}}{\partial R_\alpha}\mathbf{C}^I(\mathbf{R}). \tag{2.19}$$

Note that Eq. (2.19) is not the Hellmann–Feynman theorem[78,79] to which it bears a formal resemblance, since in Eq. (2.19) it is not the hamiltonian *operator* $H^e(\mathbf{r};\mathbf{R})$ but rather the hamiltonian *matrix* $\mathbf{H}$ that is being differentiated.

This expression will be shown to have the same form as a CI gradient but with transition density matrices replacing standard density matrices[80,81] in the appropriate expressions.[36,37] To see this note that the CI vectors, $\mathbf{C}^I(\mathbf{R})$, are constrained to be orthonormal at all geometries, that is,

$$\mathbf{C}^J(\mathbf{R})^\dagger \mathbf{C}^I(\mathbf{R}) - \delta_{IJ} = 0. \tag{2.20}$$

Taking the dot product of Eq. (2.17) with $\mathbf{C}^I(\mathbf{R})$ and using Eq. (2.20) yields

$$E_\alpha^I(\mathbf{R}) \equiv \frac{\partial E^I(\mathbf{R})}{\partial R_\alpha} = \mathbf{C}^I(\mathbf{R})^\dagger \frac{\partial \mathbf{H}}{\partial R_\alpha} \mathbf{C}^I(\mathbf{R}), \tag{2.21}$$

which has the same form as Eq. (2.19).

To demonstrate explicitly the connection between Eqs. (2.19) and (2.21) it is convenient to express the CI *energy* in terms of one- and two-electron integrals, CI coefficients, and coupling constants:

$$E^I(\mathbf{R}) \equiv \mathbf{C}^I(\mathbf{R})^\dagger \mathbf{H} \mathbf{C}^I(\mathbf{R}) \tag{2.22a}$$

$$= \sum_{\lambda,\mu} C_\lambda^I(\mathbf{R}) H_{\lambda\mu} C_\mu^I(\mathbf{R}) \tag{2.22b}$$

$$= \sum_{\lambda,\mu} \left[\sum_{i,j} C_\lambda^I h_{ij} \tilde{k}_{ij}^{\lambda\mu} C_\mu^I + \sum_{i,j,k,l} C_\lambda^I g_{ijkl} \tilde{K}_{ijkl}^{\lambda\mu} C_\mu^I \right] + V_N \tag{2.22c}$$

$$= \sum_{i,j} \sum_{\lambda,\mu} h_{ij} C_\lambda^I \tilde{k}_{ij}^{\lambda\mu} C_\mu^I + \sum_{i,j,k,l} \sum_{\lambda,\mu} g_{ijkl} C_\lambda^I \tilde{K}_{ijkl}^{\lambda\mu} C_\mu^I + V_N \tag{2.22d}$$

$$= \sum_{i,j} h_{ij} \gamma_{ij}^{II} + \sum_{i,j,k,l} g_{ijkl} \Gamma_{ijkl}^{II} + V_N \tag{2.22e}$$

where h_{ij} and g_{ijkl} are, respectively, the one-electron integrals and two-electron integrals in the molecular-orbital (MO) basis, V_N is the nuclear repulsion energy, $\tilde{k}_{ij}^{\lambda\mu}$ and $\tilde{K}_{ijkl}^{\lambda\mu}$ are coupling constants used to construct the hamiltonian matrix element $H_{\lambda\mu}(\mathbf{R}) = \langle \psi_\lambda(\mathbf{r};\mathbf{R}) | H^e \psi_\mu(\mathbf{r};\mathbf{R}) \rangle_\mathbf{r}$, from the MO integrals, and γ_{ij}^{II} and Γ_{ijkl}^{II} are one-electron and two-electron density matrices, respectively, again in the MO basis.

Thus, from Eqs (2.21) and (2.22), $E_\alpha^I(\mathbf{R})$ is given by

$$\frac{\partial E^I(\mathbf{R})}{\partial R_\alpha} = \sum_{i,j} \left(\frac{\partial}{\partial R_\alpha} h_{ij} \right) \gamma_{ij}^{II} + \sum_{i,j,k,l} \left(\frac{\partial}{\partial R_\alpha} g_{ijkl} \right) \Gamma_{ijkl}^{II} + \frac{\partial V_N}{\partial R_\alpha} \tag{2.23}$$

where $\partial V_N / \partial R_\alpha$ is the derivative of the nuclear repulsion energy. Similarly from eq. (2.19) the first derivative nonadiabatic coupling term is given by

$$^{CI}f_\alpha^{JI}(\mathbf{R}) = \Delta E_{IJ}(\mathbf{R})^{-1} \left\{ \sum_{i,j} \left(\frac{\partial}{\partial R_\alpha} h_{ij} \right) \gamma_{ij}^{JI} + \sum_{i,j,k,l} \left(\frac{\partial}{\partial R_\alpha} g_{ijkl} \right) \Gamma_{ijkl}^{JI} \right\}, \tag{2.24}$$

where $\Delta E_{JI}(\mathbf{R}) \equiv E^J(\mathbf{R}) - E^I(\mathbf{R})$ and *transition density* matrix elements γ_{ij}^{JI} and Γ_{ijkl}^{JI} occur in place density matrix elements, γ_{ij}^{II} and Γ_{ijkl}^{II} which appear in the energy gradient expression [Eq. (2.23)].

Since the molecular orbitals are linear combinations of atomic orbitals [see Eq. (2.11)], the derivatives of the molecular orbitals involve two terms:

$$\frac{\partial}{\partial R_\alpha}\phi_i(r;\mathbf{R}) = \sum_p \left[\left(\frac{\partial}{\partial R_\alpha}\tau_{pi}(\mathbf{R})\right)\chi_p(r;\mathbf{R}) + \tau_{pi}(\mathbf{R})\left(\frac{\partial}{\partial R_\alpha}\chi_p(r;\mathbf{R})\right)\right], \quad (2.25)$$

where the derivative of $\tau(\mathbf{R})$, the MO coefficients, is defined as follows

$$\frac{\partial}{\partial R_\alpha}\tau_{pi}(\mathbf{R}) = \sum_j \tau_{pj}(\mathbf{R})U^\alpha_{ji}(\mathbf{R}) \tag{2.26a}$$

$$\equiv \tau^\alpha_{pi}(\mathbf{R}). \tag{2.26b}$$

Thus the derivative of $\mathbf{h}$ and $\mathbf{g}$, the MO integrals, also involve two terms

$$\frac{\partial}{\partial R_\alpha}h_{ij}(\mathbf{R}) = h^\alpha_{ij}(\mathbf{R}) + h^{U^\alpha}_{ij}(\mathbf{R}) \tag{2.27a}$$

$$\frac{\partial}{\partial R_\alpha}g_{ijkl}(\mathbf{R}) = g^\alpha_{ijkl}(\mathbf{R}) + g^{U^\alpha}_{ijkl}(\mathbf{R}), \tag{2.27b}$$

where the superscript α on a quantity in the MO basis indicates that the quantity in question is constructed from the derivative of the atomic integrals (here h^α_{pq} and g^α_{pqrs}) so that we have

$$h_{ij} = \sum_{p,q} \tau_{pi}h_{pq}\tau_{qj}, \tag{2.28a}$$

$$g_{ijkl} = \sum_{p,q,r,s} \tau_{pi}\tau_{qj}g_{pqrs}\tau_{rk}\tau_{sl}, \tag{2.28b}$$

while the corresponding derivative quantities are given by

$$h^\alpha_{ij} = \sum_{p,q} \tau_{pi}h^\alpha_{pq}\tau_{qj} \tag{2.29a}$$

$$g^\alpha_{ijkl} = \sum_{p,q,r,s} \tau_{pi}\tau_{qj}g^\alpha_{pqrs}\tau_{rk}\tau_{sl}. \tag{2.29b}$$

The remaining terms, $h^{U^\alpha}_{ij}$ and $g^{U^\alpha}_{ijkl}$ are constructed from $\mathbf{U}^\alpha(\mathbf{R})$ in the following manner:

$$h^{U^\alpha}_{ij} = \sum_m [U^\alpha_{mi}h_{mj} + h_{im}U^\alpha_{mj}], \tag{2.30a}$$

$$g_{ijkl}^{U^x} = \sum_m [U_{mi}^\alpha g_{mjkl} + U_{mj}^\alpha g_{imkl} + g_{ijml}U_{mk}^\alpha + g_{ijkm}U_{ml}^\alpha]. \qquad (2.30b)$$

The CI gradient then can be expressed as follows,

$$\frac{\partial E^I}{\partial R_\alpha} = E_I^\alpha + \sum_{i,j} \frac{\partial E^I}{\partial U_{ij}} \frac{\partial U_{ij}}{\partial R_\alpha} + \frac{\partial V_N}{\partial R_\alpha} \qquad (2.31a)$$

$$= \sum_{\lambda,\mu} [C_\lambda^I H_{\lambda\mu}^\alpha C_\mu^I + C_\lambda^I H_{\lambda\mu}^{U^x} C_\mu^I] + \frac{\partial V_N}{\partial R_\alpha} \qquad (2.31b)$$

$$\equiv E_I^\alpha + E_I^{U^x} + \frac{\partial V_N}{\partial R_\alpha} \qquad (2.31c)$$

where

$$E_I^\alpha \equiv \sum_{p,q} \gamma_{pq}^{II} h_{pq}^\alpha + \sum_{p,q,r,s} \Gamma_{pqrs}^{II} g_{pqrs}^\alpha \qquad (2.32a)$$

$$E_I^{U^x} = \sum_{i,j} L_{ij}^{II} U_{ij}^\alpha. \qquad (2.32b)$$

In Eq. (2.32a) the density matrix elements have been transformed to the atomic-orbital (AO) basis in order to avoid transforming the AO derivative integrals to the MO basis for each degree of freedom. In Eq. (2.32b) the Lagrangian,[13] L_{ij}^{II}, has been introduced where

$$L_{mi}^{II} = \frac{\partial E^I}{\partial U_{mi}} \qquad (2.33a)$$

$$= 2\sum_j \gamma_{ij}^{II} h_{mj} + 4\sum_{j,k,l} \Gamma_{ijkl}^{II} g_{mjkl}. \qquad (2.33b)$$

In an analogous manner Eq. (2.24) gives the CI contribution to the first derivative nonadiabatic coupling matrix element as

$$^{CI}f_\alpha^{JI} = \Delta E_{JI}^{-1}[f_{JI}^\alpha + f_{JI}^{U^x}] \equiv \Delta E_{JI}^{-1}[^{CI}\tilde{f}_\alpha^{JI}(\mathbf{R})], \qquad (2.31c')$$

where

$$f_{JI}^\alpha = \sum_{p,q} \gamma_{pq}^{JI} h_{pq}^\alpha + \sum_{p,q,r,s} \Gamma_{pqrs}^{JI} g_{pqrs}^\alpha, \qquad (2.32a')$$

$$f_{JI}^{U^x} = \sum_{i,j} L_{ij}^{JI} U_{ij}^\alpha, \qquad (2.32b')$$

and the transition Lagrangian is given by

$$L_{mi}^{JI} = 2 \sum_j \gamma_{ij}^{JI} h_{mj} + 4 \sum_{j,k,l} \Gamma_{ijkl}^{JI} g_{mjkl}. \tag{2.33b'}$$

Equations (2.31c) and (2.31c') *provide the unified approach for the evaluation of CI gradients* (Eq. (2.31)] *and the CI contribution to the first-order nonadiabatic coupling elements* [Eq. (2.31c')].

The evaluation of Eqs. (2.32b) and (2.32b') can be further simplified by avoiding the explicit determination of $U^\alpha(R)$.[82-85] However, discussion of this procedure is deferred until after an expression for the CSF contribution to the first derivative nonadiabatic coupling matrix elements has been developed as the same approach can also be employed to reduce the computational effort needed to construct this term.

We now consider the CSF contribution to the first derivative nonadiabatic coupling matrix element. A CSF, ψ_λ, is an antisymmetric symmetry adapted sum of products of molecular orbitals:

$$\psi_\lambda(\mathbf{r}; \mathbf{R}) \equiv P^a \prod_i \phi_{\lambda_i}(\mathbf{r}_i; \mathbf{R}), \tag{2.34}$$

so that its derivative has the form

$$\frac{\partial}{\partial R_\alpha} \psi_\lambda(\mathbf{r}; \mathbf{R}) = \sum_j P^a \prod_{i(\neq j)} \phi_{\lambda_i}(r_i) \frac{\partial}{\partial R_\alpha} \phi_{\lambda_j}(r_j). \tag{2.35}$$

Thus, the overlap between a CSF and the derivative of a CSF can be represented as the matrix element of an one-electron operator,

$$\left\langle \psi_\lambda(\mathbf{r}; \mathbf{R}) \left| \frac{\partial}{\partial R_\alpha} \psi_\mu(\mathbf{r}; \mathbf{R}) \right\rangle_\mathbf{r} = \langle \psi_\lambda(\mathbf{r}; \mathbf{R}) | D_\alpha \psi_\mu(\mathbf{r}; \mathbf{R}) \rangle_r, \tag{2.36}$$

where

$$D_\alpha(\mathbf{r}) = \sum_i d_\alpha(r_i) \tag{2.37}$$

and

$$d_\alpha(r)\phi_k(r; \mathbf{R}) \equiv \frac{\partial}{\partial R_\alpha} \phi_k(r; \mathbf{R}). \tag{2.38}$$

The matrix element of a one-electron operator can be obtained as the trace of the one-electron density with the appropriate integrals, thus

$$^{\text{CSF}}f_\alpha^{JI}(\mathbf{R}) = \sum_{i,j} \gamma_{ij}^{JI} d_{ij}^\alpha, \tag{2.39}$$

where γ_{ij}^{JI} is the *square* one-electron transition density matrix and $\mathbf{d}^\alpha(\mathbf{R})$ is an antisymmetric matrix with matrix elements

$$d_{ij}^\alpha(\mathbf{R}) = \left\langle \phi_i(r; \mathbf{R}) \left| \frac{\partial}{\partial R_\alpha} \phi_j(r; \mathbf{R}) \right\rangle_r. \tag{2.40}$$

The antisymmetry of $\mathbf{d}^\alpha(\mathbf{R})$ is a consequence of the orthonormality of the molecular orbitals,

$$\langle \phi_i(r; \mathbf{R}) | \phi_j(r; \mathbf{R}) \rangle_r = \delta_{ij}. \tag{2.41}$$

Here the adjective "square" has been emphasized in reference to the one-particle transition density matrix. The one-particle transition density matrix is in general not symmetric, that is, the full or square matrix must be retained. However, in most electronic structure applications the associated one-electron integrals, for example $\mathbf{h}$, are symmetric, permitting the off-diagonal density matrix element to be stored in folded or triangular form. Since $\mathbf{d}^\alpha$ is not symmetric, it it necessary to construct and store the transition density matrix in its unfolded or square form.

From Eqs. (2.25) and (2.26) d_{ij}^α is comprised of two terms:

$$d_{ij}^\alpha(\mathbf{R}) = \sum_{p,q} \tau_{pi}(\mathbf{R}) \left\langle \chi_p(r; \mathbf{R}) \left| \frac{\partial}{\partial R_\alpha} \chi_q(r; \mathbf{R}) \right\rangle_r \tau_{qj}(\mathbf{R}) + \sum_m \langle \phi_i(r; \mathbf{R}) | \phi_m(r; \mathbf{R}) \rangle_r U_{mj}^\alpha \tag{2.42a}$$

$$\equiv \sigma_{ij}^\alpha(\mathbf{R}) + U_{ij}^\alpha(\mathbf{R}) \tag{2.42b}$$

so that

$$^{\text{CSF}}f_\alpha^{JI}(\mathbf{R}) = \sum_{i,j} \gamma_{ij}^{JI}(\sigma_{ij}^\alpha + U_{ij}^\alpha). \tag{2.43}$$

Thus, the CSF contribution to the first derivative nonadiabatic coupling matrix element requires only two new quantities, d_{ij}^α and the *square* γ_{ij}^{JI}, which do not appear in the formulas for a CI gradient.

Combining Eqs. (2.31c′) and (2.43) we obtain the following expression for the first-order nonadiabatic coupling matrix element:

$$f_\alpha^{JI}(\mathbf{R}) = \Delta E_{JI}^{-1} \left(\sum_{p,q} \gamma_{pq}^{JI} h_{pq}^\alpha + \sum_{p,q,r,s} \Gamma_{pqrs}^{JI} g_{pqrs}^\alpha \right) + \sum_{i,j} [(\Delta E_{JI}^{-1} L_{ij}^{JI} + \gamma_{ij}^{JI}) U_{ij}^\alpha + \gamma_{ij}^{JI} \sigma_{ij}^\alpha] \tag{2.44a}$$

$$\equiv \Delta E_{JI}^{-1}\left(\sum_{p,q} \gamma_{pq}^{JI} h_{pq}^{\alpha} + \sum_{p,q,r,s} \Gamma_{pqrs}^{JI} g_{pqrs}^{\alpha}\right) + f_{\alpha}^{JI,U}(\mathbf{R}) + \sum_{i,j} \gamma_{ij}^{JI}\sigma_{ij}^{\alpha}. \qquad (2.44b)$$

B. Molecular-Orbital Derivatives

We now turn to the evaluation of the U_{ij}^{α}, which appear in Eqs. (2.32) or (2.44). In order to implement the formalism of Section II A a common set of molecular orbitals must be employed in the description of each pair of the states under consideration. As noted earlier, in the present approach the common set of orbitals is obtained from a state-averaged MCSCF (SA-MCSCF) procedure. The SA-MCSCF energy is a weighted sum of the energy of the individual states,

$$E_{\text{SAMC}} = \sum_I W_I \mathbf{C}_{\text{MC}}^I(\mathbf{R})^\dagger \mathbf{H}(\mathbf{R})\mathbf{C}_{\text{MC}}^I(\mathbf{R}). \qquad (2.45)$$

The derivatives of the MO coefficients U_{ij}^{α} are obtained from the system of linear equations that results from differentiating the equations which define the parameters, orbitals, and CI coefficients, which minimize E_{SAMC}. These equations are known as the coupled-perturbed state-averaged MCSCF,[38] CP-SAMCSCF, equations. However, for the purposes of this discussion, it is convenient to consider the equations that arise for a single-state MCSCF procedure, the CP-MCSCF[13] equations, and defer our discussion of the CP-SAMCSCF equations to Appendix C.

1. Coupled-Perturbed MCSCF Equations

The conditions defining the MCSCF parameters, the MCSCF variational conditions, are as follows[45,67,86]:

$$\frac{\partial}{\partial \Delta_{ij}} E_{\text{MC}} \equiv GO_{ij} \qquad (2.46a)$$

$$= L_{ij}^{\text{MC}} - L_{ji}^{\text{MC}} \qquad (2.46b)$$

$$= 0, \qquad (2.46c)$$

$$\frac{\partial}{\partial C_{\text{MC},\lambda}^I} E_{\text{MC}} \equiv GC_{\lambda}^I \qquad (2.47a)$$

$$= -2[(\mathbf{H}_{\text{MC}} - E_{\text{MC}}^I)\mathbf{C}_{\text{MC}}^I]_\lambda \qquad (2.47b)$$

$$= 0, \qquad (2.47c)$$

where Δ_{ij} is a *unique* element of the antisymmetric matrix, $\tilde{\Delta}$, used to generate a unitary transformation, $e^{-\tilde{\Delta}}$ of the orthonormal molecular orbitals. Only the nonredundant elements of $\tilde{\Delta}$, that is, those elements corresponding to orbital mixings that can change the energy, are included in Eq. (2.46). A similar set of parameters ξ_{ij}, where $\tilde{\xi}$ is an antisymmetric matrix, may be defined to generate a unitary transformation of the CI eigenfunctions of the MCSCF hamiltonian. However, a more efficient MCSCF procedure is obtained if these CI variations are expressed directly in the CSF basis[86] and Eq. (2.47) reflects this choice of basis.

Before differentiating the MCSCF variational conditions to obtain the CP-MCSCF equations, it is necessary to consider the contributions to U_{ij}^{α} that arise from the molecular-orbital orthonormality constraints [Eq. (2.41)]. Differentiating this equation gives

$$\frac{\partial}{\partial R_{\alpha}}(\langle \phi_i(r; \mathbf{R})|\phi_j(r; \mathbf{R})\rangle_r - \delta_{ij}) = 0 \tag{2.48a}$$

$$= \frac{\partial}{\partial R_{\alpha}}\left(\sum_{p,q} \tau_{pi}(\mathbf{R})\langle \chi_p(r; \mathbf{R})|\chi_q(r; \mathbf{R})\rangle_r \tau_{qj}(\mathbf{R}) \right) \tag{2.48b}$$

$$\equiv \frac{\partial}{\partial R_{\alpha}} \sum_{p,q} [\tau_{pi}(\mathbf{R})S_{pq}(\mathbf{R})\tau_{qj}(\mathbf{R})], \tag{2.48c}$$

which yields

$$U_{ij}^{\alpha}(\mathbf{R}) + S_{ij}^{\alpha}(\mathbf{R}) + U_{ji}^{\alpha}(\mathbf{R}) = 0 \tag{2.49}$$

where

$$S_{ij}^{\alpha}(\mathbf{R}) = \sum_{p,q} \tau_{pi}(\mathbf{R})\left(\frac{\partial}{\partial R_{\alpha}}\langle \chi_p(r; \mathbf{R})|\chi_q(r; \mathbf{R})\rangle_r \right)\tau_{qj}(\mathbf{R}) \tag{2.50a}$$

$$\equiv \sum_{p,q} \tau_{pi}(\mathbf{R})S_{pq}^{\alpha}(\mathbf{R})\tau_{qj}(\mathbf{R}). \tag{2.50b}$$

The constraint equation [Eq. (2.49)] is incorporated by defining $\tilde{\Delta}_{ij}^{\alpha}$ and T_{ij}^{α} such that

$$U_{ij}^{\alpha}(\mathbf{R}) = \tilde{\Delta}_{ij}^{\alpha}(\mathbf{R}) + T_{ij}^{\alpha}(\mathbf{R}), \tag{2.51}$$

where $\tilde{\Delta}_{ij}^{\alpha}$ is an antisymmetric matrix whose elements will be determined by solving the CP-MCSCF equations and T_{ij}^{α} has the following structure:

$$T_{ij}^{\alpha} = -S_{ij}^{\alpha} \quad \text{for } i < j \tag{2.52a}$$

$$= -\tfrac{1}{2}S_{ii}^{\alpha} \quad \text{for } i = j \tag{2.52b}$$

$$= 0 \qquad \text{for } i > j. \tag{2.52c}$$

Equation (2.51) embodies the fact that, as the positions of the nuclei are changed, the orbitals evolve in response to the variational conditions and also in response to the molecular-orbital orthonormality constraints. The latter contribution arises because the overlap matrix in the atomic orbital basis, $S_{pq}(\mathbf{R})$, is a function of the nuclear coordinates.

Differentiating the MCSCF variational conditions gives

$$\frac{\partial}{\partial R_{\alpha}} GO_{ij} = GO_{ij}^{\alpha} + \sum_{k,l} \frac{\partial GO_{ij}}{\partial T_{kl}} \frac{\partial T_{kl}}{\partial R_{\alpha}} + \sum_{kl} \frac{\partial GO_{ij}}{\partial \Delta_{kl}} \frac{\partial \Delta_{kl}}{\partial R_{\alpha}} + \sum_{\lambda} \frac{\partial GO_{ij}}{\partial C_{MC,\lambda}^{I}} \frac{\partial C_{MC,\lambda}^{I}}{\partial R_{\alpha}} \tag{2.53a}$$

$$= GO_{ij}^{\alpha} + \sum_{k,l} \frac{\partial GO_{ij}}{\partial T_{kl}} T_{kl}^{\alpha} + \sum_{kl} \frac{\partial GO_{ij}}{\partial \Delta_{kl}} \Delta_{k_{l}}^{\alpha} + \sum_{\lambda} \frac{\partial GO_{ij}}{\partial C_{MC,\lambda}^{I}} C_{MC,\lambda}^{I\alpha} \tag{2.53b}$$

$$= GO_{ij}^{\alpha} + GO_{ij}^{T^{\alpha}} + \sum_{kl} \frac{\partial GO_{ij}}{\partial \Delta_{kl}} \Delta_{kl}^{\alpha} + \sum_{\lambda} \frac{\partial GO_{ij}}{\partial C_{MC,\lambda}^{I}} C_{MC,\lambda}^{I\alpha} \tag{2.53c}$$

$$= 0, \tag{2.53d}$$

$$\frac{\partial}{\partial R_{\alpha}} GC_{\lambda} = GC_{\lambda}^{\alpha} + \sum_{k,l} \frac{\partial GC_{\lambda}}{\partial T_{kl}} T_{kl}^{\alpha} + \sum_{kl} \frac{\partial GC_{\lambda}}{\partial \Delta_{kl}} \Delta_{kl}^{\alpha} + \sum_{\lambda} \frac{\partial GC_{\lambda}}{\partial C_{MC,\lambda}^{I}} C_{MC,\lambda}^{I\alpha} \tag{2.54a}$$

$$= GC_{\lambda}^{\alpha} + GC_{\lambda}^{T^{\alpha}} + \sum_{kl} \frac{\partial GC_{\lambda}}{\partial \Delta_{kl}} \Delta_{kl}^{\alpha} + \sum_{\lambda} \frac{\partial GC_{\lambda}}{\partial C_{MC,\lambda}^{I}} C_{MC,\lambda}^{I\alpha} \tag{2.54b}$$

$$= 0. \tag{2.54c}$$

In Eqs. (2.53)–(2.54) and throughout the remainder of this chapter the compound summation index kl denotes a sum that extends over the unique nonredundant elements of an antisymmetric matrix. This is to be distinguished from the summation over k, l, which denotes a (double) sum over all the elements in a matrix. In eqs. (2.53) and (2.54) the quantities GO_{ij}^{α} and GC_{λ}^{α} are defined as in Eqs. (2.46) and (2.47) except that the derivative integrals $\mathbf{h}^{\alpha}$ and $\mathbf{g}^{\alpha}$ replace the undifferentiated quantities $\mathbf{h}$ and $\mathbf{g}$ in their evaluation. Equations (2.53) and (2.54) are equivalent to the following system of linear equations, the CP-MCSCF equations:

$$
\begin{pmatrix}
\dfrac{\partial E_{MC}^2}{\partial \Delta^2} & \dfrac{\partial E_{MC}^2}{\partial \Delta \partial C} \\[2ex]
\dfrac{\partial E_{MC}^2}{\partial C \partial \Delta} & \dfrac{\partial E_{MC}^2}{\partial C^2}
\end{pmatrix}
\begin{pmatrix}
\Delta^\alpha \\[2ex]
C_{MC}^{I^\alpha}
\end{pmatrix}
= -
\begin{pmatrix}
\mathbf{GO}^\alpha + \mathbf{GO}^{T^\alpha} \\[2ex]
\mathbf{GC}^\alpha + \mathbf{GC}^{T^\alpha}
\end{pmatrix}
\tag{2.55a}
$$

where Eqs. (2.46) and (2.47) have been used to reexpress derivatives of gradients **GO** and **GC** as second derivatives of E_{MC}. This equation is abbreviated as follows

$$
\mathscr{H}\delta^\alpha = -(\mathbf{G}^\alpha + \mathbf{G}^{T^\alpha}).
\tag{2.55b}
$$

The left-hand side of Eqs. (2.55) is the same hessian that naturally arises in a fully quadratic MCSCF procedure. It is the need to solve Eq. (2.55a), or its equivalent discussed subsequently, that makes a fully quadratically convergent SA-MCSCF procedure not just a convenience but an essential component of the methodology discussed in this chapter.

Using Eq. (2.44) the $\mathbf{U}^\alpha(\mathbf{R})$ are seen to enter the expression for the first derivative nonadiabatic coupling matrix elements in the following manner:

$$
f_\alpha^{JI,U}(\mathbf{R}) = \sum_{i,j} (\Delta E_{JI}(\mathbf{R})^{-1} L_{ij}^{JI}(\mathbf{R}) + \gamma_{ij}^{JI}(\mathbf{R})) U_{ij}^\alpha(\mathbf{R})
\tag{2.56a}
$$

$$
\equiv \sum_{i,j} [\Delta E_{JI}(\mathbf{R})^{-1} L_{ij}^{JI}(\mathbf{R}) + \gamma_{ij}^{JI}(\mathbf{R})][\tilde{\Delta}_{ij}^\alpha(\mathbf{R}) + T_{ij}^\alpha(\mathbf{R})]
\tag{2.56b}
$$

$$
\equiv f_\alpha^{JI,\Delta}(\mathbf{R}) + f_\alpha^{JI,T}(\mathbf{R}).
\tag{2.56c}
$$

The contribution to this term from the derivative of the overlap matrix, T_{ij}^α, requires very little computational effort and will not be considered further. In evaluating $f_\alpha^{JI,\Delta}(\mathbf{R})$, the variational contribution, it would appear that the CP-MCSCF equations must be solved for each internal degree of freedom R_α, in order to obtain Δ_{ij}^α. However, the repeated solution of the CP-MCSCF equations can be avoided using the Z-vector method of Handy and Schaefer.[82] Using the antisymmetry of $\tilde{\Delta}_{ij}^\alpha$, $f_\alpha^{JI,\Delta}(\mathbf{R})$ can be rewritten as

$$
f_\alpha^{JI,\Delta}(\mathbf{R}) = \sum_{ij} \{\Delta E_{JI}^{-1}(L_{ij}^{JI} - L_{ji}^{JI}) + \gamma_{ij}^{JI} - \gamma_{ji}^{JI}\}\Delta_{ij}^\alpha
\tag{2.57a}
$$

$$
\equiv \sum_{ij} l_{ij}^{JI}\Delta_{ij}^\alpha,
\tag{2.57b}
$$

where the sum now extends only over the unique elements of Δ^α. This expression can be rewritten by formally inverting Eq. (2.55) to give

$$f_\alpha^{JI,\Delta}(\mathbf{R}) = -\begin{pmatrix} \mathbf{l}^{JI} \\ \\ \mathbf{0} \end{pmatrix}^\dagger \begin{pmatrix} \dfrac{\partial \mathbf{E}_{MC}^2}{\partial \Delta^2} & \dfrac{\partial \mathbf{E}_{MC}^2}{\partial \Delta \partial \mathbf{C}} \\ \\ \dfrac{\partial \mathbf{E}_{MC}^2}{\partial \mathbf{C} \partial \Delta} & \dfrac{\partial \mathbf{E}_{MC}^2}{\partial \mathbf{C}^2} \end{pmatrix}^{-1} \begin{pmatrix} \mathbf{GO}^\alpha + \mathbf{GO}^{T\alpha} \\ \\ \mathbf{GC}^\alpha + \mathbf{GC}^{T\alpha} \end{pmatrix} \tag{2.58a}$$

$$= -\mathbf{l}^{JI\dagger}\mathscr{H}^{-1}(\mathbf{G}^\alpha + \mathbf{G}^{T\alpha}). \tag{2.58b}$$

To evaluate Eq. (2.58b) it is sufficient to solve one equation of the form (2.55), that being

$$\mathscr{H}\mathbf{X}^{JI} = -\mathbf{l}^{JI} \tag{2.59}$$

so that Eq. (2.58b) reduces to

$$f_\alpha^{JI,\Delta}(\mathbf{R}) = \mathbf{X}^{JI\dagger}\mathbf{G}^\alpha + \mathbf{X}^{JI\dagger}\mathbf{G}^{T\alpha}. \tag{2.60}$$

Each of the contributions to $f_\alpha^{JI,\Delta}(\mathbf{R})$ in Eq. (2.60) can be represented as a sum of contributions from the orbital and CI portions of the gradients:

$$f_\alpha^{JI,\Delta}(\mathbf{R}) = \mathbf{X}_{orb}^{JI}{}^\dagger\mathbf{GO}^\alpha + \mathbf{X}_{orb}^{JI}{}^\dagger\mathbf{GO}^{T\alpha} + \mathbf{X}_{CI}^{JI\dagger}\mathbf{GC}^\alpha + \mathbf{X}_{CI}^{JI\dagger}\mathbf{GC}^{T\alpha}. \tag{2.61}$$

It is important to bear in mind that the term "CI contribution" used in the context of this discussion refers to the CI expansion employed in the MCSCF calculation.

The principal computational effort in the evaluation of Eq. (2.61) is now the construction of the gradients, $\mathbf{GO}^\alpha$ and $\mathbf{GC}^\alpha$, in the *molecular-orbital* basis. This requires a transformation of the derivative integrals from the atomic-orbital basis to the molecular-orbital basis for each nuclear degree of freedom R_α. However, as first noted by Rice and Amos,[83] this costly step can also be avoided since it is possible to evaluate the traces in Eq. (2.61) directly in the atomic-orbital basis. This point is developed in Appendix A. Furthermore the orbital and CI contributions that depend on $\mathbf{T}^\alpha(\mathbf{R})$ [the second and fourth terms in Eq. (2.61)] can be efficiently evaluated using techniques developed to compute MCSCF second derivatives. The efficient construction of these terms is also discussed in Appendix A. *With these computational economies the evaluation of an energy gradient or nonadiabatic coupling matrix element requires a fraction of the time needed to perform the multireference CI calculation.*

2. Additional Constraints on the Molecular Orbitals

In an MCSCF procedure there are generally three classes of molecular orbitals: (1) those that are doubly occupied in all CSFs, the core or inactive orbitals; (2) those that are not occupied in any of the CSFs, the virtual orbitals, and (3) those that are partially occupied in at least one of the CSFs, the active orbitals. The inactive or core orbitals as well as the virtual orbitals are not uniquely defined by the MCSCF variational conditions, since a rotation among the orbitals within either of these spaces does not change the MCSCF energy. Similarly, if a complete-active-space (CAS)[87-89] treatment is used in the MCSCF problem, the MCSCF energy is also invariant to rotations among the active orbitals. However, in many calculations core orbitals and core-correlating orbitals are excluded from the multireference CI calculations. In some cases, limitations on computer resources necessitate further truncation of the orbitals and, perhaps, some selection of reference configurations in the CI calculation.

In these instances the multireference CI energy (but not the MCSCF energy) is dependent on the definition of the individual orbitals. In this situation the orbitals in these spaces must be uniquely defined. This can be accomplished in the following manner. The core and virtual orbitals can be defined by diagonalizing a Fock operator in these spaces. It is convenient to choose this Fock operator to be the closed-shell Fock operator corresponding to the core orbitals in the MCSCF procedure. In this case the corresponding density matrix is geometry independent (in the molecular-orbital basis).[90] This choice of Fock operator is also particularly convenient for the truncation of virtual orbitals.[91] In the case of the active orbital subspace, a natural orbital[92] transformation can be used to define the unique set of active orbitals.

These additional rotations result in new antisymmetric contributions to U_{ij}^{α}. The lack of invariance of the CI energy (or the nonadiabatic coupling matrix element) to these rotations is manifest by a Lagrangian matrix (and transition density matrix) with nonsymmetric diagonal blocks corresponding to these subspaces. If the Lagrangian or transition density were symmetric in these subblocks, the additional antisymmetric contributions would not contribute to the traces in Eqs. (2.32b) or (2.44).

These new antisymmetric contributions to $\mathbf{U}^{\alpha}(\mathbf{R})$ are obtained by differentiating the equations used to uniquely define these orbital spaces. The approach for the determination of these antisymmetric contributions is similar whether one diagonalizes a Fock operator (for the core and virtual orbitals) or a density matrix (as in the case of the active orbital rotation) to uniquely define the orbitals in some subspace. The contributions to $\mathbf{U}^{\alpha}(\mathbf{R})$ from a core/virtual orbital space rotation is given in Appendix B, while the contribution from an active space orbital rotation is presented in Appendix D.

C. Second Derivative Nonadiabatic Coupling

In this subsection the determination of the second derivative nonadiabatic coupling matrix elements, $h_{\alpha\beta}^{JI}(\mathbf{R})$ and $k_{\alpha\beta}^{JI}(\mathbf{R})$, is considered. As noted previously these matrix elements, (1) provide mass-dependent modifications to Born–Oppenheimer potential energy surfaces, (2) couple potential energy surfaces, and (3) are related to the existence of rigorous diabatic bases[93] as discussed in Section II F. The second derivative matrix elements are related to the first derivative matrix elements by [recall Eq. (2.8)]:

$$\frac{\partial}{\partial R_\beta} f_\alpha^{JI}(\mathbf{R}) = \left\langle \frac{\partial}{\partial R_\beta} \Psi_J(\mathbf{r}; \mathbf{R}) \middle| \frac{\partial}{\partial R_\alpha} \Psi_I(\mathbf{r}; \mathbf{R}) \right\rangle_\mathbf{r} + \left\langle \Psi_J(\mathbf{r}; \mathbf{R}) \middle| \frac{\partial^2}{\partial R_\alpha \partial R_\beta} \Psi_I(\mathbf{r}; \mathbf{R}) \right\rangle_\mathbf{r}$$

$$\tag{2.62a}$$

$$\equiv k_{\beta\alpha}^{JI}(\mathbf{R}) + h_{\beta\alpha}^{JI}(\mathbf{R}). \tag{2.62b}$$

For $J = I$, this equation reduces to

$$k_{\beta\alpha}^{II}(\mathbf{R}) = -h_{\beta\alpha}^{II}(\mathbf{R}). \tag{2.63}$$

From Eqs. (2.5a) and (2.7b) these terms give rise to positive definite corrections to an individual Born–Oppenheimer potential energy surface.

As discussed below, $k_{\alpha\beta}^{JI}(\mathbf{R})$ can be evaluated using analytic gradient methods. Thus Eq. (2.62) provides a method for evaluating $h_{\alpha\beta}^{JI}(\mathbf{R})$ for $J \neq I$ using only analytic gradient techniques provided a divided difference is used to evaluate the left-hand side of Eq. (2.62). This approach obviates the need to develop the additional technology required to implement analytic second derivatives methods. We will discuss both finite difference (first derivative) and analytic (second derivative) methods for the evaluation of $h_{\alpha\beta}^{JI}(\mathbf{R})$ and $k_{\alpha\beta}^{JI}(\mathbf{R})$.

1. Evaluation of $k_{\alpha\beta}^{JI}$ (R) Using Analytic Gradient Techniques

The second derivative nonadiabatic coupling matrix element $k_{\alpha\beta}^{JI}(\mathbf{R})$ is a sum of three terms:

$$k_{\alpha\beta}^{JI}(\mathbf{R}) = \left\langle \frac{\partial}{\partial R_\alpha} \Psi_J(\mathbf{r}; \mathbf{R}) \middle| \frac{\partial}{\partial R_\beta} \Psi_I(\mathbf{r}; \mathbf{R}) \right\rangle_\mathbf{r} \tag{2.64a}$$

$$= \left\langle \sum_\mu \left[\left(\frac{\partial}{\partial R_\alpha} C_\mu^J(\mathbf{R}) \right) \psi_\mu(\mathbf{r}; \mathbf{R}) + C_\mu^J(\mathbf{R}) \frac{\partial}{\partial R_\alpha} \psi_\mu(\mathbf{r}; \mathbf{R}) \right] \right.$$

$$\left. \cdot \middle| \sum_\lambda \left[\left(\frac{\partial}{\partial R_\beta} C_\lambda^I(\mathbf{R}) \right) \psi_\lambda(\mathbf{r}; \mathbf{R}) + C_\lambda^I(\mathbf{R}) \frac{\partial}{\partial R_\beta} \psi_\lambda(\mathbf{r}; \mathbf{R}) \right] \right\rangle_\mathbf{r}$$

$$= {}^{CI}k_{\alpha\beta}^{JI}(\mathbf{R}) + {}^{CSF}k_{\alpha\beta}^{JI}(\mathbf{R}) + {}^{CSF-CI}k_{\alpha\beta}^{JI}(\mathbf{R}), \tag{2.64b}$$

where

$$^{CI}k_{\alpha\beta}^{JI}(\mathbf{R}) = \sum_\lambda \left(\frac{\partial}{\partial R_\alpha} C_\lambda^J(\mathbf{R}) \right) \left(\frac{\partial}{\partial R_\beta} C_\lambda^I(\mathbf{R}) \right), \tag{2.65a}$$

$$
{}^{\text{CSF}}k_{\alpha\beta}^{JI}(\mathbf{R}) = \sum_{\lambda,\mu} C_\lambda^J(\mathbf{R}) C_\mu^I(\mathbf{R}) \left\langle \left(\frac{\partial}{\partial R_\alpha} \psi_\lambda(\mathbf{r}; \mathbf{R}) \right) \middle| \left(\frac{\partial}{\partial R_\beta} \psi_\mu(\mathbf{r}; \mathbf{R}) \right) \right\rangle_r, \quad (2.65b)
$$

and

$$
{}^{\text{CI-CSF}}k_{\alpha\beta}^{JI}(\mathbf{R}) = \sum_{\lambda,\mu,i,j} \left[\left(\frac{\partial}{\partial R_\alpha} C_\lambda^J \right) C_\mu^I \tilde{k}_{ij}^{\lambda\mu} d_{ij}^\beta - C_\lambda^J \left(\frac{\partial}{\partial R_\beta} C_\mu^I \right) \tilde{k}_{ij}^{\lambda\mu} d_{ij}^\alpha \right]
$$

$$
\equiv \sum_{i,j} [\gamma_{ij}^{J'I} d_{ij}^\beta - \gamma_{ij}^{JI'} d_{ij}^\alpha]. \quad (2.65c)
$$

The minus sign in eq. (2.65c) arises from the previously noted antisymmetry of d_{ij}^α.

a. Evaluation of ${}^{\text{CSF}}k^{JI}(\mathbf{R})$. In evaluating the CSF contribution to $k_{\beta\alpha}^{JI}(\mathbf{R})$, ${}^{\text{CSF}}k^{JI}(\mathbf{R})$, the fact that two orbitals have been differentiated must be considered. This gives rise to a contribution from the square two-particle transition density matrix in addition to a contribution from the square one-particle transition density matrix. In particular,

$$
{}^{\text{CSF}}k_{\alpha\beta}^{JI}(\mathbf{R}) = \sum_{i,j} \gamma_{ij}^{JI} e_{ij}^{\alpha\beta} - \sum_{i,j,k,l} \Gamma_{ijkl}^{JI} f_{ijkl}^{\alpha\beta}, \quad (2.66)
$$

where

$$
e_{ij}^{\alpha\beta}(\mathbf{R}) = \left\langle \frac{\partial}{\partial R_\alpha} \phi_i(\mathbf{r}; \mathbf{R}) \middle| \frac{\partial}{\partial R_\beta} \phi_j(\mathbf{r}; \mathbf{R}) \right\rangle_r \quad (2.67a)
$$

$$
= \sum_k U_{ki}^\alpha U_{kj}^\beta + \sum_{p,q} \left[\tau_{pi} \left\langle \frac{\partial}{\partial R_\alpha} \chi_p(\mathbf{r}; \mathbf{R}) \middle| \frac{\partial}{\partial R_\beta} \chi_q(\mathbf{r}; \mathbf{R}) \right\rangle_r \tau_{qj} \right.
$$

$$
\left. + \tau_{pi} \left\langle \frac{\partial}{\partial R_\alpha} \chi_p(\mathbf{r}; \mathbf{R}) \middle| \chi_q(\mathbf{r}; \mathbf{R}) \right\rangle_r \tau_{qj}^\beta + \tau_{pi}^\alpha \left\langle \chi_p(\mathbf{r}; \mathbf{R}) \middle| \frac{\partial}{\partial R_\beta} \chi_q(\mathbf{r}; \mathbf{R}) \right\rangle_r \tau_{qj} \right],
$$

$$
(2.67b)
$$

where τ_{pi}^α has been defined in Eq. (2.26),

$$
f_{ijkl}^{\alpha\beta}(\mathbf{R}) = - \left\langle \frac{\partial}{\partial R_\alpha} \phi_i(\mathbf{r}; \mathbf{R}) \middle| \phi_j(\mathbf{r}; \mathbf{R}) \right\rangle_r \left\langle \chi_k(\mathbf{r}; \mathbf{R}) \middle| \frac{\partial}{\partial R_\beta} \phi_l(\mathbf{r}; \mathbf{R}) \right\rangle_r
$$

$$
- \left\langle \phi_i(\mathbf{r}; \mathbf{R}) \middle| \frac{\partial}{\partial R_\beta} \phi_j(\mathbf{r}; \mathbf{R}) \right\rangle_r \left\langle \frac{\partial}{\partial R_\alpha} \phi_k(\mathbf{r}; \mathbf{R}) \middle| \phi_l(\mathbf{r}; \mathbf{R}) \right\rangle_r \quad (2.68a)
$$

$$= \left\langle \phi_i(r; \mathbf{R}) \left| \frac{\partial}{\partial R_\alpha} \phi_j(r; \mathbf{R}) \right. \right\rangle_r \left\langle \phi_k(r; \mathbf{R}) \left| \frac{\partial}{\partial R_\beta} \phi_l(r; \mathbf{R}) \right. \right\rangle_r$$

$$+ \left\langle \phi_i(r; \mathbf{R}) \left| \frac{\partial}{\partial R_\beta} \phi_j(r; \mathbf{R}) \right. \right\rangle_r \left\langle \phi_k(r; \mathbf{R}) \left| \frac{\partial}{\partial R_\alpha} \phi_l(r; \mathbf{R}) \right. \right\rangle_r \tag{2.68b}$$

$$= d_{ij}^\alpha(\mathbf{R}) d_{kl}^\beta(\mathbf{R}) + d_{ij}^\beta(\mathbf{R}) d_{kl}^\alpha(\mathbf{R}) \tag{2.68c}$$

and where use has been made of the antisymmetry of d_{ij}^α. The quantity $f_{ijkl}^{\alpha\beta}$ will also arise in the expression for $h_{\alpha\beta}^{JI}(\mathbf{R})$. The fact that $f_{ijkl}^{\alpha\beta}$ is a product of one-electron integrals greatly facilitates the evaluation of $^{\mathrm{CSF}}k^{JI}(\mathbf{R})$. In this case all the d_{ij}^α needed to construct $f_{ijkl}^{\alpha\beta}$ can be held in memory so that the square two-particle density matrix need not be stored, rather the contributions to Eq. (2.66a) are evaluated directly. Equation (2.66a) requires the use of the "square" two-particle transition density matrix, since the symmetry of the $f_{ijkl}^{\alpha\beta}$ with respect to the permutation of indices

$$f_{ijkl}^{\alpha\beta} = -f_{jikl}^{\alpha\beta} = -f_{ijlk}^{\alpha\beta} = f_{jilk}^{\alpha\beta} \tag{2.69}$$

is different from that of the two electron integrals g_{ijkl}.

b. Evaluation of $^{\mathrm{CI}}k^{JI}(\mathbf{R})$ and $^{\mathrm{CSF-CI}}k^{JI}(\mathbf{R})$: The Coupled Perturbed-CI Equations. The evaluation of $^{\mathrm{CI}}k^{JI}(\mathbf{R})$ and $^{\mathrm{CSF-CI}}k^{JI}(\mathbf{R})$ requires the derivatives of the CI coefficients, which are obtained by solving the coupled-perturbed CI (CP-CI) equations. These equations were introduced previously Eq. (2.16) but not discussed in detail. We have

$$\frac{\partial}{\partial R_\alpha} [\mathbf{H} \mathbf{C}^I(\mathbf{R}) - E^I(\mathbf{R}) \mathbf{C}^I(\mathbf{R}) = 0, \tag{2.16}$$

which gives a system of linear equations for the requisite vector,

$$\frac{\partial}{\partial R_\alpha} \mathbf{C}^I \equiv \mathbf{V}_\alpha^I,$$

$$(\mathbf{H} - E^I(\mathbf{R})) \frac{\partial}{\partial R_\alpha} \mathbf{C}^I(\mathbf{R}) = - \left[\frac{\partial}{\partial R_\alpha} \{ \mathbf{H} - E^I(\mathbf{R}) \} \right] \mathbf{C}^I(\mathbf{R}) \tag{2.17a}$$

$$= - \left[\mathbf{H}^\alpha(\mathbf{R}) + \mathbf{H}^{U^z}(\mathbf{R}) + \frac{\partial}{\partial R_\alpha} E^I(\mathbf{R}) \right] \mathbf{C}^I(\mathbf{R}) \tag{2.17b}$$

$$= - \mathbf{P}[\mathbf{H}^\alpha(\mathbf{R}) + \mathbf{H}^{U^z}(\mathbf{R})] \mathbf{C}^I(\mathbf{R}) \tag{2.17c}$$

where $\mathbf{P}$ is the projection operator,

$$\mathbf{P} = 1 - \mathbf{C}^I\mathbf{C}^{I\dagger}, \tag{2.70}$$

and as discussed following Eq. (2.27) $\mathbf{H}^\alpha$ is the hamiltonian matrix constructed from the integrals in Eq. (2.29), while $\mathbf{H}^{U^\alpha}$ is the hamiltonian matrix constructed from the integrals in Eq. (2.30). As a consequence the solution of the CP-CI equations requires the prior solution of the CP-MCSCF equations in order to evaluate $\mathbf{U}^\alpha$. Given the solutions to the CP-CI equations, $\mathbf{V}^I_\alpha$, it is straightforward to use Eq. (2.65a) to construct $^{CI}k^{JI}(\mathbf{R})$. Similarly, the evaluation of $^{CSF-CI}k^{JI}(\mathbf{R})$ becomes analogous to the evaluation of $^{CSF}f^{JI}(\mathbf{R})$ [see Eq. (2.43)] with

$$^{CSF-CI}k^{JI}_{\alpha\beta}(\mathbf{R}) = {}^{CSF}f^{J^\alpha I}_\beta(\mathbf{R}) - {}^{CSF}f^{JI^\beta}_\alpha(\mathbf{R}), \tag{2.71}$$

where J^α in Eq. (2.71) implies that $\mathbf{V}^J_\alpha$ replaces $\mathbf{C}^J$ in the evaluation of the

transition density matrices in Eq. (2.43).

2. Analytic Evaluation of $\langle \phi_j | (\partial^2/\partial R_\alpha \partial R_\beta)\phi_i \rangle$

In order to evaluate $h^{JI}_{\alpha\beta}(\mathbf{R})$ analytically we begin with an expression for the second derivatives of the molecular orbitals, $\langle \phi_j | (\partial^2/\partial R_\alpha \partial R_\beta)\phi_i \rangle$. This is obtained as follows:

$$
\begin{aligned}
\frac{\partial^2}{\partial R_\alpha \partial R_\beta} &\phi_i(r; \mathbf{R}) \\
&= \sum_p \left[\tau^{\alpha\beta}_{pi}(\mathbf{R})\chi_p(r; \mathbf{R}) + \tau^\alpha_{pi}(\mathbf{R})\chi^\beta_p(r; \mathbf{R}) + \tau^\beta_{pi}(\mathbf{R})\chi^\alpha_p(r; \mathbf{R}) + \tau_{pi}(\mathbf{R})\chi^{\alpha\beta}_p(r; \mathbf{R}) \right] \\
&= \sum_k \phi_k(r; \mathbf{R})U^{\alpha\beta}_{ki}(\mathbf{R}) + \sum_{k,p} \left[\chi^\beta_p(r; \mathbf{R})\tau_{pk}(\mathbf{R})U^\alpha_{ki}(\mathbf{R}) + \chi^\alpha_p(r; \mathbf{R})\tau_{pk}(\mathbf{R})U^\beta_{ki}(\mathbf{R}) \right] \\
&\quad + \sum_p \chi^{\alpha\beta}_p(r; \mathbf{R})\tau_{pi}(\mathbf{R}) \\
&= \sum_k \phi_k(r; \mathbf{R})U^{\alpha\beta}_{ki}(\mathbf{R}) + \sum_p \left[\chi^\beta_p(r; \mathbf{R})\tau^\alpha_{pi}(\mathbf{R}) + \chi^\alpha_p(r; \mathbf{R})\tau^\beta_{pi}(\mathbf{R}) + \chi^{\alpha\beta}_p(r; \mathbf{R})\tau_{pi}(\mathbf{R}) \right] \\
&\equiv \phi^{\alpha\beta}_i(r; \mathbf{R}), \tag{2.72}
\end{aligned}
$$

where

$$\chi^\alpha_p(r; \mathbf{R}) \equiv \frac{\partial}{\partial R_\alpha} \chi_p(r; \mathbf{R})$$

and [recall (2.26)]

$$\frac{\partial \tau_{pi}}{\partial R_\alpha} = \sum_j \tau_{pj} U_{ji}^\alpha \equiv \tau_{pi}^\alpha \tag{2.73a}$$

$$\frac{\partial \tau_{pi}^\alpha}{\partial R_\beta} \equiv \tau_{pi}^{\beta\alpha} = \sum_j \tau_{pj} U_{ji}^{\alpha\beta}. \tag{2.73b}$$

Note that as a result of these definitions

$$\frac{\partial}{\partial R_\beta} \mathbf{U}^\alpha \neq \mathbf{U}^{\beta\alpha}.$$

Thus, from Eq. (2.72)

$$\left\langle \phi_j(r; \mathbf{R}) \left| \frac{\partial^2}{\partial R_\alpha \partial R_\beta} \right| \phi_i(r; \mathbf{R}) \right\rangle_r = U_{ji}^{\alpha\beta} + \sum_{p,q} [\tau_{pj}\sigma_{pq}^\beta \tau_{qi}^\alpha + \tau_{pj}\sigma_{pq}^\alpha \tau_{qi}^\beta + \tau_{pj}\sigma_{pq}^{\alpha\beta}\tau_{qi}] \tag{2.74a}$$

$$\equiv U_{ji}^{\alpha\beta}(\mathbf{R}) + \sigma_{ji}^{\alpha\beta}(\mathbf{R}) \tag{2.74b}$$

$$\equiv d_{ij}^{\alpha\beta}(\mathbf{R}) \tag{2.74c}$$

where

$$\sigma_{pq}^\beta(\mathbf{R}) = \left\langle \chi_p(r; \mathbf{R}) \left| \frac{\partial}{\partial R_\beta} \chi_q(r; \mathbf{R}) \right. \right\rangle_r, \tag{2.75a}$$

$$\sigma_{pq}^{\alpha\beta}(\mathbf{R}) = \left\langle \chi_p(r; \mathbf{R}) \left| \frac{\partial^2}{\partial R_\alpha \partial R_\beta} \chi_q(r; \mathbf{R}) \right. \right\rangle_r, \tag{2.75b}$$

In Eqs. (2.74) and (2.75) the second derivative overlap matrix element $d_{ij}^{\alpha\beta}$ has been cast into a form that is analogous to the derivative overlap matrix element d_{ij}^α defined in Eq. (2.42).

In order to obtain $\mathbf{U}^{\alpha\beta}(\mathbf{R})$, the second derivative CP-SAMCSCF equations must be solved. As in the case of the first derivative CP-SAMCSCF, the left-hand side of the second derivative CP-SAMCSCF equations is the full hessian of the SA-MCSCF problem. However, the right-hand side is hessian of the SA-MCSCF problem. However, the right-hand side is considerably more complicated. As in the case of the first derivative CP-SAMCSCF equations, it contains contributions from the variational conditions as well as from the orthogonality equations. In addition, it contains contributions from the solution of the first derivative CP-SAMCSCF equations. Some of the ideas involved in the second derivative CP-SAMCSCF equations are illustrated in the treatment of the second derivative orthogonality constraints, which follows. More details concerning the structure of these equations are found in Appendix E.

The second derivative of the orthonormality condition,

$$\frac{\partial^2}{\partial R_\alpha \partial R_\beta}(\langle \phi_i(r; \mathbf{R})|\phi_j(r; \mathbf{R})\rangle_r - \delta_{ij}) = \frac{\partial^2}{\partial R_\alpha \partial R_\beta} \sum_{p,q} [\tau_{pi}(\mathbf{R})S_{pq}(\mathbf{R})\tau_{qj}(\mathbf{R})]$$

$$= 0 \qquad (2.76a)$$

yields

$$U_{ji}^{\alpha\beta} + S_{ij}^{\alpha\beta} + U_{ij}^{\alpha\beta} = -\sum_k [U_{ki}^\alpha S_{kj}^\beta + U_{ki}^\beta S_{kj}^\alpha + S_{ik}^\alpha U_{kj}^\beta + S_{ik}^\beta U_{kj}^\alpha + U_{ki}^\alpha U_{kj}^\beta + U_{ki}^\beta U_{kj}^\alpha]$$

$$(2.76b)$$

In analogy with Eq. (2.49), Eq. (2.76b) is satisfied if $U_{ij}^{\alpha\beta}$ is defined as a sum of an antisymmetric matrix and an overlap dependent term,

$$U_{ij}^{\alpha\beta}(\mathbf{R}) = \tilde{\Delta}_{ij}^{\alpha\beta}(\mathbf{R}) + T_{ij}^{\alpha\beta}(\mathbf{R}), \qquad (2.77a)$$

where

$$\tilde{\Delta}_{ij}^{\alpha\beta}(\mathbf{R}) = -\tilde{\Delta}_{ji}^{\alpha\beta}(\mathbf{R}) \qquad (2.77b)$$

and

$$T_{ij}^{\alpha\beta} = -(S_{ij}^{\alpha\beta} + \sum_k [U_{ki}^\alpha S_{kj}^\beta + U_{ki}^\beta S_{kj}^\alpha + S_{ik}^\alpha U_{kj}^\beta + S_{ik}^\beta U_{kj}^\alpha + U_{ki}^\alpha U_{kj}^\beta + U_{ki}^\beta U_{kj}^\alpha]),$$

$$\text{for} \quad i < j, \quad (2.78a)$$

$$T_{ij}^{\alpha\beta} = -\tfrac{1}{2}S_{ii}^{\alpha\beta} - \sum_k [U_{ki}^\alpha S_{ki}^\beta + U_{ki}^\beta S_{ki}^\alpha + U_{ki}^\alpha U_{ki}^\beta + U_{ki}^\beta U_{ki}^\alpha], \quad \text{for } i = j \quad (2.78b)$$

$$T_{ij}^{\alpha\beta} = 0, \qquad\qquad\qquad\qquad\qquad \text{for } i > j. \quad (2.78c)$$

As noted previously the determination of $\Delta_{ij}^{\alpha\beta}$ from the second derivative CP-SAMCSCF equations parallels the determination of Δ_{ij}^α from the first derivative CP-SAMCSCF equations. The details of the treatment of the second-order CP-SAMCSCF equations can be found in Appendix E and Ref. 38. Note, however, that as in the evaluation of $\mathbf{f}^{JI}(\mathbf{R})$ the Z-vector method can be used to avoid solving the second order CP-SAMCSCF equations for $\Delta_{ij}^{\alpha\beta}$.

3. Analytic Evaluation of $h_{\alpha\beta}^{JI}(\mathbf{R})$

Given the formal expression for $d_{ij}^{\alpha\beta}$, it is possible to complete the description of the analytic evaluation of $h_{\alpha\beta}^{JI}(\mathbf{R})$ using only previously introduced concepts.

Since

$$\frac{\partial^2}{\partial R_\alpha \partial R_\beta} \Psi_I(\mathbf{r}; \mathbf{R}) = \sum_\lambda \left[\left(\frac{\partial^2}{\partial R_\alpha \partial R_\beta} C^I_\lambda(\mathbf{R}) \right) \psi_\lambda(\mathbf{r}; \mathbf{R}) + \left(\frac{\partial}{\partial R_\alpha} C^I_\lambda(\mathbf{R}) \right) \left(\frac{\partial}{\partial R_\beta} \psi_\lambda(\mathbf{r}; \mathbf{R}) \right) \right.$$

$$\left. + \left(\frac{\partial}{\partial R_\beta} C^I_\lambda(\mathbf{R}) \right) \left(\frac{\partial}{\partial R_\alpha} \psi_\lambda(\mathbf{r}; \mathbf{R}) \right) + C^I_\lambda \left(\frac{\partial^2}{\partial R_\alpha \partial R_\beta} \psi_\lambda(\mathbf{r}; \mathbf{R}) \right) \right],$$

(2.79)

we again find $h^{JI}_{\alpha\beta}(\mathbf{R})$ to be a sum of three terms:

$$h^{JI}_{\alpha\beta}(\mathbf{R}) = {}^{\mathrm{CI}}h^{JI}_{\alpha\beta}(\mathbf{R}) + {}^{\mathrm{CSF}}h^{JI}_{\alpha\beta}(\mathbf{R}) + {}^{\mathrm{CI-CSF}}h^{JI}_{\alpha\beta}(\mathbf{R}),$$

(2.80)

where

$$^{\mathrm{CI}}h^{JI}_{\alpha\beta}(\mathbf{R}) = \sum_\lambda C^J_\lambda(\mathbf{R}) \left(\frac{\partial^2}{\partial R_\alpha \partial R_\beta} C^I_\lambda(\mathbf{R}) \right),$$

(2.81a)

$$^{\mathrm{CSF}}h^{JI}_{\alpha\beta}(\mathbf{R}) = \sum_{\lambda,\mu} C^J_\lambda(\mathbf{R}) C^I_\mu(\mathbf{R}) \left\langle \psi_\lambda(\mathbf{r}; \mathbf{R}) \left| \left(\frac{\partial^2}{\partial R_\alpha \partial R_\beta} \psi_\mu(\mathbf{r}; \mathbf{R}) \right) \right\rangle_{\mathbf{r}} \right.,$$

(2.81b)

$$^{\mathrm{CI-CSF}}h^{JI}_{\alpha\beta}(\mathbf{R}) = \sum_{\lambda,\mu,i,j} \left[C^J_\lambda \left(\frac{\partial}{\partial R_\alpha} C^I_\mu \right) \tilde{k}^{\lambda\mu}_{ij} d^\beta_{ij} + C^J_\lambda \left(\frac{\partial}{\partial R_\beta} C^I_\mu \right) \tilde{k}^{\lambda\mu}_{ij} d^\alpha_{ij} \right],$$

(2.81c)

where $\tilde{\mathbf{k}}^{\lambda\mu}$ is the one particle coupling constant matrix introduced in Eq. (2.22).

a. Evaluation of $^{\mathrm{CI}}\mathbf{h}^{IJ}(\mathbf{R})$. From Eq. (2.81a) the evaluation of the $^{\mathrm{CI}}h^{JI}(\mathbf{R})$ formally involves the solution to the second-order coupled-perturbed CI equations, which are obtained from

$$\frac{\partial^2}{\partial R_\alpha \partial R_\beta} [\mathbf{H}\mathbf{C}^I(\mathbf{R}) - E^I(\mathbf{R})\mathbf{C}^I(\mathbf{R})] = 0.$$

(2.82)

However, just as $^{\mathrm{CI}}f^{JI}(\mathbf{R})$ [Eqs. (2.18) and (2.19)] can be obtained without solving the first-order CP-CI equations, $^{\mathrm{CI}}h^{JI}(\mathbf{R})$ can be obtained without solving the second-order CP-CI equations, Eq. (2.82). Multiplying the second-order CP-CI equations by $\mathbf{C}^J(\mathbf{R})^\dagger$ gives for ${}^{CI}h^{JI}(\mathbf{R})$:

$$
\begin{aligned}
{}^{CI}h^{JI}_{\alpha\beta}(\mathbf{R}) = (E^J - E^I)^{-1} \Bigg| & \mathbf{C}^J(\mathbf{R})^\dagger \frac{\partial^2 \mathbf{H}(\mathbf{R})}{\partial R_\alpha \partial R_\beta} \mathbf{C}^I(\mathbf{R}) \\
& - \left(\mathbf{C}^J(\mathbf{R})^\dagger \frac{\partial \mathbf{C}^I(\mathbf{R})}{\partial R_\alpha} \right) \frac{\partial E^I}{\partial R_\beta} - \left(\mathbf{C}^J(\mathbf{R})^\dagger \frac{\partial \mathbf{C}^I(\mathbf{R})}{\partial R_\beta} \right) \frac{\partial E^I}{\partial R_\alpha} \\
& + \left(\mathbf{C}^J(\mathbf{R})^\dagger \frac{\partial \mathbf{H}}{\partial R_\alpha} \frac{\partial \mathbf{C}^I(\mathbf{R})}{\partial R_\beta} \right) + \left(\mathbf{C}^J(\mathbf{R})^\dagger \frac{\partial \mathbf{H}}{\partial R_\beta} \frac{\partial \mathbf{C}^I(\mathbf{R})}{\partial R_\alpha} \right) \Bigg|
\end{aligned}
$$

$$\text{(2.83a)}$$

$$
= \Delta E_{JI}^{-1} \left\{ \mathbf{C}^{J\dagger} \frac{\partial^2 \mathbf{H}}{\partial R_\alpha \partial R_\beta} \mathbf{C}^I - {}^{CI}f^{JI}_\alpha E^I_\beta - {}^{CI}f^{JI}_\beta E^I_\alpha + {}^{CI}\tilde{f}^{JI^\beta}_\alpha + {}^{CI}\tilde{f}^{JI^\alpha}_\beta \right\}.
$$

$$\text{(2.83b)}$$

Evaluation of ${}^{CI}h^{JI}(\mathbf{R})$ is then analogous to evaluating a CI second derivative just as the evaluation of ${}^{CI}f^{JI}(\mathbf{R})$ is analogous to evaluating a CI first derivative. Note that in Eq. (2.83b)

$$
{}^{CI}\tilde{f}^{JI^\beta}_\alpha(\mathbf{R}) = \mathbf{C}^J(\mathbf{R})^\dagger \frac{\partial \mathbf{H}}{\partial R_\alpha} \frac{\partial \mathbf{C}^I(\mathbf{R})}{\partial R_\beta} = \mathbf{C}^J(\mathbf{R})^\dagger (\mathbf{H}^\alpha + \mathbf{H}^{U^\alpha}) \mathbf{V}^I_\beta \qquad \text{(2.84)}
$$

can now be evaluated in the MO basis, since $\partial \mathbf{H}/\partial R_\alpha$ is needed to solve the first-order CP-CI equations. From Eq. (2.83b) it is seen that the only new term that must be considered is

$$
\mathbf{C}^J(\mathbf{R})^\dagger \frac{\partial^2 \mathbf{H}(\mathbf{R})}{\partial R_\alpha \partial R_\beta} \mathbf{C}^I(\mathbf{R}).
$$

We note that

$$
\frac{\partial^2 \mathbf{H}}{\partial R_\alpha \partial R_\beta} = \mathbf{H}^{\alpha\beta} + \mathbf{H}^{\alpha, U^\beta} + \mathbf{H}^{\beta, U^\alpha} + \mathbf{H}^{U^\alpha, U^\beta} + \mathbf{H}^{U^{\alpha\beta}}, \qquad \text{(2.85)}
$$

where in analogy with eq. (2.31b) the superscripts denote the type of integral used in constructing the matrix elements. Thus, the contribution to Eq. (2.83) resulting from the first term in Eq. (2.85) has the form of eq. (2.32a') with the second derivative integrals $\mathbf{h}^{\alpha\beta}$ and $\mathbf{g}^{\alpha\beta}$ replacing the first derivative integrals $\mathbf{h}^\alpha$ and $\mathbf{g}^\alpha$. The contribution of the last term in Eq. (2.85) is of the form of Eq. (2.32b') with $U^{\alpha\beta}$ replacing U^α, that is,

$$
\mathbf{C}^{J\dagger} \mathbf{H}^{U^{\alpha\beta}} \mathbf{C}^I = \sum_{i,j} L^{JI}_{ij} U^{\alpha\beta}_{ij}. \qquad \text{(2.86)}
$$

The contribution from term two (and similarly term three) in Eq. (2.85) can be written as

$$\mathbf{C}^{J\dagger}\mathbf{H}^{\alpha,U^\beta}\mathbf{C}^I = \sum_{i,j} L_{ij}^{JI,\alpha}U_{ij}^\beta, \tag{2.87}$$

where $L_{ij}^{JI,\alpha}$ is the transition Lagrangian [Eq. (2.33b')] constructed with transition density matrices and first derivative integrals [Eqs. (2.30)]. The fourth term in Eq. (2.85) $\mathbf{H}^{U^\alpha,U^\beta}$ involves integrals of the form

$$h_{ij}^{U^\alpha U^\beta} = \sum_{m,n} [U_{mi}^\alpha h_{mn} U_{nj}^\beta + U_{mi}^\beta h_{mn} U_{nj}^\alpha], \tag{2.88a}$$

$$g_{ijkl}^{U^\alpha U^\beta} = \sum_{m,n} (U_{mi}^\alpha U_{nj}^\beta + U_{mi}^\beta U_{nj}^\alpha) g_{mnkl} + \cdots. \tag{2.88b}$$

This term can then be efficiently evaluated in terms of Lagrangians:

$$\mathbf{C}^{J\dagger}\mathbf{H}^{U^\alpha,U^\beta}\mathbf{C}^I = \sum_{i,j} L_{ij}^{JI,U^\alpha}U_{ij}^\beta - \sum_{i,j,m} L_{ij}^{JI}U_{im}^\beta U_{mj}^\alpha, \tag{2.89}$$

where L_{ij}^{JI,U^α} is a Lagrangian constructed from $\mathbf{h}$ and $\mathbf{g}$ integrals and $\mathbf{U}^\alpha$ transformed density matrices, for example, $\gamma^{JI,U^\alpha} \equiv \mathbf{U}^\alpha\gamma^{JI}$. This type of transformed transition density matrix is given explicitly in Appendix A.

b. *Evaluation of* $^{CSF-Cl}h^{JI}(\mathbf{R})$. The expression for $^{CSF-Cl}h^{JI}(\mathbf{R})$ is analogous to that found for $^{CSF-Cl}k^{JI}(\mathbf{R})$ [Eq. (2.65c)]. We find

$$^{CSF-Cl}h_{\alpha\beta}^{JI}(\mathbf{R}) = \sum_{i,j} [\gamma_{ij}^{JI^\alpha}d_{ij}^\beta + \gamma_{ij}^{JI^\beta}d_{ij}^\alpha] \tag{2.90a}$$

$$= {}^{CSF}f_\alpha^{J,I^\beta}(\mathbf{R}) + {}^{CSF}f_\beta^{J,I^\alpha}(\mathbf{R}). \tag{2.90b}$$

Equations (2.65c) and (2.90) are quite similar with the principal difference being that in the contribution to $^{CSF-Cl}h^{JI}(\mathbf{R})$, Eq. (2.90), only the first-order CP-CI solutions associated with state I are used to construct the one-particle transition density matrix and both contributions have the same phase.

c. *Evaluation of* $^{CSF}h^{JI}(\mathbf{R})$. From Eqs. (2.74) and (2.81b) $^{CSF}h^{JI}(\mathbf{R})$ is given by

$$^{CSF}h_{\alpha\beta}^{JI}(\mathbf{R}) = \sum_{i,j} \gamma_{ij}^{JI}d_{ij}^{\alpha\beta} + \sum_{i,j,k,l} \Gamma_{ijkl}^{JI}f_{ijkl}^{\alpha\beta} \tag{2.91a}$$

$$= \sum_{i,j} \gamma_{ij}^{JI}\tilde{\Delta}_{ij}^{\alpha\beta} + \sum_{i,j} \gamma_{ij}^{JI}(T_{ij}^{\alpha\beta} + \sigma_{ij}^{\alpha\beta}) + \sum_{i,j,k,l} \Gamma_{ijkl}^{JI}f_{ijkl}^{\alpha\beta} \tag{2.91b}$$

which is similar in form to Eq. (2.66) for $^{CSF}k^{JI}$. In addition it is clear from Eqs. (2.77) and (2.91b) that the Z-vector method can be used to avoid the solution of the second derivative CP-SAMCSCF equations for $\Delta_{ij}^{\alpha\beta}$ just as it was used in the determination of the first derivative coupling to avoid the explicit evaluation of Δ_i^α.

D. Body-Fixed-Frame Methods

In Sections II B and II C it was not necessary to specify the precise nature of R_α, the coordinates used to perform the differentiation. Thus, R_α could represent an internal nuclear motion, an overall nuclear rotation or translation, or an arbitrary atom-centered displacement. Because of the nature of the atomic-orbital basis functions, which are atom-centered cartesian gaussian functions, it is convenient to evaluate the derivative quantities initially with respect to atom-centered cartesian displacements. Derivatives with respect to these space-fixed coordinates are then transformed into the internal motion, nuclear rotation, nuclear translation coordinate system, which is referred to as the body-fixed[94-96] coordinate system and is more appropriate for dynamical treatments. In this subsection we consider the computational economies that can be achieved by *a priori* incorporation of the simplifications which are obtained in the body-fixed coordinate system.

Three coordinate systems are considered: (1) the space-fixed frame (SFF), with coordinates, $\mathbf{r}, \mathbf{R}$; (2) the center-of-mass fixed frame (CMFF), with coordinates, $\mathbf{q}; \mathbf{Q}, \mathbf{C}$; and (3) the body-fixed frame (BFF), with coordinates $\mathbf{w}, \mathbf{W}, \mathbf{B}, \mathbf{C}$. Here $\mathbf{r}, \mathbf{q}, \mathbf{w}$ denote the $3N^e$ coordinates of the N^e electrons in the SFF, CMFF, and BFF, respectively, $\mathbf{R}$ denotes the $3N$ SFF coordinates of the N nuclei, $\mathbf{Q}$ denotes $3N - 3$ nuclear coordinates exclusive of translation, $\mathbf{W}$ denotes $3N - 6$ (or $3N - 5$) internal coordinates, $\mathbf{C}$ denotes the coordinates of the center-of-mass of the *entire* system and $\mathbf{B}$ denotes the orientation of the molecule (body-fixed axis). The interrelation of these three coordinate systems is well known.[94-96] For diatomic systems, which are to be considered in this section and for which the present approach obtains its maximal advantage, we collect the relevant results below. The definition of the internal coordinate systems that will be needed are

$$\mathbf{C} = \left(\sum_i m_e \mathbf{r}^i + M_1 \mathbf{R}^1 + M_2 \mathbf{R}^2 \right) \Big/ M^T, \qquad (2.92)$$

$$\mathbf{q}^i = \mathbf{r}^i - \mathbf{O}, \qquad (2.93)$$

$$\mathbf{O} = (M_1 \mathbf{R}^1 + M_2 \mathbf{R}^2)/M, \qquad (2.94)$$

$$\mathbf{Q} = \mathbf{R}^2 - \mathbf{R}^1, \qquad (2.95)$$

$$
\begin{pmatrix} w^i_x \\ w^i_y \\ w^i_z \end{pmatrix} = \begin{pmatrix} -\sin\phi & \cos\phi & 0 \\ -\cos\theta\cos\phi & -\cos\theta\sin\phi & \sin\theta \\ \sin\theta\cos\phi & \sin\theta\sin\phi & \cos\theta \end{pmatrix} \begin{pmatrix} q^i_x \\ q^i_y \\ q^i_z \end{pmatrix}, \tag{2.96}
$$

$$
W = |\mathbf{R}^2 - \mathbf{R}^1|. \tag{2.97}
$$

The orientation of the molecule is defined by $\mathbf{B} = (\theta, \phi)$, where θ, ϕ are the spherical polar coordinates of $\mathbf{Q}$ and the choice of axis system in Eq. (2.96) follows Kronig.[94] Here $M^T = M_1 + M_2 + m_e N^e$ is the total system mass and $M = M_1 + M_2$ is the total nuclear mass. The total kinetic energy operator in the SFF, CMFF, and BFF is, respectively,

$$
\hat{T}^{\text{SFF}} = \hat{T}^r_e - \sum_k \frac{\Delta^R_k}{2M_k}, \tag{2.98}
$$

$$
\hat{T}^{\text{CMFF}} = \hat{T}^q_e + \hat{T}_C + \frac{\hat{\mathbf{P}}^q \cdot \hat{\mathbf{P}}^q}{2M} - \frac{1}{2\mu}\Delta^Q, \tag{2.99}
$$

$$
\hat{T}^{\text{BFF}} = \hat{T}^w_e + \hat{T}_C + \frac{\hat{\mathbf{P}}^w \cdot \hat{\mathbf{P}}^w}{2M} + \hat{T}_W + \hat{T}_{ROT}, \tag{2.100}
$$

where

$$
\hat{T}_W = -\frac{1}{2\mu}\frac{1}{W}\frac{\partial^2}{\partial W^2} W, \tag{2.101}
$$

$$
\hat{T}_{ROT} = -\frac{1}{2\mu W^2}\left[\frac{1}{\sin\theta}\left(\frac{\partial}{\partial\theta} - i\hat{L}^w_x\right)\sin\theta\left(\frac{\partial}{\partial\theta} - i\hat{L}^w_x\right) \right.
$$
$$
\left. + \frac{1}{\sin^2\theta}\left[\frac{\partial}{\partial\phi} - i(\cos\theta\hat{L}^w_z + \sin\theta\hat{L}^w_y)^2\right] \right], \tag{2.102}
$$

$$
\hat{\mathbf{P}}^q = -i\sum_j \nabla^q_j, \tag{2.103}
$$

$$
\hat{\mathbf{L}}^w = -i\sum_j w_j \times \nabla^w_j, \tag{2.104}
$$

$$
\hat{T}_C = \frac{-1}{2M^T}\Delta^C, \tag{2.105}
$$

$$
\hat{T}^r_e = \frac{-1}{2}\sum_j \Delta^r_j, \tag{2.106}
$$

with $\mu^{-1} = M_1^{-1} + M_2^{-1}$ and $\Delta \equiv \nabla \cdot \nabla$. In the preceding, atomic units have been used and when specifying an operator the following convention has been followed. A superscript variable name is used to indicate the variable name used in the formal operator, while the subscript index denotes the corresponding particle index, for example,

$$\nabla_j^q \equiv \left(\frac{\partial}{\partial q_x^j}, \frac{\partial}{\partial q_y^j}, \frac{\partial}{\partial q_z^j} \right).$$

The P^2 term in Eqs. (2.99) and (2.100) is referred to as the mass polarization term.

The electronic Schrödinger equation $(T_e + V - E^I(\mathbf{R}))\psi_I(\mathbf{r}; \mathbf{R}) = 0$ is similar in the three coordinate systems. The difference is in the dependence of Ψ_I on the nuclear coordinates; in the SFF $\Psi_I = \Psi_I(\mathbf{r}; \mathbf{R})$; in the CMFF $\Psi_I = \Psi_I(\mathbf{q}; \mathbf{Q})$, and in the BFF $\Psi_I = \Psi_I(\mathbf{w}; W)$. Thus, for a diatomic molecule the electronic wavefunction depends parametrically on six nuclear coordinates in the SFF, three nuclear coordinates in the CMFF, and only in the BFF does the electronic wavefunction depend parametrically on a single coordinate, the internuclear distance, W.

From these definitions several valuable equalities can be derived. From the inverse of Eqs. (2.92)–(2.95) and the chain rule we have

$$M^2\Delta^Q = M_2^2\Delta_1^R + M_1^2\Delta_2^R - 2M_1 M_2 \nabla_1^R \cdot \nabla_2^R \qquad (2.107)$$

in the space spanned by the Ψ_I. Then from Eqs. (2.98) and (2.99) the operator equivalence

$$\sum_{i,j} \nabla_i^q \cdot \nabla_j^q = \Delta_1^R + \Delta_2^R + 2\nabla_1^R \cdot \nabla_2^R \qquad (2.108)$$

obtains so that

$$MP^{JI}(\mathbf{Q}) = \sum_{j\alpha} k_{j\alpha,j\alpha}^{JI}(\mathbf{R}) + \sum_{\alpha} [k_{1\alpha,2\alpha}^{JI}(\mathbf{R}) + k_{2\alpha,1\alpha}^{JI}(\mathbf{R})], \qquad (2.109)$$

where $j = 1, 2$ and $\alpha = x, y, z$, the index $j\alpha$ refers to the SFF variable R_α^j, and

$$MP^{JI}(\mathbf{Q}) = -\left\langle \Psi_J \left| \sum_{i,j} \nabla_i^q \cdot \nabla_j^q \right| \Psi_I \right\rangle. \qquad (2.110)$$

In deriving Eq. (2.109), Eq. (2.8)

$$h_{\alpha\beta}^{JI}(\mathbf{R}) = \frac{\partial}{\partial R_\alpha} f_\beta^{JI}(\mathbf{R}) - k_{\alpha\beta}^{JI}(\mathbf{R})$$

and the antisymmetry of $f_\beta^{JI}(\mathbf{R}), f_\beta^{JI}(\mathbf{R}) = -f_\beta^{IJ}(\mathbf{R})$, which ·is valid for real-valued Ψ_I, have been used. Equation (2.109) provides a powerful diagnostic for the derivative methods discussed in this chapter. The evaluation of $k_{\alpha\beta}^{JI}(\mathbf{R})$ requires the complete derivative apparatus discussed in Section II C, including the solution of the CP-SAMCSCF equations, and the CP–CI equations. The matrix element $MP^{JI}(\mathbf{Q})$, on the other hand, is independent of the derivative apparatus. Thus, the verification of Eq. (2.109) provides a stringent test of the algorithms involved.

Another useful diagnostic relation can be obtained by comparing the CMFF and BFF expressions for the kinetic energy operator with Ψ_I restricted to be Σ states. In that subspace we have the operator equivalence

$$\frac{\hat{\mathbf{L}}^w \cdot \hat{\mathbf{L}}^w}{W^2} + \frac{1}{W}\frac{\partial^2}{\partial W^2}\, W = \Delta^Q, \tag{2.111}$$

so that

$$\left(\frac{M}{W}\right)^2 L_y^{2JI}(W) = \sum_j \{M_k^2 k_{jx,jx}^{JI}(\mathbf{Q})\} - M_1 M_2 [k_{1x,2x}^{JI}(\mathbf{Q}) + k_{2x,1x}^{JI}(\mathbf{Q})], \tag{2.112}$$

where $j = 1, 2$, and $k = 3 - j$,

$$L_y^{2JI}(W) = \langle \Psi_J(\mathbf{w}; W)|(\hat{L}_y^w)^2|\Psi_I(\mathbf{w}; W)\rangle_\mathbf{w} \tag{2.113}$$

and $\hat{L}^w$ is evaluated with the origin at the center-of-mass of the molecule [$\mathbf{O}$ in Eq. (2.94)]. As for Eq. (2.109) the value of Eq. (2.112) as a diagnostic results from its relating a quantity independent of nuclear displacement derivatives to quantities that depend explicitly on such derivatives.

The physical content of Eqs. (2.109) and (2.112) can be seen as follows. Denote the operator that generates a translation of all the nuclei along the space-fixed axes as $\hat{t}^N$ and denote as $\hat{p}_y^N$ the operator that generates a rotation of the nuclei about the body-fixed y axis located at the nuclear center of mass, that is,

$$\hat{t}^N = \sum_{k=1}^{2} \nabla_k^R$$

and

$$\hat{p}_y^N = \sum_{k=1}^{2} (R_z^k - O_z)\frac{\partial}{\partial R_z^k}.$$

Then Eqs. (2.109) and (2.113) can be reexpressed in terms of $\hat{t}^N$ and $\hat{p}_y^N$ as

$$MP^{JI}(\mathbf{Q}) = -\sum_\alpha \langle t_\alpha^N \Psi_J | t_\alpha^N \Psi_I \rangle, \tag{2.114}$$

$$L_y^{2^{JI}}(W) = -\langle p_y^N \Psi_J | p_y^N \Psi_I \rangle, \tag{2.115}$$

where $\alpha = x, y, z$. Thus, Eqs. (2.109) and (2.112) are seen to relate derivatives of Born–Oppenheimer electronic wavefunctions with respect to noninternal nuclear degrees of freedom to matrix elements of electronic operators.

In Section III we will require the Born–Oppenheimer diagonal correction for a $^1\Sigma^+$ state. From Eq. (2.100) this is given by

$$H^{II}(R) = \frac{MP^{II}(R)}{2M} + \frac{L^{2,II}(R)}{2\mu W^2} - \frac{K_R^{II}(R)}{2\mu}, \tag{2.116}$$

where R is the value of the internuclear distance and

$$K_R^{II}(R) = \left\{ \frac{M_2}{M_1} k_{1z,1z}^{II}(R) + \frac{M_1}{M_2} k_{2z,2z}^{II}(R) - 2k_{1z,2z}^{II}(R) \right\} \tag{2.117}$$

and

$$L^{2^{II}}(R) = L_x^{2^{II}}(R) + L_y^{2^{II}}(R) + L_z^{2^{II}}(R). \tag{2.118}$$

In Eq. (2.116) derivatives with respect to only two space-fixed coordinates, R_z^1 and R_z^2, are required. Each derivative preserves the original molecular symmetry axis. This should be compared with the direct evaluation of Eq. (2.4). There derivatives with respect to six nuclear coordinates are required of which four (two unique) break the original molecular symmetry axis. Thus, if Eq. (2.4) is used directly a lower point group (C_S), involving a larger, more costly, CSF space is required to characterize the CI wavefunction. This problem is exacerbated since the CPCI equations, the most costly step in the derivative procedure, must be solved for four degrees of freedom in this larger CSF space instead of the two degrees of freedom in the smaller CSF space required if Eq. (2.116) is used.

Similar economies obtain for triatomic systems. In this case of the nine nuclear displacements required to evaluate Eq. (2.4) directly in the SFF only the six that preserve the original plane of symmetry are required in the BFF treatment. Also, C_s symmetry can be used in the characterization of the Born–Oppenheimer electronic wavefunctions. These computational efficiencies should facilitate the characterization of electronically nonadiabatic interactions in triatomic systems. This capability complements the recent theoretical work on the vibronic states of X_3 systems in the presence of a Jahn–Teller intersection by Mead, Truhlar, and coworkers.[97–99] by providing the capability to determine the basic quantities required in their formalism.

Note that the economies indicated above are obtained without any additional algorithm development beyond that required to determine the requisite one-electron integrals, provided the algorithms required for the evaluation of nuclear wavefunctions derivatives are already available. The requisite matrix elements, $MP^{JI}(\mathbf{Q})$ defined in Eq. (2.110) and $L^{2^{JI}}(W)$ defined in Eq. (2.113) are each of the form

$$A^{JI}(\mathbf{R}) \equiv \langle \Psi_J(\mathbf{r}; \mathbf{R}) | \sum_i \hat{\mathbf{a}}_i \cdot \sum_j \hat{\mathbf{a}}_j | \Psi_I(\mathbf{r}; \mathbf{R}) \rangle_{\mathbf{r}}, \tag{2.119a}$$

$$= \langle \Psi_J(\mathbf{r}; \mathbf{R}) | \sum_i \hat{\mathbf{a}}_i^2 + \sum_{i \neq j} \hat{\mathbf{a}}_i \cdot \hat{\mathbf{a}}_j | \Psi_I(\mathbf{r}; \mathbf{R}) \rangle_{\mathbf{r}}, \tag{2.119b}$$

where $\hat{\mathbf{a}}$ is a vector of antihermetian operators. Thus,

$$A^{JI}(\mathbf{R}) = \sum_{i,j} \gamma_{ij}^{JI} \langle \phi_i | \hat{a}^2 | \phi_j \rangle_{\mathbf{r}} + 2 \sum_{i,j,k,l} [\Gamma_{ijkl}^{JI} \langle \phi_i | \hat{a} | \phi_j \rangle_{\mathbf{r}} \langle \phi_k | \hat{a} | \phi_l \rangle_{\mathbf{r}}], \tag{2.120}$$

where γ_{ij}^{JI} and Γ_{ijkl}^{JI} are the square one- and two-particle transition density matrices required for the evaluation of $f_\alpha^{JI}(\mathbf{R})$, $k_{\alpha\beta}^{JI}(\mathbf{R})$, and $h_{\alpha\beta}^{JI}(\mathbf{R})$. These density matrices are normalized such that

$$E^I = \sum_{i,j} \gamma_{ij}^{II} h_{ij} + \sum_{i,j,k,l} \Gamma_{ijkl}^{II} g_{ijkl}. \tag{2.121}$$

Finally note that in Eq. (2.116) the values of $K_R^{II}(W)$ and $MP^{II}(\mathbf{Q})$ are independent of nuclear masses, while the value of $L^{2^{II}}(W)$ is mass dependent through the definition of $\mathbf{O}$ [Eq. (2.94)]. Hence, $L^{2^{II}}(W)$ must be reevaluated for distinct mass combinations. This, however, is not a significant limitation since the computational effort required to evaluate $L^{2^{II}}(W)$ is minor compared to that required to evaluate $k^{II}(R)$.

E. Avoided and Actual Surface Crossings: Direct Evaluation of the Energy Difference Gradient

In Section II A it was shown that the energy gradient $E_\alpha^I(\mathbf{R})$ and the first derivative nonadiabatic coupling matrix element $f_\alpha^{JI}(\mathbf{R})$ could be evaluated using a unified density matrix driven procedure. The energy gradients facilitate the determination of extrema, in particular, minima and saddle points, on individual potential energy surfaces. The $f_\alpha^{JI}(\mathbf{R})$, on the other hand, are an intersurface property. They are most relevant, that is, electronically nonadiabatic processes are most likely to be important, in regions of nuclear coordinate space for which two (or more) potential energy surfaces are in close proximity or actually touch or cross. Such regions correspond not, in general, to an extremum on a single potential energy surface, but rather to an extremum, a minimum, in $\Delta E_{IJ}(\mathbf{R})^2$, the square of the separation of two potential energy surfaces.

In this subsection it is shown how the unified density matrix driven procedure of Section II A can be extended to directly evaluate the energy difference gradient $\partial \Delta E_{IJ}(\mathbf{R})/\partial R_\alpha$. This algorithm will form the basis for a Newton–Raphson-based procedure for determining allowed/avoided surface crossings.

1. The Energy Difference Gradient

Avoided/actual surface crossings represent solutions of the equation

$$\frac{\partial}{\partial R_\alpha} \Delta E_{IJ}(\mathbf{R})^2 = 2g_\alpha^{IJ}(\mathbf{R})\Delta E_{IJ}(\mathbf{R}) \equiv G_\alpha^{IJ}(\mathbf{R}) = 0, \quad \text{for all } \alpha, \quad (2.122)$$

with $\Delta E_{IJ}(\mathbf{R}) \neq 0$ corresponding to an avoided surface crossing and $\Delta E_{IJ}(\mathbf{R}) = 0$ corresponding to an actual surface crossing. The condition $\Delta E_{IJ}(\mathbf{R}) = 0$ has also been discussed by Koga and Morokuma[100] in relation to the minimum energy crossing problem. Here, $g_\alpha^{IJ}(\mathbf{R}) \equiv E_\alpha^I(\mathbf{R}) - E_\alpha^J(\mathbf{R})$ and the energy gradient $E_\alpha^J(\mathbf{R})$ is defined by $E_\alpha^I(\mathbf{R}) \equiv \partial E^I(\mathbf{R})/\partial R_\alpha$. Thus, $g_\alpha^{IJ}(\mathbf{R})$ is seen to represent the difference between the slopes of the potential energy surfaces. It can therefore be determined from two independent evaluations of the energy gradient. However, the requisite computational effort will be reduced considerably if $g_\alpha^{IJ}(\mathbf{R})$ can be evaluated directly. This can be accomplished using the one- and two-particle *difference density matrices* $\Delta \gamma^{IJ}$ and $\Delta \Gamma^{IJ}$ defined in terms of the standard[80,81] one- and two-particle density matrices density matrices γ^{II} and Γ^{II} by $\Delta \gamma^{IJ} = \gamma^{II} - \gamma^{JJ}$ and $\Delta \Gamma^{IJ} = \Gamma^{II} - \Gamma^{JJ}$. Key to achieving this reduction is the observation that evaluation of $g_\alpha^{IJ}(\mathbf{R})$ is formally identical to the evaluation of $^{CI}\tilde{f}_\alpha^{IJ}(\mathbf{R})$ provided the difference density matrices replace transition density matrices in Eqs. (2.32') and (2.33b'). In particular we have

$$g_\alpha^{IJ}(\mathbf{R}) = \Delta E_{IJ}^\alpha(\mathbf{R}) + \Delta E_{IJ}^{U\alpha}(\mathbf{R}), \quad (2.31c'')$$

where

$$\Delta E_{IJ}^\alpha = \sum_{p,q} \Delta \gamma_{pq}^{IJ} h_{pq}^\alpha + \sum_{p,q,r,s} \Delta \Gamma_{pqrs}^{IJ} g_{pqrs}^\alpha \quad (2.32a'')$$

and

$$\Delta E_{IJ}^{U\alpha} = \sum_{i,j} \Delta L_{ij}^{IJ} U_{ij}^\alpha. \quad (2.32b'')$$

Here the difference Lagrangian[13] ΔL_{im}^{IJ} is given by

$$\Delta L_{mi}^{IJ} = 2 \sum_j \Delta \gamma_{ij}^{IJ} h_{mj} + 4 \sum_{j,k,l} \Delta \Gamma_{ijkl}^{IJ} g_{mjkl}. \quad (2.33b'')$$

As noted in Section II A if the standard one- and two- particle density matrices γ^{II} and $\mathbf{\Gamma}^{II}$ replace the difference density matrices in the equations for the energy difference gradient [Eqs. (2.31c″), (2.32″), and (2.33b″)], then these equations [Eqs. (2.31c), (2.32), and (2.33b)] yield the energy gradient $E_\alpha^I(\mathbf{R})$. Thus Eqs. (2.31c), (2.32), and (2.33b) provide a unified approach for the evaluation of *three* classes of derivatives, energy gradients (using standard density matrices), energy difference gradients (using difference density matrices), and (the CI contribution to) first derivative nonadiabatic coupling matrix elements (using transition density matrices).

2. Locating Actual/Avoided Crossings of Potential Energy Surfaces

The solution of Eq. (2.122) can be accomplished using a Newton–Raphson procedure

$$\mathbf{F}^{IJ}(\mathbf{R}_0)\boldsymbol{\delta}(\mathbf{R}_0) = -\,\mathbf{G}^{IJ}(\mathbf{R}_0), \tag{2.123}$$

where $\boldsymbol{\delta}(\mathbf{R}_0) = \mathbf{R} - \mathbf{R}_0$ and $\mathbf{F}^{IJ}(\mathbf{R}_0)$ is the second derivative or hessian matrix given by

$$F_{\alpha\beta}^{IJ}(\mathbf{R}_0) = \frac{\partial}{\partial R_\alpha}\,G_\beta^{IJ}(\mathbf{R}_0) \tag{2.124a}$$

$$= g_\alpha^{IJ}(\mathbf{R}_0)g_\beta^{IJ}(\mathbf{R}_0) + \Delta E_{IJ}(\mathbf{R}_0)\frac{\partial}{\partial R_\alpha}\,g_\beta^{IJ}(\mathbf{R}_0). \tag{2.124b}$$

The partial derivative in Eq. (2.124b) can be evaluated using forward or centered divided differences

$$\frac{\partial}{\partial R_\alpha}\,g_\beta^{IJ}(\mathbf{R}_0) \cong [g_\beta^{IJ}(\mathbf{R}_0 + \varepsilon\mathbf{I}^\alpha) - g_\beta^{IJ}(\mathbf{R}_0)]/\varepsilon \tag{2.125a}$$

or

$$\frac{\partial}{\partial R_\alpha}\,g_\beta^{IJ}(\mathbf{R}_0) \cong [g_\beta^{IJ}(\mathbf{R}_0 + \varepsilon\mathbf{I}^\alpha) - g_\beta^{IJ}(\mathbf{R}_0 - \varepsilon\mathbf{I}^\alpha)]/2\varepsilon. \tag{2.125b}$$

Since Eq. (2.123) is formally equivalent to that for solving $E_\alpha^I(\mathbf{R}) = 0$ using the identification $G_\alpha^{IJ}(\mathbf{R})\leftrightarrow E_\alpha^I(\mathbf{R})$, existing surface walking procedures[101] designed to locate extrema on a potential energy surface can be used to locate extrema on the energy difference hypersurface $\Delta E_{IJ}(\mathbf{R})^2$. However, searches on the energy difference hypersurface are significantly more complicated than the analogous searches on the energy hypersurface [$E^I(\mathbf{R})$]. While extrema on a potential energy surface occur at isolated points, minima on the energy difference hypersurface need not be isolated. According to the multidimensional extension[61-64] of the von Neumann–Wigner noncrossing rule[65] for a system with M internal degrees of freedom, crossings of potential energy surfaces of the same overall symmetry are permitted, but not guaranteed, to occur on hypersurfaces of dimension $M - 2$. Thus, for triatomic or larger systems for which $M - 2 > 0$, solutions to Eq. (2.123) for which $\Delta E_{IJ}(\mathbf{R}) = 0$ and/or $\Delta E_{IJ}(\mathbf{R}) \neq 0$ may exist.

Solutions to Eq. (2.122) may be obtained in spaces of reduced dimensionality by restricting the number of degrees of freedom included in the Newton–Raphson search algorithm. Although straightforward, the reduced dimensionality procedure is quite important. Nonadiabatic interactions are frequently highly localized. In triatomic and larger polyatomic systems, a systematic means for locating regions of large nonadiabatic effects is desirable. In triatomic systems, by eliminating from the Newton–Raphson geometry optimization procedure the degree of freedom corresponding to an approximate reaction coordinate (ξ) one obtains a *seam of actual/avoided crossings* parametrized by ξ. It is in the vicinity of this seam that one expects nonadiabatic effects to be preeminent. Finally recall that the actual crossing of two states of the same symmetry is permitted, not required, by the noncrossing rule. Thus, an allowed crossing seam may merge into an avoided crossing seam when the solution permitted by the noncrossing rule ceases to exist. In this regard, the fact that both classes of solutions can be obtained from the same set of equations, Eqs. (2.122)–(2.124) is particularly convenient.

F. On the Existence of Rigorous Diabatic Bases

When nonadiabatic effects are large, as in the vicinity of actual surface crossings [see Eq. (2.44)], the adiabatic representation may be inconvenient for characterizing the nuclear dynamics. In this instance it is conceptually and computationally convenient to define a new electronic basis, the diabatic basis,[93,102-105] $\Psi_I^d(\mathbf{r}; \mathbf{R})$, which is a unitary transformation of the adiabatic electronic basis such that

$$f_\alpha^{JI,d}(\mathbf{R}) \equiv \left\langle \Psi_J^d(\mathbf{r}; \mathbf{R}) \left| \frac{\partial}{\partial R_\alpha} \Psi_I^d(\mathbf{r}; \mathbf{R}) \right. \right\rangle_{\mathbf{r}} = 0. \qquad (2.126)$$

The existence of a rigorous diabatic basis[103] has been a matter of considerable interest and some controversy in recent years.[93,106-108] While it is not the purpose of this review to discuss the subtleties of the existence of rigorous and/or approximate diabatic (quasidiabatic)[108,109] bases or the intrinsic limitations of the truncated Born–Oppenheimer expansion, it is useful to observe that the methods introduced in Section II C to determine $\mathbf{k}^{JI}(\mathbf{R})$ can be used to consider, from a computational perspective, the conditions discussed by Mead and Truhlar[93] for the existence of a rigorous diabatic basis.

We illustrate this point by examining the multidimensional two-electronic-state problem where the electronically diabatic basis is given by

$$\begin{pmatrix} \Psi_I^d(\mathbf{r}; \mathbf{R}) \\ \Psi_J^d(\mathbf{r}; \mathbf{R}) \end{pmatrix} = \begin{pmatrix} \cos\theta(\mathbf{R}) & -\sin\theta(\mathbf{R}) \\ \sin\theta(\mathbf{R}) & \cos\theta(\mathbf{R}) \end{pmatrix} \begin{pmatrix} \Psi_I(\mathbf{r}; \mathbf{R}) \\ \Psi_J(\mathbf{r}; \mathbf{R}) \end{pmatrix} \tag{2.127}$$

and where $\theta(\mathbf{R})$ is to be determined. Inserting Eq. (2.127) into the requirement Eq. (2.126) gives the system of equations

$$\frac{\partial}{\partial R_\alpha}\theta(\mathbf{R}) = -f_\alpha^{JI}(\mathbf{R}), \tag{2.128}$$

which is solved by partial integration. In order for $\theta(\mathbf{R})$ to be uniquely defined the condition for an exact derivative must hold:

$$\left(\frac{\partial^2}{\partial R_\alpha \partial R_\beta} - \frac{\partial^2}{\partial R_\beta \partial R_\alpha}\right)\theta(\mathbf{R}) = 0, \tag{2.129}$$

which from Eqs. (2.126) and (2.8) becomes

$$k_{\alpha\beta}^{JI}(\mathbf{R}) = k_{\beta\alpha}^{JI}(\mathbf{R}). \tag{2.130}$$

Thus, the methodology used to determine the $\mathbf{k}^{JI}(\mathbf{R})$ introduced in Section II C can be used to study computationally the existence of a rigorous diabatic basis. Note that if in the two-state problem the CSF basis consists of two terms, then Eq. (2.130) is a consequence of the antisymmetry of $\mathbf{f}^{JI}(\mathbf{R})$. However, if the number of CSFs is greater than two, $\mathbf{k}^{JI}(\mathbf{R})$ will in general not be symmetric.[93] Finally, note that for diatomic systems for which there is only one internal degree of freedom Eq. (2.130) becomes trivial.

III.   APPLICATIONS

In this section applications of the techniques introduced in Section II to problems of a chemical nature are presented. Recently, several groups have used techniques based on divided difference procedures[43-47] to evaluate nonadiabatic interactions for MCSCF and limited CI wavefunctions and used these methods to study nonadiabatic effects in regions of allowed (conical intersections) and avoided crossings. Other groups have used approximate diabatization procedures[47,110-112] to consider electronically nonadiabatic effects. However, in this chapter we will focus attention on applications that employ the analytic derivative techniques presented in this work. Since these computational techniques employ large-scale direct CI wavefunctions [developed from (state-averaged) MCSCF reference spaces], we will be able to consider the nonadiabatic interactions using wavefunctions that yield reliable potential energy surfaces. This aspect of our treatment is quite important, since the magnitude of the nonadiabatic coupling is a sensitive function of the separation of the potential energy surfaces in question. Furthermore, since our expressions for the nonadiabatic coupling matrix elements are exact, that is, are limited only by the accuracy of the MCSCF/CI Born–Oppenheimer wavefunctions themselves, it is possible to use the present techniques to treat some of the more subtle aspects of the breakdown of the Born–Oppenheimer approximation including the adiabatic correction to a Born–Oppenheimer potential energy surface. This aspect of molecular electronic structure has previously been largely (see, however, Refs. 58–60) the province of very specialized treatments for small diatomic systems.[52,95,113]

We begin by considering an example of the adiabatic correction to the Born–Oppenheimer potential energy curve (PEC) for the $X^1\Sigma^+$ state of LiH. The adiabatic correction to this PEC is a matter of some concern, since independent estimates of this quantity using advanced experimental[114,115] and theoretical techniques[52] had produced qualitatively different results. We will then turn to an example of radiationless energy transfer resulting in the quenching of an electronically excited state. In this latter study we will emphasize the electronic structure aspects of the quenching process and will be concerned with mechanistic aspects of the process rather than providing extensive tabulations of data. Our treatment will focus on the determination of the region(s) of coordinate space in which nonadiabatic effects are significant and the nature of the nonadiabatic coupling in these regions. The expected accuracy of the treatment employed will be carefully documented and the accuracy of key "asymptotic" quantities established by comparison with experimental data. The calculations reported in this work were carried out on dedicated minicomputer systems (Perkin–Elmer 3230 and Alliant FX/40 systems) and therefore should be considered to be at a level routinely available using the present techniques.

The electronic quenching problem to be considered

$$\text{He} + \text{H}_2(B\,^1\Sigma_u^+) \rightarrow \text{HeH}_2(2\,^1\text{A}') \rightarrow \text{He} + \text{H}_2(X\,^1\Sigma_g^+)$$

has been motivated by recent experimental[116–118] and theoretical work.[119–121] We will be concerned with the possibility of observing the metastable excited state of the HeH_2 moiety, which previous theoretical results have suggested to be stable[122] as well as with the mechanism of the overall quenching reaction. In addition to the intrinsic interest of the chemistry of the He–$\text{H}_2(B)$ quenching, this system provides a laboratory for studying a reaction in which a region of large nonadiabatic effects *cannot* be located on the basis of symmetry arguments and simultaneously provides an example of a system in which the multidimensional extension[61-64] of the von Neumann–Wigner noncrossing rule[65] is essential to understanding the chemistry being considered.

A. The Adiabatic Correction to the Born–Oppenheimer Potential Energy Curve for the $X\,{}^1\Sigma^+$ State of Lithium Hydride

The total second derivative nonadiabatic coupling matrix element [see Eqs (2.5) and (2.7)],

$$H^{JI}(\mathbf{R}) = \left\langle \Psi_J(\mathbf{r};\mathbf{R}) \left| \sum_{i,w} \frac{-1}{2M_i} \frac{\partial^2}{\partial R_w^{i2}} \Psi_I(\mathbf{r};\mathbf{R}) \right\rangle_{\mathbf{r}} \right. \tag{3.1}$$

where $w = x, y, z$ and i indexes the N nuclei, both modifies ($J = I$) and couples ($J \neq I$) Born–Oppenheimer potential energy surfaces. Again from Eqs. (2.5) and (2.7) the $I = J$ term can be rewritten as

$$H^{II}(\mathbf{R}) = -K^{II}(\mathbf{R}) = -\sum_{l,w} k^{II}_{lw,lw}(\mathbf{R}), \tag{3.2}$$

where

$$k^{II}_{iw,jw}(\mathbf{R}) = \left\langle \frac{\partial}{\partial R_w^i} \Psi_I(\mathbf{r};\mathbf{R}) \left| \frac{\partial}{\partial R_w^j} \Psi_I(\mathbf{r};\mathbf{R}) \right\rangle_{\mathbf{r}} \right. \tag{3.3}$$

and is referred to as the adiabatic correction to the Born–Oppenheimer potential energy surface. Under favorable circumstances, that is, when the Born–Oppenheimer potential energy surface is sufficiently isolated so that interstate coupling can be neglected, the adiabatic correction can be inferred from spectroscopic studies of isotopically substituted species.

As discussed in Section II D, science Eq. (3.1) includes derivatives with respect to all nuclear coordinates, it contains nonzero contributions from derivatives with respect to overall nuclear translation and rotation. Unlike derivatives with respect to internal coordinates, evaluation of derivatives of the Born–Oppenheimer wavefunction with respect to nuclear translations and rotations do not require derivative technology but rather can be expressed in terms of *electronic* mass polarization and angular momentum operators. In particular for the case of a ${}^1\Sigma^+$ state of a diatomic molecule orientated along the z axis with internuclear distance R, the adiabatic correction is given by [see Eqs. (2.116)–(2.118)]

$$H^{II}(R) = \frac{MP^{II}(R)}{2M} + \frac{L^{2\,II}(R)}{2\mu W^2} - \frac{K_R^{II}(R)}{2\mu}, \tag{3.4}$$

where

$$K_R^{II}(R) = \frac{M_2}{M_1} k^{II}_{1z,1z}(R) + \frac{M_1}{M_2} k^{II}_{2z,2z}(R) - 2k^{II}_{1z,2z}(R), \tag{3.5}$$

$$MP^{JI}(R) = -\langle \Psi_J(\mathbf{r}; R) | \sum_{i,j} \nabla_i^r \cdot \nabla_j^r | \Psi_I(\mathbf{r}; R) \rangle_{\mathbf{r}}, \tag{3.6}$$

$$L^{2''}(R) = L_x^{2''}(R) + L_y^{2''}(R) + L_z^{2''}(R), \tag{3.7a}$$

$$L_k^{2JI}(R) = \langle \Psi_J(\mathbf{r}; R) | \sum_i l_k(r_i) \cdot \sum_j l_k(r_j) | \Psi_j(\mathbf{r}; R) \rangle_{\mathbf{r}}. \tag{3.7b}$$

$\mathbf{I}(r_i)$ is the orbital angular momentum operator for the ith electron relative to the center of mass of the nuclei, μ, the reduced mass, satisfies $\mu^{-1} = M_1^{-1} + M_2^{-1}$ and the total nuclear mass $M = M_1 + M_2$. In Section II D the computational efficiency that can be achieved through the use of Eq. (3.4) rather than Eq. (3.2) for the evaluation of $H^{II}(\mathbf{R})$ was noted.

Here Eq. (3.4) will be used to evaluate the adiabatic correction for the $X^1\Sigma^+$ state of LiH. The situation in 1986 is summarized in Table I columns 2 and 3.[40] This table presents as a function of R the mass independent[52,56,57] quantities $U^H(R)$ and $U^{Li}(R)$, which are related to the adiabatic correction as follows:

$$H(X^1\Sigma^+, X^1\Sigma^+, R) = U^{Li}(R)/M(^xLi) + U^H(R)/M(^yH). \tag{3.8}$$

Here $M(^xLi)$ and $M(^yH)$ denote the mass of the ^xLi isotope of lithium and the ^yH isotopes of hydrogen and the right-hand side of Eq. (3.8) shows the mass dependence of $H(X^1\Sigma^+, X^1\Sigma^+, R)$, which is suppressed in the notation on the left-hand side of that equation. We have also written $H^{II}(R) \equiv H(I, I, R)$.

The experimental estimate of U^H (column 3) was obtained by Chen, Harding, Stwalley, and Vidal (CHSV)[115] from an inverted perturbation approach (IPA)[114] analysis of the $A^1\Sigma^+ \to X^1\Sigma^+$ emission for four isotopes, $^xLi^yH$, $x = 6, 7$ and $y = 1, 2$, of lithium hydride. This analysis assumes that the only modification to the Born–Oppenheimer PEC comes from Eq. (3.8), that is, it neglects interstate interactions. The theoretical values were obtained by Bishop and Cheung (BC)[52] and represent a direct evaluation of $H(I, I, R)$. BC performed their calculations using an axis system attached to the geometric center of the molecule and specialized CI wavefunctions involving an elliptical orbital basis. The Born–Oppenheimer wavefunctions of BC provide the lowest (best) variational energies reported to date for the $X^1\Sigma^+$ state of LiH. The $U^H(R)$ are plotted in Fig. 1.

From Table I and Fig. 1 it is seen that for internuclear distances (R) such that $R \leqslant r_e(X^1\Sigma^+) = 3.03a_0$ the adiabatic correction derived from experiment by CHSV and that computed by BC are in good qualitative agreement. Both decrease monotonically with R. Note that CHSV did not report U^{Li}, which to within the precision of their model is independent of R. As discussed in

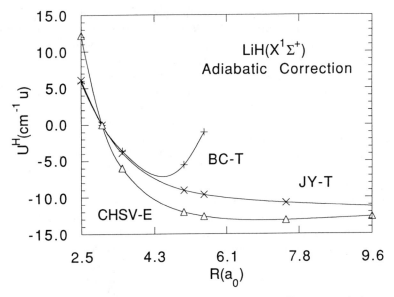

Figure 1. Mass independent contribution from hydrogen (U^H) to the $X\,^1\Sigma^+$ potential energy curve for LiH. Δ is the experimental result from Ref. 115; $+$ is the theoretical result from Ref. 52; and x is the result using the methods of this chapter from Ref. 40.

TABLE I
Analysis of the Adiabatic Correction: Decomposition into Mass Independent Terms

	U^{Ha}			U^{Lia}	
R^b	JY^c	BC^d	$CHSV^e$	JY^c	BC^d
2.5	6.1	5.8	12.2	7.2	7.3
3.0	0.0	0.0	0.0	0.0	0.0
3.5	-3.9	-3.6	-5.99	-3.7	-3.8
5.0	-9.0	-5.5	-12.0	-5.0	-3.8
5.5	-9.6	-1.0	-12.6	-2.9	8.2
7.5	-10.7		-13.1	7.0	
9.6			-12.7		
15.0	-12.2				

^{a}In cm$^{-1}\times$ u
^{b}In a_0.
^{c}This work relative to value at $R = 3.0$, $U^H = 72.4$, or $U^{Li} = 907.7$.
^{d}Reference 52 relative to value at $R = 3.0$, $U^H = 70.5$, or $U^{Li} = 924.0$.
^{e}Reference 115 relative to interpolated value at $R = 3.0$, $U^H = 0.10$.

Ref. 40 observable properties, such as the isotope dependence of $r_e(X\ ^1\Sigma^+)$, are considerably less sensitive to $U^{Li}(R)$ owing to the larger mass of lithium and smaller percentage change (compared to analogous changes for hydrogen) in mass when ^7Li is replaced by ^6Li. For R appreciably greater than $r_e(X\ ^1\Sigma^+)$ qualitative differences in the functional form of $U^H(R)$ are found. The $U^H(R)$ deduced by CHSV remains essentially a monotonically decreasing function of R, while that of BC begins to increase sharply (see Fig. 1). In order to address this discrepancy, evaluation of the adiabatic correction was undertaken[40] using the present techniques and a multireference CI wavefunction designed to provide a reliable representation of the wavefunctions over the entire range of internuclear distances required in this study.

The wavefunctions used in these calculations were second-order CI (SOCI) wavefunctions relative to an active space, which consists of all CSF's resulting from the distribution of four electrons among the $(1-4)\sigma$ orbitals. The molecular orbitals used to construct these CSFs were obtained from a complete active-space, two-state, state-averaged MCSCF procedure with weight vector $\mathbf{W} = (0.55, 0.45)$, using the above indicated active space. This treatment allows for proper dissociation of the $X\ ^1\Sigma^+$ state and also helps to provide an even handed treatment of the ionic $(1\sigma^2 2\sigma^2)$ and neutral $(1\sigma^2 2\sigma 3\sigma)$ contribution to the wavefunction. The molecular orbitals were in turn expanded in terms of (standard) extended atom centered gaussian basis sets. The lithium basis is the $(11s7p|5s4p)$ basis of McLean and Chandler[123] augmented with two d polarization functions ($\alpha = 0.4, 0.1$). The hydrogen basis is the $(9s|4s)$ gaussian basis of Siegbahn and Liu[124] augmented with two p polarization functions ($\alpha = 1.4, 0.25$). Using this basis set the SOCI expansion consists of 3798 CSF's in C_{2v} symmetry.

The results of the present calculations are summarized in Table I (columns 1 and 4) and $U^H(R)$ is plotted in Fig. 1. It is seen that the present results agree with those of BC for $R \lesssim r_e(X\ ^1\Sigma^+)$, that is, the region in which theory and experiment agree. Agreement with BC for $U^{Li}(R)$ is also obtained in this region. However, for $R > r_e(X\ ^1\Sigma^+)$ the $U^H(R)$ reported here remains a monotonically decreasing function of R. Thus our results disagree with those of BC but are in qualitative accord with those of CHSV. The improved agreement between theory and experiment is most likely due to the ability of the present calculations to describe all internuclear separations considered in this study with comparable accuracy. In this regard it is important to emphasize that the present approach uses standard rather than specialized CI wavefunctions, so that the considerable computational experience available for this type of wavefunction can be employed in designing treatments of polyatomic systems that cannot be readily studied with specialized wavefunctions. Such considerations will be employed in future treatments of the LiH system designed to resolve the remaining discrepancies between the present theoretical results and experiment.

B.   The $He + H_2(X\,^1\Sigma_u^+) \to He + H_2(X\,^1\Sigma_g^+)$ Reaction

The captioned reaction has a long and controversial history. The original experimental studies of the quenching of $H_2(B\,^1\Sigma_u^+)$ by helium were performed by Moore's group[116,117] in the early 1970s. This work motivated theoretical studies, which sought to explain the observed quenching in terms of (allowed) surface crossings. However, these attempts proved unsuccessful.[125–127] Subsequently, Farantos et al.[128] found an avoided crossing for general C_s approach. This finding was rationalized by Nicolaides and Zdetsis[129] in terms of a concept they introduced termed maximum ionicity excited state (MIES) theory. More recently, Perry and Yarkony[120] showed that the avoided crossing determined by Farantos et al. was actually part of an extended region of nuclear coordinate space for which the two potential energy surfaces in question are in close proximity.

Theoretical studies of the dynamics of the quenching process have also been reported, including the original trajectory surface hopping[30] study of Farantos[130] and the very recent nonadiabatic wavepacket treatment from Lester's group.[121] These dynamical studies employed a potential energy surface originally developed by Farantos[131] using the Sorbie–Murrel[132] approach based on the *ab initio* data of Farantos et al.[128] and Romelt et al.[126]

While this previous work has resulted in a clearer picture of quenching of $H_2(B)$ by helium several important questions remain unresolved. The Farantos surface is known to have significant deficiencies.[118,119] Perhaps as a result, the trajectory surface hopping studies of Farantos[131] yield a quenching cross section significantly smaller than the experimental value.[116,117,130] Furthermore, it has recently been suggested that the MIES charge-transfer structure, $(HeH)^+-H^-$, which is composed of two stable charged moieties and exists as a local minimum on the $2^1A'$ PES may be observable experimentally.[122] To consider the possibility of observing the $(HeH)^+-H^-$ charge transfer structures on the $2^1A'$ PES and to understand the mechanism of the radiationless quenching, we reexamine the $1,2^1A'$ potential energy surfaces and the nonadiabatic interactions which couple them using the methods of Section II.

The numerical results discussed in this section are taken from the original study of Perry and Yarkony and a more recent treatment of Manaa and Yarkony.[69] The characterization of the electronic wavefunctions follows Perry and Yarkony. Three, entrance channel, internal coordinates will be used to characterize nuclear configurations, r, the hydrogen–hydrogen distance, R, the distance between the center of mass of H_2 and the helium atom, and γ the angle between the H_2 axis and the line connecting helium to the center of mass of H_2. Nuclear configurations will be specified by the ordered triple $\mathbf{R} \equiv (R, r, \gamma)$ with R, r in atomic units and γ in degrees. The wavefunctions are SOCI wavefunctions relative to a four-electron–four-orbital reference space. The active orbitals (orbitals in the reference space) are determined from a SA-MCSCF procedure for the $1, 2^1 A'$ states with weight vector $\mathbf{W} = (0.55, 0.45)$. The molecular orbitals are expanded in terms of the extended contracted gaussian basis of Romelt et al. (Ref. 126, Table 1) He($7s2p1d/5s2p1d$) and H($7s3p/5s3p$), which was augmented with two diffuse s functions on helium, $\alpha = 0.07$, 0.01 and one diffuse s function on hydrogen, $\alpha = 0.02$. Using this basis the SOCI space contains 11,410 CSFs.

To consider the expected reliability of this level of treatment, it is useful to compare the predicted spectroscopic properties at the He + H_2 asymptote with the available experimental data. Using the $1^1 A'$ and $2^1 A'$ wavefunctions, respectively, the following reactant channel spectroscopic properties were obtained for $H_2(X^1\Sigma_g^+)$ and $H_2(B^1\Sigma_u^+)$. For $H_2(X^1\Sigma_g^+)$ we find $r_e = 0.7464(0.74144)$ Å, $\omega_e = 4328(4401.2)$ cm^{-1} and for $H_2(B^1\Sigma_u^+)$ we find $r_e = 1.2661(1.2928)$ Å, $\omega_e = 1424(1358.09)$ cm^{-1} and $T_e = 91227(91700)$ cm^{-1}, where the experimental values[133] are given parenthetically. For reference the SOCI energies for the $1^1 A'$ and $2^1 A'$ wavefunctions, $E(1^1 A')$ and $E(2^1 A')$, at the (near) equilibrium geometries $\mathbf{R} = (50, 1.415, 90)$ and $\mathbf{R} = (50, 2.393, 90)$ are, respectively, $E(1^1 A') = -4.065804$ a.u. (-260.82763 kcal/mol), $E(2^1 A') = -3.601826$ a.u. (29.40017 kcal/mol) and $E(1^1 A') = -3.996224$ a.u. (-216.82244 kcal/mol), $E(2^1 A') = -3.650107$ a.u. (0.0 kcal/mol). Here the parenthetical values are measured relative to $E(2^1 A')$ at $\mathbf{R} = (50, 2.393, 90)$, which will serve as the standard reference point in this work. It is significant to note that the choice of weight vector, which clearly favors the interaction region, does not significantly alter the results in the reactant channel. Using $\mathbf{W} = (0.95, 0.05)$ for the characterization of the $1^1 A'$ state and $\mathbf{W} = (0.05, 0.95)$ for the $2^1 A'$ state gives the following spectroscopic properties for $H_2(X^1\Sigma_g^+)$, $r_e = 0.7474$ Å, $\omega_e = 4323$ cm^{-1} and for $H_2(B^1\Sigma_u^+)$, $r_e = 1.2686$ Å, $\omega_e = 1412$ cm^{-1} and $T_e = 91152$ cm^{-1}.

Very recently Pibel et al.[118] have estimate the entrance channel barrier on the $2^1 A'$ potential energy surface from experimental quenching data and an adiabatic collision model. They report an upper bound (assuming the applicability of their adiabatic quenching model) to the barrier of 250 ± 80 cm^{-1}. Using the analytic gradient methodology of Section II, Perry and Yarkony located a saddle point in the entrance channel at $\mathbf{R} = (2.650, 2.520, 64.178)$ with a corresponding classical barrier of 520 cm^{-1}(1.49 kcal/mol) (see Table II). This result is in significantly better agreement with experiment than all previous theoretical treatments including the most recent previous treatment which yielded a classical barrier of 1200 cm^{-1}.[118,119] Recent preliminary studies show that at least a portion of the existing discrepancy

TABLE II
Energies[a] for the $1^1A'$ and $2^1A'$ States from SOCI Wavefunctions[b]

r	R	γ	$E(1^1A')$	$E(2^1A')$
Extrema				
2.520	2.690	64.178	−178.72226	1.49288
3.807	1.625	45.882	−50.69712	−38.99064

r	$R(r)$	$\gamma(r)$	$E(1^1A')$	$E(2^1A')$
Crossing Seam				
5.750000	2.313281	21.053830	4.28081	4.28093
5.250000	2.127251	25.584322	−11.70477	−11.70472
4.250000	1.750089	36.861334	−34.00658	−34.00646
3.812000	1.594403	43.100100	−38.00010	−37.99991
3.730000	1.566372	44.364160	−38.14103	−38.14099
3.500000	1.489807	48.061934	−37.18900	−37.18892
3.000000	1.333254	56.712661	−25.52514	−25.52509
2.700000	1.243611	62.065596	−8.90662	−8.90655
2.550000	1.198878	64.696541	3.37081	3.37092
2.400000	1.115354	67.250642	19.07133	19.07133

[a] In kcal/mol relative to $E(2^1A') = -3.650107$ at (50, 2.393, 90).
[b] R, r in a_0; γ in degrees.

between the experimental and theoretical barriers can be attributed to limitations in the atomic orbital basis set used by Perry and Yarkony.

These comparisons demonstrate that the treatment employed in this subsection provides a reliable description of both the ground- and excited-state surfaces. This treatment can then be expected to provide realistic representations of the key features to be discussed here, the actual crossing seam and the derivative couplings in the vicinity of this seam.

As a first step in understanding the mechanism of the quenching of $H_2(B\,^1\Sigma_u^+)$ the methodology outlined in Section II E is used to show that the region of close approach reported by Perry and Yarkony is in fact the consequence of an *actual crossing seam* between the $1^1A'$ and $2^1A'$ potential energy surfaces. According to the noncrossing rule an actual crossing seam of dimension 1 *may*, and as just noted does, exist for this system. This seam will be parametrized by the single coordinate r, the H_2 distance. Thus, for each value of the parameter r the solution of $G_\alpha^{IJ}(\mathbf{R}) = 0$ [Eq. (2.122)] is sought in the subspace spanned by R and γ. The seam is then represented by the triple of internal coordinates $\mathbf{R}(r) \equiv [R(r), r, \gamma(r)]$ for which $\Delta E_{IJ}(\mathbf{R}) = 0$. The seam $\mathbf{R}(r)$ is presented in Table II and the functions $R(r)$ and $\gamma(r)$ are plotted in Fig. 2. The energy along the seam, $E[\mathbf{R}(r)]$, is plotted in Fig. 3. For the

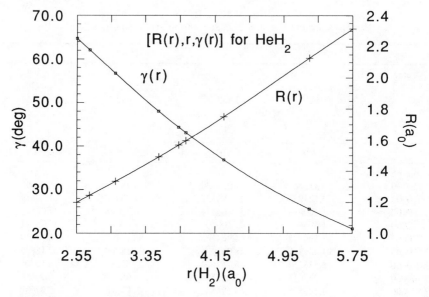

Figure 2. Plot of parameters $\gamma(r)$ and $R(r)$ for energetically accessible portion of seam representing the actual crossing of the $1, 2\,^1A'$ states of HeH_2.

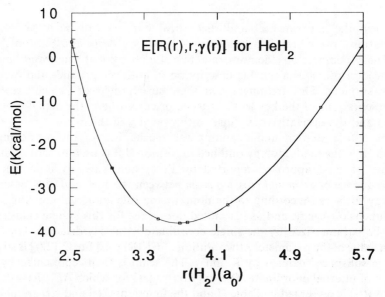

Figure 3. Plot of (common) energy along energetically accessible portion of seam representing the actual crossing of the $1, 2\,^1A'$ states of HeH_2.

purposes of these calculations the two states were considered to be degenerate when $\Delta E_{IJ}(\mathbf{R})$ was less than 0.0002 kcal/mole. (The corresponding changes in the geometrical parameters were on the order of 10^{-6}, distances in atomic units, γ in radians.) From Table II and Fig. 3 it is seen that a seam of actual crossings, *exoergic* with respect to the $H_2(B\,^1\Sigma_u^+)$ state asymptote, exists for $r \approx [2.60, 5.70]$. This portion of the actual crossing seam is thus relevant to the low-energy-scattering experiments reported by Pibel et al.[118] For the range of r presented in the table we find $\mathbf{R}(r) = 0.55986 + 0.20568r + 0.017417r^2$ and $\gamma(r) = 126.27 - 28.676r + 1.8058r^2$. Note that $\gamma \neq 0°$ or $90°$ so that this is *not* a high-symmetry section of these potential energy surfaces.

The electronic character of the wavefunctions in the vicinity of the seam has been analyzed in terms of the molecular dipole moment by Perry and Yarkony. In that work it was shown that along the seam significant charge reorganization occurs involving $(HeH)^+ - H^-$ and $He^0 - H_2^0$ moieties. The dipole moment does not vary systematically with r. Rather for fixed r, it is a sensitive function of R and γ with for example the charge transfer structure $(HeH)^+ - H^-$, corresponding to either the $1\,^1A'$ or the $2\,^1A'$ state. The zwitterionic character of the $2\,^1A'$ wavefunction in the vicinity of the seam was originally noted by Farantos et al.[128] and interpreted in terms of MIES theory.

It is interesting to note that the seam of actual surface crossings lies within the general vicinity of the global minimum on the $2\,^1A'$ potential energy surface. This point is located at $\mathbf{R} = (1.625, 3.807, 45.882)$ (see Table II) and is stable by 38.99 kcal/mol, with respect to the $He - H_2(B\,^1\Sigma_u^+)$ asymptote. The location of this extremum is in reasonable accord with the *ab initio* data of Farantos et al.[128] The charge distribution of this wavefunction was also analyzed in terms of the molecular dipole moment by Perry and Yarkony. Again this analysis shows that the wavefunction for the $2\,^1A'$ state at this geometry has a large dipole moment consistent with the $(HeH)^+ - H^-$ structure.

The degeneracy of the $1, 2\,^1A'$ potential energy surfaces makes it clear that nonadiabatic effects will be significant in the vicinity of the seam described previously. In order to determine the range of nuclear configurations for which nonadiabatic effects remain appreciable, that is, the breadth of the seam, and to facilitate adiabatic states treatments of the dynamics of the quenching reaction the first derivative nonadiabatic coupling matrix elements, $\mathbf{f}(\mathbf{R})$, are required.[30,33] This data has been reported elsewhere.[120] Here we focus on a computational study of the structure of the $\mathbf{f}(\mathbf{R})$ at the point $\mathbf{R}_1 = (1.594416, 3.812, 43.099850)$, the point of nearest degeneracy of the $1, 2\,^2A'$ potential energy surfaces reported by Perry and Yarkony. (Using the methodology discussed in Section II E, this approximate crossing point of Perry and Yarkony for which the $1, 2\,^2A'$ potential energy surfaces are degenerate to 0.001 kcal/mole was refined to give the result reported in Table II.)

The partial derivatives required of the determination of the $\mathbf{f(R)}$ will be taken with the center of mass of the triatomic system held fixed. We employ 1 and 4 as the mass (in atomic mass units) of hydrogen and helium, respectively. Below the ordered triple $(f_R(\mathbf{R}), f_r(\mathbf{R}), f_\gamma(\mathbf{R})/R)$ will be denoted as $\mathbf{f(R)}$ or $\mathbf{f}$ when the geometry dependence is suppressed. Determination of the relative phases of the components of $\mathbf{f(R)}$ is a matter of some concern. The use of the analytic derivative approach facilitates this determination. In this case the relative phases of the $\mathbf{f(R)}$ at a single point are uniquely determined from the phases of the *single* pair of CI wavefunctions. The relative phase of the $\mathbf{f(R)}$ at neighboring points will be determined from continuity and the requirement that when transversing an avoided/actual crossing the convention $[\Psi_a, \Psi_b] \to [-\Psi_b, \Psi_a]$, should hold. Here the ordered pair, $\Psi \equiv [\Psi_1, \Psi_2]$ denotes the wavefunctions for the two adiabatic states $\Psi_1 = \Psi(1^1A')$ and $\Psi_2 = \Psi(2^1A')$. The preceding phase convention is consistent with the two state representation of the adiabatic states Ψ_1, Ψ_2 in terms of, geometry independent, diabatic states Ψ_1^d, Ψ_2^d

$$\begin{pmatrix} \Psi_1 \\ \Psi_2 \end{pmatrix} = \begin{pmatrix} \cos\phi & \sin\phi \\ -\sin\phi & \cos\phi \end{pmatrix} \begin{pmatrix} \Psi_1^d \\ \Psi_2^d \end{pmatrix}, \tag{3.9}$$

where ϕ goes from 0 to $\pi/2$ as the avoided crossing is traversed.

The nonadiabatic coupling matrix elements are given in Table III. In this table $\mathbf{f(R)}$ is reported as a function of the displacement $\delta\mathbf{R}$ from the reference point $\mathbf{R}_1$ with $\mathbf{R} = (R_1 + \delta R, r_1 + \delta r, \gamma_1 + \delta\gamma) \equiv \mathbf{R}_1 + \delta\mathbf{R}$. For $\delta\mathbf{R} = [0, \delta r, 0]$, with $-\varepsilon < \delta r < \varepsilon$, $f_r(\mathbf{R})$ has a constant sign and achieves a large but finite extremum. However, each $f_\alpha(\mathbf{R})$ for $\alpha \neq r$, that is, the f's for the "extra" degrees of freedom, resembles the derivative of $f_r(\mathbf{R})$ evincing two extrema and a sign variation. A similar situation obtains in the R and γ directions. This functional dependence, which is carefully documented in Table III, reflects the following observations. In the vicinity of $\mathbf{R}_1$ the adiabatic state wavefunctions are dominated by two CSFs, $\psi_1 = 1a'^2 2a'^2$ and $\psi_2 = 1a'^2 2a' 3a'$, which are constructed from molecular orbitals given approximately by $1a' = \sigma(\text{HeH}^1)$, $2a' = 1s(\text{H}^2)$, and $3a' = \sigma^*(\text{HeH}^1)$. Thus, ψ_1 represents a charge-transfer CSF, while ψ_2 is less polar. At $\delta\mathbf{R} = [0, \delta r, 0]$, $\Psi = [\Psi_a, \Psi_b]$ while at $\delta\mathbf{R} = [0, -\delta r, 0]_1$, $\Psi \cong [-\Psi_b, \Psi_a]$, where Ψ_a, Ψ_b are largely, geometry dependent, linear combinations of the approximately geometry independent ψ_1 and ψ_2. Thus, in this instance the adiabatic states Ψ are expressed, in a qualitative sense, as linear combinations of diabatic states ψ_1, ψ_2 as in Eq. (3.9). The behavior of $f_r(\mathbf{R})$ is then a consequence of the phase convention in Eq. (3.9), which is commonly applied in one-dimensional problems.

The $\mathbf{f(R)}$ presented in Table III, which are couplings due to internal modes, change by orders of magnitude over a small range of nuclear configurations. However, the coupling due to an overall rotation in the molecular plane (ω), that is,

$$f_\omega \equiv \left\langle \Psi(2^1 A') \left| \frac{\partial}{\partial \omega} \Psi(1^1 A') \right\rangle_\mathbf{r} \right.$$

is expected to be small and approximately constant in the vicinity of a given reference point. This expectation follows from Eq. (3.9) and the antisymmetry of $\partial/\partial \omega$. In this study the value of f_ω is obtained computationally as the difference of two (large) numbers whose sum is f_y. The fact that the expected behavior is observed, that is, we find $f_\omega \cong 0.33$ for all points near and including reference point 1, lends credence to the accuracy of the numerical techniques used in this study and facilitates assignment of the relative phase of the $\mathbf{f(R)}$ at neighboring points.

The actual crossing seam reported in this work has important implications for the chemistry of the $He + H_2(B\,^1\Sigma_u^+)$ system. This feature of the $1, 2^1 A'$ potential energy surfaces should significantly influence the quenching of $H_2(B\,^1\Sigma_u^+)$ by He and alter expectations of the lifetime of the state corresponding to the global minimum on the $2^1 A'$ potential energy surface. It is anticipated that the lower entrance channel barrier noted previously and the actual crossing seam reported here will serve to increase the calculated quenching cross section. The large nonadiabatic coupling between the $1, 2^1 A'$ wavefunctions in the vicinity of the global minimum on the $2^1 A'$ PES will serve to decrease the predicted lifetime of this "state." To facilitate dynamical studies of these questions additional work on the $1, 2^1 A'$ potential energy surfaces and the nonadiabatic interactions is in progress.

IV. FUTURE DIRECTIONS

The theoretical study of electronically nonadiabatic processes has seen considerable progress in recent years. Progress has been made on several fronts. In this chapter the important advances in the computation of derivative coupling matrix elements [36-38,40] has been emphasized. However, considerable progress has also been made in the direct construction of approximate diabatic or quasidiabatic electronic states. Several approaches have emerged including approximate diabatic states which (1) preserve a particular molecular property[110,134] or attribute of the molecular wavefunction,[47] (2) are defined in terms of a particular set of atomlike orbitals as

TABLE III

Nonadiabatic Coupling Between the $1^1A'$ and $2^1A'$ States from SOCI Wavefunctions at $\mathbf{R} = (1.594, 3.812, 43.000)$

R^a	r^a	γ^a	$E(2^1A')^b$	ΔE^c	f_R^d	f_r^d	f_γ/R
0.000000	0.000000	0.00000000	−37.99919	0.00127	−27135.346	−6992.423	38524.596
0.000000	−0.010000	0.00000000	−37.40330	0.59509	−66.496	0.031	54.018
0.000000	−0.001000	0.00000000	−37.93973	0.06063	−650.026	−3.083	535.099
0.000000	−0.000500	0.00000000	−37.96948	0.03092	−1272.124	−11.854	1060.837
0.000000	−0.000250	0.00000000	−37.98435	0.01608	−2439.463	−43.975	2084.326
0.000000	−0.000050	0.00000000	−37.99624	0.00421	−9126.144	−643.064	8926.139
0.000000	−0.000025	0.00000000	−37.99772	0.00274	−13795.096	−1525.916	14799.221
0.000000	−0.000010	0.00000000	−37.99861	0.00185	−19725.637	−3315.092	23834.471
0.000000	0.000010	0.00000000	−37.99976	0.00073	−38795.315	−21197.610	81499.626
0.000000	0.000025	0.00000000	−38.00034	0.00046	65813.769	−52242.544	69804.675
0.000000	0.000050	0.00000000	−38.00042	0.00179	22684.831	−3539.130	−10068.162
0.000000	0.000250	0.00000000	−38.00045	0.01364	2912.402	−61.128	−2220.143
0.000000	0.000500	0.00000000	−38.00047	0.02850	1389.510	−13.962	−1095.038
0.000000	0.001000	0.00000000	−38.00049	0.05820	679.012	−3.348	−543.250
0.000000	0.010000	0.00000000	−37.99971	0.59322	66.348	0.031	−53.833
−0.001000	0.000000	0.00000000	−37.93898	0.08074	−6.995	357.969	−839.170
−0.000250	0.000000	0.00000000	−37.98485	0.01971	−114.196	1479.422	−3397.583
−0.000050	0.000000	0.00000000	−37.99702	0.00357	−3474.939	8303.718	−16767.263
−0.000025	0.000000	0.00000000	−37.99845	0.00174	−14538.792	15500.978	−24762.853
0.000025	0.000000	0.00000000	−37.99889	0.00292	−5214.683	−8233.459	23655.494
0.000050	0.000000	0.00000000	−37.99842	0.00487	−1873.094	−5437.095	14346.099
0.000250	0.000000	0.00000000	−37.99444	0.02105	−100.204	−1348.574	3262.143
0.001000	0.000000	0.00000000	−37.97931	0.08210	−6.765	−349.597	820.750
0.000000	0.000000	−0.06875493	−37.68511	0.26784	−147.877	50.530	0.767
0.000000	0.000000	−0.03437747	−37.84240	0.13439	−295.162	100.113	3.380
0.000000	0.000000	−0.06887549	−37.96800	0.02762	−1444.889	462.384	82.239

0.000000	-0.00343775	0.000000	-37.98368	0.01428	-2809.623	836.548	307.597
0.000000	-0.00057296	0.000000	-37.99670	0.00326	-12474.690	1785.870	5914.947
0.000000	-0.00042972	0.000000	-37.99734	0.00273	-14834.128	1531.408	8430.741
0.000000	-0.00028648	0.000000	-37.99798	0.00221	-18072.390	776.133	12841.392
0.000000	-0.00014324	0.000000	-37.99860	0.00172	-22472.812	-1275.623	21253.837
0.000000	0.00014324	0.000000	-37.99972	0.00097	-23817.111	-20335.018	67304.606
0.000000	0.00021486	0.000000	-37.99994	0.00090	-13731.420	-27910.806	76986.996
0.000000	0.00024351	0.000000	-38.00001	0.00090	-8422.223	-29906.273	77383.779
0.000000	0.00028018	0.000000	-38.00010	0.00092	-1379.106	-31127.843	74547.303
0.000000	0.00028648	0.000000	-38.00012	0.00092	-218.553	-31172.475	73710.375
0.000000	0.00030023	0.000000	-38.00015	0.00094	2162.579	-31101.433	71609.640
0.000000	0.00032945	0.000000	-38.00021	0.00097	6769.409	-30369.872	66143.839
0.000000	0.00035810	0.000000	-38.00026	0.00102	10352.800	-29077.934	60187.051
0.000000	0.00057296	0.000000	-38.00054	0.00159	17236.488	-16378.498	24641.623
0.000000	0.00343775	0.000000	-38.00260	0.01248	3115.316	-1243.279	403.479
0.000000	0.00687549	0.000000	-38.00493	0.02580	1522.456	-563.945	94.177
0.000000	0.03437747	0.000000	-38.02342	0.13261	298.604	-104.276	3.461
0.000000	0.06875493	0.000000	-38.04621	0.26618	149.006	-51.651	0.777

^{a} δR, δr, in a_0; $\delta\gamma$ in degrees.
^{b} In kcal/mol relative to $E(2^1 A') = -3.650107$ a.u. obtained at $\mathbf{R} = (50, 2.393, 90)$.
^{c} $\Delta E = E(2^1 A') - E(1^1 A')$ in kcal/mol.

has been done by Levy and coworkers[111] using polarized atomic orbitals, or (3) are based on block diagonalization techniques.[108,109]

The recently introduced gradient driven techniques for locating actual[69] and/or avoided[48] crossings discussed in this chapter should have a significant impact on the field. By enabling efficient determination of regions of significant electronic nonadiabaticity for systems in which molecular point group system is not a determining factor, these techniques have the potential to extend considerably the range of tractable systems.

New benchmark systems are likely to emerge. The $He-H_2$ system discussed in this chapter is a likely candidate. As a three-atom, four-electron system, it is an ideal candidate for study using any electronic structure technique. The recently identified[69] actual crossing seam discussed in this chapter, which is associated with significant charge reorganization, should make this system an important test case for approximate diabatization procedures. Interest in this sytem will be further stimulated by recent detailed experimental studies of the electronic quenching cross sections[118] and wavepacket studies of the dynamics of the quenching process.[121] Other likely candidates for benchmark systems may be found in the charge-transfer area, which has long provided fertile ground for the study of electronically nonadiabatic processes. The $H^+ + O_2 \rightarrow H + O_2^+$ system which has again been the object of theoretical[135,136] and experimental studies[137] is such a benchmark candidate.

Vibronic spectroscopy of electronically excited states that are perturbed or even predissociated by avoided curve crossings is another area in which important theoretical contributions can be expected. Systems exhibiting valence–Rydberg interactions such as the states of N_2[22] and O_2[23,24] noted in the Introduction are prime candidates in this regard. In this area the second derivative methodology for the evaluation of $k_{\alpha\beta}^{IJ}(\mathbf{R})$ discussed in Section II C and applied in Section III A to the LiH system[40] can be expected to find its principal application. The experimental determinations of the second derivatives couplings such as those available in N_2[22] will provide unique opportunities for experimental–theoretical synergism. Another motivating factor will be use of resonance-enhanced multiphoton ionization spectroscopy, which regularly probes excited molecular states that are strongly perturbed by valence–Rydberg couplings.[138]

The progress in electronic structure capabilities has been and will continue to be complemented by the considerable progress that has been made in dynamical techniques using both time-independent formalisms including the RIOSA approach of Baer and coworkers,[139-141] the nonadiabatic log-derivative[142] propagator method of Alexander et al.[143] and time-dependent formalisms in particular nonadiabatic wavepacket techniques.[21,144,145]

These advances in electronic structure and quantum dynamics treatments of electronically nonadiabatic processes can be expected to provide a synergism which will drive this field in the near future.

APPENDIX A: EVALUATION OF TERMS OF THE FORM Tr(MU) IN THE ATOMIC-ORBITAL BASIS

In this appendix the evaluation of Eq. (2.61) in the AO basis is considered.[83-85] Consider first the contribution from terms 1 and 3 in that equation:

$$\mathbf{X}^{JI\dagger}\mathbf{G}^\alpha = \mathbf{X}_{orb}^{JI}{}^\dagger\mathbf{G}\mathbf{O}^\alpha + \mathbf{X}_{CI}^{JI\dagger}\mathbf{G}\mathbf{C}^\alpha. \tag{A.1}$$

To evaluate the second (CI) contribution to (A.1) we note that

$$\mathbf{X}_{CI}^{JI\dagger}\mathbf{G}\mathbf{C}^\alpha = \sum_{\lambda,\mu} 2X_{CI,\lambda}^{JI}(E_{MC} - H)_{\lambda\mu}^\alpha C_{MC,\mu}^I \tag{A.2a}$$

$$= -\sum_{\lambda,\mu} 2X_{CI,\lambda}^{JI}H_{\lambda\mu}^\alpha C_{MC,\mu}^I \tag{A.2b}$$

From Eq. (A.2b) we see that it is only necessary to construct one $[\gamma_{ij}^{X_{CI},I_{MC}}]$ and two $[\Gamma_{ijkl}^{X_{CI},I_{MC}}]$ particle transition density matrices between $\mathbf{X}_{CI}^{JI}$ and $\mathbf{C}_{MC}^I$ and then use standard techniques to transform these matrices to the AO basis to yield $\gamma_{pq}^{X_{CI},I_{MC}}$ and $\Gamma_{pqrs}^{X_{CI},I_{MC}}$. These density matrices are then traced with derivative integrals [see Eq. (A.4c)] to evaluate (A.2).

A similar expression is obtained for the orbital contribution to $\mathbf{X}^{JI\dagger}\mathbf{G}\mathbf{O}^\alpha$. The orbital contribution is first expressed as the trace of a Lagrangian and an antisymmetric matrix:

$$\mathbf{X}_{orb}^{JI}{}^\dagger\mathbf{G}\mathbf{O}^\alpha = \sum_{ij} X_{ij}^{JI} GO_{ij}^\alpha \tag{A.3a}$$

$$= \sum_{ij} X_{ij}^{JI}(L_{ij}^{MC,\alpha} - L_{ji}^{MC,\alpha}) \tag{A.3b}$$

$$= \sum_{i,j} \tilde{X}_{ij}^{JI} L_{ij}^{MC,\alpha} \tag{A.3c}$$

where $\tilde{X}_{ij}^{JI}$ is the antisymmetric matrix with unique elements X_{ij}^{JI}. The Lagrangian is then expanded and the sums reordered to obtain the following expression:

$$\sum_{m,i} L_{mi}^{MC,\alpha}\tilde{X}_{mi}^{JI} = \sum_{m,i}\left[\tilde{X}_{mi}^{JI}2\left(\sum_j \gamma_{ij}^{MC,II}h_{mj}^\alpha + \sum_{j,k,l} 2\Gamma_{ijkl}^{MC,II}g_{mjkl}^\alpha\right)\right] \tag{A.4a}$$

$$= \sum_{p,q}\left[\sum_{m,i,j}(2\tilde{X}_{mi}^{JI}\tau_{pm}\tau_{qj}\gamma_{ij}^{MC,II})\right]h_{pq}^\alpha$$

$$+ \sum_{p,q,r,s} \left[\sum_{m,i,j,k,l} (4\tilde{X}^{JI}_{mi} \tau_{pm} \tau_{qj} \Gamma^{MC,II}_{ijkl} \tau_{rk} \tau_{sl}) \right] g^{\alpha}_{pqrs}$$

$$\equiv \sum_{p,q,i,j} \left(2(\tau\tilde{X}^{JI}_{orb})_{pi} \tau_{qj} \gamma^{MC,II}_{ij} \right) h^{\alpha}_{pq}$$

$$+ \sum_{p,q,r,s,i,j,k,l} \left(4(\tau\tilde{X}^{JI}_{orb})_{pi} \tau_{qj} \Gamma^{MC,II}_{ijkl} \tau_{rk} \tau_{sl} \right) g^{\alpha}_{pqrs} \tag{A.4b}$$

$$\equiv \sum_{p,q} \gamma^{\tilde{X}_{orb},IMC}_{pq} h^{\alpha}_{pq} + \sum_{p,q,r,s} \Gamma^{\tilde{X}_{orb},IMC}_{pqrs} g^{\alpha}_{pqrs} \tag{A.4c}$$

where in Eq. (A.4b) the sums over i, j are performed first in order to *define* new "density matrices" in the AO basis given in (A.4c). The orbital (A.4c) and CI (A.2b) contributions are then combined to obtain:

$$\mathbf{X}^{JI\dagger}\mathbf{G}^{\alpha} = \sum_{p,q} \gamma^{X,IMC}_{pq} h^{\alpha}_{pq} + \sum_{p,q,r,s} \Gamma^{X,IMC}_{pqrs} g^{\alpha}_{pqrs} \tag{A.5}$$

where

$$\gamma^{X,IMC}_{pq} = \gamma^{X_{orb},IMC}_{pq} + \gamma^{X_{CI},IMC}_{pq} \tag{A.6a}$$

and

$$\Gamma^{X,IMC}_{pqrs} = \Gamma^{X_{orb},IMC}_{pqrs} + \Gamma^{X_{CI},IMC}_{pqrs} \tag{A.6b}$$

Note from Eq. (A.4c) and the discussion following (A.2) that the definitions of the two density matrices that contribute to Eq. (A.6a) [or (A.6b)] differ.

Finally, consider the T^{α} contribution to $f^{JI,\Delta}_{\alpha}(\mathbf{R})$ defined in Eq. (2.58b) and reexpressed in Eq. (2.61). This term does not require a great deal of computational effort as efficient methods exist to build $\mathbf{GO}^{T^{\alpha}}$ and $\mathbf{GC}^{T^{\alpha}}$. However, the T^{α} contribution is best obtained as follows:

$$\mathbf{X}^{JI\dagger}_{CI}\mathbf{GC}^{T^{\alpha}} = \sum_{i,j} L^{X,IMC}_{ij} T^{\alpha}_{ij} \tag{A.7}$$

and

$$\mathbf{X}^{JI\dagger}_{orb}\mathbf{GO}^{T^{\alpha}} = \sum_{i,j} X^{JI}_{ij} L^{T^{\alpha}}_{ij} \tag{A.8a}$$

$$= \sum_{i,j} L^{X^{JI}}_{ij} T^{\alpha}_{ij} - \sum_{i,j,m} L_{mj} X^{JI}_{mk} T^{\alpha}_{ki}. \tag{A.8b}$$

Thus, the generation of the Lagrangian, $\mathbf{L}^{X^{JI}}$, has no dependence on the number of nuclear degrees of freedom, α. The economy afforded by reorganizing Eq. (A.3c) in this manner was noted earlier by Page et al.[13] in their work on MCSCF second derivatives. For the MCSCF Lagrangian employed in these equations, $\mathbf{L}^{XJI}$ need not be constructed by contracting h_{ij}^X and g_{ijkl}^X integrals with density matrices, rather we build $\mathbf{L}^{XJI}$ by contracting a $\mathbf{Y}^{MC,I}$ matrix,

$$Y_{mi,nj}^{MC,I} = 2\gamma_{ij}^{MC,II} h_{mn} + \sum_{k,l} [4\Gamma_{ijkl}^{MC,II} g_{mnkl} + 8\Gamma_{ikjl}^{MC,II} g_{mknl}] \tag{A.9}$$

which was used to build the MCSCF orbital hessian and by transforming the Lagrangian.

$$L_{mi}^{MC,X,I} = \sum_{n,j} Y_{mi,nj}^{MC,I} \tilde{X}_{nj} + \sum_{j} L_{ji}^{MC,II} \tilde{X}_{mj}. \tag{A.10}$$

APPENDIX B: CONTRIBUTION TO DERIVATIVE QUANTITIES FROM INEQUIVALENT CORE AND VIRTUAL ORBITALS

In the discussion of orbital constraints in Section II B 2 it was noted that when an incomplete virtual orbital space or a redefined fully occupied orbital space is used in the CI calculations, additional contributions to the first derivative quantities arise. In this appendix equations are presented that permit evaluation of these contributions.

As noted in Section II B 2 it is convenient to require that the core and virtual orbitals be eigenfunctions of the core Fock operator, $\hat{F}_c$, where

$$\hat{F}_c = \hat{h} + \sum_{i}^{core} (2\hat{J}_i - \hat{K}_i). \tag{B.1}$$

This leads to the requirement

$$\frac{\partial}{\partial R_\alpha} \left\langle \phi_m(r; \mathbf{R}) | \hat{F}_c \phi_n(r; \mathbf{R}) \right\rangle_r = 0, \tag{B.2}$$

where $\phi_m(r; \mathbf{R})$ and $\phi_n(r; \mathbf{R})$ belong to different sets of virtual orbitals or to different sets of core orbitals. Carrying out the differentiation yields the following equations for the the unique elements $\Lambda_{ij}(\mathbf{R})$ of the antisymmetric matrix $\tilde{\Lambda}_{ij}^\alpha(\mathbf{R})$

$$(\langle \phi_n(r; \mathbf{R}) | \tilde{F}_c \phi_n(r; \mathbf{R}) \rangle_r - \langle \phi_m(r; \mathbf{R}) | \tilde{F}_c \phi_m(r; \mathbf{R}) \rangle_r) \Lambda_{mn}^\alpha$$
$$= -\sum_k [U_{km}^\alpha \langle \phi_k(r; \mathbf{R}) | \tilde{F}_c \phi_n(r; \mathbf{R}) \rangle_r + \langle \phi_m(r; \mathbf{R}) | \tilde{F}_c \phi_k(r; \mathbf{R}) \rangle_r U_{kn}^\alpha]$$
$$- \sum_{k,j} [U_{kj}^\alpha \{4g_{mknj} - (g_{mknj} + g_{mjnk})\}] - \langle \phi_m(r; \mathbf{R}) | \tilde{F}_c^\alpha \phi_n(r; \mathbf{R}) \rangle_r, \tag{B.3}$$

where the sum over j is restricted to core orbitals, $\hat{F}_c^\alpha$ represents the Fock operator in Eq. (B.1) constructed from first derivative integrals [Eq. (2.29)], and the derivative of the MO coefficients $\bar{U}_{ij}^\alpha(\mathbf{R})$ is given by

$$\bar{U}_{ij}^\alpha(\mathbf{R}) = \tilde{\Lambda}_{ij}^\alpha(\mathbf{R}) + \tilde{\Delta}_{ij}^\alpha(\mathbf{R}) + T_{ij}^\alpha(\mathbf{R}) \tag{B.4a}$$

$$\equiv \tilde{\Lambda}_{ij}^\alpha(\mathbf{R}) + U_{ij}^\alpha(\mathbf{R}). \tag{B.4b}$$

Solving for Λ_{ij}^α gives

$$\Lambda_{mn}^\alpha = (\varepsilon_m^c - \varepsilon_n^c)^{-1} \left\{ \sum_k [U_{km}^\alpha \langle \phi_k(r; \mathbf{R}) | \hat{F}_c \phi_n(r; \mathbf{R}) \rangle_r + \langle \phi_m(r; \mathbf{R}) | \hat{F}_c \phi_k(r; \mathbf{R}) \rangle_r U_{kn}^\alpha] \right.$$

$$\left. + \sum_{k,j} [U_{kj}^\alpha \{4g_{mknj} - (g_{mknj} + g_{mjnk})\}] - \langle \phi_m(r; \mathbf{R}) | \hat{F}_c^\alpha \phi_n(r; \mathbf{R}) \rangle_r \right\}. \tag{B.5}$$

In Eq. (B.5) the diagonal elements of the $\hat{F}_c$ have been denoted as orbital energies ε_m^c.

As in Section II B 1 avoiding the transformation of the derivative integrals from the AO basis to the MO basis is the key to the efficient evaluation of these contributions to the energy gradient or first derivative nonadiabatic coupling matrix elements. To avoid the repeated transformation of the derivative integrals from the AO basis to the MO basis, it is necessary to eliminate the explicit use of $\mathbf{U}^\alpha(\mathbf{R})$, which require the solution of the CP-SAMCSCF equations in the MO basis. By expanding U_{km}^α [see Eq. (2.51)] we have a general equation of the form

$$\Lambda_{mn}^\alpha = \sum_{k,l} [Q_{mn,kl} \tilde{\Delta}_{kl}^\alpha + Q_{mn,kl} T_{kl}^\alpha] + (\varepsilon_m^c - \varepsilon_n^c)^{-1} (F_c^\alpha)_{mn}, \tag{B.6}$$

where $\mathbf{Q} = \mathbf{Q}^F + \mathbf{Q}^G$

$$Q_{mn,kl}^F = (\varepsilon_m^c - \varepsilon_n^c)^{-1} \{F_{kn} \delta_{lm} + F_{mk} \delta_{ln}\} \tag{B.7a}$$

$$Q_{mn,kl}^G = (\varepsilon_m^c - \varepsilon_n^c)^{-1} \{4g_{mnkl} - g_{mknl} - g_{mlnk}\}, \quad \text{where } l \text{ is a core orbital,} \tag{B.7b}$$

$$Q_{mn,kl}^F = Q_{mn,kl}^G = 0, \quad \text{where } l \text{ is an active orbital.} \tag{B.7c}$$

We then define

$$Z_{mn,kl} = Q_{mn,kl} - Q_{mn,lk} \tag{B.8}$$

to obtain

$$\Lambda_{mn}^\alpha = \sum_{kl} Z_{mn,kl} \Delta_{kl}^\alpha + \sum_{k,l} Q_{mn,kl} T_{kl}^\alpha + (\varepsilon_m^c - \varepsilon_n^c)^{-1} (F_c^\alpha)_{mn}, \tag{B.9}$$

where the compound index kl now is restricted to the nonredundant orbital rotations of the MCSCF problem. Each term in Eq. (B.9) results in a contribution to $f^{JI}(\mathbf{R})$ as in Eq. (2.56) formally:

$$f_\alpha^{JI,\wedge}(\mathbf{R}) = f_{\Delta^\alpha}^{JI,\wedge}(\mathbf{R}) + f_{T^\alpha}^{JI,\wedge}(\mathbf{R}) + f_{F^\alpha}^{JI,\wedge}(\mathbf{R}). \tag{B.10}$$

The term that depends on T^α requires very little computational effort. We find

$$f_{T^\alpha}^{JI,\lambda}(\mathbf{R}) = \sum_{mn,k,l} l_{mn}^{JI} Q_{mn,kl} T_{kl}^\alpha \tag{B.11a}$$

$$\equiv \sum_{k,l} Q_{kl}^l T_{kl}^\alpha, \tag{B.11b}$$

where the sum over mn is restricted to nonredundant orbital pairs in either the virtual orbital space or the core orbital space and l_{mn}^{JI} is defined in Eq. (2.57). It is also straightforward to evaluate the derivative Fock operator, $(F_c^\alpha)_{mn}$, contribution to this term in the AO basis as

$$f_{F^\alpha}^{JI,\wedge}(\mathbf{R}) = \sum_{mn} l_{mn}^{JI} (F_c^\alpha)_{mn} \tag{B.12a}$$

$$= \sum_{mn,p,q} l_{mn}^{JI} \tau_{pm} \tau_{qn} \{ h_{pq}^\alpha + \gamma_{rs}^{\text{core}} [2g_{pqrs}^\alpha - \tfrac{1}{2}(g_{prqs}^\alpha + g_{psqr}^\alpha)] \} \tag{B.12b}$$

$$\equiv \sum_{p,q} \gamma_{pq}^l \{ h_{pq}^\alpha + \gamma_{rs}^{\text{core}} [2g_{pqrs}^\alpha - \tfrac{1}{2}(g_{prqs}^\alpha + g_{psqr}^\alpha)] \}, \tag{B.12c}$$

where $\gamma_{rs}^{\text{core}}$ is the one-particle density matrix for the core orbitals in the AO basis. Finally, consider the contribution from the term $f_{\Delta^\alpha}^{JI,\wedge}(\mathbf{R})$. Since Δ^α is never evaluated explicitly, this contribution cannot be evaluated directly as in Eq. (B.11). Formally, the contribution to the first-order nonadiabatic coupling arising from this term is

$$f_{\Delta^\alpha}^{JI,\wedge}(\mathbf{R}) = \sum_{mn,kl} l_{mn}^{JI} Z_{mn,kl} \Delta_{kl}^\alpha \tag{B.13a}$$

$$= \sum_{kl} \xi_{kl}^{JI} \Delta_{kl}^\alpha. \tag{B.13b}$$

Equation (B.13b) is analogous in form to Eq. (2.57b). Thus, its contribution can be obtained from the Z-vector method. Because these orbital contributions to $f_{\Delta^\alpha}^{JI,\wedge}(\mathbf{R})$ in Eq. (B.13b) are linear, the contributions from Eq. (B.13b) and Eq. (2.57b) are obtained by adding ξ_{mn}^{JI} to l_{mn}^{JI} and proceeding as in Eq. (2.58).

APPENDIX C: COUPLED PERTURBED STATE-AVERAGED MCSCF EQUATIONS

In this appendix, we consider the changes in the expressions for a first-order nonadiabatic coupling matrix element or a CI gradient that are required when a state-averaged MCSCF procedure rather than an MCSCF procedure is used to generate the molecular orbitals. Once again, we wish to evaluate the derivative integral contribution from these new terms in the AO basis. The SA-MCSCF energy was presented in Eq. (2.45) in the text. From the SA-MCSCF energy expression

$$E_{\text{SAMC}}(\mathbf{R}) = \sum_I W_I \mathbf{C}_{\text{MC}}^I(\mathbf{R})^\dagger \mathbf{H}(\mathbf{R}) \mathbf{C}_{\text{MC}}^I(\mathbf{R}), \tag{2.45}$$

we can easily deduce the SA-MCSCF variational conditions,

$$\frac{\partial}{\partial \Delta_{ij}} E_{\text{SAMC}} \equiv GO_{ij} = \sum_I W_I GO_{ij}(I, I) \tag{C.1a}$$

$$= L_{ij}^{\text{SAMC}} - L_{ji}^{\text{SAMC}} \tag{C.1b}$$

$$= 0 \tag{C.1c}$$

$$\frac{\partial}{\partial \eta_{IJ}} E_{\text{SAMC}} \equiv G\eta_{IJ} \tag{C.2a}$$

$$= -2(W_I - W_J) \mathbf{C}_{\text{MC}}^J{}^\dagger \mathbf{H}_{\text{MC}} \mathbf{C}_{\text{MC}}^I \tag{C.2b}$$

$$= 0 \tag{C.2c}$$

$$\frac{\partial}{\partial C_\lambda^I} E_{\text{SAMC}} \equiv GC_\lambda^I \tag{C.3a}$$

$$= -2W_I (\mathbf{H}_{\text{MC}} - E_{\text{MC}}^I) \mathbf{C}_{\text{MC}}^I \tag{C.3b}$$

$$= 0 \tag{C.3c}$$

where $GO_{ij}(I, I)$ is given in Eq. (2.46a) (as GO_{ij}), Eq. (C.2) must hold for each unique pair IJ of states in Eq. (2.45) and Eq. (C.3) must hold for all states in the complement of the "occupied-state" space in Eq. (2.45). The state-averaged MCSCF lagrangian is constructed from Eq. (2.33) with a density that is the weighted sum of the density of the occupied states,

$$\gamma_{ij}^{\text{SAMC}} = \sum_I W_I \gamma_{ij}^{\text{MC},II}, \tag{C.4a}$$

$$\Gamma_{ijkl}^{\text{SAMC}} = \sum_I W_I \Gamma_{ijkl}^{\text{MC},II}. \tag{C.4b}$$

In this appendix the term "occupied states" refers to those states included in Eq. (2.45). Whereas in the MCSCF procedure there is only one class of CI variational parameters in the SA-MCSCF procedure there are two classes of CI variational parameters. The mixings between the CI vectors that occur in SA-MCSCF energy expression are represented in the eigenvector basis [Eq. (C.2)] and the mixings of the "occupied" CI vectors with the "unoccupied" CI vectors are handled in the CSF basis [Eq. (C.3)]. Differentiating these variational expressions with respect to nuclear displacements generates the coupled-perturbed SA-MCSCF equations,

$$
\begin{bmatrix}
\dfrac{\partial E^2}{\partial \Delta^2} & \dfrac{\partial E^2}{\partial \eta \partial C} & \dfrac{\partial E^2}{\partial \Delta \partial C} \\[2ex]
\dfrac{\partial E^2}{\partial \eta \partial \Delta} & \dfrac{\partial E^2}{\partial \eta^2} & \dfrac{\partial E^2}{\partial \eta \partial C} \\[2ex]
\dfrac{\partial E^2}{\partial C \partial \Delta} & \dfrac{\partial E^2}{\partial C \partial \eta} & \dfrac{\partial E^2}{\partial C^2}
\end{bmatrix}
\begin{bmatrix}
\Delta^\alpha \\[2ex] \eta^\alpha \\[2ex] C^\alpha
\end{bmatrix}
= -
\begin{bmatrix}
\mathbf{GO}^\alpha + \mathbf{GO}^{T\alpha} \\[2ex]
\mathbf{G}\eta^\alpha + \mathbf{G}\eta^{T\alpha} \\[2ex]
\mathbf{GC}^\alpha + \mathbf{GC}^{T\alpha}
\end{bmatrix}.
\tag{C.5}
$$

Following our earlier discussions, these terms contribute to the first-order nonadiabatic coupling matrix elements through through the following equation,

$$
f_\alpha^{JI,\Delta}(R) =
\begin{bmatrix}
\mathbf{l}_0^{JI} \\[2ex] 0 \\[2ex] 0
\end{bmatrix}^\dagger
\begin{bmatrix}
\dfrac{\partial E^2}{\partial \Delta^2} & \dfrac{\partial E^2}{\partial \eta \partial C} & \dfrac{\partial E^2}{\partial \Delta \partial C} \\[2ex]
\dfrac{\partial E^2}{\partial \eta \partial \Delta} & \dfrac{\partial E^2}{\partial \eta^2} & \dfrac{\partial E^2}{\partial \eta \partial C} \\[2ex]
\dfrac{\partial E^2}{\partial C \partial \Delta} & \dfrac{\partial E^2}{\partial C \partial \eta} & \dfrac{\partial E^2}{\partial C^2}
\end{bmatrix}^{-1}
\begin{bmatrix}
\mathbf{GO}^\alpha + \mathbf{GO}^{T\alpha} \\[2ex]
\mathbf{G}\eta^\alpha + \mathbf{G}\eta^{T\alpha} \\[2ex]
\mathbf{GC}^\alpha + \mathbf{GC}^{T\alpha}
\end{bmatrix},
\tag{C.6}
$$

which we again abbreviate as

$$
= \mathbf{l}^{JI\dagger} \mathscr{H}^{-1}(\mathbf{G}^\alpha + \mathbf{G}^{T\alpha}).
\tag{C.7}
$$

We then solve a linear equation that is independent of the number of nuclear degrees of freedom,

$$
\mathscr{H} X^{JI} = \mathbf{l}^{JI}.
\tag{C.8}
$$

Inserting the solution of this equation into Eq. (C.5) yields the following expression for the variational contribution to the first-order nonadiabatic coupling matrix element:

$$f_\alpha^{JI,\Delta}(\mathbf{R}) = \mathbf{X}_{orb}^{JI\,\dagger}\mathbf{GO}^\alpha + \mathbf{X}_{orb}^{JI\,\dagger}\mathbf{GO}^{T^x} + \mathbf{X}_\eta^{JI\dagger}\mathbf{G}\boldsymbol{\eta}^\alpha + \mathbf{X}_\eta^{JI\dagger}\mathbf{G}\boldsymbol{\eta}^{T^x}$$
$$+ \mathbf{X}_{CI}^{JI\dagger}\mathbf{GC}^\alpha + \mathbf{X}_{CI}^{JI\dagger}\mathbf{GC}^{T^x}. \tag{C.9}$$

We see from Eq. (C.3) that the new term that arises in a state-averaged MCSCF calculation can be expressed as

$$\mathbf{X}_\eta^{JI\dagger}\mathbf{G}\boldsymbol{\eta}^\alpha + \mathbf{X}_\eta^{JI\dagger}\mathbf{G}\boldsymbol{\eta}^{T^x} = \sum_{KL} X_{\eta KL}^{JI}\{-2(W_K - W_L^L)\mathbf{C}_{MC}^{K\,\dagger}(\mathbf{H}_{MC}^\alpha + \mathbf{H}_{MC}^{T^x})\mathbf{C}_{MC}^L\}. \tag{C.10}$$

The derivative integral contribution to this term can now be computed in the AO basis if one generates the following density matrices,

$$\gamma_{pq}^{X^{JI},SAMC} = \sum_{KL} X_{\eta KL}^{JI}\{-2(W_K - W_L)\}\gamma_{pq}^{MC,KL}, \tag{C.11}$$

$$\Gamma_{pqrs}^{X^{JI},SAMC} = \sum_{KL} X_{\eta KL}^{JI}\{-2(W_K - W_L)\}\Gamma_{pqrs}^{MC,KL}. \tag{C.12}$$

APPENDIX D: CONTRIBUTION TO DERIVATIVE QUANTITIES FROM INEQUIVALENT ACTIVE ORBITALS

In Section II B 2 it was noted that when an incomplete virtual orbital space or a redefined fully orbital space is used in the CI calculations additional contributions to the first derivative quantities arise. In Appendix B equations were presented that permit evaluation of these contributions. In this appendix that discussion is expanded to include the case where the MCSCF or SA-MCSCF active orbitals must be constrained. In this development the commonly applied constraint that the active orbitals be eigenfunctions (natural orbitals) of the MCSCF one-particle density is considered. This constraint is necessary, for example, if selected configurations from a CAS MCSCF space are used to define the reference space for a CI calculation. In that case the CI energy is not invariant to rotations among the active orbitals.

A. MCSCF Active Orbitals

The equations for the derivatives of the independent, active orbital rotations in a CAS calculation were derived by Page et al.[13],

$$\Lambda_{ij}^\alpha = \frac{2\gamma_{ij}^{MC,I,I^x}}{\lambda_i - \lambda_j}, \tag{D.1}$$

where γ_{ij}^{MC,I,I^z} is the one-particle transition density constructed from the CI vector from the MCSCF calculation $[\mathbf{C}_{MC}^I]$ and the CI portion of the solution to the CP-MCSCF equations, $[\mathbf{C}_{MC}^{I^z}]$ and λ_i is an eigenvalue of the one-particle density matrix. Again as in Section II B 2 efficient determination of this quantity requires the elimination of explicit reference to the solutions of the CP-MCSCF equations $\mathbf{C}_{MC}^{I^z}$. To accomplish this Eq. (D.1) is rewritten as a matrix times the CP-MCSCF solutions,

$$\Lambda_{ij}^\alpha = \sum_{\mu,\beta} \left(\frac{2C_{MC,\mu}^I \tilde{k}_{\mu\beta}^{ij}}{\lambda_i - \lambda_j} \right) C_{MC,\beta}^{I^z} \tag{D.2a}$$

$$= \sum_\beta Q_{ij,\beta} C_{MC,\beta}^{I^z} \tag{D.2b}$$

where the indices i,j refer to active space orbitals. Thus, the contribution to $f_{\Lambda^z}^{JI}(\mathbf{R})$ can be written as (compare with Eq. (B.14)]

$$f_{\Lambda^z}^{JI}(\mathbf{R}) = \sum_{ij,\mu} l_{ij}^{JI} Q_{ij,\mu} C_{MC,\mu}^{I^z} \tag{D3.a}$$

$$= \sum_\mu q_\mu^{JI} C_{MC,\mu}^{I^z}, \tag{D.3b}$$

where the sum over the compound index ij is restricted to active space rotations. Since $\mathbf{C}_{MC}^{I^z}$ represents the solution to the CP-MCSCF equations Eq. (D.3) is evaluated using the Z-vector method with q_μ^{JI} added to the CI portion of l^{JI} in Eqs. (2.58) and (2.59) producing the following vector:

$$\mathbf{l}_{\Lambda^z}^{JI} = \begin{pmatrix} \mathbf{l}^{JI} \\ \mathbf{q}^{JI} \end{pmatrix} \tag{D.4}$$

B. State-Averaged-MCSCF Active Orbitals

When the active orbitals are taken to be the natural orbitals of the state-averaged density matrix, Eq. (C.4), then Eq. (D.1) must be modified. The appropriate equation in this case is

$$\Lambda_{ij}^\alpha = \sum_K \frac{2W_K \gamma_{ij}^{MC,K,K^z}}{\lambda_i - \lambda_j} + \sum_{KL} \frac{(W_K - W_L)2\gamma_{ij}^{MC,K,L}}{\lambda_i - \lambda_j} \eta_{KL}^\alpha. \tag{D.5}$$

Then following the procedure in part A results in an $\mathbf{l}^{JI}$ vector, the analog of Eq. (D.4), which has the following form:

$$\mathbf{l}_{\Lambda^{\alpha}}^{JI} = \begin{bmatrix} \mathbf{l}^{JI} \\ \mathbf{q}_{\eta} \\ q^{K} \\ \vdots \\ q^{L} \end{bmatrix} \tag{D.6}$$

where

$$q_{\eta_{KL}} = \sum_{ij} l_{ij}^{JI} \frac{[2\gamma_{ij}^{MC,KL}(W_K - W_L)]}{\lambda_i - \lambda_j} \tag{D.7}$$

and

$$q_{\mu}^{K} = \sum_{\beta,ij} l_{ij}^{JI} \left(\frac{2C_{MC,\beta}^{K} \tilde{k}_{ij}^{\beta\mu}}{\lambda_i - \lambda_j} \right) W_K \tag{D.8a}$$

$$= \sum_{ij} l_{ij}^{JI} Q_{ij,\mu}^{K} W_K \tag{D.8b}$$

where the sum over ij runs over the nonredundant active-orbital indices.

APPENDIX E: THE SECOND DERIVATIVE CP-SAMCSCF EQUATIONS

In this appendix the second derivative coupled perturbed state-averaged MCSCF (CP-SAMCSCF) equations are described. Here for brevity derivatives with respect to CI degrees of freedom are expressed in the eigenstate basis rather than in the CSF/eigenstate basis as was done in Section II B 1 and Appendix C. The ideas developed in those subsections can be used to reformulate the expression presented here into the more computationally convenient mixed CSF/eigenstate representation of Appendix C.

As in the case of the first derivative CP-SAMCSCF equations discussed in Appendix C, the second derivative of the state-averaged MCSCF equations can be expressed as a system of linear equation in which the left-hand side is the hessian of the state-averaged MCSCF problem. However, the right-hand side of this equation is considerably more complex than is the case for the first derivative CP-SAMCSCF equations. The right hand side involves quantities constructed from derivative integrals, overlap constraints, and the results of the first derivative CP-MCSCF equations. In particular we find

$$
\begin{bmatrix}
\dfrac{\partial \mathbf{E}^2}{\partial \Delta^2} & \dfrac{\partial \mathbf{E}^2}{\partial \Delta \partial \xi} \\[2ex]
\dfrac{\partial \mathbf{E}^2}{\partial \xi \partial \Delta} & \dfrac{\partial \mathbf{E}^2}{\partial \xi^2}
\end{bmatrix}
\begin{bmatrix}
\Delta^{\alpha\beta} \\[2ex]
\xi^{\alpha\beta}
\end{bmatrix}
= -
\begin{bmatrix}
\mathbf{G3}^{\alpha\beta} \\[2ex]
\mathbf{G4}^{\alpha\beta}
\end{bmatrix}
\tag{E.1}
$$

where the state-averaged MCSCF energy E_{SAMC} has been abbreviated E. The terms on the right-hand side of Eq. (E.1) can be expressed as follows:

$$
\begin{aligned}
\mathbf{G3} &= \mathbf{GO}^{\alpha\beta} + \mathbf{GO}^{U^{\alpha},U^{\beta}} + \mathbf{GO}^{\alpha,U^{\beta}} + \mathbf{GO}^{\beta,U^{\alpha}} + \mathbf{GO}^{T^{\alpha\beta}} + \mathbf{GO}^{Y^{\alpha\beta}} \\
&\quad + 2\sum_{I} W_{I}[\mathbf{GO}(I^{\alpha}, I^{\beta}) + \mathbf{GO}^{\alpha}(I, I^{\beta}) + \mathbf{GO}^{\beta}(I, I^{\alpha}) \\
&\quad + \mathbf{GO}^{U^{\alpha}}(I, I^{\beta}) + \mathbf{GO}^{U^{\beta}}(I, I^{\alpha})],
\end{aligned}
\tag{E.2a}
$$

$$
\begin{aligned}
\mathbf{G4} &= \mathbf{GC}^{\alpha\beta} + \mathbf{GC}^{U^{\alpha},U^{\beta}} + \mathbf{GC}^{\alpha,U^{\beta}} + \mathbf{GC}^{\beta,U^{\alpha}} + \mathbf{GC}^{T^{\alpha\beta}} + \mathbf{GC}^{Y^{\alpha\beta}} \\
&\quad + [W_{I} - W_{J}][\mathbf{GC}(I^{\alpha}, J^{\beta}) + \mathbf{GC}(I^{\beta}, J^{\alpha}) \\
&\quad + \mathbf{GC}^{\alpha}(I^{\beta}, J) + \mathbf{GC}^{\alpha}(I, J^{\beta}) + \mathbf{GC}^{U^{\alpha}}(I^{\beta}, J) + \mathbf{GC}^{U^{\alpha}}(I, J^{\beta}) \\
&\quad + \mathbf{GC}^{\beta}(I^{\alpha}, J) + \mathbf{GC}^{U^{\beta}}(I^{\alpha}, J) + \mathbf{GC}^{\beta}(I, J^{\alpha}) + \mathbf{GC}^{U^{\alpha}}(I, J^{\alpha})].
\end{aligned}
\tag{E.2b}
$$

As in the case of the first derivative CP-MCSCF equations [see Eq. (2.53)], the second derivative molecular-orbital orthogonality terms, $\mathbf{GO}^{T^{\alpha\beta}}$ and $\mathbf{GC}^{T^{\alpha\beta}}$ arise from the definitions of $\tilde{\Delta}_{ij}^{\alpha\beta}$ and $T_{ij}^{\alpha\beta}$ in Eqs. (2.76) and (2.77). The representative terms $\mathbf{GO}^{Y^{\alpha\beta}}, \mathbf{GC}^{Y^{\alpha\beta}}, G\tilde{C}_{JI}(J^{\beta}, I^{\alpha}), GC_{JI}^{\alpha}(J, I^{\beta})$, and $\mathbf{GO}(I^{\alpha}, I^{\beta})$ in Eqs. (E.2) require further discussion. As discussed in detail in Ref. 38 in deriving these expressions differentiations are initially performed in the CSF basis and then transformed to the CI or eigenstate basis.

$Y^{\alpha\beta}$ contributions arise from the second derivative of the CI-orthonormality condition, Eq. (2.20). This gives

$$
\sum_{\lambda} \left(\frac{\partial^2}{\partial R_{\alpha} \partial R_{\beta}} C_{\mathrm{MC},\lambda}^{J} \right) C_{\mathrm{MC},\lambda}^{I} = \tilde{\xi}_{JI}^{\alpha\beta} + Y_{JI}^{\alpha\beta}
\tag{E.3}
$$

where $\tilde{\xi}_{JI}^{\alpha\beta}$ is an antisymmetric matrix whose unique elements $\xi_{JI}^{\alpha\beta}$ appear on the the left-hand side of Eq. (E.1) and

$$
Y_{JI}^{\alpha\beta} = -\sum_{K} (\tilde{\xi}_{KI}^{\alpha} \tilde{\xi}_{KJ}^{\beta} + \tilde{\xi}_{KI}^{\beta} \tilde{\xi}_{KJ}^{\alpha}), \quad J < I,
\tag{E.4a}
$$

$$Y_{JI}^{\alpha\beta} = -\sum_K \tilde{\xi}_{KI}^{\alpha} \tilde{\xi}_{KI}^{\beta}, \qquad\qquad J = I, \qquad\qquad \text{(E.4b)}$$

$$Y_{JI}^{\alpha\beta} = 0, \qquad\qquad J > I. \qquad\qquad \text{(E.4c)}$$

The derivatives of the $\mathbf{C}^J$ are given by

$$\frac{\partial}{\partial R_\alpha} C_\lambda^J(\mathbf{R}) = \sum_K C_\lambda^K(\mathbf{R}) \tilde{\xi}_{KJ}^{\alpha}, \qquad\qquad \text{(E.5a)}$$

$$\frac{\partial^2}{\partial R_\alpha \partial R_\beta} C_\lambda^J(\mathbf{R}) = \sum_K C_\lambda^K(\mathbf{R})(\tilde{\xi}_{KJ}^{\alpha\beta} + Y_{KJ}^{\alpha\beta}). \qquad\qquad \text{(E.5b)}$$

For $K, J > L$, $\xi_{KJ}^{\alpha\beta} = \xi_{KJ}^{\alpha} = 0$ since these parameters are not determined from the state-averaged MCSCF equations. Here L is the number of states included in the averaging procedure so that $W_J = 0$ for $J > L$.

Using the definition of $\mathbf{GO}_{ij}$ from Eqs. (2.46) and (C.1), $\mathbf{GO}_{ij} = \sum_I W_I(L_{ij}^{II} - L_{ji}^{II})$, together with the definitions of γ^{II} and Γ^{II} in terms of the structure factors $\tilde{k}^{II}$ and $\tilde{K}^{II}$ in Eqs. (2.22d) and (2.22e), we find

$$GO_{ij}^{Y^{\alpha\beta}} = 2\sum_{I,J} W_I(L_{ij}^{JI} - L_{ji}^{JI}) Y_{JI}^{\alpha\beta}, \qquad\qquad \text{(E.6)}$$

$$GO_{ij}^{\beta}(I, I^\alpha) = \sum_K \tilde{\xi}_{KI}^{\alpha} (L_{ij}^{\beta,KI} - L_{ji}^{\beta,KI}), \qquad\qquad \text{(E.7)}$$

where

$$L_{mi}^{\alpha,JI} = 2\sum_j \gamma_{ij}^{\text{MC},JI} h_{mj}^{\alpha} + 4\sum_{j,k,l} \Gamma_{ijkl}^{\text{MC},JI} g_{mjkl}^{\alpha}. \qquad\qquad \text{(E.8)}$$

Similarly since GC_{JI} is given by

$$GC_{JI} = (W_J - W_I)\left[\sum_{i,j} \gamma_{ij}^{\text{MC},JI} h_{ij} + \sum_{i,j,k,l} \Gamma_{ijkl}^{\text{MC},JI} g_{ijkl}\right] \qquad\qquad \text{(E.9)}$$

we have

$$GC_{JI}^{\beta}(J, I^\alpha) = (W_J - W_I)\sum_K\left[\sum_{i,j} \gamma_{ij}^{\text{MC},JK} h_{ij}^{\beta} + \sum_{i,j,k,l} \Gamma_{ijkl}^{\text{MC},JK} g_{ijkl}^{\beta}\right]\tilde{\xi}_{KI}^{\alpha}, \qquad\qquad \text{(E.10)}$$

$$GC_{JI}(J^\beta, I^\alpha) = (W_J - W_I)\sum_{\lambda,\mu} C_{\text{MC},\lambda}^{J^\alpha}\left[\sum_{i,j} \tilde{k}_{ij}^{\text{MC},\lambda\mu} h_{ij} + \sum_{i,j,k,l} \tilde{K}_{ijkl}^{\text{MC},\lambda\mu} g_{ijkl}\right]C_{\text{MC},\mu}^{I^\beta}$$

$$= (W_J - W_I)\sum_K [\tilde{\xi}_{KJ}^{\alpha} \tilde{\xi}_{KI}^{\beta} E_{\text{MC}}^K] \qquad\qquad \text{(E.11)}$$

and

$$GC_{JI}^{Y^{\alpha\beta}} = (W_J - W_I)(Y_{IJ}^{\alpha\beta}E_{\mathrm{MC}}^I + Y_{JI}^{\alpha\beta}E_{\mathrm{MC}}^J).$$ (E.12)

Once the right-hand side of Eq. (E.1) has been constructed, the same techniques used to treat the first derivative CP-SAMCSCF equations can be applied to the solution and simplification of Eq. (E.1).

References

1. J. Gerrat and I. M. Mills, *J. Chem. Phys.* **49** 1719 (1968).

2. P. Pulay, *Mol. Phys.* **17**, 197 (1969).

3. S. Bratoz, *Colloq. Intern. CNRS (Paris)* **82**, 287 (1958).

4. D. M. Bishop and M. Randic, *J. Chem. Phys.* **44**, 2480 (1966).

5. J. A. Pople, R. Krishnan, H. B. Schegel, and J. S. Binkley, *Int. J. Quant. Chem. Symposium* **13**, 225 (1979).

6. B. R. Brooks, W. D. Laidig, P. Saxe, J. D. Goddard, Y. Yamaguchi, and H. F. Schaefer, *J. Chem. Phys.* **72**, 4652 (1980).

7. Y. Osamura, Y. Yamaguchi, and H. F. Schaefer, *J. Chem. Phys.* **75**, 2919 (1981).

8. Y. Osamura, Y. Yamaguchi, and H. F. Schaefer, *J. Chem. Phys.* **77**, 383 (1982).

9. J. E. Rice, R. D. Amos, N. C. Handy, T. J. Lee, and H. F. Schaefer, *J. Chem. Phys.* **85**, 963 (1986).

10. T. J. Lee, N. C. Handy, J. E. Rice, A. C. Scheiner, and H. F. Schaefer, *J. Chem. Phys.* **85**, 3930 (1986).

11. T. Takada, M. Dupuis, and H. F. King, *J. Chem. Phys.* **75**, 332 (1981).

12. P. Jorgensen and J. Simons, *J. Chem. Phys.* **79**, 334 (1983).

13. M. Page, P. Saxe, G. F. Adams, and B. H. Lengsfield, *J. Chem. Phys.* **81**, 434 (1984).

14. P. Pulay, *Modern Theoretical Chemistry*, H. F. Schaefer, ed., Plenum Press, New York, 1977, Vol. 4.

15. P. Jorgensen and J. Simons, *Geometrical Derivatives of Energy Surfaces and Molecular Properties*, Reidel, Dordrecht, 1982.

16. Y. Osamura, Y. Yamaguchi, and H. F. Schaefer, *Theor. Chim Acta* **72**, 71 (1987).

17. Y. Osamura, Y. Yamaguchi, and H. F. Schaefer, *Theor. Chim Acta* **72**, 93 (1987).

18. T. Helgaker and P. Jorgensen, *Advances in Quantum Chemistry*, Academic Press, New York, 1988.

19. P. Pulay, *Adv. Chem. Phys.* **69**, 241 (1987).

20. W. J. Hehre, L. Radom, P. R. Schleyer, and J. A. Pople, *Ab Initio Molecular Orbital Theory*, Wiley-Interscience, New York, 1986.

21. A. E. Orel and K. C. Kulander, *Chem. Phys. Lett.* **146**, 428 (1988).

22. D. Stahel, M. Leoni, and K. Dressler, *J. Chem. Phys.* **79**, 2541 (1979).

23. W. J. van der Zande, W. Koot, J. R. Peterson, and Los, *Chem. Phys. Lett.* **140**, 175 (1987).

24. W. J. van der Zande, W. Koot, J. Los, and J. R. Peterson, *J. Chem. Phys.* **89**, 7658 (1988).

25. H. Koppel, W. Domcke, and L. S. Cederbaum, *Adv. Chem. Phys.* **57**, 59 (1984).

26. R. L. Whetten, G. S. Ezra, and E. R. Grant, *Ann. Revs. Phys. Chem.* **36**, 277 (1985).

27. J. W. Zwanziger and E. R. Grant, *J. Chem. Phys.* **87**, 2954 (1987).

28. M. V. Berry, *Proc. R. Soc. Lond. Ser. A* **392**, 45 (1984).

29. B. Simon, *Phys. Rev. Lett.* **51**, 2167 (1983).

30. J. C. Tully, *Modern Theoretical Chemistry*, Plenum, New York, 1976.

31. B. C. Garrett and D. G. Truhlar, *Theoretical Chemistry Advances and Perspectives*, Academic Press, New York, 1981.

32. Z. H. Top and M. Baer, *J. Chem. Phys.* **66**, 1363 (1977).

33. M. Baer, in *Theory of Chemical Reaction Dynamics*, M. Baer ed., Chemical Rubber, Boca Raton, 1985 Vol. 2, Chapter 4.

34. R. J. Buenker, G. Hirsch, S. D. Peyerimhoff, P. J. Bruna, M. Romelt, M. Bettendorff, and C. Petrongolo, *Current Aspects of Quantum Chemistry*, Elsevier, New York, 1981.

35. M. Desouter-Lecomte, C. Galloy, J. C. Lorquet, and M. V. Pires, *J. Chem. Phys.* **71**, 3661 (1979).

36. B. H. Lengsfield, P. Saxe, and D. R. Yarkony, *J. Chem. Phys.* **81**, 4549 (1984).

37. P. Saxe, B. H. Lengsfield, and D. R. Yarkony, *Chem. Phys. Lett.* **113**, 159 (1985).

38. B. H. Lengsfield and D. R. Yarkony, *J. Chem. Phys.* **84**, 348 (1986).

39. P. Saxe and D. R. Yarkony, *J. Chem. Phys.* **86**, 321 (1987).

40. J. O. Jensen and D. R. Yarkony, *J. Chem. Phys.* **89**, 3853 (1988).

41. I. Shavitt, *Modern Theoretical Chemistry*, Plenum Press, New York, 1976, Vol. 3.

42. H. J. Werner, in *Advances in Chemical Physics*, K. P. Lawley, ed., Wiley, New York, 1987, Vol. 69, p. 1.

43. D. Dehareng, X. Chapuisat, J. C. Lorquet, C. Galloy, and G. Raseev, *J. Chem. Phys.* **78**, 1246 (1983).

44. C. Galloy and J. C. Lorquet, *J. Chem. Phys.* **67**, 4672 (1971).

45. H. J. Werner and W. Meyer, *J. Chem. Phys.* **74**, 5802 (1981).

46. R. J. Buenker, G. Hirsch, S. D. Peyerimhoff, P. J. Bruna, J. Romelt, M. Bettendorff, and C. Petrongolo, *Current Aspects of Quantum Chemistry*, Elsevier, New York, 1982.

47. H. Werner, B. Follmeg, and M. H. Alexander, *J. Chem. Phys.* **89**, 3139 (1988).

48. D. R. Yarkony, *J. Chem. Phys.* **92**, 2457 (1990).

49. D. R. Yarkony, *J. Phys. Chem.* **94**, 5572 (1990).

50. D. M. Bishop and L. M. Cheung, *Phys. Rev. A* **18**, 1846 (1978).

51. D. M. Bishop and L. M. Cheung, *J. Mol. Spectrosc.* **75**, 462 (1979).

52. D. M. Bishop and L. M. Cheung, *J. Chem. Phys.* **78**, 1396 (1983).

53. R. D. Bardo and M. Wolfsberg, *J. Chem. Phys.* **67**, 593 (1977).

54. R. D. Bardo and M. Wolfsberg, *J. Chem. Phys.* **68**, 2686 (1978).

55. R. D. Bardo, L. I. Kleinman, A. W. Raczkowski, and M. Wolfsberg, *J. Chem. Phys.* **69**, 1106 (1978).

56. L. I. Kleinman and M. Wolfsberg, *J. Chem. Phys.* **60**, 4740 (1974).

57. L. I. Kleinman and M. Wolfsberg, *J. Chem. Phys.* **60**, 4749 (1974).

58. P. Pulay and H. Sellers, *Chem. Phys. Lett.* **103**, 463 (1984).

59. H. Sellers, *Chem. Phys. Lett.* **108**, 339 (1984).

60. N. C. Handy, Y. Yamaguchi, and H. F. Schaefer, *J. Chem. Phys.* **84**, 4481 (1986).

61. G. Herzberg and H. C. Longuet-Higgins, *Disc. Faraday Soc.* **35**, 77 (1963).

62. H. C. Longuet-Higgins, *Proc. R. Soc. Lond. A* **344**, 147 (1975).

63. J. Katriel and E. R. Davidson, *Chem. Phys. Lett.* **76**, 259 (1980).

64. C. A. Mead and D. G. Truhlar, *J. Chem. Phys.* **84**, 1055 (1986).

65. J. von Neumann and E. Wigner, *Physik. Z.* **30**, 467 (1929).

66. J. Hinze, *J. Chem. Phys.* **559**, 6424 (1973).

67. B. H. Lengsfield, *J. Chem. Phys.* **77**, 4073 (1982).

68. R. Shepard, in *Advances in Chemical Physics*, K. P. Lawley ed., J. Wiley, New York, 1987. Vol. 69, p. 63.

69. M. R. Manaa and D. R. Yarkony, *J. Chem. Phys.* **93**, 4473 (1990).

70. A. A. Frost, *J. Phys. Chem.* **72**, 289 (1968).

71. C. Eaker, *J. Chem. Phys.* **93**, 8073 (1990).

72. C. W. Eaker and C. A. Parr, *J. Chem. Phys.* **64**, 1322 (1976).

73. F. O. Ellision, *J. Am. Chem. Soc.* **85**, 3540 (1963).

74. J. Romelt, *Int. J. Quant. Chem.* **24**, 627 (1983).

75. R. N. Diffenderfer and D. R. Yarkony, *J. Phys. Chem.* **86**, 5098 (1982).

76. M. R. Hoffman, D. J. Fox, J. F. Gaw, Y. Osamura, Y. Yamaguchi, R. S. Grev, G. Fitzgerald, H. F. Schaefer, P. J. Knowles, and N. C. Handy, *J. Chem. Phys.* **80**, 2660 (1984).

77. B. H. Lengsfield, in *Proceedings of the NATO Workshop on Geometrical Derivatives of Energy Surfaces and Molecular Properties*, Sonderborg, Denmark, 1984.

78. J. Hellmann, *Einfuhrung in die Quantenchemie*, Deuticke, Leipzig, 1937.

79. R. P. Feynman, *Phys. Rev.* **56**, 340 (1939).

80. R. McWeeny and B. T. Sutcliffe, *Methods of Molecular Quantum Mechanics*, Academic Press, London, 1969.

81. E. R. Davidson, *Reduced Density Matrices in Quantum Chemistry*, Academic Press, New York, 1976.

82. N. C. Handy and H. F. Schaefer, *J. Chem. Phys.* **81**, 5031 (1984).

83. J. E. Rice and R. D. Amos, *Chem. Phys. Lett.* **122**, 585 (1985).

84. T. J. Lee and W. Allen, *J. Chem. Phys.* **87**, 7062 (1987).

85. R. Shepard, *Int. J. Quant. Chem.* **31**, 33 (1987).

86. B. Liu and B. H. Lengsfield, *J. Chem. Phys.* **75**, 478 (1981).

87. P. Siegbahn, A. Heiberg, B. Roos, and B. Levy, *Phys. Scr.* **21**, 323 (1980).

88. B. O. Roos, P. R. Taylor, and P. E. M. Siegbahn, *Chem. Phys.* **48**, 157 (1980).

89. B. O. Roos, *Int. J. Quantum Chem. Symp.* **14**, 175 (1980).

90. D. R. Yarkony, *J. Chem. Phys.* **90**, 1657 (1989).

91. C. W. Bauschlicher, *J. Chem. Phys.* **72**, 880 (1980).

92. P. O. Löwdin, *Phys. Rev.* **97**, 1474 (1955).

93. C. A. Mead and D. G. Truhlar, *J. Chem. Phys.* **77**, 6090 (1982).

94. R. de L. Kronig, *Band Spectra and Molecular Structure*, Macmillan, New York, 1929.

95. W. Kolos, *Advances in Quantum Chemistry*, Academic, New York, 1970.

96. B. T. Sutcliffe, *Quantum Dynamics of Molecules, in NATO Advanced Study Institute Series B: Physics*, R. G. Wooley, Plenum, New York, 1980, Vol. 57.

97. C. A. Mead, *J. Chem. Phys.* **78**, 807 (1983).

98. T. C. Thompson, D. G. Truhlar, and C. A. Mead, *J. Chem. Phys.* **82**, 2392 (1985).

99. T. C. Thompson and C. A. Mead, *J. Chem. Phys.* **82**, 2408 (1985).

100. N. Koga and K. Morokuma, *Chem. Phys. Lett.* **119**, 371 (1985).

101. For a recent review of surface-walking procedures see H. Schlegel, *Adv. Chem. Phys.* **67**, 249 (1987).

102. A. D. McLachlan, *Mol. Phys.* **4**, 417 (1961).

103. F. T. Smith, *Phys. Rev.* **179**, 111 (1969).

104. W. Lichten, *Phys. Rev.* **131**, 339 (1963).

105. W. Lichten, *Phys. Rev.* **164**, 164 (1967).

106. M. Baer, *Chem. Phys. Lett.* **35**, 112 (1975).

107. M. Baer, *Chem. Phys.* **15**, 49 (1976).

108. T. Pacher, C. A. Mead, L. S. Cederbaum, and H. Koppel, *J. Chem. Phys.* **91**, 7057 (1989).

109. T. Pacher, L. S. Cederbaum, and H. Koppel, *J. Chem. Phys.* **89**, 7367 (1988).

110. C. W. Bauschlicher and S. R. Langhoff, *J. Chem. Phys.* **89**, 4246 (1988).

111. M. C. Montabanel-Bacchus, G. Chambaud, B. Ley, and P. Mille, *Journal de Chimie Physique* **80**, 425 (1983).

112. P. Archirel and B. Levy, *Chem. Phys.* **106**, 51 (1986).

113. L. Wolniewicz and K. Dressler, *J. Chem. Phys.* **88**, 3861 (1988).

114. C. R. Vidal and W. C. Stwalley, *J. Chem. Phys.* **77**, 883 (1982).

115. Y. C. Chen, D. R. Harding, W. C. Stwalley, and C. R. Vidal, *J. Chem. Phys.* **85**, 2436 (1986).

116. D. Atkins, E. H. Fink, and C. B. Moore, *J. Chem. Phys.* **52**, 1604 (1970).

117. E. H. Fink, D. L. Atkins, and C. B. Moore, *J. Chem. Phys.* **56**, 900 (1972).

118. C. D. Pibel, K. L. Carlton, and C. B. Moore, *J. Chem. Phys.* **93**, 323 (1990).

119. R. M. Grimes, W. A. Lester, and M. Dupuis, *J. Chem. Phys.* **84**, 5437 (1986).

120. J. K. Perry and D. R. Yarkony, *J. Chem. Phys.* **89**, 4945 (1988).

121. P. Pernot, F. M. Grimes, W. A. Lester Jr., and C. Cerjan, *Chem. Phys. Lett.* **163**, 297 (1989).

122. S. C. Farantos and J. Tennyson, *J. Chem. Phys.* **82**, 2163 (1985).

123. A. D. McLean and G. Chandler, unpublished results; the lithium basis is reproduced in D. R. Yarkony, *Int. J. Quant. Chem.* **31**, 91 (1987).

124. P. Siegbahn and B. Liu, *J. Chem. Phys.* **68**, 2457 (1978).

125. H. F. Schaefer, P. Wallach, and C. F. Bender, *J. Chem. Phys.* **56**, 1219 (1972).

126. J. Romelt, S. D. Peyerimhoff, and R. J. Buenker, *Chem. Phys.* **34**, 403 (1979).

127. J. Romelt, S. D. Peyerimhoff, and R. J. Buenker, *Chem. Phys.* **41**, 133 (1979).

128. S. C. Farantos, G. Theodorakopoulos, and C. A. Nicolaides, *Chem. Phys. Lett.* **100**, 163 (1983).

129. C. A. Nicolaides and A. Zdetsis, *J. Chem. Phys.* **80**, 1900 (1984).

130. S. C. Farantos, *Mol. Phys.* **54**, 835 (1985).

131. S. C. Farantos, J. N. Murrel, and S. Carter, *Chem. Phys. Lett.* **108**, 367 (1984).

132. K. S. Sorbie and J. N. Murrel, *Mol. Phys.* **80**, 1900 (1984).

133. K. P. Huber and G. Herzberg, *Molecular Spectra and Molecular Structure IV. Constants of Diatomic Molecules*, Van Nostrand Reinhold, New York, 1979.

134. H. Werner and W. Meyer, *J. Chem. Phys.* **74**, 5794 (1981).

135. D. Grimbert, B. Lassier-Govers, and V. Sidis, *Chem. Phys.* **124**, 187 (1988).

136. V. Sidis, D. Gimbert, M. Sizun, and M. Baer, *Chem. Phys. Lett.* **163**, 19 (1989).

137. M. Noll and J. P. Toennies, *J. Chem. Phys.* **85**, 3313 (1986).
138. P. J. H. Tjossem, T. A. Cool, D. A. Webb, and E. R. Grant, *J. Chem. Phys.* **88**, 617 (1988).
139. M. Baer and H. Nakamura, *J. Chem. Phys.* **87**, 465 (1987).
140. M. Baer, G. Niedner, and J. P. Toennies, *J. Chem. Phys.* **88**, 1461 (1988).
141. M. Baer, G. Niedner-Schatteburt, and J. P. Toennies, *J. Chem. Phys.* **91**, 4169 (1989).
142. B. R. Johnson, *J. Comput. Phys.* **13**, 445 (1973).
143. M. H. Alexander, G. Parlant, and T. H. Hemmer, *J. Chem. Phys.* **91**, 2388 (1989).
144. J. Alvarellos and H. Metiu, *J. Chem. Phys.* **88**, 4957 (1988).
145. R. Heather, X. P. Jiang, and H. Metiu, *J. Chem. Phys.* **90**, 2555 (1989).

DIABATIC POTENTIAL ENERGY SURFACES FOR CHARGE-TRANSFER PROCESSES

V. SIDIS

*Laboratoire des Collisions Atomiques et Moléculaires,
Université de Paris-Sud, ORSAY Cedex, France*

CONTENTS

State-Selected and State-to-State Ion–Molecule Reaction Dynamics, Part 2: Theory, Edited by Michael Baer and Cheuk-Yiu Ng. Advances in Chemical Physics Series, Vol. LXXXII. ISBN 0-471-53263-0 © 1992 John Wiley & Sons, Inc.

I. INTRODUCTION

An electron transfer between the impinging partners of a molecular encounter generally causes changes of both their state of charge and (ro-)vibrational excitation. This process is termed (*ro-*)*vibronic charge transfer*:

$$A^+ + BC(v_0) \leftrightarrow A + BC^+(v_+), \tag{1a}$$

but the nomenclature applies as well to processes of the type

$$A^- + BC(v_0) \leftrightarrow A + BC^-(v_-) \tag{1b}$$

and

$$A(*) + BC(v_0) \leftrightarrow A^\pm + BC^\mp(v_\pm). \tag{1c}$$

In the preceding reactions A may be an atom or a molecule and the double arrow indicates that the reaction may be read from the left to the right or *vice versa*.

A. The Quasi Molecular Model

(Ro)vibronic charge transfer at small relative velocities ($v \lesssim 10^{-1}$ a.u. $= 2.18\,10^7$ cm/s) of the impinging molecules is but one of the many heavy-particle collision processes whose theoretical investigation may be undertaken within the framework of a *quasimolecular* model. The basic idea of this model is to view a slow heavy-particle collision as a process of temporary formation and then breakup of a sort quasimolecular system $(ABC)^{\pm,0}$ built from all the nuclei and electrons of the colliding partners.[1] One may thereby effect the well-known Born and Oppenheimer[2-4] *separation of electronic and nuclear motions*. As is well known this fundamental step rests on the electronic to nuclear mass ratio ($m_{el}/m_{nuc} \lesssim 1/2000$), which makes electrons move much faster than nuclei at comparable energies.

A collision problem thenceforth splits into two parts: (1) determination of electronic wavefunctions and energies for fixed nuclei and (2) treatment of the nuclear motion in the average potentials thus generated by the electrons.

B. Insufficiency of the BO Approximation

The mentioned approach is often confused with the Born and Oppenheimer (BO) *approximation*,[2,5] which consists of representing the total wavefunction of the whole quasimolecular system as a *single product*: an electronic times a nuclear wavefunction. A BO electronic wavefunction represents an *eigenstate* of the electronic hamiltonian (H_{el}), which, in view of the clamped nuclei approximation (1), is obtained from the (actual) total hamiltonian of the system by dropping out *all* nuclear kinetic energy terms (T_{nuc}). Since different BO electronic wavefunctions are obtained for different arrangements of the clamped nuclei, they depend not only on the electronic coordinates but also on those of the nuclei, albeit parametrically so. Likewise, the corresponding *eigenvalues* of the electronic hamiltonian depend on the nuclear coordinates; these electronic energies are intended to represent the potentials that govern the nuclear motions. This is the means by which the invaluable notion of *electronic potential energy surface is* introduced.

The basic reason for proceeding as just described rests on an *adiabatic hypothesis*. Electrons in a molecule are likely to move so fast compared to nuclei that they may be assumed to readjust "instantaneously" and continuously to the slowly varying nuclear field. In a molecular collision problem, such an adiabatic behavior,[6] would amount to forcing the quasimolecular system $(ABC)^{\pm,0}$ to remain in the *same electronic state throughout the collision*, namely, the one that *correlates* with the initial electronic state of the reactants by continuous changes of the nuclear geometry as imposed by the motion of the collision partners when they approach and then recede.

The conditions of validity of the BO approximation have been analyzed in many works.[7–9] Suffice it to say that a necessary condition for its applicability is the *smallness* of coupling terms arising between BO products from the *nuclear kinetic energy operators*. Owing to the appearance of these operators as first- and second-order derivatives with respect to the nuclear coordinates, this requirement is fulfilled when both the electronic wavefunctions vary slowly with nuclear displacements and when the latter motions are slow. Further scrutiny of the mentioned condition[7–9] reveals that the BO approximation based on a single-product wavefunction is likely to fail when the considered electronic state gets close in energy to other states as is the case of degeneracies or near degeneracies.

The BO approximation applies best to the ground state of a stable molecule when it is well removed energetically from higher lying states and when the nuclear motions are confined to small displacements around their equilibrium geometry. It has successfully been applied to the description of pure elastic scattering[10] and resonant charge transfer[11] in diatomic collisional systems.

Moreover, it has made it possible to investigate theoretically many reactive $(A + BC \rightarrow AB + C)$ and nonreactive collision processes (vibrational and rotational excitation) by viewing their dynamics as a *classical* evolution of nuclei along a single adiabatic potential energy surface.[12-15]

There are many cases, however, where the strict *electronically adiabatic* view conveyed by the BO approximation breaks down; to cite a few: nonradiative transitions in polyatomic molecules,[16,17] the Jahn–Teller effect,[18,19] vibronic coupling effects in molecular spectroscopy,[20] predissociation and perturbations in the spectra of diatomic molecules,[21] and inelastic as well as charge-transfer processes in collisions of atoms and molecules.[22-27] Breakdown of the BO approximation is present in the case of vibronic charge transfer of interest to us here. First, the transferred electron is located around a different partner in the beginning and at the end of the collision; thence, the description of the initial and final asymptotic conditions of a charge-transfer process implies the explicit involvement of at least two electronic states of the $(ABC)^{\pm,0}$ quasimolecule. Second, it is well known, from both textbook perturbation theory and the detailed investigation of two-state models[22,28-30] that collision-induced transitions between two states, at small relative velocities, are most likely to occur when the spacing between their energy levels is small and/or is comparable with the relevant coupling. Third, it is precisely in the regions of nuclear coordinate space where such closeness conditions are met that BO electronic wavefunctions often display their largest variations. It is thus obvious that the just depicted characteristics of a nonadiabatic phenomenon impose a "correlated" description of nuclei and electrons; this is readily achieved by a superposition of BO product wavefunctions,* which constitute the basis of the *adiabatic representation*.[3,5] Although it is at this very point that the notion of diabatic state is traditionally brought in, we will defer it until another important notion is recalled, namely, that of sudden behavior.[6]

The conditions that make a sudden behavior appear in a quantal system are very similar to those (already discussed) which determine adiabatic behavior, namely, the separation between slow and rapid motions or between the associated long and short characteristic periods, respectively. If a quantal system undergoes a perturbation over much shorter time scales than its characteristic periods, measured as representative reciprocal energy spacings, then it stays in the same state as it was initially prepared in. This will be the case in the course of a collision when the typical time taken by the system

---

*The mentioned superposition bears some resemblance with the description of "electron correlations" in atomic and molecular physics using superpositions of configurations built as (antisymmetrized) products of one-electron wavefunctions.

to change its nuclear geometry is much shorter than the reciprocal energy spacings that actually depend on that geometry. Let us now imagine a two-state system whose adiabatic potential energy surfaces $(E_2 > E_1)$ are *well separated* except in a small domain of the nuclear coordinates where they exhibit a narrow pinching with an energy gap ΔE. If the system, initially prepared in the lowest adiabatic state ψ_1^i, traverses the pinching zone in a shorter time than ΔE^{-1}, then it stays in the very ψ_1^i state until its exit whereon it projects onto the adiabatic states $\psi_{1,2}^f$. When $|\langle \psi_1^i | \psi_2^f \rangle| \approx 1$, the pinching corresponds to an *avoided-curve-crossing*[31] situation. Sudden behavior in this case makes state ψ_1^i *correlate* with ψ_2^f and vice versa $(\psi_2^i \leftrightarrow \psi_1^f)$. Such a "boring through" characterizes diabatic behavior (etymologically $\delta\iota\alpha\beta\alpha\tau\epsilon\sigma$: which may traverse). It is well known (and will be shown again in Sections III and V) that in the mentioned avoided-crossing case the electronic (adiabatic) BO wavefunctions display large (not to say huge) variations with nuclear geometry in the very pinching zone. As will also be discussed later on such rapid variations cause both conceptual and computational difficulties. The notion of diabatic behavior just introduced bears the idea of an alternative representation spanned by basis states whose *dependence on nuclear geometry is much smoother than that of BO wavefunctions when such critical situations occur.*

C. Historical Background

The notion of diabatic states has been a matter of repeated rediscovery over the past 60 years.

The widespread literature on the subject attributes their fatherhood to Lichten[32] and first formal definition to Smith.[33] Almost everyone acknowledges, however, that they were implicit in studies performed in the early 1930s on nonadiabatic transitions in two-state models (whatever curve crossing[28,29] or noncrossing[30] models). Actually, both the terminology and definition of these states in their modern acceptance appeared as early as 1935 in a work by Hellmann and Sirkin[34] on "the anomalously small steric factors in chemical kinetics." Like many of their contemporaries these authors were concerned among other things with the curve-crossing problem in ionic covalent reactions,[28b,35] which they studied using a semiclassical approach where the electrons of the quasimolecular collisional system are treated quantally whereas the nuclei are treated classically.[1] Their analysis of the problem led them to introduce diabatic states defined as linear combinations of eigenstates of the electronic hamiltonian (H_{el}) having *negligible couplings via the nuclear momentum* operators. Diagonal matrix elements of H_{el} between such states thereby represented *diabatic potential energy curves* that could run freely through avoided adiabatic crossings.[31] Corresponding off-diagonal matrix elements of H_{el} would induce electronic transitions between two diabatic states. This pioneering work has, however, passed unnoticed up to the mid 1970s.

The suggestion to abandon BO electronic states in favor of alternative wavefunctions displaying as slow a dependence on nuclear displacements as possible reappeared in the early 1960s in two contributions by Hobey and McLachlan[7] and McLachlan[36] who dealt with "the wavefunctions of electronically degenerate states and the dynamic Jahn–Teller effect in polyatomic systems." The required transformation and its conditions of existence were investigated nearly as is done today to define diabatic states. Yet, the field was not mature enough to grasp at this occasion of building a theory of slow nonadiabatic molecular collisions based on a special nonadiabatic representation of the electronic hamiltonian. As pointed out in Ref. 37," in the mid 1960s one could barely treat electronic transitions in low and medium energy collisions between the simplest atomic partners (e.g., H and He). The state of the theory was such that it was believed that the detailed study of molecular collision processes would remain for long an exclusive theme of experimental physics." In addition it was not well perceived except in pinpoint contributions[38–40] how beneficial a molecular description of heavy particle collisions would be.

In a series of articles dealing with atomic collision problems Lichten[32,41,42] triggered a sort of "thought Renaissance" in the field by promoting four governing ideas. First, molecular physics may be amalgamated into collision theory with the objective of understanding electronic transitions; these manifest themselves either as anomalies in elastic scattering and resonant charge transfer or as inelastic processes (including charge transfer and ionization). Second, processes exist in many-electron systems *that cannot be accounted for by an adiabatic theory* thereby forcing one to abandon it (or at least some of its prescriptions) in favor of a quasiadiabatic theory that emphasizes the description of molecular states in terms of configuration-state function (CSF) built from *molecular orbitals* (MO).[43,44] Third, such states termed *diabatic* may freely undergo *crossings*; diabatic curve crossings are readily predicted from *correlation diagrams* connecting (one-electron) MO and (many-electron) CSF energy levels at infinite and zero interparticle distances (separated- and fused-collision-partner limits, respectively).[44] Fourth, a hint at the importance of electron transitions at the thus revealed crossings could thenceforth be estimated using results of the celebrated Landau–Zener[28,29] model. An important merit of Lichten in this context has been to stress the physicochemical content of the diabatic states he was just introducing. He pointed out that in the curve-crossing problem the diabatic states are not just convenient mathematical constructs but actually convey specific information on characters the system is likely to retain during its evolution (as, for example, a MO configuration). A prolongation of this original approach has been the emergence of the so-called MO promotion model[41] that aimed at the prediction of *diabatic MO crossings* (in the newly introduced sense) in symmetric[41] and asymmetric[45] diatomic systems.* Lichten's masterly ideas have not had immediate impact on the quantum-chemistry community both because of their qualitative nature and the emphasis they were putting on special rudimentary MO and single-configuration descriptions of the diabatic states. For nearly two decades such states thereby dragged the bad reputation of being inaccurate, and useful (if at all) for energetic collisions between two atomic systems. Yet both Lichten's ideas and the rapid development of the so-called "collision spectroscopy"[51,52] in the late 1960s provoked a strong incitement for the search of a formal definition of diabatic states, which would hopefully be as rigorous as that of the well-established adiabatic states and would allow their determination by state-of-the art quantum-chemistry techniques.

Three contributions actually provided the basic formalism sought.[33,53−56] The earlier prescriptions of Refs. 7, 34, and 36 were thus rediscovered in part by Levine et al.[53] and Smith[33] who suggested that one define diabatic electronic states as those that avoid possible rapid variations of the adiabatic states with respect to nuclear coordinates. Smith's general analysis of the conditions to be fulfilled by the unitary adiabatic-to-diabatic transformation achieving this requirement made his proposal to be retained as the standard definition of diabatic states. A quite different formulation based on the Feshbach projection operator technique was put forward by O'Malley.[54−56] Originally, the proposals of Smith and O'Malley looked quite distinct. Later practice revealed that they actually pursue the same goal[27]: the removal of badly behaved off-diagonal matrix elements of the nuclear momentum operator between electronic BO states. The former does so in an *a posteriori* way, that is, after having determined these matrix elements, whereas the latter does it in an *a priori* way, that is, by controlled transformation of a basis in which matrix elements of nuclear momentum operator are knwon beforehand (e.g., by construction) to be null or negligible.

---

*This model rests on the idea that adiabatic MO of many-electron systems as provied by molecular structure calculations[46] often display *avoided crossings* that one may tentatively put in correspondence with *true crossings* in systems made of a single electron and two bare nuclei. Its basic argument is that both inner shells and Rydberg shells of diatomic systems mimic those of one-electron diatomic molecules. Thus, diabatic MO should maintain to some extent the same behavior and should thus exhibit the same sort of level crossings. The latter crossings happen to occur owing to the special symmetry of the two-center Coulomb field[47−50] and obey rules[50] that are actually exploited in the MO promotion model. In many-electron systems electron transitions are induced at those crossings by the screening effects, which cause the departure from the pure Coulomb field. It is precisely those effects that are responsible for the avoidance of the mentioned crossings in the adiabatic view.

The contributions of Smith[33] and O'Malley[53] have caused significant progress in the study of the dynamics of nonadiabatic transitions in diatomic systems during the 1970s. Yet, many fewer attempts were made during the same period to apply similar ideas to more general polyatomic collisional systems. One reason for such a lack of interest might have been the widespread use of the so-called trajectory surface hopping (TSH) description of molecular collisions.[25,57] The propagation of classical trajectories, in the TSH model, to describe the evolution of the nuclear motions is quite frequently achieved using adiabatic potential energy surfaces; yet, the explicit handling of cumbersome couplings between the related states in numerical work is never truly needed in TSH, since it does not care for a rigorous description of an electronic transition *per se*. Therefore, adiabatic states seemed to be sufficient. Moreover, despite a lengthy literature on the drawbacks of adiabatic electronic bases in nonadiabatic conditions, the stubborn nonsensical idea that they still uniquely provide the most rigorously founded description of the dynamics kept being perpetuated.

Nevertheless, the progressive undertaking of quantal descriptions of nonadiabatic molecular collisions[26,37,58-60] led to the readvocation of the use of a diabatic representation instead of the ordinary adiabatic one. Baer[59] (and later Chapuisat et al.[61]) thereby extended Smith's[33] idea to polyatomic systems.

Over a decade an abundant literature grew concerned with those definitions of diabatic states[23,24,61-67] and with methods of determining them in practice using state-of-the-art quantum-chemistry techniques. The present review intends to survery some of these methods with specific focus on those which help dealing with vibronic charge transfer.

II. PRELIMINARIES

A. Coordinates

We consider the general case of a polyatomic system made of the $N_{nuc} = N_{nuc}^{\mathcal{T}} + N_{nuc}^{\mathcal{P}}$ nuclei and $N_{el} = N_{el}^{\mathcal{T}} + N_{el}^{\mathcal{P}}$ electrons of the colliding target $\mathcal{T}$ and projectile $\mathcal{P}$. The quantum-mechanical hamiltonian of this system in the laboratory reference frame (in atomic units)

$$H_{lab} = -\frac{1}{2}\left(\sum_I^{N_{nuc}} m_I^{-1} \Delta_I + \sum_i^{N_{el}} \Delta_i \right) + \mathcal{V} \qquad (2)$$

is first transformed by separating off its center-of-mass (c.m.) motion. An appropriate set of $3(N_{nuc} + N_{el}) - 3$ relative coordinates is then chosen to describe the particles movements in the c.m. frame.

Let us first consider the nuclei. Ordinarily, coordinate sets that enable one to write the nuclear kinetic energy operator in the form of sums of individual relative kinetic energy operators are preferred. This is readily achieved iteratively by a well-known method; one defines the first three relative coordinates as components of the vector joining an arbitrary pair of nuclei. A second relative vector is next formed by joining a third nucleus to the c.m. of the initial pair, the third relative vector is then obtained by joining a fourth nucleus to the c.m. of the previous three, and so on. Alternatively, one may apply this procedure to subsets of nuclei forming aggregates and subsequently use it again to form interaggregate relative vectors. In a collision process that does not involve any chemical bond rearrangement between the impinging molecules [Eqs. (1a)–(1c)] one obvious interaggregate distance of this sort is the vector $\mathbf{R}$ joining the nuclear c.m. of the target ($\mathcal{T}$) to that of the projectile ($\mathcal{P}$). The kinetic energy operator T_{nuc} in the c.m. frame then writes:

$$T_{\text{nuc}} = -\frac{1}{2\mu}\Delta_{\mathbf{R}} + T_{\text{nuc}}^{\mathcal{T}} + T_{\text{nuc}}^{\mathcal{P}}, \qquad (3)$$

where the last two terms are nuclear kinetic energy operators of the target and projectile in their own c.m. frame and μ is their reduced mass:

$$\mu^{-1} = m_{\mathcal{T}}^{-1} + m_{\mathcal{P}}^{-1}. \qquad (4)$$

The previously mentioned (Jacobi-type) procedure to sequentially build sets of relative coordinates is seldom (if ever) applied to electrons. Rather, it is a usual practice to refer them to a common origin lying on the $\mathbf{R}$ axis at a distance $\mathcal{D}_{\mathcal{P}}R$ from the projectile and $\mathcal{D}_{\mathcal{T}}R$ from the target ($\mathcal{D}_{\mathcal{P}} + \mathcal{D}_{\mathcal{T}} = 1$). With such a choice of electronic coordinates (noted $\mathbf{q}_i$), the kinetic energy T_{el} of the N_{el} electrons in the c.m. frame is

$$T_{\text{el}} = -\frac{1}{2}\sum_i \Delta_{\mathbf{q}_i} - \left(\frac{\mathcal{D}_{\mathcal{P}}}{m_{\mathcal{T}}} - \frac{\mathcal{D}_{\mathcal{T}}}{m_{\mathcal{P}}}\right)\mathbf{\nabla}_{\mathbf{R}} \cdot \sum_i \mathbf{\nabla}_{\mathbf{q}_i}$$

$$+ \left(\frac{\mathcal{D}_{\mathcal{P}}^2}{m_{\mathcal{T}}} - \frac{\mathcal{D}_{\mathcal{T}}^2}{m_{\mathcal{P}}}\right)\left(\sum_i \mathbf{\nabla}_{\mathbf{q}_i}\right)^2. \qquad (5a)$$

Clearly the choice

$$\mathcal{D}_{\mathcal{P}}/\mathcal{D}_{\mathcal{T}} = m_{\mathcal{T}}/m_{\mathcal{P}} \qquad (5b)$$

cancels the second term in T_{el} and identifies the origin of electronic coordinates with the nuclear c.m. G. The third term in T_{el} represents a mass polarization contribution; owing to the electron-to-mass ratio, this term is negligible with respect to the first term and is usually dropped.

The potential $\mathscr{V}$ depends only on the nucleus–nucleus, electron–nucleus, and electron–electron interparticle distances and is thereby not affected by the lab–c.m. transformation. Neither is it affected by a rotation of the c.m. reference frame. This gives the freedom to express the hamiltonian in the so-called body-fixed (BF) frame, which rotates with the **R** axis about G as the collision proceeds. The azimuthal (Φ) and polar (Θ) spherical components of **R** in the fixed reference frame constitute the first two Euler angles of the discussed rotation. The third Euler angle noted δ is associated with a rotation of the system as a wole about the **R** axis. (This angle is of no relevance in linear systems.) One is thus left in the BF frame with $3(N_{nuc} + N_{el}) - e$ internal coordinates, e equals 6, in general, except for linear systems in which case it amounts to 5.

Internal nuclear coordinates actually represent *deformations* of the nuclear geometry. They may be split into two sets $(3N_{nuc}^{\mathscr{T}} - e^{\mathscr{T}}) + (3N_{nuc}^{\mathscr{P}} - e^{\mathscr{P}})$ coordinates represent vibration-type motions of the individual $\mathscr{T}$ and $\mathscr{P}$ aggregates. The remaining $e^{\mathscr{T}} + e^{\mathscr{P}} - e$ coordinates represent, respectively, the relative $\mathscr{T} - \mathscr{P}$ motion along **R**, a relative $\mathscr{T} - \mathscr{P}$ (twist) rotation about **R**, and two independent rotations for each of $\mathscr{T}$ and $\mathscr{P}$.* Defining a polar axis $\mathbf{Z}_{\mathscr{T},\mathscr{P}}$ attached to each aggregate $\mathscr{T}, \mathscr{P}$ enables one to characterize these rotations by the angles $\gamma_{\mathscr{T},\mathscr{P}}$ and $\delta_{\mathscr{T},\mathscr{P}}$; the former two specify rotations about axes perpendicular to the respective $(\mathbf{R}, \mathbf{Z}_{\mathscr{T},\mathscr{P}})$ planes, whereas the latter two correspond to rotations about each of the chosen $\mathbf{Z}_{\mathscr{T},\mathscr{P}}$ axes.

The dependence of the potential $\mathscr{V}$ upon nuclear coordinates occurs only through the $3N_{nuc} - e$ internal degrees of freedom. Therefore, as will appear shortly, *they are the only coordinates to be considered when dealing with adiabatic versus diabatic choices.*

Let us illustrate the matter somewhat. For an atom–atom collision, the sole internal deformation coordinate system is R. Consideration of an atom–diatom system adds up two new internal deformation coordinates, for example, the diatom bond distance r and the $(\mathbf{R}, \mathbf{r})$ angle γ. In a diatom–diatom system three additional coordinates arise, namely, the r', γ' pair as just defined for one diatom, plus the dihedral angle $\eta_{\mathscr{T}\mathscr{P}}$ between the $(\mathbf{R}, \mathbf{r})$ and $(\mathbf{R}, \mathbf{r}')$ planes. For an atom–triatom collision one has, in addition to R:

1. The three internal coordinates determining the shape of the triatomic aggregate [for example, the bond distance d of a diatomic pair, the distance r of the third atom to the diatom c.m., and the angle $(\mathbf{d}, \mathbf{r})$].

---

*When $\mathscr{P}$ or $\mathscr{T}$ is an atom, the latter two rotations and the twisting motions are irrelevant. When $\mathscr{P}$ or $\mathscr{T}$ is a linear molecule, a rotation about its axis is irrelevant.

2. Two overall rotations of the triatomic plane with respect to **R** [e.g., one rotation characterized by the angle γ_{tri} takes place around an axis perpendicular to the $(\mathbf{r}, \mathbf{R})$ plane and another one characterized by the angle δ_{tri}, takes place around **r** itself].

Obviously, such choices are not mandatory. They are often suggested by the physics of the considered process. For instance, if, as a result of a $BCD^+ + A$ charge-transfer collision the triatomic BCD neutral dissociates[68] into a unique atom–diatom arrangement $B + CD$, then the preceding coordinate set (with $\mathbf{r}_{B,CD}$ and $\mathbf{d}_{C,D}$) is a natural one. On the other hand, if the other $BC + D$ and/or $BD + C$ dissociation arrangements are also possible, coordinate sets designed to treat rearrangement processes (as hyperspherical[69] or Eckart[70] coordinates) should be used. At the other extreme, the choice of internal nuclear coordinates in the form of small nuclear displacements with respect to a reference nuclear configuration[71] may be adequate to the treatment of cases when bulky targets and/or projectiles undergo small deformations with respect to their equilibrium geometry throughout the collision.

B. The Hamiltonian in the BF Frame

There are many ways in which body-fixed reference frames may be chosen. Except if stated otherwise, we will consider the set of internal electronic: $\{\boldsymbol{\rho}_i\}_{i=1,N_{el}}$ and Jacobi-type nuclear: $R, r_{\mathscr{T}}, \gamma_{\mathscr{T}}, \{\mathbf{r}_\alpha\}_{\alpha=1,N_{nuc}-3}$ coordinates in the rotating BF frame, which is characterized by the three Euler angles Φ, Θ, δ. The z axis in the BF frame lies along **R** and the BF $(\mathbf{Gz}, \mathbf{Gx})$ plane contains a (Jacobi-type) internal nuclear vector $\mathbf{r}_{\mathscr{T}}$ of the target aggregate (Fig. 1). We have:

$$H = T_{nuc} + H_{el}, \tag{6}$$

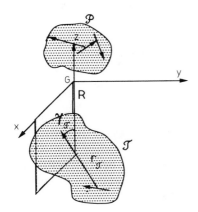

Figure 1. The body-fixed reference frame. Dotted areas represent the target ($\mathscr{T}$) and projectile ($\mathscr{P}$) aggregates. G is the $\mathscr{T} - \mathscr{P}$ center of mass. Vectors in the dotted areas represent schematically Jacobi-type coordinates. One of these vectors ($\mathbf{r}_{\mathscr{T}}$) in the target aggregate is chosen to define the $(\mathbf{Gz}, \mathbf{Gx}) \equiv (\mathbf{r}_{\mathscr{T}}, \mathbf{R})$ plane. The three Euler angles (not shown) that determine the position of the body-fixed frame with respect to the space-fixed frame are as usual the two spherical angles of vector $\mathbf{R} \equiv \mathbf{Gz}$ and a rotation of angle δ of the $(\mathbf{Gx}, \mathbf{Gz})$ plane about **R**.

where

$$H_{el} = -\frac{1}{2}\sum_{i=1}^{N_{el}}\Delta_{\rho_i} + \mathscr{V}(\{\rho_i\}, R, r_{\mathscr{F}}, \gamma_{\mathscr{F}}, \{r_\alpha\})$$ (7)

and[60,72]

$$
\begin{aligned}
T_{nuc} = &-\frac{1}{2\mu R^2}\left\{\frac{\partial}{\partial R}R^2\frac{\partial}{\partial R} - (K^2 - 2K_z) - (\Gamma^2 - 2\Gamma_z) - \frac{1}{\sin^2\gamma_{\mathscr{F}}}(K_z - \Gamma_z)^2\right.\\
&-\frac{\partial^2}{\partial\gamma_{\mathscr{F}}^2} - \cot\gamma_{\mathscr{F}}\frac{\partial}{\partial\gamma_{\mathscr{F}}} + (K_+ - \Gamma_+)\left[-\cot\gamma_{\mathscr{F}}(K_z - \Gamma_z) - \frac{\partial}{\partial\gamma_{\mathscr{F}}}\right]\\
&\left.+ (K_- - \Gamma_-)\left[-\cot\gamma_{\mathscr{F}}(K_z - \Gamma_z) + \frac{\partial}{\partial\gamma_{\mathscr{F}}}\right] + \Gamma_+K_- + \Gamma_-K_+\right\}\\
&-\frac{1}{2\mu_{\mathscr{F}}r_{\mathscr{F}}^2}\left[\frac{\partial}{\partial r_{\mathscr{F}}}r_{\mathscr{F}}^2\frac{\partial}{\partial r_{\mathscr{F}}} - \frac{1}{\sin^2\gamma_{\mathscr{F}}}(K_z - \Gamma_z)^2 - \frac{\partial^2}{\partial\gamma_{\mathscr{F}}^2} - \cot\gamma_{\mathscr{F}}\frac{\partial}{\partial\gamma_{\mathscr{F}}}\right]\\
&-\sum_{\alpha=1}^{N_{nuc}-3}\frac{1}{2\mu_\alpha}\Delta_{r_\alpha},
\end{aligned}
$$ (8)

where

$$K^2 = -\frac{\partial^2}{\partial\Theta^2} - \cot\Theta\frac{\partial}{\partial\Theta} - \frac{1}{\sin^2\Theta}\left[\frac{\partial^2}{\partial\Phi^2} + \frac{\partial^2}{\partial\delta^2} - 2\cos\Theta\frac{\partial^2}{\partial\Phi\partial\delta}\right],$$ (9a)

$$K_\pm = \exp(\mp i\delta)\left(-i\cot\Theta\frac{\partial}{\partial\delta} + \frac{i}{\sin\Theta}\frac{\partial}{\partial\Phi} \pm \frac{\partial}{\partial\Theta}\right),$$ (9b)

$$K_z = -i\frac{\partial}{\partial\delta},$$ (9c)

$$\Gamma = m + L, \quad m = -i\sum_{\alpha=1}^{N_{nuc}-3}r_\alpha \wedge \nabla_{r_\alpha}, \quad L = -i\sum_{j=1}^{N_{el}}\rho_j \wedge \nabla_{\rho_j},$$ (10a)

$$\Gamma_\pm = \Gamma_x \pm i\Gamma_y$$ (10b)

All partial derivatives in the preceding equations are to be effected holding the other coordinates fixed in the body-fixed reference frame. If spin–orbit interaction is to be taken into account in $\mathscr{V}$, then the electronic orbital angular momentum term in Γ [Eqs. (8) and (10a)] is to be replaced by the total $J = L + S$ electronic angular momentum. The appearance of the electronic angular momentum terms in T_{nuc} is readily understood as a

Coriolis-type interaction arising from the description of the system in a rotating frame. Likewise, the appearance of $\mathfrak{m}$ components has the same origin. Obviously, $\mathfrak{m}$ does not appear for atom–atom or atom–diatom collisional systems.

The stationary quantum-mechanical description of the $\mathcal{T} - \mathcal{P}$ quasi-molecular collision system stems from the Schrödinger equation:

$$(T_{\text{nuc}} + H_{\text{el}})\Psi = E\Psi, \tag{11}$$

where E is the total (conservative) energy of the system and Ψ is its total wavefunction. As is frequently done when treating correlated motions of particles in a many-body system, one may first write Ψ in the form of an expansion over wavefunction products: each factor in these products describes independent motions or independent groups of correlated motions. Accordingly, Ψ may be sought in the form:

$$\Psi = \sum_k D_k(\Phi, \Theta, \delta) \sum_n \mathcal{N}_{kn}^{\text{nuc}}(R, r_{\mathcal{T}}, \gamma_{\mathcal{T}}, \{\mathbf{r}_\alpha\}) \varphi_n^{\text{el}} \tag{12}$$

$\mathcal{N}_{kn}^{\text{nuc}}$ and φ_n^{el} describe, respectively, the collective internal nuclear and electronic motions, whereas D_k describes overall rotations of given spatial configurations of electrons and nuclei.

The problem of defining and using adiabatic or diabatic representations has to do with the actual choice of φ_n^{el} basis functions. As will be seen shortly, these functions are universally chosen to depend on the electronic coordinates $\{\boldsymbol{\rho}_j\}$ *and possibly* on $R, r_{\mathcal{T}}, \gamma_{\mathcal{T}}, \{\mathbf{r}_\alpha\}$ but *not* on the Euler angles Φ, Θ, δ; hence, the operators $K^2, K_{z,\pm}$ in T_{nuc} act exclusively on D_k. The latter functions may thus be chosen as the coefficients of the irreducible representation of the rotation group.[73] In that case the label k stands collectively for there indices K, M_K, Ω:

$$D_k = D_{M_K\Omega}^K \tag{13a}$$

$$K^2 D_{M_K\Omega}^K = K(K+1)D_{M_K\Omega}^K, \quad K_z D_{M_K\Omega}^K = \Omega D_{M_K\Omega}^K, \tag{13b}$$

$$K_\pm D_{M_K\Omega}^K = \sqrt{K(K+1) - \Omega(\Omega-1)} D_{M_K\Omega\mp 1}^K \tag{13c}$$

which may be used when inserting Ψ into Eqs. (6), (8), and (11). In addition the $L^2, L_{z,\pm}$ operators act solely on φ_n^{el}. Terms in T_{nuc} depending exclusively on combinations of the latter six operators are related to rotations and electrorotational interactions. These terms will altogether be noted $\mathcal{R}$. So that T_{nuc} appears in the form:

$$T_{\text{nuc}}$$

$$= -\frac{1}{2\mu R^2}\left\{\frac{\partial}{\partial R}R^2\frac{\partial}{\partial R} - \Upsilon_1 - \Upsilon_2 + \mathfrak{m}_+ K_- + \mathfrak{m}_- K_+\right.$$

$$+\left[\cot\gamma_{\mathscr{F}}(K_+\mathfrak{m}_z + \mathfrak{m}_+ K_z - \mathfrak{m}_+\mathfrak{m}_z - \mathfrak{m}_+ L_z - L_+\mathfrak{m}_z) - (K_+ - \Gamma_+)\frac{\partial}{\partial\gamma_{\mathscr{F}}}\right]$$

$$+\left.\left[\cot\gamma_{\mathscr{F}}(K_-\mathfrak{m}_z + \mathfrak{m}_- K_z - \mathfrak{m}_-\mathfrak{m}_z - \mathfrak{m}_- L_z - L_-\mathfrak{m}_z) + (K_- - \Gamma_-)\frac{\partial}{\partial\gamma_{\mathscr{F}}}\right]\right\}$$

$$-\frac{1}{2\mu_{\mathscr{F}}r_{\mathscr{F}}^2}\left[\frac{\partial}{\partial r_{\mathscr{F}}}r_{\mathscr{F}}^2\frac{\partial}{\partial r_{\mathscr{F}}} - \Upsilon_2\right] - \sum_{\alpha=1}^{N_{\text{nuc}}-3}\frac{1}{2\mu_\alpha}\Delta_{\mathbf{r}_\alpha} + \mathscr{R}, \tag{14}$$

where:

$$\Upsilon_1 = \mathbf{m}^2 + 2\mathbf{m}\cdot\mathbf{L} - 2\mathfrak{m}_z \tag{15}$$

$$\Upsilon_2 = \frac{\partial^2}{\partial\gamma_{\mathscr{F}}^2} + \cot\gamma_{\mathscr{F}}\frac{\partial}{\partial\gamma_{\mathscr{F}}} + \frac{1}{\sin^2\gamma_{\mathscr{F}}}(-2K_z\mathfrak{m}_z + 2\mathfrak{m}_z L_z + \mathfrak{m}_z^2). \tag{16}$$

III. THE ADIABATIC REPRESENTATION

As recalled in Section I, the electronic basis set $\{\varphi_n^{\text{el},a}\}$ characterizing this representation stems from the Born and Oppenheimer approximation; it is generated for each *fixed geometry of the nuclei* $(R, r_{\mathscr{F}}, \gamma_{\mathscr{F}}, \{\mathbf{r}_\alpha\})$ by:

$$H_{\text{el}}\varphi_n^{\text{el},a}(\{\boldsymbol{\rho}_i\}; R, r_{\mathscr{F}}, \gamma_{\mathscr{F}}, \{\mathbf{r}_\alpha\}) = E_n^a(R, r_{\mathscr{F}}, \gamma_{\mathscr{F}}, \{\mathbf{r}_\alpha\})\varphi_n^{\text{el},a}(\{\boldsymbol{\rho}_i\}; R, r_{\mathscr{F}}, \gamma_{\mathscr{F}}, \{\mathbf{r}_\alpha\}), \tag{17}$$

where indices following a semicolon specify parametric dependences of the electronic wavefunctions. Using this definition in the total Schrödinger equation (11) yields

$$[T_{\text{nuc}} + E_n^a(R, r_{\mathscr{F}}, \gamma_{\mathscr{F}}, \{\mathbf{r}_\alpha\}) - E]F_n^a = \sum_m \Xi_{nm}F_n^a, \tag{18}$$

with

$$F_n^a = \sum_k D_k(\Phi, \Theta, \delta)\mathscr{N}_{kn}^a(R, r_{\mathscr{F}}, \gamma_{\mathscr{F}}, \{\mathbf{r}_\alpha\}). \tag{19}$$

The adiabatic representation thereby compels us to evaluate the Ξ_{nm} coupling terms. Putting aside electrorotational (Coriolis-type) interactions in $\mathscr{R}$ [Eq.

(14)], the so-called internal deformation-type *nonadiabatic* coupling matrix elements contributing to Ξ_{nm} are of the form

$$T_{nm}^{(I)} = \langle \varphi_n^{\text{el},a} | \frac{\partial}{\partial Q_I} | \varphi_m^{\text{el},a} \rangle, \tag{21}$$

$$T_{nm}^{(IJ)} = \langle \varphi_n^{\text{el},a} | \frac{\partial^2}{\partial Q_I \partial Q_J} | \varphi_m^{\text{el},a} \rangle, \tag{22}$$

where Q_I $(I = 1, 3N_{\text{nuc}} - e)$ is any of $R, r_{\mathcal{F}}, \gamma_{\mathcal{F}}$ or any of the three components of a vector in the set $\{\mathbf{r}_\alpha\}$. It is readily established that

$$T_{nm}^{(IJ)} = \frac{\partial T_{nm}^{(J)}}{\partial Q_I} + \sum_\lambda T_{n\lambda}^{(I)} T_{\lambda m}^{(J)} \tag{23a}$$

$$= T_{nm}^{(JI)}. \tag{23b}$$

Moreover, using the commutation relation

$$\left[\frac{\partial}{\partial Q_I}, H_{\text{el}} \right] = \frac{\partial H_{\text{el}}}{\partial Q_I} = \frac{\partial \mathscr{V}}{\partial Q_I}, \tag{24}$$

one gets, for $m \neq n$,

$$T_{nm}^{(I)} = -[E_n^a - E_m^a]^{-1} \langle \varphi_n^{\text{el},a} | \frac{\partial \mathscr{V}}{\partial Q_I} | \varphi_m^{\text{el},a} \rangle \tag{25a}$$

$$= -T_{mn}^{(I)*}. \tag{25b}$$

These properties reveal the major drawback of adiabatic basis functions, namely, *the importance of $T_{nm}^{(I)}$ and $T_{nm}^{(IJ)}$ matrix elements and thus that of Ξ_{nm} coupling terms when the energy difference between two potential energy hypersurfaces is small.* Even worse, these matrix elements may diverge or become discontinuous in polyatomic systems $(N_{\text{nuc}} \geqslant 3)$ near crossings of potential energy hypersurfaces[31,58,74-77] (see Section V). Although methods exist to calculate $T_{nm}^{(I)}$ matrix elements,[76-85] it is obvious that the numerical work involved in their determination and *effective use* in close-coupling treatments of reaction dynamics is a quite tedious task when they become singular. Actual calculations of $T_{nm}^{(R)}, T_{nm}^{(r_{\mathcal{F}})}$, and $T_{nm}^{(\gamma_{\mathcal{F}})}$ matrix elements have been calculated for a few triatomic and some tetra-atomic systems.[77-86] Still, in one category of work the nonadiabatic matrix elements were provided for restricted samples of nuclear geometries and were only used in qualitative discussions.[78,79,81-84] In another category of work they were just used to

pinpoint regions of strong nonadiabatic coupling were sound albeit rudimentary recipes provided transition probabilities from one electronic state
to another.[25,75,86-87]* Finally in a third category of works they were
immediately transformed away by a change of representation[58,60,77] before
undertaking the scattering calculations.

IV. THE DIABATIC REPRESENTATION

It is important to recall that the actual collision problem consists of solving
Eq. (11) using an expansion of the system's wavefunction in the form given
by Eq. (12). Obviously, there is absolutely no obligation to chose φ_n^{el} as a
solution of Eq. (17).

Formally, a strictly diabatic basis set $\{\varphi_n^{el,d}\}$ may be defined by the
condition[7,33,34,36,59]

$$\langle \varphi_n^{el,d} | \frac{\partial}{\partial Q_I} | \varphi_m^{el,d} \rangle = 0, \quad \forall n \neq m, \forall Q_I, \tag{26}$$

which removes at once the previously mentioned drawbacks of the adiabatic
representation. With this definition, the stem of coupled equations equivalent
to Eq. (18) is

$$[T_{\text{nuc}} + E_n^d(R, r_{\mathcal{F}}, \gamma_{\mathcal{F}}, \{\mathbf{r}_\alpha\}) - E]F_n^d = \sum_m (H_{nm} + \mathscr{C}_{nm})F_m^d, \tag{27}$$

where orthonormality ($\langle \varphi_m^{el,d} | \varphi_n^{el,d} \rangle = \delta_{mn}$) has been assumed and

$$E_n^d = \langle \varphi_n^{el,d} | H_{el} | \varphi_n^{el,d} \rangle, \tag{28}$$

$$H_{mn} = (1 - \delta_{mn}) \langle \varphi_m^{el,d} | H_{el} | \varphi_n^{el,d} \rangle, \tag{29}$$

$$\mathscr{C}_{mn} = \int d\boldsymbol{\rho}_1 \cdots d\boldsymbol{\rho}_{N_{el}} \varphi_m^{el,d*} [\mathscr{R}, \varphi_n^{el,d}], \tag{30a}$$

$$[\mathscr{R}, \varphi_n^{el,d}] = (\mathscr{R}\varphi_n^{el,d}) - \varphi_n^{el,d}\mathscr{R}. \tag{30b}$$

F_n^d is defined by Eq. (19) with the replacement of the superscript a by d.

---

*Two exceptions, however, are the works on the $Na(^2P) + H_2$ quenching by Blais and
Truhlar[88] and Eaker[89] who did use $\tau_{nm}^{(I)}$ matrix elements in time-dependent coupled equations
to determine transition probabilities for a particular use in the TSH[25] method.

Yet, Eq. (26) is a trivial definition;[63,90,91] indeed, since, in general,

$$\frac{\partial}{\partial Q_I}|\varphi_n^{el}\rangle = \sum_m |\varphi_m^{el}\rangle\langle\varphi_m^{el}|\frac{\partial}{\partial Q_I}|\varphi_n^{el}\rangle, \tag{31}$$

it entails

$$\frac{\partial}{\partial Q_I}|\varphi_n^{el,d}\rangle = |\varphi_n^{el,d}\rangle\langle\varphi_n^{el,d}|\frac{\partial}{\partial Q_I}|\varphi_n^{el,d}\rangle. \tag{32}$$

In addition differentiation of the orthonormality condition yields

$$\langle\varphi_m^{el,d}|\frac{\partial}{\partial Q_I}|\varphi_n^{el,d}\rangle = -\langle\varphi_n^{el,d}|\frac{\partial}{\partial Q_I}|\varphi_m^{el,d}\rangle^* \tag{33}$$

which shows that $\langle\varphi_n^{el,d}|\partial/\partial Q_I|\varphi_n^{el,d}\rangle$ in Eq. (32) is a purely imaginary number. Hence, one immediately realizes that Eq. (26) yields an electronic basis, which, except for phase factors, *does not depend on any of the $3N_{nuc} - e$ internal nuclear coordinates*. The mentioned phase factors depend exclusively on the internal nuclear coordinates and may thus be ignored since they may be incorporated in F_n^d. The strict definition (26) of diabatic states is obviously useless for any practical purpose. Indeed, since it does not refer to any simplifying physicochemical property of the system, it implies that reference is permanently being made to a complete basis set of functions representing $3N_{el}$ motions. Useful definitions should thereby achieve condition (26) approximately by restricting it to certain *finite electronic subspaces*, and/or to *limited domains of internal nuclear coordinates* and/or by requiring that $\langle\varphi_n^{el,d}|\partial/\partial Q_I|\varphi_m^{el,d}\rangle$ be *small enough to be neglected*.

A.   On the Transformation of an Arbitrary Representation to a Strictly Diabatic Representation

We will examine here whether strictly diabatic bases may be defined by an orthogonal transformation $\mathbb{C}$ of an arbitrary basis in a finite subspace.

Let us assume that a basis of N_{bas} electronic functions $\{\varphi_i^{el}\}$ has been selected to treat a given nonadiabatic molecular collision process. This may be the adiabatic basis of Section III or not. If an equivalent diabatic basis of N_{bas} function $\{\varphi_n^{el,d}\}$ exists, it obeys the orthogonal transformation equation

$$\varphi_n^{el,d} = \sum_{i=1}^{N_{bas}} \varphi_i^{el}C_{in} \tag{34}$$

Fulfilment of Eq. (26) in this basis entails[33,59,63]:

$$\frac{\partial \mathbb{C}}{\partial Q_I} + \mathbb{T}^{(I)}\mathbb{C} = 0, \quad \forall Q_I. \tag{35}$$

The elements of matrix $\mathbb{T}^{(I)}$ are defined as in Eq. (21), without imposing the superscript a; those of matrix $\mathbb{C}$ are defined in Eq. (34). The necessary and sufficient conditions for a solution of this equation to exist are[59,60]

$$\frac{\partial \mathbb{T}^{(I)}}{\partial Q_J} - \frac{\partial \mathbb{T}^{(J)}}{\partial Q_I} = [\mathbb{T}^{(I)}, \mathbb{T}^{(J)}], \quad \forall I, J. \tag{36}$$

With Eqs. (35) and (36) $\mathbb{C}$ formally* obeys the integral equation[59,92]

$$
\begin{aligned}
\mathbb{C}(Q_1, Q_2, &\ldots, Q_{3N_{\text{nuc}}-e}) \\
= \mathbb{C}&(Q_1^0, Q_2^0, \ldots, Q_{3N-e}^0) \\
&+ \int_{Q_1}^{Q_1^0} dQ_1' \, \mathbb{T}^{(1)}(Q_1', Q_2^0, \ldots, Q_{3N_{\text{nuc}}-e}^0)\mathbb{C}(Q_1', Q_2^0, \ldots, Q_{3N_{\text{nuc}}-e}^0) \\
&+ \int_{Q_2}^{Q_2^0} dQ_2' \, \mathbb{T}^{(2)}(Q_1, Q_2', \ldots, Q_{3N_{\text{nuc}}-e}^0)\mathbb{C}(Q_1, Q_2', \ldots, Q_{3N_{\text{nuc}}-e}^0) \\
&+ \int_{Q_{3N_{\text{nuc}}-e}}^{Q_{3N_{\text{nuc}}-e}^0} dQ_{3N_{\text{nuc}}-e}' \, \mathbb{T}^{(3N_{\text{nuc}}-e)}(Q_1, Q_2, \ldots, Q_{3N_{\text{nuc}}-e}') \\
&\times \mathbb{C}(Q_1, Q_2, \ldots, Q_{3N_{\text{nuc}}-e}').
\end{aligned} \tag{37}
$$

The latter equation means that to propagate $\mathbb{C}$ from some chosen point $\{Q_I^0\}$ to $\{Q_I\}$ one has to integrate with respect to Q_I at fixed $Q_2^0, \ldots, Q_{3N_{\text{nuc}}-e}^0$, then with respect to Q_2 at fixed $Q_1, Q_3, \ldots, Q_{3N_{\text{nuc}}-6}^0$, and so on, and finally with respect to $Q_{3N_{\text{nuc}}-e}$ at fixed $Q_1, Q_2, \ldots, Q_{3N_{\text{nuc}}-e-1}$ (see Ref. 92).

Yet analysis of the conditions (36) showed that they may not be fulfilled in general.[63] The basic argument lies in the fact that the variation of $\mathbb{T}^{(I)}$ with $Q_1, Q_2 \cdots Q_{3N_{\text{nuc}}-e}$ involves contributions from basis states lying outside the considered subspace of dimension N_{bas}; indeed, it is readily established that

$$\frac{\partial \mathbb{T}^{(I)}}{\partial Q_J} - \frac{\partial \mathbb{T}^{(J)}}{\partial Q_I} = [\mathbb{T}^{(I)}, \mathbb{T}^{(J)}] + \mathbb{O}^{(I,J)} \tag{38}$$

---

*Equations (35) and (37) are but formal definitions. Indeed the numerical work[92] that would be required for arbitrary polyatomics, in general, and for singular $\mathbb{T}^{(I)}$ matrices in particular, would be prohibitively cumbersome.

with the matrix elements of $\mathbb{O}^{(I,J)}$ defined as

$$
O_{nm}^{(I,J)} = \sum_{k=N_{bas}+1}^{\infty} \left(\langle \varphi_n^{el} | \frac{\partial}{\partial Q_I} | k \rangle \langle k | \frac{\partial}{\partial Q_J} | \varphi_m^{el} \rangle - \langle \varphi_n^{el} | \frac{\partial}{\partial Q_J} | k \rangle \langle k | \frac{\partial}{\partial Q_I} | \varphi_m^{el} \rangle \right),
$$

(39)

where the $|k\rangle$'s form an orthonormal basis set in complementary subspace. Only in exceptional cases can the matrices $\mathbb{O}^{(I,J)}$ be made to vanish. Hence, the transformation of an arbitrary $\{Q_I\}$-dependent finite basis (in particular, a *basis of* N_{bas} adiabatic states) into an equivalent *strictly diabatic* basis [Eq. (34)] is impossible in general. This does not mean, however, that the objective of constructing finite nontrivial basis sets that may get rid of the drawbacks of adiabatic bases is out of reach. Construction of such bases is the subject of the next sections.

B. Nontrivial Strictly Diabatic Bases

Instead of attempting from the outset to obtain diabatic states by an orthogonal transformation of a given $\{Q_I\}$-dependent basis (e.g., the adiabatic basis) as is done in Section IV A, let us proceed the other way around. Let us start from a complete strictly diabatic basis $\{|d\rangle\}$; according to Section IV this is a "crude" $\{Q_I\}$-independent basis. Of course, such a basis is defined except for a constant unitary transformation. Let us assume that this constant transformation has been properly chosen (on physicochemical grounds) and that the complete space may be split into two subspaces $\mathbf{S}^r$ and $\mathbf{S}^p$. For example, focusing on charge-transfer processes, this partitioning may be conceived as a way that helps distinguish between two specific characters: reactant-type $(\mathbf{S}^r)$ or product-type $(\mathbf{S}^p)$. We will discuss later how one manages in practice to realize this partitioning (see Sections VI A, VI C 2a). One may then solve independently the electronic Schrödinger equation *in each subspace*

$$
P^L H_{el} P^L X_n^L(\{\boldsymbol{\rho}_i\}; \{Q_I\}) = E_n^L(\{Q_I\}) X_n^L(\{\boldsymbol{\rho}_i\}; \{Q_I\}),
$$

(40)

where $P^L = \sum_d |d^L\rangle\langle d^L|$ is the projector onto subspace $\mathbf{S}^L$ ($L=r$ or p) and $\{Q_I\}_{I=1,3N_{nuc}-e}$ represents the chosen set of internal nuclear coordinates (e.g., $R, r_{\mathscr{T}}, \gamma_{\mathscr{T}}, \{\mathbf{r}_\alpha\}$). It is obvious that

$$
\langle X_k^r | \frac{\partial}{\partial Q_I} | X_n^p \rangle \equiv 0, \quad \forall Q_I.
$$

(41)

Since the use of the projectors prevent any state in $\mathbf{S}^r$ to mix with any state in $\mathbf{S}^p$. It is then possible to pick out one eigenstate of Eq. (40) in *each* subspace $\mathbf{S}^r$ and $\mathbf{S}^p$ to form a *nontrivial strictly diabatic* 2×2 *subspace*. The just described procedure is of course readily generalized to more than two subspaces. In addition, since

$$\left[\frac{\partial}{\partial Q_I}, P^L\right] \equiv 0, \quad \forall L,$$

(42)

it follows that

$$\left\langle X_k^L \middle| \frac{\partial}{\partial Q_I} \middle| X_n^M \right\rangle = \delta_{LM}\left(\delta_{kn}\frac{\partial E_n^L}{\partial Q_I} + (E_n^L - E_k^L)^{-1}\left\langle X_k^L \middle| \frac{\partial H_{el}}{\partial Q_I} \middle| X_n^L \right\rangle\right).$$

(43)

This property shows that X_n^L states that are energetically well separated in each subspace have weak dependences upon the internal nuclear coordinates (as is ordinarily admitted for acceptable BO states). If this is the case, then merging the two bases: $\{|X_k^r\rangle\}$ and $\{|X_n^p\rangle\}$ provides one type of quasidiabatic representation. If, on the contrary, two (or more) $|X_k^L\rangle$'s in $\mathbf{S}^L$ bring about the same troubles as those which make the adiabatic representation unpractical, then $\mathbf{S}^L$ should be fragmented into two (or more) crude diabatic subspaces and the procedure restarted.

The preceding scheme is inspired by a suggestion that was put forward by O'Malley.[54,56] It provides states that are simultaneously quasiadiabatic and quasidiabatic. They are quasiadiabatic via Eq. (40) and they are quasidiabatic in the sense that they are not affected by troublesome $\partial/\partial Q_I$ matrix elements; in fact, they just retain the weak $\partial/\partial Q_I$ matrix elements, which are universally accepted when the BO approximation applies. Expressed differently, the described procedure does just what the BO approximation (when valid) does except that it does it *within* subspaces and it does exactly what a strictly diabatic treatment is intended to do except that it does it *between* subspaces. It rests of course on bringing out criteria to explicitly build the subspaces. This point together with the upgrading of the method with the objective of achieving state-of-the-art chemical accuracy are practical preoccupations which are dealt with in Section VI.

V. CHARACTERISTIC TWO-STATE MODEL CASES

It is instructive to have in mind the behavior of typical interactions in the adiabatic and diabatic representations. These are best illustrated by two-state-model cases. Since ion–molecule charge transfer is of primary concern to us here, we will restrict the discussion to related examples inspired by actual cases. For simplicity, reference will be made to atom–diatom collisional systems; accordingly, the set of internal nuclear coordinates is R, r, and γ (Fig. 2).

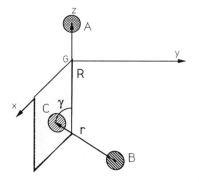

Figure 2. Same as Fig. 1 for the case when the projectile is an atom (A) and the target a diatomic molecule (BC).

On the basis of the discussions in the preceding section (see also Section VI) we may assume that two (strictly or nearly) diabatic states may be constructed one of which (X^r) owns a reactant-type character (say $\mathscr{P}^+ + \mathscr{T}$), whereas the other (X^p) owns a product-type character (say $\mathscr{P} + \mathscr{T}^+$). For simplicity, we may consider that $X^{r,p}$ are energetically the lowest eigenstates of the corresponding Eq. (40). We thereby omit the subscript of these states. In the considered 2×2 subspace one may construct the equivalent $\varphi^a_{1,2}$ adiabatic basis which obtains from the $X^{r,p}$ set according to:

$$\varphi^a_1 = X^r \cos\theta + X^p \sin\theta, \tag{44a}$$

$$\varphi^a_2 = -X^r \sin\theta + X^p \cos\theta, \tag{44b}$$

where the angle θ is determined from

$$\langle \varphi^a_1 | H_{el} | \varphi^a_2 \rangle \equiv 0 = \tfrac{1}{2}(E^p - E^r)\sin 2\theta + H^{pr}\cos 2\theta \tag{45}$$

with

$$H^{pr} = \langle X^p | H_{el} | X^r \rangle = H^{rp} \tag{46a}$$

$$E^L = \langle X^L | H_{el} | X^L \rangle \qquad (L = p, r) \tag{46b}$$

One thus readily obtains:

$$\theta = \tfrac{1}{2}\arctan[2H^{pr}(E^r - E^p)^{-1}] \tag{47}$$

and

$$T^{(I)}_{21} = \langle \varphi^a_2 | \frac{\partial}{\partial Q_I} | \varphi^a_1 \rangle = \frac{(E^r - E^p)\partial H^{rp}/\partial Q_I - H^{rp}\partial(E^r - E^p)/\partial Q_I}{(E^r - E^p)^2 + 4(H^{rp})^2}. \tag{48}$$

The $\langle X^{\mathrm{p}}|\partial/\partial Q_I|X^{\mathrm{r}}\rangle$ term does not appear in the expression of $\tau_{21}^{(I)}$ owing to the assumed (strictly or nearly) diabatic nature of the $X^{\mathrm{p,r}}$ basis.

One important characteristic of $\mathscr{P}^+ + \mathscr{T} \to \mathscr{P} + \mathscr{T}^+$ charge-transfer problems is the exponential behavior of the diabatic interaction H^{rp} at large intermolecular distances.[93–97] It may be represented roughly by the functional dependence[95–97] (Fig. 3)

$$H^{\mathrm{rp}}(R, r, \gamma) = A(r, \gamma)R^v \exp[-\lambda(r)R] \qquad (49)$$

Specification of the R, r, γ dependences* of the adiabatic energy difference $E^{\mathrm{r}} - E^{\mathrm{p}}$ enables one to get some insight into the behavior of the $\tau_{21}^{(I)}$ couplings in the equivalent adiabatic basis [Eq. (44)].

In many $A^+ + BC$ charge-transfer systems the energy difference $E^{\mathrm{r}} - E^{\mathrm{p}}$ at large interparticle distance R may be approximated as the difference ΔI between the ionization potential of the neutral molecule and the recombination energy of the atomic ion (or vice versa), hence

$$E^{\mathrm{r}} - E^{\mathrm{p}} \cong \Delta I(r) \qquad (50)$$

is grossly a function of r only. Let us examine cases (like $H^+ + H_2$,[57,58] $Ar^+ + H_2$,[75] $F^+ + CO$,[85] for instance) when $\Delta I(r)$ vanishes at a certain value of the molecule bond distance r_c. Near r_c we have

$$\Delta I(r) \approx \alpha(r - r_c). \qquad (51)$$

If H^{rp} does not depend too much on r and γ, one obtains with Eq. (48)

$$\tau_{21}^{(r)}(R, r) \cong -\frac{\alpha H^{\mathrm{rp}}}{\alpha^2(r - r_c)^2 + 4(H^{\mathrm{rp}})^2}, \qquad (52)$$

$$\tau_{21}^{(R)}(R, r) \cong \frac{\alpha(r - r_c)(v/R - \lambda)H^{\mathrm{rp}}}{\alpha^2(r - r_c)^2 + 4(H^{\mathrm{rp}})^2}. \qquad (53)$$

For *fixed R* $\tau_{21}^{(r)}$ and $\tau_{21}^{(R)}$ are seen to be Lorentzian functions (Fig. 4). The full width at half-maximum of $\tau_{21}^{(r)}$ behaves as $4H^{\mathrm{rp}}/\alpha$. Hence, as R increases, $|\tau_{21}^{(r)}|$ gets thinner and larger and tends to a "δ-function" in the limit $R \to \infty$ (Fig. 4a). The function $\tau_{21}^{(R)}$ at fixed R, vanishes ar $r_0 = r_c$, it displays a positive

---

*The variable r representing the molecule bond distance and the superscript r labeling the reactant channel should not be confused. When r appears in brackets as a superscript [see Eq. (52)], it corresponds to the former meaning.

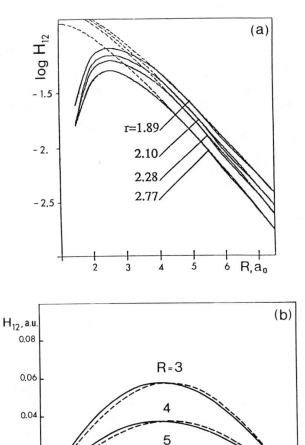

Figure 3. Illustration of the functional dependences of the H^{rp} interaction in the $H^+ + O_2$ $(X\,^3\Sigma_g^-) \to H(1\,^2S) + O_2^+ (X\,^2\Pi_g)$ system. (a) Exponential R dependence of the H^{rp} interaction for some fixed values of r and $\gamma = 45°$. Solid lines: calculations of Ref. 97; dashed lines: corresponding $A(r, \gamma) \exp[-\lambda(r)R]$ fit. (b) γ dependence of H^{rp} for some fixed values of R and $r = 2.282a_0$. Solid lines: calculations of Ref. 97; dashed lines: $P_2^1(\cos \gamma)$ function adjusted to the maximum of the calculated H^{rp} interaction. (Both figures are taken from Ref 97.)

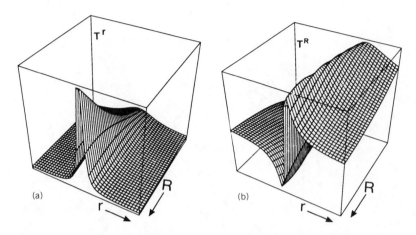

Figure 4.   Typical shapes of $T_{21}^{(r)}$ and $T_{21}^{(R)}$ matrix elements in a 2×2 adiabatic basis $(\varphi_{1,2}^a)$. The corresponding diabatic energy difference behaves as $\alpha(r - r_c)$ and the associated diabatic coupling H^{rp} as $A \exp(-\lambda R)$.

and a negative peak at $r_M = r_c \pm 2H^{rp}/\alpha$. These peaks get bigger, thinner, and closer to each other as R increases. Concurrently, as R increases, the diabatic to adiabatic rotation angle θ [Eq. (47)] varies more and more abruptly near $r = r_c$ and tends to a step function in the limit $R \to \infty$ (see e.g., Ref. 58). It may also be noted that, for $v = 0$ and fixed r, $T_{21}^{(r)}$ as a function of R displays on either side of r_c a bell shape with a maximum at $|H^{rp}| = |E^r - E^p|/2$. This maximum is characteristic of exponential models[93a,b]; it transforms into a peak that sharpens and moves toward large R as $r \to r_c$ (Fig. 4b). The behavior of $T_{21}^{(R)}$ as a function of R remains qualitatively the same for $v \neq 0$ near $r = r_c$.

The rapid and large variations of $T_{21}^{(r)}$ and $T_{21}^{(R)}$ [Eqs. (52) and (53)] at large R values obviously arises from the vanishing of H^{rp} at $r = r_c$ when $R \to \infty$. Yet another circumstance may cause the vanishing of H^{rp}, namely, a γ dependence of the preexponential factor of Eq. (49) as

$$A(r, \gamma) = \bar{A}(r) P_l^m(\cos \gamma) \tag{54}$$

where P_l^m is a Legendre function (see, e.g., Fig. 3b). Let us consider, for example, the case $l = 1$, $m = 0$ corresponding to a $\cos \gamma$ dependence of H^{rp}. Assuming again a negligible dependence of $\bar{A}$ upon r, one obtains in addition to Eqs. (52) and (53)

$$T_{21}^{(\gamma)} = \frac{\alpha(r - r_c)\tan H^{rp}}{\alpha^2(r - r_c)^2 + 4(H^{rp})^2}. \tag{55}$$

One thus sees that the previously mentioned rapid variation and sharp peaking of $T_{21}^{(R,r \text{ or } \gamma)}$ as well as the sudden variation of θ arise near $r \cong r_c$ when $\gamma \to \pi/2$.

Let us next examine a case when (like in the $Na + H_2$,[81,83] $Cs + O_2$,[98] or $H^+ + O_2$[97] collisional systems) r_c is an increasing function of R. For simplicity consider the linear case:

$$r_c(R) = \mathscr{A} + \mathscr{B}R. \tag{56a}$$

Here, the $r - p$ crossing may be viewed as occurring not only at $r_c(R)$ for fixed R but also at

$$R_c(r) = \mathscr{B}^{-1}(r - \mathscr{A}) \tag{56b}$$

for fixed r, that is,

$$E^r - E^p \cong \alpha[r - r_c(R)] = \alpha\mathscr{B}[R_c(r) - R]. \tag{56c}$$

For this case $T_{21}^{(r)}$ remains the same as given by Eq. (52), whereas Eq. (53) transforms into

$$T_{21}^{(R)} = \frac{\alpha H^{rp}[(r - r_c)(v/R - \lambda) + \mathscr{B}]}{\alpha^2(r - r_c)^2 + 4(H^{rp})^2}, \quad r_c = r_c(R). \tag{57}$$

For $H^{rp} \neq 0$, $T_{21}^{(R)}$ now vanishes at $r_0 = r_c + \mathscr{B}R/(\lambda R - v)$ and exhibits its two maxima at

$$r_M = r_0 \pm \left[\left(\frac{\mathscr{B}R}{\lambda R - v}\right)^2 + \left(\frac{2H^{rp}}{\alpha}\right)^2\right]^{1/2}. \tag{58}$$

When $H^{rp} \to 0$, either because $R \to \infty$ [Eq. (49)] or because γ approaches a zero of $P_l^m(\cos\gamma)$ [Eq. (54)], the maximum that moves toward $r_c + 2\mathscr{B}R[\lambda R - v]^{-1}$ dies off, whereas that approaching r_c tends to a "δ-function."

The rudimentary cases that have been cited help in getting acquainted with some characteristic variations of $T_{21}^{(l)}$ coupling matrix elements for some simplistic (albeit realistic) $E^r - E^p$ and H^{rp} functional dependences. Beyond their "educational" virtues, they are particularly interesting because they correspond to a blend of familiar dynamical models that have all been worked out by implicitly using a strictly diabatic representation. In effect, the considered curve crossing in the r variable, for fixed R and no dependence upon γ, corresponds to the assumptions of the Landau–Zener[28,29] model. The case when r_c is independent of R corresponds (for fixed r and γ) to the exponential model of Demkov[93a] and Nikitin.[93b] On the other hand, the case when r_c depends linearly on R corresponds (for fixed r and γ) to the linear–exponential model of Nikitin.[93b] Finally, the case of a $\cos \gamma$ dependence of H^{rp} corresponds, for fixed R, when γ varies on both sides of $\pi/2 - \gamma \cong \pi/2 \pm \varepsilon$ ($\varepsilon \ll 1$)—to the linear model of a conical intersection in the r, ε coordinates. It is thus seen that *the construction of diabatic states offers the possibility of emphasizing the characteristics of such models in actual problems.*

VI. PRACTICAL CONSTRUCTION OF DIABATIC STATES

A difficulty one actually meets in practice when attempting to directly construct quasidiabatic states is that the replacement of the "$=$" sign in Eq. (26) by the "$\simeq$" sign does not help building general algorithms to derive the transformation matrix $\mathbb{C}$ that achieves the modified definition.

The alternative guidelines presented in Section IV B may serve as a practical procedure if one could first generate a *managable* strictly diabatic basis. It has been repeatedly pointed out that a *finite* adiabatic basis that would have been obtained for some given arrangement $\{Q_I^0\}$ of the nuclei could constitute a strictly diabatic basis when *used unaltered for any other nuclear arrangement* Q_I. Yet, it has quite as much been recognized that such a basis (termed crude adiabatic)[99,100] is not useful in practice primarily because its ability to properly describe actual deformations of electron clouds rapidly degrades when the nuclei move away from the reference nuclear arrangement $\{Q_I^0\}$. In other words, diagonalization of the H_{el} matrix in crude adiabatic bases of realistic sizes would yield poor descriptions of the true adiabatic states away from $\{Q_I^0\}$. The crude adiabatic starting point is thus particularly ill suited for collision problems where the R coordinate (at least) varies between typical atomic dimensions ($\approx 1a_0$) and infinity. It is precisely for such reasons that normal electronic structure calculations on polyatomic systems make use of electronic basis sets attached to every new set of (fixed) positions of the nuclei.

In practice, a general electronic wavefunction φ_n^{el} [Eq. (12)] is written as a linear combination of configuration state functions (CSF); each CSF is expressed as a fixed superposition of Slater determinants built from N_{el} spin orbitals.[46] An orbital in those Slater determinants is expanded over a finite set of nucleus-centered one-electron basis functions. Clearly the *utmost diabatic states* in this context should only involve *constant linear combinations* of CSF's and basis orbitals. Strictly diabatic states would in addition impose that those basis orbitals remain fixed in the chosen reference frame (Section II B); however, as already discussed for crude adiabatic states, this is generally not an acceptable option. Right at this point procedures suggest themselves that depend on the class of problem under consideration. These are now examined in turn.

A. Preservation of CSF's: states differing by two orbitals

A CSF represents a set of orbitals of the system having *identifiable labels* and being ascribed *specific occupation numbers*. Usually, in the absence of spin–orbit coupling, CSF's are built as eigenstates of the S^2 and S_z total electronic spin operators of the system. When H_{el} is invariant under the operations of a symmetry point group, the CSF's may be chosen to transform according to its irreducible representations (unless some conflict may arise with the objective of building diabatic states).

1. Case Study

It was originally pointed out by Sidis and Lefebvre-Brion[101] that since any $\partial/\partial Q_I$ operator *acts on a Slater determinant* as a one electron operator (i.e., as: $\sum_{j=1}^{N_{el}}(\partial/\partial Q_I)_j$) *its matrix elements exactly vanish* between CSF's representing states that are (at least) doubly excited with respect to each other.* Numerous examples of curve-crossing patterns have been identified as implying the coming together of strictly diabatic CSF's of this sort.[32,101–109] In particular, collisional systems involving He^+ ions and atomic[101,102,104,105] or molecular neutrals[106–109] offer nice illustrations of related electron transition processes. For instance, charge-exchange excitation

$$He^+(1s) + BC \rightarrow He(1s^2) + BC^{+*} \tag{59}$$

results from *two-electron rearrangement* processes whereby one electron of the neutral fills the He vacancy while another electron is expelled toward an excited orbital of the BC^+ molecular ion[106,108] (Fig. 5). These Auger-type processes labeled DII (Diabatic II)[104,110] are known to occur at curve crossing between a "promoted" core-excited valence state and a Rydberg series. In what follows the salient features of such interactions are illustrated taking the elementary $He^+–H_2$ system as an example.[109] The electronic configuration in the entrance state of the collision is asymptotically ($R \rightarrow \infty$) $1s_{He}1\sigma_{gH_2}^2$; it transforms at finite R into $1a'2a'^2$, which corresponds to an excited $^2A'$ state of the triatomic quasimolecular system. This state is predicted[107,108] and indeed found[109] to undergo a series of crossings with states owning the configuration: $1a'^2 na'$ ($n > 2$). These crossings roughly

*Certain classes of singly excited CSF's have the same property as for example: $\langle(\zeta^2)_{singlet}\eta|\partial/\partial Q_I|(\zeta\xi)_{triplet}\eta\rangle = 0$, since: $|(\zeta^2)_{singlet}\eta\rangle = |\zeta\zeta\eta|$ and $|(\zeta\xi)_{triplet}\eta\rangle = 6^{-1/2}(2|\zeta\xi\bar\eta| - |\zeta\bar\xi\eta| - |\bar\zeta\xi\eta|)$.

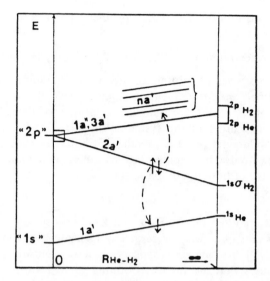

Figure 5. Schematic MO correlation diagram for the $(He - H_2)^+$ collisional system in C_s symmetry indicating how the so-called Diabatic II mechanism proceeds. As a result of the adiabatic treatment of the $a'_{1sHe} - a'_{1\sigma_g H_2}$ interaction the $2a'$ MO is promoted and gets closer in energy to higher lying na' MO. When the matching condition (60) is fulfilled, an electron from the $2a'^2$ shell fills the $1a'$ vacancy brought by the He^+ ion, while the other electron is expelled to one of the outer na' MO. (The correlation of the MO levels in the $R = 0$ limit in this diagram is of the quasi-atom–atom type; actual adiabatic and diabatic correlations are discussed in refs. 107–108 and 111, respectively.) (The figure shown is taken from Ref. 107.)

occur when

$$2\varepsilon_{2a'} \cong \varepsilon_{1a'} + \varepsilon_{na'} \qquad (60)$$

where ε_ϕ is the energy of orbital ϕ (defined as the energy needed to extract the electron out of this orbital in the mentioned configurations). $\varepsilon_{na'}$ is either negative or positive and the $1a'^2 na'$ configuration thereby describes the Rydberg series converging to $1a'^2$ and the associated continuum.[102,109]

The common set of orthonormal orbitals that constitute the diabatic CSF's for the preceding two-electron rearrangement processes (59) should fulfill a few requirements stemming from what follows.

1. Owing to the large energy gap separating the a'_{1sHe} and $a'_{1\sigma_g H_2}$ orbitals (5–11 eV for realistic molecule bond distances), their mutual interaction should be treated *adiabatically* up to keV collision energies. In other words,

at low collision energies ($E < 1$ keV) the

$$He^+(1s\,^2S) + H_2(1\sigma_g^2\,X\,^1\Sigma_g^+) \rightarrow He(1s^2\,^1S) + H_2^+(1\sigma_g\,X\,^2\Sigma_g^+) \qquad (61)$$

charge-transfer process is rather unlikely. This will generally be the case for a broad class of molecules other than H_2. If this is not the case then a more careful treatment, such as that discussed in Sections VI B and VI C, is required. As a result of the adiabatic treatment of the $a'_{1sHe} - a'_{1\sigma_g H_2}$ interaction the energy of the resulting $1a'$ MO decreases while that of the $2a'$ MO increases. This energy level splitting with decreasing atom–molecule distance is the very mechanism that causes the matching condition (60) to occur (Fig. 5).

2. Actually, the previously mentioned MO-energy level movement does something more; it induces crossings (labeled DI: diabatic I)[104,108,110] between the "promoted" $2a'$ MO and orbitals like $a'_{1\sigma_u H_2}$ that lie asymptotically higher in energy. These crossings take place at rather short distances of approach $R < 1.5a_0$ $(\forall r, \gamma)$[111] and may be safely neglected if one is focusing on soft collision problems (otherwise, procedures, such as those touched upon in Section VI D, ought to be considered). With *those restrictions in mind* both valence and Rydberg-type orbitals of $(HeBC)^+$ systems may be treated *adiabatically* with a view to investigating processes like those specified in Eq. (59).

The actual building of quasidiabatic states may then proceed along the same lines as those proposed in Ref. 109 (which should be consulted for details). One first manages to build a Fock-type operator for an SCF scheme in order to obtain simultaneously

$$\langle 1a'^2 2a' | H_{el} | 1a' 2a'^2 \rangle = 0 \qquad (62)$$

and

$$\varepsilon_{2a'} - \varepsilon_{1a'} = \langle 1a' 2a'^2 | H_{el} | 1a' 2a'^2 \rangle - \langle 1a'^2 2a' | H_{el} | 1a'^2 2a' \rangle. \qquad (63)$$

The virtue of those conditions is that the $1a'$–$2a'$ interaction is treated adiabatically (as discussed in paragraph 1) and that $\varepsilon_{1a'}$ and $\varepsilon_{2a'}$ represent the binding energies of an electron attached to the $1a'2a'$ configuration of the $(HeH_2)^{++}$-ion core. Next, one may generate the na' $(n > 2)$ MO as improved virtual orbitals (IVO).[112,113] This is actually achieved by requiring the available orbitals in the space orthogonal to the $1a'$ and $2a'$ MO to describe an electron in the field of the $(1a')^2$ closed-shell configuration of the $(HeH_2)^{++}$-ion core. Consequently, such IVO's obey

$$\langle 1a'^2 na' | H_{el} | 1a'^2 ma' \rangle \equiv 0 \qquad (n \neq m > 2) \qquad (64)$$

among other properties.[109] Finally, following Section IV B, two *independent CI calculations* are carried out: one in the $\{(1a'^2 2a'), (1a'^2 na')\}$ subspace and the other one in the $\{(1a'2a'^2), (1a'2a'na'), (1a'n\phi m\phi)\}$ subspace (where $\phi = a'$ or a'' and $n\phi$ or $m\phi \neq 1a', 2a'$). Results of these calculations are shown in Figs. 6 and 7.

The described computational scheme may be extended to more complicated systems. As advocated in Section IV B it enables one to work out in a well-defined way accurate "quasi-adiabatic–diabatic" states. If still higher accuracy is desired, one may resort to the method described in Section VI C 2b. The use of methods based on quasidegenerate perturbation theory has also been proposed as a device to improve quasidiabatic states obtained by the aforementioned partitioning technique.[20,100,115,116]

2. *Philosophy*

Before proceeding further it is worth stressing some particular aspects that have emerged during our discussion. The generation of satisfactory diabatic states as was done previously has been made possible by a precise specification

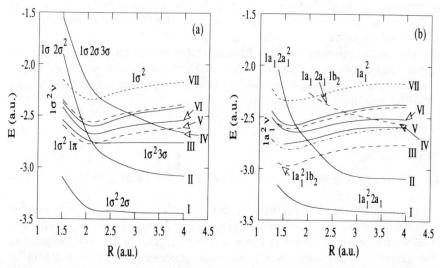

Figure 6. Cuts of a few diabatic potential energy hypersurfaces of the $(He–H_2^+)$ molecular ion. The H–H distance is fixed at $r = 1.4a_0$. (a) $\gamma = 0$. Solid curves: $^2\Sigma^+$ states; dashed curves: $^2\Pi$ states; dotted curve: limit of the $1\sigma^2 v$ series. (b) $\gamma = \pi/2$. solid curves: $^2A_1$, dashed curves: $^2B_2$ states; dashed–dotted curve: $^2B_1$ states; dotted curve: limit of the $1a_1^2 v$ series. In both figures the states are labeled by their dominant configuration. The asymptotes are as follows. (I) $He(1s^2\,^1S) + H_2^+(1s\sigma_g)$, (II) $He^+ + H_2(X\,^1\Sigma_g^+)$, (III) $He(1s^2\,^1S) + H_2^+(2p\sigma_u)$, (IV) $He^+ + H_2(b\,^3\Sigma_u^+)$, (V) $He(1s^2\,^1S) + H_2^+(2p\pi_u)$, (VI) $He(1s^2\,^1S) + H_2^+(2s\sigma_g)$. (This figure is adapted from Ref. 109.)

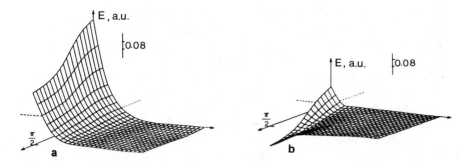

Figure 7. Perspective view of diabatic potential energy surfaces of the $(He-H_2)^+$ molecular ion. (a) "Core-excited" $(1a'2a'^2)^2A'$ state correlating with the $(1\sigma2\sigma^2)^2\Sigma^+$ state and $(1a_12a_1^2)^2A_1$ states of Figs. 6a and 6b, respectively. (b) The $(1a'^23a')^2A'$ member of the $(1a'^2v)$ series correlating with the $(1\sigma^23\sigma)^2\Sigma^+$ and the $(1a_1^21b_2)^2B_2$ states of Figs. 6a and 6b, respectively. The views are plotted for $1.5a_0 \leq R \leq 6a_0$ and $0 \leq \gamma \leq \pi/2$ with steps $\Delta R = 0.18a_0$ and $\Delta\gamma = 5°$. (The figure is taken from Ref. 109.)

of the nonadiabatic process (59). The actual energy gaps between the three classes of orbitals—$1a', 2a'$, and na'—has made it possible to decide that, in the range of interparticle distances that is accessible in soft collisions, those orbitals may be treated *adiabatically* without introducing any troublesome $\partial/\partial Q_I$ matrix element. In these conditions it is possible to design an appropriate LCAO-MO-SCF scheme to generate the *adiabatic MO* that serve constructing *diabatic CSF* prototypes. Those CSF are subsequently used to define the projectors onto CI subspaces characterized by the state of excitation of the $(HeBC)^{++}$ ion core.

Another interesting feature one may stress is that the crossing of a given pair of states often results from the *independent adiabatic* treatment of some interactions; this is well illustrated by the promotion of the $2a'$ MO discussed previously. (Fig. 5).[107,108,111]

B. Preservation of Separated-Partner Characters: Orbitals

The discussion of the preceding section opens the problem of constructing diabatic states for processes involving *one-electron* transitions, which in turn brings on the question of *diabatic orbitals*. This problem arises precisely in electron-transfer reactions [viz. Eqs. (1a)–(1c)]. It is well known that the occurrence of such reactions in a low-energy collision is favored by the proximity of the reactant and product energy levels. In those cases, the critical region of nonadiabatic behavior is generally located at rather *large intermolecular* distances.[93] This critical region actually divides the R range into two domains: one in which the active electron belongs *exclusively* to the target or projectile (outer domain) and one in which it belongs to both (inner domain). The maximum of the $T_{21}^{(R)}$ matrix element in Section V, for the Demkov-type case (see end of Section V), is the pointer of that critical region. Obviously, if one wants the system's wave function to vary as little as possible when traversing it, one ought to preserve either the *prevalent characters* of the outer domain or those of the inner domain. Inasmuch as CSF's naturally constitute many-electron diabatic prototypes, we now turn to express the preceding ideas in terms of CSF's built from appropriate orbitals. In the outer domain the relevant orbitals should be *something like the separated-partner orbitals*. These are defined as orbitals of the isolated collision partners that are brought together without distortion at each of the considered nuclear geometries. On the other hand, since MO's like those discussed in Section VI A are adiabatic orbitals, the only possibilities left for the inner domain would be something like constant linear combinations of separated-partner orbitals. Clearly, the former option which is specifically designed for the description of the initial and final states of reactions (1a)–(1c) is to be preferred.

A difficulty one has to face when using CSF's made of separated-partner orbitals is their *nonorthogonality* at finite intermolecular distances R. This difficulty is altogether ignored in calculations based on the so-called diatomics in molecules method (DIM).[25,117] The reason invoked for doing so is twofold. First, paraphrasing Tully,[25b] "actual specification of the basis function is *never required in DIM*. This is a great advantage because it introduces substancial flexibility into the basis functions. They can be considered to expand, contract, distort in a way that preserves symmetry in order to optimally adapt to any particular molecular environment." Secondly, those unspecified basis functions are nonetheless conferred the property of having weak dependences upon nuclear geometry[25,117]: in other words, they are assumed from the outset to be quasidiabatic. Although qualitatively sound arguments have been put forward to support such assumptions, they still constitute a ticklish issue.[24,118,119]

1. The Case of Two Orbitals in Atom–Atom Systems

To have a first glimpse of the matter, let us first examine a two-state atomic case and consider only the *two atomic orbitals* between which the "active" electron is being exchanged: viz., $\phi_{\mathscr{T}}$ and $\phi_{\mathscr{P}}$ centered on the targer ($\mathscr{T}$) and projectile ($\mathscr{P}$), respectively. The objective of defining wavefunctions, which describe an electron exclusively attached to $\mathscr{T}$ or $\mathscr{P}$ is frustrated by the existence of a nonvanishing overlap integral:

$$\omega = \langle \phi_{\mathscr{T}} | \phi_{\mathscr{P}} \rangle. \tag{65}$$

As a by-product one obtains

$$\frac{d\omega}{dR} = \left\langle \frac{\partial \phi_{\mathscr{T}}}{\partial R} \middle| \phi_{\mathscr{P}} \right\rangle + \left\langle \phi_{\mathscr{T}} \middle| \frac{\partial \phi_{\mathscr{P}}}{\partial R} \right\rangle \tag{66a}$$

$$= \mathscr{D}_{\mathscr{T}} \left\langle \frac{\partial \phi_{\mathscr{T}}}{\partial z} \middle| \phi_{\mathscr{P}} \right\rangle - \mathscr{D}_{\mathscr{P}} \left\langle \phi_{\mathscr{T}} \middle| \frac{\partial \phi_{\mathscr{P}}}{\partial z} \right\rangle \tag{66b}$$

$$= \left\langle \frac{\partial \phi_{\mathscr{T}}}{\partial z} \middle| \phi_{\mathscr{P}} \right\rangle \tag{66c}$$

$$= - \left\langle \phi_{\mathscr{T}} \middle| \frac{\partial \phi_{\mathscr{P}}}{\partial z} \right\rangle, \tag{66d}$$

where z is the component of the electron coordinate ρ along the R axis and $\mathscr{D}_{\mathscr{T},\mathscr{P}}$ are defined in Eq. (5b). The appearence of $\partial/\partial z$ and $\mathscr{D}_{\mathscr{T},\mathscr{P}}$ in these equations is related to the relations

$$\rho_{\mathscr{T}} = \rho + \mathscr{D}_{\mathscr{T}} \mathbf{R}, \tag{67a}$$

$$\rho_{\mathscr{P}} = \rho - \mathscr{D}_{\mathscr{P}} \mathbf{R}, \tag{67b}$$

which define electron coordinates referred to $\mathscr{T}$ and $\mathscr{P}$, respectively, and entail

$$\mathbf{V}_{\mathbf{R}} \equiv \mathbf{V}_{\mathbf{R}}|_{\mathbf{G}} = \mathbf{V}_{\mathbf{R}}|_{\mathscr{T}} + \mathscr{D}_{\mathscr{T}} \mathbf{V}_{\rho} = \mathbf{V}_{\mathbf{R}}|_{\mathscr{P}} - \mathscr{D}_{\mathscr{P}} \mathbf{V}_{\rho}, \tag{68a}$$

$$\frac{\partial}{\partial R} \equiv \frac{\partial}{\partial R}\bigg|_{G} = \frac{\partial}{\partial R}\bigg|_{\mathscr{T}} + \mathscr{D}_{\mathscr{T}} \frac{\partial}{\partial z} = \frac{\partial}{\partial R}\bigg|_{\mathscr{P}} - \mathscr{D}_{\mathscr{P}} \frac{\partial}{\partial z}. \tag{68b}$$

The latter equation relates the different expressions of the $\partial/\partial R$ operator that obtain when effecting the differentiation with $G, \mathscr{T}$, or $\mathscr{P}$ held fixed. Our aim, then, is to look for an orthogonalization method that achieves at once a cancellation of the off-diagonal matrix element of $\partial/\partial R$. To do so, let us first orthogonalize arbitrarily $\phi_{\mathscr{P}}$ to $\phi_{\mathscr{T}}$ using a Schmidt procedure:

$$|\tilde{\phi}_{\mathscr{P}}\rangle = \mathscr{N}(|\phi_{\mathscr{P}}\rangle - \omega|\phi_{\mathscr{T}}\rangle), \tag{69a}$$

$$|\tilde{\phi}_{\mathscr{T}}\rangle \equiv |\phi_{\mathscr{T}}\rangle, \tag{69b}$$

with

$$\mathscr{N} = (1 - \omega^2)^{-1/2}. \tag{70}$$

Clearly with Eq. (68b) we get ($\phi_{\mathscr{P}}$ and $\phi_{\mathscr{T}}$ being real functions)[119]

$$\langle \tilde{\phi}_{\mathscr{P}} | \frac{\partial}{\partial R} | \tilde{\phi}_{\mathscr{T}} \rangle = - \langle \tilde{\phi}_{\mathscr{T}} | \frac{\partial}{\partial R} | \tilde{\phi}_{\mathscr{P}} \rangle \qquad (71a)$$

$$= \mathscr{N} \mathscr{D}_{\mathscr{T}} \frac{d\omega}{dR} \qquad (71b)$$

$$= \mathscr{D}_{\mathscr{T}} \frac{d}{dR} \arcsin \omega, \qquad (71c)$$

which vanishes only if $\mathscr{D}_{\mathscr{T}} = 0$, that is, $G \equiv \mathscr{T}$ (case of infinitely heavy target).[94] For other cases we still have at our disposal an orthogonal transformation between $|\tilde{\phi}_{\mathscr{T}}\rangle$ and $|\tilde{\phi}_{\mathscr{P}}\rangle$, namely,

$$|\tilde{\tilde{\phi}}_{\mathscr{T}}\rangle = \cos\theta |\tilde{\phi}_{\mathscr{T}}\rangle + \sin\theta |\tilde{\phi}_{\mathscr{P}}\rangle, \qquad (72a)$$

$$|\tilde{\tilde{\phi}}_{\mathscr{P}}\rangle = - \sin\theta |\tilde{\phi}_{\mathscr{T}}\rangle + \cos\theta |\tilde{\phi}_{\mathscr{P}}\rangle, \qquad (72b)$$

such that

$$\frac{d\theta}{dR} = - \langle \tilde{\phi}_{\mathscr{T}} | \frac{\partial}{\partial R} | \tilde{\phi}_{\mathscr{P}} \rangle, \qquad (73)$$

which entails

$$\langle \tilde{\tilde{\phi}}_{\mathscr{P}} | \frac{\partial}{\partial R} \tilde{\tilde{\phi}}_{\mathscr{T}} \rangle = - \langle \tilde{\tilde{\phi}}_{\mathscr{T}} | \frac{\partial}{\partial R} | \tilde{\tilde{\phi}}_{\mathscr{P}} \rangle = 0. \qquad (74)$$

Equation (73) is easily integrated with (71c) to give

$$\theta = - \mathscr{D}_{\mathscr{T}} \arcsin \omega. \qquad (75)$$

In the particular case $\mathscr{D}_{\mathscr{T}} \simeq 1$, that is, $\mathscr{D}_{\mathscr{P}} \simeq 0$ and $G \simeq P$ (which corresponds to a much heavier projectile than target), one finds, as could have been anticipated, that the proper atomiclike diabatic orbitals obtain from the Schmidt orthogonalization of $\phi_{\mathscr{T}}$ to $\phi_{\mathscr{P}}$ [i.e., Eq. (69) with the interchange of $\mathscr{T}$ and $\mathscr{P}$]. In the case of equal target and projectile masses: $\mathscr{D}_{\mathscr{T}} = \mathscr{D}_{\mathscr{P}} = \frac{1}{2}$, the sought diabatic orbitals are found to result from a Löwdin symmetric orthogonalization[119]:

$$|\tilde{\tilde{\phi}}_{\mathscr{T}}\rangle = \mathscr{N}[C_{+}|\phi_{\mathscr{T}}\rangle + C_{-}|\phi_{\mathscr{P}}\rangle], \qquad (76a)$$

$$|\tilde{\tilde{\phi}}_{\mathscr{P}}\rangle = \mathscr{N}[C_{-}|\phi_{\mathscr{T}}\rangle + C_{+}|\phi_{\mathscr{P}}\rangle], \qquad (76b)$$

$$C_{\pm} = \tfrac{1}{2}[(1 - \omega)^{1/2} \pm (1 + \omega)^{1/2}]. \tag{76c}$$

The preceding elementary example shows that *a relation exists between the orthogonalization procedure that is intended to furnish the atomiclike diabatic orbitals and the origin of electronic coordinates that is held fixed while effecting the $\partial/\partial R$ differentiation.*[119] For instance, in a collision between a heavy target and a light projectile ($\mathscr{D}_{\mathscr{T}} \cong 0$), the arbitrary use of a symmetric orthogonalization of $\phi_{\mathscr{T}}$ and $\phi_{\mathscr{P}}$ will let an off-diagonal matrix element of $\partial/\partial R|_{G \cong \mathscr{T}}$ remain; the absolute value of this matrix element is: $|(\mathscr{N}/2)d\omega/dR|$. Contrary to what is sometimes thought, the orthogonalization of separated-partner orbitals to obtain diabatic orbitals is not arbitrary; the interdependence of orthogonalization and diabatization determines the choice.

2. The Case of a Few Orbitals in Atom–Atom Systems: A Hint at the Electron Translation Factor Problem

Consideration of charge-exchange problems involving more than a single active orbital for each collision partner reveals, via Eq. (68), a difficulty that was not immediately perceptible in the preceding examples. This is illustrated by an elementary three-state case involving two orthonormal orbitals $|\phi_{\mathscr{T}1}\rangle$, $|\phi_{\mathscr{T}2}\rangle$ of the isolated target and one orbital $|\phi_{\mathscr{P}}\rangle$ of the projectile. The general formulation of the problem is the same as in the preceding case except that, owing to the relation

$$\left\langle \phi_{\mathscr{T}1} \left| \frac{\partial \phi_{\mathscr{T}2}}{\partial R} \right\rangle = \mathscr{D}_{\mathscr{T}} \left\langle \phi_{\mathscr{T}1} \left| \frac{\partial \phi_{\mathscr{T}2}}{\partial z} \right\rangle \right. \tag{77}$$

that stems from Eq. (68), the $\partial/\partial R$ matrix element does not vanish between (nondegenerate) atomic *orbitals of the same collision partner* obeying dipole-transition selection rules. This feature rocks the whole theoretical structure built up until this point, since it implies that the scattering equations cannot be decoupled when the collision partners are infinitely separated. The appearance of such spurious couplings is inherent to all methods, whether adiabatic, diabatic, or mixed, that attempt to describe the dynamical evolution of arbitrary quasimolecular systems using expansions over orbital basis sets *attached to fixed positions of the nuclei.* Its origin lies in the fact that those "clamped basis sets" try to describe electron clouds that are more or less *attached to nuclei that are actually moving.*[23,24,120]

a. Travelling Orbitals. To convince oneself that the just mentioned diagnosis is correct, it suffices to note that, in a reference frame where the center of mass of the nuclei is at rest (Section II), an isolated atomic orbital that *translates* with the target (resp. the projectile) is actually described by

$$\phi_{\mathscr{T}}^{\text{trans}} = \phi_{\mathscr{T}}(\boldsymbol{\rho}_{\mathscr{T}})\exp\left(-i\mathscr{D}_{\mathscr{T}}\frac{\mathbf{k}}{\mu}\cdot\boldsymbol{\rho}\right) \tag{78a}$$

(or:

$$\phi_{\mathscr{P}}^{\text{trans}} = \phi_{\mathscr{P}}(\boldsymbol{\rho}_{\mathscr{P}})\exp\left(i\mathscr{D}_{\mathscr{P}}\frac{\mathbf{k}}{\mu}\cdot\boldsymbol{\rho}\right) \tag{78b}$$

resp.), where $\mathbf{k}$ is the linear momentum associated with the $\mathbf{R}$ motion. "Clamped basis sets" lack the plane wave factors of Eqs. (78); those factors, which account for the translation of electrons along with the nuclei to which they belong, are named electron translation factors (ETF). It is easily verified that

$$[-\tfrac{1}{2}\Delta_{\boldsymbol{\rho}} + \mathscr{V}_{\mathscr{T}}(\boldsymbol{\rho}_{\mathscr{T}})]\phi_{\mathscr{T}}^{\text{trans}} - \frac{i\mathbf{k}}{\mu}\cdot\mathbf{\nabla}_{\mathbf{R}}|_G\phi_{\mathscr{T}}^{\text{trans}}$$

$$= \exp\left(-i\mathscr{D}_{\mathscr{T}}\frac{\mathbf{k}}{\mu}\cdot\boldsymbol{\rho}\right)\left[-\tfrac{1}{2}\Delta_{\boldsymbol{\rho}} + \mathscr{V}_{\mathscr{T}}(\boldsymbol{\rho}_{\mathscr{T}}) + \mathscr{D}_{\mathscr{T}}^2\frac{k^2}{2\mu}\right]\phi_{\mathscr{T}}(\boldsymbol{\rho}_{\mathscr{T}}) \tag{79}$$

(and a similar equation for $\mathscr{P}$). Hence, when $\mathscr{T}$ (or $\mathscr{P}$) is just translating, troublesome terms like those appearing in Eq. (77) do not occur with $\phi_{\mathscr{T}}^{\text{trans}}$ (or $\phi_{\mathscr{P}}^{\text{trans}}$): no $\partial/\partial z$ term appears in the r.h.s. of Eq. (79). This is the result of an exact cancellation between $-i\mu^{-1}\mathbf{k}\cdot\mathbf{\nabla}_{\mathbf{R}}|_G\phi_{\mathscr{T}}^{\text{trans}}$ and a term arising from $-\Delta_{\boldsymbol{\rho}}\phi_{\mathscr{T}}^{\text{trans}}/2$. Thus, it seems advisable to use basis sets that incorporate ETF's[23,24,120] not only to remove the mentioned spurious terms but also to account correctly for momentum transfer, which is tributary of a proper representation of the overall motions of electrons with the centers they belong to.[120] The presently available calculations, which make use of expansions of electronic wavefunctions over traveling orbitals as those defined in Eqs. (78), are actually confined to energetic collisions of one- or two-electron atomic systems that are constrained to move classically with constant velocity along straight line trajectories.[122,123] The major difficulty in using such expansions lies in the obligation to redo the whole calculation of H_{el} and T_{nuc} matrix element each time the velocity ($\mathbf{k}/\mu$) of the nuclei is changed. Aside from computational complications[123] such an approach ruins the conventional concept of potential energy surfaces: the potentials now depend on the nuclear velocity.

b.   Common ETF's. To overcome the previously mentioned difficulties, it was suggested[124] to resort to so-called common ETF's[23,123,124] whose purpose is to retain essentially the same H_{el} matrix as that which would be built from "clamped atomic basis sets" (except possibly for second-order terms in the nuclear velocity and $1/\mu$ corrections to energy levels)[23] while removing spurious couplings by correcting the effect of the $-i\mathbf{V_R}$ operator. This is achieved by replacing $-\mathscr{D}_{\mathscr{F}}$ or $\mathscr{D}_{\mathscr{P}}$ in Eqs. (78a) and (78b), respectively, by a common switching function $f(\boldsymbol{\rho}, \mathbf{R})$, which tends to $-\mathscr{D}_{\mathscr{F}}$ or $\mathscr{D}_{\mathscr{P}}$ when $\boldsymbol{\rho}$ approaches $\mathscr{D}_{\mathscr{F}}\mathbf{R}$ or $-\mathscr{D}_{\mathscr{P}}\mathbf{R}$, respectively.[23,124] In this context the matrix elements of the $-i\mathbf{V_R}$ operator are replaced by those of $-i\mathbf{V_R} + \mathbf{A}$, where[23]

$$\mathbf{A} = -i[\mathbf{V}_{\boldsymbol{\rho}}(f\boldsymbol{\rho})\cdot\mathbf{V}_{\boldsymbol{\rho}} + \tfrac{1}{2}\mathbf{V}_{\boldsymbol{\rho}}^2(f\boldsymbol{\rho})]. \tag{80}$$

Still, such an approach suffers from the arbitrariness in the choice of both the form and the parameters of the switching function $f(\boldsymbol{\rho}, \mathbf{R})$.[123] Although variational procedures have been proposed to reduce the mentioned arbitrariness,[125-127] the considerable increase in computational work they actually imply thwarted the spreading out of such attempts. Although the derivation of the theory with common ETF's stems from semiclassical ideas, a formulation has been proposed that reconciles it with quantum mechanics.[128] Were it not for the lack of basic principle to derive f from, the latter formulation would offer real possibilities to settle many problems like those [related to Eq. (68)] that have triggered the preceding lengthy but unavoidable digression. To our knowledge, no attempt has been made so far to introduce variants of those ideas in quantal and/or semiclassical treatments of state-to-state vibronic processes in low-energy collisions of molecular systems.

c. Pragmatic Approach. The philosophy one may extract from the preceding subsection is that a treatment based on "clamped orbital" expansions is still acceptable at low energies *provided one corrects the couplings related to nuclear momentum operators like $\partial/\partial R$.*

Inspection of Eqs. (68) and (79) indicates that *those corrections essentially amount to transferring the origin of electronic coordinates from the c.m. G to $\mathscr{F}$ (or $\mathscr{P}$) when evaluating matrix elements between two orbitals centered on $\mathscr{F}$ (or $\mathscr{P}$ resp.).* To make this more transparent, let $\phi_{\mathscr{F}_1}$ and $\phi_{\mathscr{F}_2}$ be eigenstates of an isolated one-electron operator $h_{\mathscr{F}}$:

$$h_{\mathscr{F}} = -\tfrac{1}{2}\Delta_{\boldsymbol{\rho}_{\mathscr{F}}} + \mathscr{V}_{\mathscr{F}}(\boldsymbol{\rho}_{\mathscr{F}}), \tag{81}$$

with eigenvalues $\varepsilon_{\mathscr{F}_1}$ and $\varepsilon_{\mathscr{F}_2}$ (resp.). Equation (80) is easily shown to yield[128]

$$\langle \phi_{\mathscr{F}_1}|\mathbf{A}|\phi_{\mathscr{F}_2}\rangle = i\langle\phi_{\mathscr{F}_1}|[h_{\mathscr{F}}, f\boldsymbol{\rho}]|\phi_{\mathscr{F}_2}\rangle \tag{82a}$$

$$= i(\varepsilon_{\mathscr{F}_1} - \varepsilon_{\mathscr{F}_2})\langle\phi_{\mathscr{F}_1}|f\boldsymbol{\rho}|\phi_{\mathscr{F}_2}\rangle. \tag{82b}$$

Since, by definition, the switching function should approximately equal $-\mathscr{D}_{\mathscr{T}}$ (at large $\mathbf{R}$) for wavefunctions like $\phi_{\mathscr{T}_1}$ and $\phi_{\mathscr{T}_2}$ that are mostly located around center $\mathscr{T}$, we have

$$\langle \phi_{\mathscr{T}_1} | \mathbf{A} | \phi_{\mathscr{T}_2} \rangle \approx - i\mathscr{D}_{\mathscr{T}} (\varepsilon_{\mathscr{T}_1} - \varepsilon_{\mathscr{T}_2}) \langle \phi_{\mathscr{T}_1} | \mathbf{\rho} | \phi_{\mathscr{T}_2} \rangle \tag{83a}$$

$$\approx i\mathscr{D}_{\mathscr{T}} \langle \phi_{\mathscr{T}_1} | \mathbf{V}_{\rho} | \phi_{\mathscr{T}_2} \rangle. \tag{83b}$$

Considering, in particular, the z component of that matrix element, it is immediately seen that $\langle \phi_{\mathscr{T}_1} | A_z - i(\partial/\partial R) | \phi_{\mathscr{T}_2} \rangle$ should consititute a negligible contribution. Indeed,

$$\langle \phi_{\mathscr{T}_1} | A_z - \frac{i\partial}{\partial R} | \phi_2 \rangle = \langle \phi_{\mathscr{T}_1} | - i \frac{\partial}{\partial R} \Big|_{\mathscr{T}} | \phi_{\mathscr{T}_2} \rangle + \langle \phi_{\mathscr{T}_1} | A_{z,\text{mod}} | \phi_{\mathscr{T}_2} \rangle. \tag{84}$$

The first term in this equation is zero and the second term

$$\langle \phi_{\mathscr{T}_2} | A_{z,\text{mod}} | \phi_{\mathscr{T}_2} \rangle = \langle \phi_{\mathscr{T}} | A_z - i\mathscr{D}_{\mathscr{T}} \frac{\partial}{\partial z} | \phi_{\mathscr{T}_2} \rangle \tag{85}$$

is negligible in view of Eq. (83b). Similar results are of course readily obtained for $\mathscr{P}$-centered orbitals.

For matrix elements involving orbitals that belong to different collision partners, there was apparently no ambiguity up to Section VI B 1. It is however seen at this point that consideration of a common ETF modifies the $\partial/\partial R$ coupling in a fashion that specifically depends on the way the switching function behaves *between the collision partners*. This is also seen from the relation

$$\langle \phi_{\mathscr{T}} | \mathbf{A} | \phi_{\mathscr{P}} \rangle = i \langle \phi_{\mathscr{T}} | (\varepsilon_{\mathscr{T}} + \mathscr{V}_{\mathscr{P}} - \varepsilon_{\mathscr{P}} - \mathscr{V}_{\mathscr{T}}) f\rho | \phi_{\mathscr{P}} \rangle, \tag{86}$$

which is obtained along the same lines as Eq. (82). Since the definition of f may continue to be a matter of dispute for some while, one may resort to the following set of prescriptions:

1. Matrix elements of the $\mathbf{V}_{\mathbf{R}}$ operator between separated-partner orbitals that are attached to the same center are set to zero.
2. Matrix elements of the $\mathbf{V}_{\mathbf{R}}$ operator between separated-partner orbitals that belong to different centers are evaluated using coordinates referred to the center of mass G.
3. The corrective terms arising from the considerations of a specific form of the switching function f in the common ETF are considered as additional sources of *velocity-dependent* interactions. Owing to

prescription 1 and in view of Eqs. (84) and (85), one should modify the matrix elements of the corrective operator $\mathbf{A}$ according to

$$\langle \phi_{\mathcal{T}_1} | \mathbf{A}_{\text{mod}} | \phi_{\mathcal{T}_2} \rangle = \langle \phi_{\mathcal{T}_1} | \mathbf{A} - i \mathscr{D}_{\mathcal{T}} \mathbf{V}_{\rho} | \phi_{\mathcal{T}_2} \rangle \qquad (87)$$

and a corresponding modification for $\mathscr{P}$-centered separated partner orbitals, viz., $\mathscr{D}_{\mathcal{T}} \to -\mathscr{D}_{\mathcal{P}}$. For two-center $(\mathcal{T} - \mathscr{P})$ matrix element, $\mathbf{A}$ is kept untouched.

These prescriptions are seen to split the problem into two. One problem refers to prescriptions 1 and 2. It is both well defined and well behaved and naturally fits with the presentation made until Section VI B 1. Our dissertation on diabatic states will be continued on these grounds. The second problem has to do with point (3). Although it is formally well behaved and has actually received artful solutions in some works, it still depends upon one's ability to devise proper forms of switching functions in the interpartner region. It has been rightly argued[123] that this problem is similar to that posed by the optimization of basis set parameters in electronic structure calculations and should therefore be automatically solved by systematic enlargement of the basis. This is, however, something one cannot afford at present in the field of molecular collisions of many-electron systems.

d. Diabatization. With prescriptions 1 and 2, the problem of finding separated-partner-type orbitals that are simultaneously orthogonal and diabatic is amenable to a practical solution. The procedure consists of three steps. First, one starts by choosing as convenient a procedure as one wishes to orthonormalize the set of separated-partner orbitals $\{\phi\}$:

$$|\tilde{\phi}_i\rangle = \sum_k |\phi_k\rangle c_{ki} \qquad (88)$$

Second, taking prescriptions 1 and 2 into account, one constructs the matrix $\mathbb{T}^{(R)}$ of the $\partial/\partial R$ operator in the orthogonal basis $\{\tilde{\phi}\}$:

$$\mathbb{T}_{ij}^{(R)} = \langle \tilde{\phi}_i | \frac{\partial}{\partial R} | \tilde{\phi}_j \rangle = \sum_{k,l} c_{ki} \left(\langle \phi_k | \frac{\partial}{\partial R} | \phi_l \rangle c_{lj} + \langle \phi_k | \phi_l \rangle \frac{dc_{lj}}{dR} \right), \qquad (89)$$

where with our conventions, the only *nonvanishing* $\langle \phi_k | \partial \phi_l / \partial R \rangle$ matrix elements involve orbitals belonging to *different* collision partners and obey [Eqs. (66) and (68)]

$$\langle \phi_{k \in \mathcal{T}} | \frac{\partial}{\partial R} | \phi_{l \in \mathscr{P}} \rangle = \mathscr{D}_{\mathscr{P}} \frac{d}{dR} \langle \phi_k | \phi_l \rangle, \qquad (90a)$$

$$\langle \phi_{k \in \mathscr{P}} | \frac{\partial}{\partial R} | \phi_{l \in \mathscr{T}} \rangle = \mathscr{D}_{\mathscr{T}} \frac{d}{dR} \langle \phi_k | \phi_l \rangle. \tag{90b}$$

Third, one determines the rotation matrix $\mathbb{C}$ obeying Eq. (35) (for $Q_I \equiv R$) to transform away the $\mathbb{T}^{(R)}$ matrix in the final orbital basis $\{\tilde{\tilde{\phi}}\}$:

$$|\tilde{\tilde{\phi}}_m\rangle = \sum_i |\tilde{\phi}_i\rangle C_{im}. \tag{91}$$

The general solution is found to be[92]

$$\mathbb{C} = \exp(\mathbb{Z}) = \mathbb{W}\mathbb{U}\mathbb{W}^+ \tag{91a}$$

with

$$\mathbb{Z} = \int_R^\infty \mathbb{T}^{(R)} dR'. \tag{91b}$$

$\mathbb{W}$ is the unitary transformation that diagonalizes the antisymmetric matrix $\mathbb{Z}$ and $\mathbb{U}$ is defined by

$$U_{kl} = \exp(\zeta_k)\delta_{kl}, \tag{92}$$

where the ζ_k's are the purely imaginary (or zero) eigenvalues of $\mathbb{Z}$. Useful expressions of $\exp(\mathbb{Z})$ in closed form for 3×3 and 4×4 matrices may be found in Ref. 130.

All the numerical work implied in the preceding three steps is rather straightforward.

3. (Quasi-)diabatic Orbitals of the Separated-Partner Type for Atom–Diatom Systems

Let us first consider a two-state problem and specifically focus on the active orbital of the molecule ϕ_{AB} and that of the atom ϕ_C that are implied in a charge-transfer reaction. For every fixed r, γ pair (Fig. 2) these orbitals are processed exactly as done in Section VI B 1, Eqs. (69)–(75), with $\mathscr{P} \equiv A$, $\mathscr{T} \equiv BC$ and

$$\theta = -\mathscr{D}_{BC} \arcsin \omega, \tag{93a}$$

$$\omega = \langle \phi_{BC} | \phi_A \rangle. \tag{93b}$$

One then finds

$$\langle \tilde{\tilde{\phi}}_{BC} | \frac{\partial}{\partial R} | \tilde{\tilde{\phi}}_A \rangle = -\langle \tilde{\tilde{\phi}}_A | \frac{\partial}{\partial R} | \tilde{\tilde{\phi}}_{BC} \rangle \equiv 0, \quad \forall R, r, \gamma, \tag{94a}$$

$$\langle \tilde{\tilde{\phi}}_{BC} | \frac{\partial}{\partial r} | \tilde{\tilde{\phi}}_A \rangle = \mathcal{D}_{BC}(1 - \omega^2)^{-1/2} \frac{\partial \omega}{\partial r}, \tag{94b}$$

$$\langle \tilde{\tilde{\phi}}_{BC} | \frac{\partial}{\partial \gamma} | \tilde{\tilde{\phi}}_A \rangle = \mathcal{D}_{BC}(1 - \omega^2)^{-1/2} \frac{\partial \omega}{\partial \gamma}. \tag{94c}$$

This example illustrates the general rule[63] (Section IV A) stating that $\mathbb{T}^{(R)}$, $\mathbb{T}^{(r)}$, and $\mathbb{T}^{(\gamma)}$ may not be made to vanish simultaneously. Approximate simultaneous cancellation only occurs in the case of light (A) + heavy (BC) collisions when $\mathcal{D}_{BC} \approx 0$. Still, inasmuch as ϕ_A does not depend on r, we have

$$\frac{\partial \omega}{\partial r} = \left\langle \frac{\partial \phi_{BC}}{\partial r} \middle| \phi_A \right\rangle. \tag{95a}$$

To investigate the importance of the latter quantity, one may *formally* expand ϕ_A in terms of a complete set of orbitals of the diatom. This set may be chosen as consisting of ϕ_{BC} and its complementary space $\{\phi_{BC}^\perp\}$. It follows that if ϕ_{BC} is an *acceptable BO state* (in the sense that $\langle \partial \phi_{BC}/\partial R | \forall \phi_{BC}^\perp \rangle$ is negligible for the isolated molecule), then $\partial \omega/\partial r$ may be safely neglected. Otherwise, diabatization of the isolated ϕ_{BC} orbitals ought to be performed prior to the undertaking of the A–BC calculation; a survey of the proposed methods to achieve this goal is presented in Section VI D.

Let us next examine the $\partial/\partial \gamma$ matrix element. In the selected BF reference frame (Section II B, Fig. 2), ϕ_A orbitals are normally quantized relative to the **R** axis and therefore do not depend on γ. Hence,

$$\frac{\partial \omega}{\partial \gamma} = \left\langle \frac{\partial \phi_{BC}}{\partial \gamma} \middle| \phi_A \right\rangle. \tag{95b}$$

Proceeding as we have just done for the $\partial/\partial r$ matrix element, it is concluded that, if ϕ_{AB} is an *acceptable BO* state (in the sense that Coriolis coupling $\langle \partial \phi_{BC}/\partial \gamma | \forall \phi_{BC}^\perp \rangle$ associated with Λ-doubling and heterogeneous perturbations[21] is negligible), then $\partial \omega/\partial \gamma$ may be safely neglected. This conclusion may still be strengthened when the scattering equations are set up within the framework of the infinite order sudden (IOS) approximation, where γ appears only as a fixed parameter in the scattering calculation (see, for example, Ref. 37 and the chapter by Baer in this volume).

Extension of the preceding discussion and conclusions to more than two orbitals is straightforward: most of the effort in *the diabatization of isolated-partner orbitals* is to be placed on the $\partial/\partial R$ part of the nuclear derivative operator as explained in Section VI B 2 d as long as the isolated target and/or projectile orbitals meet with the conditons required to apply the BO approximation.

There is, in principle, no restriction in the number of orbitals the preceding procedure may handle. Yet a trivial example shows that in practice one will have to limit this number. Consider a light $(\mathscr{P})$ + heavy $(\mathscr{T})$ collisional system with $\mathscr{D}_{\mathscr{T}} \cong 0$; assume that one decides to describe this system using a single $\phi_{\mathscr{P}}$ orbital and as many $\phi_{\mathscr{T}_i}$ orbitals as one's supercomputer can handle. We already know from the preceding subsections that the diabatic orthonormal basis of separated-partner-type orbitals in this case consists of $\forall \phi_{\mathscr{T}_i}$ and

$$\tilde{\tilde{\phi}}_{\mathscr{P}} = \frac{\phi_{\mathscr{P}} - \sum_i \phi_{\mathscr{T}_i} \langle \phi_{\mathscr{T}_i} | \phi_{\mathscr{P}} \rangle}{(1 - \sum_i \langle \phi_{\mathscr{T}_i} | \phi_{\mathscr{P}} \rangle^2)^{1/2}}. \tag{96}$$

Clearly, if the $\mathscr{T}$ basis is very large, $\phi_{\mathscr{P}}$ is redundant. The trouble then is that in order to describe any situation where the electron lies closer to $\mathscr{P}$ than to $\mathscr{T}$ one will have to use the *whole* $\mathscr{T}$ basis. As a by-product the scattering calculation will then have to be treated accordingly; but since we are aiming at the description of state-to-state *vibronic processes*, involving not only electronic states but also manifolds of vibrational states, one immediately realizes the hugeness of the implied task. It is solely for such reasons that the previously mentioned orbital diabatization work has to be restricted to the *active orbitals*, that is, those *characterizing the states that are to be considered in the scattering calculation*.

C. Many-Electron Diabatic States for Electron-Transfer Processes

The discussions in Sections VI B have exclusively focused on *one-electron* diabatic states of the separated-partner type. When considering many-electron systems, additional features come into play, which we now proceed to examine in turn.

1. *Mutual Orthogonalization of Shells*

One feature that immediately manifests itself in many-electron systems is the hierarchized electronic shell structure of the configuration state functions (CSF).[112] It is usually better to have the sets of orbitals describing these shells mutually orthogonal. Orbital orthogonality has many advantages. First, it enforces the Pauli principle. Second, it helps discriminating between different shells and thereby it lends itself to the construction of physically motivated projection operators. Third, it is explicitly referred to when devising pseudopotential and model-potential techniques[131] to avoid energy level collapse. Last, it enables one automatically to build orthogonal many-electron states, which once again allow for the unambiguous definition of projection operators.

To describe how the mentioned shell orthogonality problem may be handled, we consider, for example, the most widely spread cases of atom–molecule charge-transfer systems. In those cases there are three main types of shells: inner, active, and remnant. These shells are usually made orthogonal in the just specified order.

The inner shell consists of the tightly bound orbitals, which constitute the cores of the separated collision partners. Frequently, the electron occupancy of the inner shell is complete. To zeroth order the inner shell occupation number may be considered to remain unaltered both in the reactants and products states. Orbitals in the cores are orthogonalized sequentially in increasing order of energy by the Schmidt procedure. Actually, for shells that remain closed, the orthogonalization procedure does not matter, since different procedures amount to effecting linear combinations of columns in the Slater determinants constituting the CSF's.

The active shell consists of a few outer (valence and low-lying Rydberg) orbitals of the separated collision partners; they are required to be orthogonal to the cores. Since the orthogonalization + diabatization procedure outlined in Sections VI B 1–3 may *be started up* with nonorthogonal or arbitrarily orthogonalized orbitals, mutual orthogonalization of the active orbitals to the cores and within the active manifold itself may be done using the Schmidt procedure *before* diabatization is performed.

All other orbitals in complementary space (Rydberg + polarization and/or other virtual orbitals)[46,112] form the remnant shell. There is no established prescription concerning the orthogonalization of these orbitals: both Schmidt and $S^{-1/2}$ techniques[132] have been proposed and used.[133,134]

Configuration-state functions built from orbitals that have been orthogonalized as indicated previously are called projected valence bond (PVB) functions[94] in reference to the basic ideas of the celebrated VB method[135] and considering the previously indicated orthogonalization–diabatization schemes (Sections VI B 1–3). By construction, PVB states differing by a single active orbital are (quasi)diabatic to the same extent as their active orbitals are.

2. *Improvement of Diabatic Prototypes*

Raw diabatic orbitals of separated-partner type and PVB-CSF built from them lack the *proper adiabatic distortions* (polarization, shell relaxation, exchange, bonding, correlation, and so on) whose description requires extensive orbital and configuration mixings (Fig. 8). These mixings can be achieved to the highest possible degree compatible with diabatization in two steps.

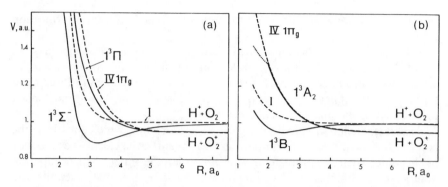

Figure 8. (a) $C_{\infty v}$ and (b) C_{2v} cuts of the $^3A''$ potential energy hypersurfaces involved in the discussion of the $H^+ + O_2(X\,^3\Sigma_g^-) \rightarrow H(1\,^2S) + O_2^+(X\,^2\Pi_g)$ reaction. Solid lines: adiabatic states; dashed lines: raw diabatic PVB-CSF. In (b), the dotted part of the $2\,^3A''$ adiabatic curve depicts a cut of a conical intersection where the symmetry of the state changes suddenly from $^3A_2$ to $^3B_1$. (This figure is taken from Ref. 97.)

a. CI in Orthogonal Subspaces. Part of the previously mentioned mixing may be readily effected in a controlled way as indicated in Section IV B. To describe the main lines of a possible procedure let us consider the charge-transfer process in Eq. (1a). We first restrict the discussion to the case when the considered process involves only the *lowest* two charge exchange CSF's of the $(ABC)^+$ system, which differ solely by their active orbital.

Inasmuch as the considered orbitals (Section VI C 1) own separated-partner labels, one may think of building the following orthogonal subspaces:

$$\text{Reactant-type:} \quad \{\text{restr } A_{\text{core}}^{+(*)}\} \quad \{\text{free } BC^{(*)}\}, \tag{97a}$$

$$\text{Product-type:} \quad \{\text{free } A^{(*)}\} \quad \{\text{restr } BC_{\text{core}}^{+(*)}\}. \tag{97b}$$

$\{\text{restr } A_{\text{core}}^{+(*)}\}$ means that any orbital other than those belonging to the orthogonalized innershell of A (as defined in Section VI C 1) is *excluded*. On the other hand, $\{\text{free } A^{(*)}\}$ means that any orbital of A may be selected at will. The nomenclature is obviously transposable to BC. One may then perform in each subspace as large a CI as one can afford. The resulting lowest root in each subspace constitutes an improved description of the related diabatic state: in particular, a *correlated description of the BC (or A) neutral in the field of the (relaxed) A^+ (resp. BC^+) ion is actually achieved.* This ensures at least that the improved diabatic states have their optimum relative energy disposition and correct long range behaviours. These are important features for a proper description of the considered charge-transfer dynamics.

As was pointed out in Section IV B the previous splitting into only two subspaces rests on the certainty that no important nonadiabatic coupling is actually affecting the lowest state in each subspace; this should at least be the case for the range of internal nuclear coordinates (R, r, γ) that is accessible to the $A-BC$ system under the considered collision conditions. Such, nonadiabatic couplings may result either from curve crossings between CSF's that differ by two orbitals, in which case we are brought back to the procedure described in Section VI A or from the closeness of CSF that differ by a single orbital from the considered ones, in which case the set of active orbitals per collision partner is to be enlarged (or its construction reconsidered, see Section VI D).

The study of diabatic ionic–covalent interactions implied by reactions like Eq. (1c) where A is an alkali atom often requires consideration of a few active orbitals and the associated CSF manifolds constituting the relevant CI subspaces. A nice example of a construction of reactant-type and product-type CI subspaces for such a case is provided by the study of Gadéa et al.[134] on the $Cs-H_2$ collisional system.

When addressing the problem of a few active orbitals per collision partner, one is confronted to the problem of deciding which orbitals ought to be chosen from the remnant shell when constructing a CI subspace for each diabatic CSF. Actually, since the remnant shell is intended to provide (among other things) the required flexibility for polarization, a possible answer would be to distribute its orbitals in consideration of dipole selection rules. If the dilemma persists, there is no other prescription than chemical intuition and experience.

An embarrassing aspect of the reactant-type and product-type partitioning of the CI space is that the CSF mixings it allows for generally turn out to be insufficient: remote states in product-type subspaces may be needed to improve the reactant states of interest and vice versa. This is patently the case in the

$$H^+ + O_2 (X\ ^3\Sigma_g^-) \rightarrow H(1\ ^2S) + O_2^+ (X\ ^2\Pi_g) \tag{98}$$

charge-exchange system.[97] The $^3A''$ diabatic potential energy surface of the $H^+ + O_2$ incident state of reaction (98) owns a well (Fig. 8).[97] This well is due not only to polarization (mixings within the reactant-type subspace) but also to exchange (mixings involving remote states in the product-type subspace), viz.[97]

$$H^+ + O_2(X\ ^3\Sigma_g^-) \leftrightarrow H(1\ ^2S) + O_2^{+*}(a\ ^4\Pi_u, A\ ^2\Pi_u, b\ ^4\Sigma_g^-, B\ ^2\Sigma_g^-). \tag{99}$$

The adiabatic treatment of these interactions is actually of importance in describing the shape of the mentioned well, which governs the rainbow scattering in this system.[136] Moreover, that shape determines the way in which the $^3A''$ diabatic states for reaction (98) come together and intersect (Fig. 8). Hence, the mentioned adiabatic treatment is seen to be of importance also for the accurate determination of the corresponding diabatic crossing locus. This matter was duly considered in Ref. 97 thanks to two clues. First, the underlying $1s_H \leftrightarrow (3\sigma_g, 1\pi_u)_{O_2}$ exchange interaction in (99) had been discussed qualitatively[137] and elementary correlation diagrams showed that when this interaction is treated adiabatically the $1s_H$ orbital energy level gets promoted and subsequently undergoes curve crossings with higher-lying orbitals, for example, $1\pi_{gO_2}$ (Fig. 9). Second, the two $^3A''$ diabatic states had to coincide with the adiabatic $1\,^3\Sigma^-$, $1\,^3\Pi$, and $1\,^3B_1$, $1\,^3A_2$ states in the limiting $C_{\infty v}$ (linear) and C_{2v} (T-shape) geometries, respectively (Fig. 8). When such indications exist they may be used to distribute the CSF in the subspaces in a more specific way than the naïve reactant-type and product-type arrangements [Eq. (97)]. Still, when devising the space partitioning, one should always keep in mind the fact that the adiabatic behavior of the system vis-à-vis certain interactions depends on the collision conditions. The above discussed $1s_H \leftrightarrow (3\sigma_g, 1\pi_u)_{O_2}$ interactions [Eq. (99)] are adiabatic in the few tens eV energy range but become nonadiabatic at keV energies.

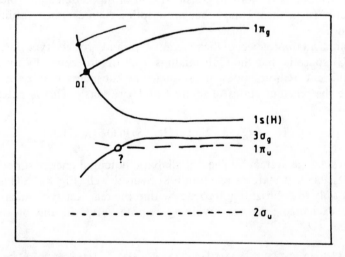

Figure 9. Schematic MO correlation diagram for the $(H-O_2)^+$ collisional system showing that when the $3\sigma_{gO_2}$ (or $1\pi_{uO_2}$) $- 1s_H$ exchange interaction is treated adiabatically the MO energy level correlated with $1s_H$ rises and crosses the MO energy level correlating with $1\pi_{gO_2}$ (This figure is adapted from Ref. 137.)

b. *Maximum Overlap between Large-Scale CI Wavefunctions and Rough Diabatic Prototypes.* For very low energies of relative motion the quasimolecular collision system is likely to behave most adiabatically except in the vicinity of real or underlying crossings of potential energy surfaces. In such conditions one's task is to determine the potential energy surfaces in the regions of adiabatic behavior with high (chemical or spectroscopic) accuracy and simultaneously manage a convenient way of handling the localized region(s) of nonadiabatic behavior. This is at the root of all attempts to obtain diabatic states by an orthogonal transformation of a small set of highly accurate adiabatic states. Actually, rotation methods based on Eq. (35) are unpractical. This is not only due to basic hindrances[63] (Section IV A) but also to the obligation one has to cope with the very acute problems that diabatic bases are intended to avoid, namely, the calculation and use of badly behaved $\mathbb{T}^{(l)}$ matrix elements. Nearly three decades have been necessary for an alternative criterion to emerge. The new proposal is due to Cimiraglia et al.[116]; its main lines are as follows. One starts by constructing a small set of n rough diabatic states $\{\varphi_k^{el,d}\}$: raw or preferably mixings of PVB–CSF's as done in Sections VI C 1 and VI C 2 a, respectively. Those states are associated with reactant and product states of the considered collision process. Besides, one determines a set of n highly accurate adiabatic states $\{\varphi_m^{el,a}\}$ correlating with the latter channel states. One projects the $\varphi_k^{el,d}$ diabatic vectors into the $n \times n$ a diabatic subspace thereby giving

$$|\pi_k^d\rangle = \sum_m |\varphi_m^{el,a}\rangle\langle\varphi_m^{el,a}|\varphi_k^{el,d}\rangle. \tag{100}$$

In general, the latter projections are neither orthogonal nor normalized. Simultaneous orthogonalization and normalization of the projected diabatic vectors by the $\mathbb{S}^{-1/2}$ technique,[97,132,134] that is,

$$|\varphi_k^{el,D}\rangle = \sum_j |\pi_j^d\rangle([\mathbb{S}^{da}]^{-1/2})_{jk}, \tag{101}$$

$$S_{jk}^{da} = \langle\pi_j^d|\pi_k^d\rangle, \tag{102}$$

provides [via Eq. (101)] the sought rotation matrix $\mathbb{C}$:

$$|\varphi_k^{el,D}\rangle = \sum_m |\varphi_m^{el,a}\rangle C_{mk}, \tag{103a}$$

$$C_{mk} = \sum_j \langle\varphi_m^{el,a}|\varphi_j^{el,d}\rangle([\mathbb{S}^{da}]^{-1/2})_{jk}. \tag{103b}$$

This is indeed the case because the transformation in Eq. (101) provides the vectors that resemble the original diabatic vectors $|\varphi_k^{\mathrm{el},d}\rangle$ most, in the sense that they maximize the overlap sum:

$$\sum_k \langle \varphi_k^{\mathrm{el},d} | \varphi_k^{\mathrm{el},D} \rangle \equiv \sum_k \langle \pi_k^d | \varphi_k^{\mathrm{el},D} \rangle \tag{104}$$

This is quite an appealing procedure both as regards its basic idea and the relatively small amount of work it requires to effect the adiabatic-to-quasi-diabatic transformation.

It is worth mentioning that the idea of the preceding process did not pop out rightaway. It was preceded by a proposal by Levy,[138] which, albeit strange at first sight, was shown later on[139] to provide the same information as the above procedure. Contrary to the preceding, Levy's method rests on the projection of the adiabatic vectors $|\varphi_m^{\mathrm{el},a}\rangle$ onto the rough diabatic subspace $\{\varphi_m^{\mathrm{el},d}\}$. (This is the feature that makes the proposal appear strange.) One thus has to force the $\{\varphi_m^{\mathrm{el},d}\}$ representation to give as accurate eigenvalues of the electronic hamiltonian as those (E_m^a) of the original adiabatic basis. This is achieved by introducing an effective (des Cloizeaux-type)[140] hamiltonian $\bar{H}_{\mathrm{el}}$ whose eigenvalues are E_m^a:

$$\bar{H}_{\mathrm{el}} = \sum_m |\varphi_m^{\mathrm{el},A}\rangle E_m^a \langle \varphi_m^{\mathrm{el},A}|, \tag{105a}$$

with

$$|\varphi_m^{\mathrm{el},A}\rangle = \sum_l |\pi_l^a\rangle ([\mathbb{S}^{ad}]^{-1/2})_{lm}, \tag{105b}$$

$$|\pi_l^a\rangle = \sum_k |\varphi_k^{\mathrm{el},d}\rangle \langle \varphi_k^{\mathrm{el},d} | \varphi_l^{\mathrm{el},a}\rangle, \tag{105c}$$

$$S_{lm}^{ad} = \langle \pi_l^a | \pi_m^a \rangle. \tag{105d}$$

The representation of the hamiltonian in the diabatic basis is then

$$H_{jk} = \langle \varphi_j^{\mathrm{el},A} | \bar{H}_{\mathrm{el}} | \varphi_k^{\mathrm{el},A} \rangle \tag{106a}$$

and it has been shown that[139]

$$H_{jk} = \langle \varphi_j^{\mathrm{el},D} | H_{\mathrm{el}} | \varphi_j^{\mathrm{el},D} \rangle. \tag{106b}$$

Although the two descriptions are equivalent the one presented first seems more natural.

As in Section VI C 2 a a few words of caution concerning the use of the preceding rotation method are necessary. It rests on the certainty that in the considered collision conditions (energy and accessible range of the internal nuclear coordinates) a set of n adiabatic states is sufficient to describe the considered electron transition processes. In other words, it requires that nonadiabatic coupling terms (if any) between states in the considered $n \times n$ subspace and complementary space may be safely neglected.

Figures 8, 10a and 10b illustrate results obtained at various stages of the construction of diabatic states for the $H^+ + O_2$ collisional system, Eq. (98). An interesting feature appears in the comparison of the results obtained with the method of Section VI C 2 a and those produced by the previous rotation method, namely, a bend of the repulsive diabatic state produced by the latter method (Fig. 10). This is precisely the sort of situation that has just been warned against in the preceding. The mentioned bending gets sharper as the geometry of the system approaches the T shape; in that neighborhood it clearly depicts a conical intersection phenomenon (Fig. 8b). Actually a bunch of avoided crossings have been found to produce that feature[97]; the states that participate in these crossings differ from the dominant $H + O_2^+$ PVB–CSF by single and double-electron excitations. In Ref. 97 the description of this crossing pattern (Fig. 11) was readily available from CI calculations in orthogonal subspaces as indicated at the end · of Section VI C 2 a. On the other hand, its description by the preceding rotation method would have required an increase of n well beyond 2. This particular example warns against some pitfalls that may sometimes lie in wait for methods which systematically attempt to get too close to true adiabatic states.

C. Extended scope

In Section VI B we have specifically focused on methods of constructing diabatic states for low-energy ion–molecule charge-transfer processes. Particular emphasis has been put on procedures that aim at enforcing the separated-partner characters in the construction of related diabatic prototypes. This has been done in view of the occurrence of those charge-transfer reactions in rather distant encounters (R larger than normal atomic and molecular dimensions). Actually, this is but one possible class of diabatic states. There are many other circumstances where different types of characters have to be brought out and preserved. Below are surveyed some of those instances that show the degree of generality of the previously mentioned ideas and procedures.

1. Bringing Out Characters

This is the part of the work where physicochemical skilfulness is required most. Characters readily show off in regions of internal nuclear coordinates,

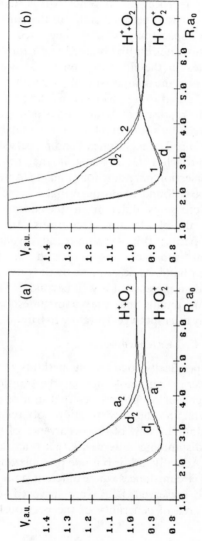

Figure 10. Comparison between $^3A''$ potential energy curves relevant to the discussion of the $H^+ + O_2(X\,^3\Sigma_g^-) \rightarrow H(1\,^2S) + O_2^+(X\,^2\Pi_g)$ reaction. The molecule bond distance is $r = 2.282a_0$ and the relative $(\mathbf{R}, \mathbf{r})$ angle is $\gamma = 45°$. (*a*) Comparison between the diabatic states ($d_{1,2}$) obtained by the rotation method of Section VIC 2 b and the corresponding adiabatic states. (*b*) Comparison between the diabatic states (1,2) obtained by diagonalization in orthogonal subspaces (Section IV B, VIC 2 a) and the $d_{1,2}$ states of Fig. 10*a*. (This figure is taken from Ref. 97.)

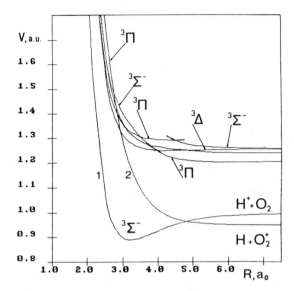

Figure 11. Example of a diabatic curve crossing pattern obtained by diagonalization in orthogonal subspaces (Sections IV B, VI C 2a in text) for the $(H-O_2)^+$ collisional system at $r = 2.282a_0$ and $\gamma = 0$. (This figure is taken from Ref. 97.)

where there is *one dominant* type of interaction.[141] A simple example is provided by the extreme opposite case to the one considered in Sections VI B 1–3, viz., that where the collision partners are united or fused. This is the united atom limit in atom–atom systems or the inserted $(D_{\infty h})$ atom limit in $A-B_2$ systems. Those cases have clearly identifiable characters that may be exploited, as, for example, *symmetry and nodal structure* of the orbitals. Likewise, real symmetry in limiting geometries (e.g., linear, isoceles and equilateral triangles—to cite but the most elementary triatomic geometries) give rise to clearcut characters. Exotic real symmetries in phase space may arise from the *existence of specific operators that commute with the electronic hamiltonian*; this is the case of the Runge–Lenz operator,[47–49] which confers special *quantum numbers* to the wavefunctions of linear one-electron polycentric systems. When no real symmetry exists, one may look for *underlying symmetries*. Particular examples are provided by isoelectronic systems involving nuclei that differ little in their atomic numbers (quasi-symmetric systems), for example, He–He, $(Li-He)^+$,[142] HeH_2.[111] Another particular example is that of *quasicoulombic systems* (deep inner shells, high Rydberg states, multiply-charged ions with few electron). Utilization of the special symmetries of real coulombic systems as quasisymmetries in non-coulombic systems[142] has been quite fruitful in the field of atomic collisions.

It has provoked the emergence of the celebrated electron-promotion model and the associated diabatic correlation diagrams based on the radial *node-conservation rule*.[45,108] For systems lacking such near-symmetries *electronic properties other than energy* may be used to bring out specific characters. Examples have only been reported for atom–atom systems. One example is the use of the rotational coupling matrix element [$\mathscr{R}$ term in Eq. (14)] between a well-behaved π_u orbital and a σ_u orbital that is independently involved in an avoided curve crossing.[141] The character put forward is large or small σ–π coupling. Another example is the use of dipole moment[144]; the character put forward is large or small dipole moment. A third example of the same style consists of using as relevant property the overlap with a suitably chosen vector[142]; the character is thereby large or small overlap. Characters associated with allowed regions of classical motion in one-electron systems (motions in potential wells of the separated or fused partners or on top of potential barriers)[145,146] have been stressed recently; in particular the use of a character like "existence of electron density on a potential saddle"[146] has provided new insight into special curve-crossing series in H_2^+-like systems. Cases also exist where the character is as simple as the valence or Rydberg nature of an orbital.[113,141] There are probably many more characters one may try to emphasize. In the limit where no character readily suggests itself there is still the possibility of postulating that it is precisely the one contained in the very system's wavefunction at a nuclear geometry selected well away from any troublesome potential surface pinching.[141] This is of course quite similar in spirit to the separated-partner character treated in Section VI B.

2. *Preservation of Characters*

As in Sections VI B and VI C, the next stage in building a diabatic state is to preserve the character thus brought out while determining the wavefunctions step by step in the considered range of internal nuclear coordinates. This may be done by letting the selected wavefunctions evolve in subspaces thereby preventing against rapid mixings.

In Sections VI B 1–3 the orbitals were kept in the form of properly orthogonalized separated-partner-type orbitals. Yet, in cases when one studies phenomena that occur at typical distances (R or r, for example) comparable with normal atomic dimensions, frozen separated-type orbitals turn out to be inappropriate. One may form linear combinations of those orbitals in a controlled way. A simple example is when one wants to prevent avoided crossings between valence and Rydberg orbital energy levels; one just performs Hartree–Fock-type calculations in separate subspaces:[113,141] *just as if the valence or Rydberg character were symmetries.* Thus, when transposed in the present orbital context, O'Malley's earlier proposal[55,56] to use projection operator techniques amounts to performing *"character-constrained"*

SCF (self-consistent-field) calculations. Some care should however be exercised in doing so since standard SCF schemes do not treat occupied and empty orbitals on the same footing; so-called IVO (improved virtual orbital) generation techniques actually help solving this problem.[112,113] Whenever desired, the resulting character-constrained-SCF MO may be improved by resorting to a rotation method similar to that discussed in Section VI C 2 b that achieves *diabatization of a few accurate adiabatic MO by maximizing their overlap with previously determined rough diabatic reference MO.*

Orbitals have been considered to some extent in the preceding because their organization in a CSF determine the character of the related diabatic many-electron wavefunction prototype. A simple example will stress the importance orbitals have in the adiabatic–diabatic problem. Consider the case where one has succeeded in building appropriate diabatic orbitals two of which, ϕ_1 and ϕ_2, cross along a certain intersection locus. The states associated with the configurations $(\cdots \phi_1^2 \phi_2^0 \cdots)$ and $(\cdots \phi_1 \phi_2 \cdots)$ will thereby cross near the ϕ_1–ϕ_2 intersection locus. But that is not all; the state representing $(\cdots \phi_1^0 \phi_2^2 \cdots)$ will cross the latter two and there is still more: the whole series $(\cdots \phi_1 \phi_2^0 \cdots \phi')(\cdots \phi_1^0 \phi_2 \cdots \phi')$ will also cross in pairs in the same neighborhoods. Clearly, if one lets the $\phi_{1,2}$ orbitals evolve adiabatically and avoid their crossing a terrible mess of multiple conical intersections will ensue merely at the single-CSF level. This situation has apparently been overlooked in the ultimate conclusion of Ref. 116.

3. Diabatic Basis Changes

It sometimes happens that the H_{el} matrix elements in a diabatic basis increase so much that their interpretative usefulness and computational convenience are lost. Elementary examples of this situation occur at small relative distances R in charge-exchange systems that come under the Demkov–Nikitin models.[93a,93b] Since a basis remains diabatic under a constant unitary transformation, the exceedingly large matrix elements may be *locally* diagonalized accordingly. This device has been used in cases when good diabatic states in one region of internal nuclear coordinates are unsuitable for another region.[147] An example is the "incorrect dissociation" of diabatic states owing π^4 or $\pi^2\sigma^2$ dominant configuration types into interacting combinations of ^1D and ^1S states or, vice versa, the "incorrect association" of diabatic states owning ^1D and ^1S configuration types into interacting combinations of π^4 and $\pi^2\sigma^2$ states.[148] Another well-known case is that of spin–orbit coupling.[22,94,104] For example, a basis that diagonalizes H_{el} solely in a region of strong electrostatic interactions between the collision partners and leaves away comparatively weak spin–orbit interactions constitutes a special class of diabatic basis. On the other hand, at large interparticle distance a basis that diagonalizes all intrapartner interactions including spin–orbit but leaves away weak interpartner electrostatic interactions is another class of diabatic basis. A constant unitary transformation will provide the right basis in the right region. It is worth pointing out that this device is actually implemented in algorithms dealing with the solution of scattering equations of the type given in Eq. (27)[149]: the integration domain is chopped into small intervals at the center of which the $\mathbb{H}$ matrix is *locally* diagonalized. Diabatic basis set changes are effected at the boundary of each domain as the system propagates.

4. Diabatic Vibronic Bases

So far we have only considered electronic diabatic states. Yet the previously discussed methods are not solely confined to those cases. For example, when considering Eq. (27) for an atom–diatom collision problem ($r_{\mathscr{F}} \equiv r$), one generally uses vibronic expansions. Those arise from the expansion of $\mathscr{N}_{kn}^{nuc}$ in Eq. (12) over vibrational bases $\{G_{v_n}\}$. One may apply the preceding procedures to those bases individually or to $\{G_{v_n}\varphi_n^{el}\}$ bases globally.

Since r is an independent internal nuclear coordinate, vibrational eigenstates of the diatomic partner in its reactant or product states constitute strictly diabatic vibrational bases $G_{v_n}^d(r)$ vis-à-vis the variations of the other two coordinates R, γ. If the latter coordinates change slowly enough, one may elect to construct an adiabatic $G_{v_n}^a(r; R, \gamma)$ basis where R, γ are fixed parameters, for example,[150]

$$\left(-\frac{d^2}{2\mu_{\mathscr{F}}\,dr^2} + E_n^d(R, r, \gamma) \right) G_{v_n}^a(r; R, \gamma) = \mathscr{E}_{nv_n}^a G_{v_n}^a(r; R, \gamma). \qquad (107)$$

Still the latter basis may be subject to troublesome rapid variation with R and γ.[151] To avoid such problems one may advantageously use the method of Section VI C 2 b: $G_{v_n}^a$ functions may be calculated exactly by solving the one-dimensional Schrödinger equation (107) numerically. An optimum diabatic basis is thereafter provided by the unitary transformation of the exact $\{G_{v_n}^a\}$ basis that achieves maximum overlap with the asymptotic $\{G_{v_n}^d\}$ basis. Products of diabatic electronic bases times diabatic vibrational bases constitute *diabatic vibronic bases*. The corresponding energy levels as functions of R and γ may freely cross. A typical crossing network of computed diabatic vibronic states[136b,c] is shown in Fig. 12; the overall trend of each series of levels is determined by that of the electronic state labeling it.

Another class of quasidiabatic vibronic states has been discussed for the

$$N_2^+(X\ ^2\Sigma_g^+, v_+) + Ar \leftrightarrow N_2(X\ ^1\Sigma_g^+, v_0) + Ar^+ \qquad (108)$$

collision in the few eV energy range.[152] It exploits the property of $N_2^+(X\ ^2\Sigma_g^+)$

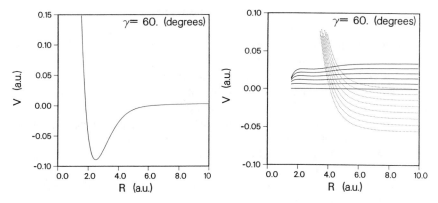

Figure 12. Right panel: example of a diabatic curve-crossing network of vibronic energy levels in the $(H-O_2)^+$ collisional system at $\gamma = 60°$. Solid lines: $H^+ + O_2(X\,^3\Sigma_g^-, v = 0, 1, \dots, 5)$ manifold; dashed lines: $H(1\,^2S) + O_2^+(X\,^2\Pi_g, v' = 0, 1, \dots, 12)$ manifold. All curves are referred to the $H^+ + O_2(X\,^3\Sigma_g^-, v = 0)$ level whose R dependence is shown in the left panel. these are samples of the input data used in the vibronic semiclassical close-coupling calculations of Ref. 136c.

and $N_2(X\,^1\Sigma_g^+)$ states of having nearly parallel energy curves as well as the property of the collision system to behave adiabatically vis-à-vis $v_+ \equiv v_0$ transitions; the mentioned quasidiabatic states are thereby obtained by mixing $G_{v_+}\varphi_{\text{left}}^{\text{el}}$ and $G_{v_0}\varphi_{\text{right}}^{\text{el}}$ states in $v_+ \equiv v_0$ pairs. (Interested readers in this example may find ample discussions in Refs. 27, 37, and 152 and corresponding references therein.)

5. Exotic Diabatic States: Hydrogenic States in Heavy–Light–Heavy Systems

It is worth pointing out that some of the preceding ideas have recently been applied to a study of the H-atom exchange process between two iodine atoms.[153] Exploiting the smallness of the hydrogen/iodine mass ratio a BO-type separation has been used which brings the IHI problem into a form bearing some resemblance with that encountered in the treatment of familiar one-electron diatomic systems. "Hydrogenic" potential energy curves, which are the analogs of electronic potential energy curves in one-electron diatomic systems, could thus be built. Diabatic energy curves have been obtained in this context using a method that amounts to selecting as characters "the H-atom localization between or outside the two iodine nuclei" and by preserving them using a diagonalization method in quasidiabatic subspaces (Sections IV B and VI C 2 a).[56]

VII. CONCLUSIONS

The notion of diabatic states is contemporaneous with that of the quasi-molecular view of atomic and molecular collisions. It has strengthened after each of its rebirths since its first emergence in the 1930s.

The essential characteristics of diabatic states is their slow dependence on parameters that determine the nature and the strength of some varying interactions. In general quasimolecular systems those parameters are some or all of the internal nuclear coordinates which determine the relative arrangements of the constituting aggregates (atoms, molecules, clusters, vibrators, rotors, and so on).

The more or less rapid variation of nuclear coordinates and their domain of spatial extension may lead to electron-transition processes. As a consequence the average field in which the nuclei evolve changes; this has the effect of modifying the velocities and spatial extensions of the nuclear motions. This in turn affects the electrons and so on. Diabatic states provide a sensible way of treating the *quantum dynamics* of those electronuclear phenomena. They are most useful when nonadiabatic couplings arising from variations of nuclear coordinates hinder the use of the BO approximation. This is particularly, but not exclusively, the case near true or avoided intersections of adiabatic potential energy surfaces. Diabatic states are useless for the detailed description of structural properties of molecular edifices with rigid nuclear armatures (except sometimes for qualitative characterization of certain states). Part of the past reluctance to employ diabatic states had to do with the fact that the relevant electronic data needed to treat the quantum dynamics of nonadiabatic molecular phenomena were supplied by molecular structure computations. Another source of distrust laid in the mathematical definition of diabatic states that tended to confuse them with trivial states having no dependence on nuclear geometry. The notion thus went into such disrepute that it was sometimes contended that any state which is not adiabatic can be considered as diabatic

The present contribution shows ample evidence that all previous objections that have been raised against diabatic states may be brushed aside. Considering the broad class of electron-transfer processes in low-energy molecular encounters the complete procedure for building well-defined diabatic states, from both mathematical and physicochemical points of view, has been described. The procedure blends in a progression basic ideas and proposals that have appeared over the past three decades. Its cornerstone is an active set of separated-partner orbitals (orthogonal to the cores) that is made orthonormal so as to achieve exact cancellation of a suitable (ETF-corrected) $\partial/\partial R$ operator. Arguments have been given to substantiate the idea that if the considered separated-partner orbitals belong to truly acceptable BO states then all other $\partial/\partial Q_I$ matrix elements ought to be satisfactorily small. Discussions have been given to serve as guidelines when the preceding condition is not fulfilled. Configuration-state functions, which are coined the name projected valence bond, are thereafter built from those orbitals (and orbitals in complementary space). Subspaces that primarily emphasize the reactantlike or productlike characters of configuration states involving the active diabatic orbitals are then generated in order to be processed by CI. The construction of those subspaces can allow for extra distortions when an adiabatic behavior of the collision system vis-à-vis interactions with remote states is anticipated. The resulting states in each subspace are obviously prevented from any undue mixing with any state in another subspace. Moreover, they incorporate utmost separated-partner correlation and mutual polarization compatible with the previously mentioned constraint. Yet this constraint may still prevent the quasidiabatic states thus determined to include further adiabatic distortions. This is allowed for in a final stage by a rotation method. A set of nearly exact adiabatic states associated with the considered reactant and products is rotated so as to maximize the overlap of the resulting states with the previously mentioned quasidiabatic prototypes.

Possible extensions of those ideas to more general nonadiabatic molecular collisions have been outlined. All calculations that are needed in the described procedure can be achieved with currently available quantum-chemistry computer codes. Just two points presently require some skilfulness: the search of the orbital character that is to be preserved and the actual form of ideal ETF that combines with it. The latter problem is still open for research. As to the former, it is not too utopian to think that in the long run the "character search" which seeds the whole procedure could well be aided by evolutionary computer-based expert systems.

References

1. N. F. Mott, *Proc. Cambridge Phil. Soc.* **27**, 553 (1931).

2. M. Born and R. Oppenheimer, *Ann. Phys.* **84**, 457 (1927).

3. M. Born, *Festchrift Gött. Nachr. Math. Phys. K1*, 1 (1951).

4. M. Born and K. Huang, *Dynamical Theory of Cristal Lattices*, Oxford University Press, New York, 1956.

5. W. Kolos, *Adv. Quant. Chem.* **5**, 99 (1970).

6. A. Messiah, *Mécanique Quantique*, Dunod, Paris, 1964, Vol. 2.

7. W. D. Hobey and A. D. Mc Lachlan, *J. Chem. Phys.* **33**, 1695 (1960).

8. R. Lefebvre and M. Garcia Sucre, *Int. J. Quant. Chem.* **IS**, 339 (1967).

9. J. H. Young, *Int. J. Quant. Chem.* **IIIS**, 607 (1970).

10. (a) R. L. Champion, L. D. Doverspike, W. G. Rich, and S. M. Bobbio, *Phys. Rev. A* **2**, 2337 (1970). (b) S. M. Bobbio, W. G. Rich, L. D. Doverspike, and R. L. Champion, *Phys. Rev.*

A **4**, 957 (1971). (c) H. P. Weise, H. P. Mittmann, A. Ding, and A. Heinglein, $Z. Naturf.$ **26a**, 1112 (1971). (d) F. A. Gianturco, G. Niedner, M. Noll, E. Semprini, F. Stefani, and J. P. Toennies, $Z. Phys. D$ **7**, 281 (1987).

11. W. Aberth, D. C. Lorents, R. P. Marchi, and F. T. Smith, $Phys. Rev. Lett.$ **14**, 776 (1965). (b) R. P. Marchi and F. T. Smith, $Phys. Rev.$ **139**, A1025 (1965).

12. M. D. Pattengill, in $Atom Molecule Collision Theory$, R. B. Bernstein, ed., Plenum, New York, 1979, p. 359.

13. W. R. Gentry, in $Atom Molecule Collision Theory$, R. B. Bernstein, ed., Plenum, New York, 1979, p. 39.

14. R. Schinke, in $Collision Theory for Atoms and Molecules$, F. A. Gianturco, ed., NATO ASI, Plenum, New York, 1989, p. 229.

15. D. G. Truhlar and J. T. Muckerman, in $Atom Molecule Collision Theory$ R. B. Bernstein, ed., Plenum, New York, 1979, p. 505.

16. S. H. Lin, $J. Chem. Phys.$ **44**, 3759 (1966).

17. M. Bixon and J. Jortner, $J. Chem. Phys.$ **48**, 715 (1968).

18. H. A. Jahn and E. Teller, $Proc. Roy. Soc. (London)$ **A161**, 220 (1937).

19. R. Engelman, $The Jahn-Teller Effect$, Wiley, New York, 1972.

20. H. Köppel, W. Domcke, and L. Cederbaum, $Adv. Chem. Phys.$ **57**, 59 (1984).

21. H. Lefebvre-Brion and R. W. Fields, $Perturbations in the Spectra of Diatomic Molecules$, Academic Press, Orlando, 1986.

22. E. E. Nikitin and S. Ya Umanskii, $Theory of Slow Atomic Collisions$, Springer Series in Chemical Physics 30, Springer-Verlag, Berlin, 1984.

23. J. B. Delos, $Rev. Mod. Phys.$ **53**, 287 (1981).

24. B. C. Garett and D. G. Truhlar, in $Theoretical Chemistry: Advances and Perspectives$, Academic Press, New York, 1981, Vol. 64, p. 215.

25. (a) J. C. Tully, in $Dynamics of Molecular Collisions$, W. H. Miller, ed., Plenum, New York, 1976, p. 217. (b) J. C. Tully $Adv. Chem. Phys.$ **42**, 63 (1980).

26. M. Baer, in $Topics in Current Physics$, Springer-Verlag, Berlin, 1983, Vol. 3, p. 117.

27. V. Sidis, in $Collision Theory for Atoms and Molecules$ F. A. Gianturco, ed., NATO ASI, Plenum, New York, 1989, p. 343.

28. L. D. Landau, $Physik Z. Sovjet Union$ **2**, 46 (1932).

29. C. Zener, $Proc. Roy. Soc. (London)$ **A137**, 396 (1932).

30. N. Rosen and C. Zener, $Phys. Rev.$ **40**, 502 (1932).

31. (a) J. Von Neumann and E. P. Wigner, $Physik Z.$ **30**, 467 (1929). (b) E. Teller, $J. Phys. Chem.$ **41**, 109 (1937). (c) G. Herzberg and H. C. Longuet-Higgins, $Discuss. Faraday Soc.$ **35**, 77 (1963).

32. W. Lichten, $Phys. Rev.$ **131**, 229 (1963).

33. F. T. Smith, $Phys. Rev.$ **179**, 111 (1969).

34. H. Hellman and J. K. Syrkin, $Acta Physica Chemica USSR$ **2**, 433, (1935).

35. (a) F. London, $Z. Phys.$ **74**, 143 (1932). (b) O. K. Rice, $Phys. Rev.$ **38**, 1943 (1931).

36. A. D. McLachlan, $Mol. Phys.$ **4**, 417 (1961).

37. V. Sidis, $Adv. At. Mol. Opt. Phys.$ **26**, 161 (1990).

38. D. R. Bates and D. A. Williams, $Proc. Phys. Soc. London$ **83**, 425 (1964).

39. W. R. Thorson, $J. Chem. Phys.$ **34**, 1744 (1961).

40. D. W. Jepsen and J. D. Hirschfelder, *J. Chem. Phys.* **32**, 1323 (1960).

41. U. Fano and W. Lichten, *Phys. Rev. Lett.* **14**, 627 (1965).

42. W. Lichten, *Phys. Rev.* **164**, 131 (1967).

43. F. Hund, *Z. Phys.* **36**, 657 (1926).

44. (a) R. S. Mulliken, *Phys. Rev.* **32**, 186 (1928). (b) R. S. Mulliken, *Rev. Mod. Phys.* **2**, 60 (1930). (c) R. S. Mulliken, *Rev. Mod. Phys.* **2**, 506 (1930).

45. M. Barat and W. Lichten, *Phys. Rev.* **A 6**, 211 (1972).

46. H. F. Schaeffer III, *The Electronic Structure of Atoms and Molecules*, Addison-Wesley, Reading, MA, 1972.

47. C. A. Coulson and A. Joseph, *Int. J. Quant. Chem.* **1**, 337 (1967).

48. H. A. Erikson and E. L. Hill, *Phys. Rev.* **75**, 29 (1949).

49. M. Kotani, K. Ohno, and K. Kayama, in *Handbook of Physics: Molecules II* S. Flugge, ed., Springer-Verlag, Berlin, 1961, Vol. XXXVII/2, p. 58.

50. S. S. Gerstein and V. D. Krivtchenkov, *Sov. Phys. JETP* **13**, 1044 (1961).

51. D. Coffey Jr., D. C. Lorents, and F. T. Smith, *Phys. Rev.* **187** 201 (1969).

52. (a) M. Barat, in *The Physics of Electronic and Atomic Collisions*, Invited Papers and Progress reports of the VIIIth ICPEAC, B. C. Cobic and M. B. Kurepa, eds., Institute of Physics, Beograd, Yugoslavia, 1973, p. 42. (b) M. Barat, in *Fundamental Processes in Energetic Atomic Collisions*, H. O. Lutz, J. S. Briggs, and H. Kleinpoppen, eds., NATO ASI, Plenum, New York, 1983, p. 389. (c) Q. C. Kessel, E. Pollack, and W. W. Smith, in *Collision Spectroscopy*, R. G. Cooks, ed., Plenum, New York, 1978, p. 147.

53. R. D. Levine, B. R. Johnson, and R. B. Bernstein, *J. Chem. Phys.* **50**, 1694 (1969).

54. T. F. O'Malley, *Phys. Rev.* **162**, 98 (1967).

55. T. F. O'Malley, *J. Chem. Phys.* **51**, 322 (1969).

56. T. F. O'Malley, *Adv. At. Mol. Phys.* **7**, 223 (1971).

57. J. C. Tully and R. K. Preston, *J. Chem. Phys.* **55**, 562 (1971).

58. (a) Z. H. Top and M. Baer, *J. Chem. Phys.* **66**, 1363 (1977). (b) Z. H. Top and M. Baer, *Chem. Phys.* **25**, 1 (1977).

59. M. Baer, *Chem. Phys. Lett.* **35**, 112 (1975).

60. M. Baer, *Chem. Phys.* **15**, 49 (1976).

61. J. B. Delos and W. R. Thorson, *J. Chem. Phys.* **70**, 1774 (1979).

62. X. Chapuisat, A. Nauts, and D. Dehareng-Dao, *Chem. Phys. Lett.* **95**, 139 (1983).

63. C. A. Mead and D. G. Truhlar, *J. Chem. Phys.* **77**, 6090 (1982).

64. T. C. Thompson, D. G. Truhlar, and C. A. Mead, *J. Chem. Phys.* **82**, 2392 (1985).

65. T. Pacher, L. S. Cederbaum, and H. Köppel, *J. Chem. Phys.* **89**, 7367 (1988).

66. T. Pacher, C. A. Mead, L. Cederbaum, and H. Köppel, *J. Chem. Phys.* **91**, 7057 (1989).

67. M. Desouter-Lecomte, D. Dehareng, and J. C. Lorquet, *J. Chem. Phys.* **86**, 1429 (1987).

68. V. Sidis, *J. Phys. Chem.* **93**, 8128 (1989).

69. (a) L. M. Delves, *Nucl. Phys.* **9**, 391 (1959); **20**, 275 (1960). (b) F. T. Smith, *J. Math. Phys.* **3**, 735 (1962). (c) R. C. Whitten and F. T. Smith, *J. Math. Phys.* **9**, 1103 (1968).

70. (a) J. Robert and J. Baudon, *J. Phys. B* **19**, 171 (1986). (b) J. Robert and J. Baudon, *J. Physique* **47**, 631 (1986).

71. E. B. Wilson Jr., J. C. Decius, and P. C. Cross, *Molecular Vibrations*, McGraw-Hill, New York, 1955.

72. J. C. Bellum and P. McGuire, *J. Chem. Phys.* **79**, 765 (1983).

73. R. N. Zare, *Angular Momentum*, Wiley, New York, 1976.

74. (a) T. Carington, *Acc. Chem. Res.* **7**, 20 (1974). (b) P. G. Mezy, *Potential Energy Hypersurfaces*, Elsevier, Amsterdam, 1987.

75. S. Chapman and R. K. Preston, *J. Chem. Phys.* **60**, 650 (1974).

76. C. Petronglo, R. J. Buenker, and S. Peyerimhoff, *Chem. Phys. Lett.* **116**, 249 (1985).

77. F. Rebentrost and W. A. Lester Jr., *J. Chem. Phys.* **64**, 3879 (1976).

78. C. Galloy and J. C. Lorquet, *J. Chem. Phys.* **67**, 4672 (1977).

79. R. J. Buenker, G. Hirsch, S. D. Peyerimhoff, P. J. Bruna, J. Römelt, M. Bettendorf, and C. Petronglo, in *Current Aspects of Quantum Chemistry*, R. Carbo, ed., Elsevier, Amsterdam, 1982, p. 8.

80. (a) B. H. Lengsfield, P. Saxe, and D. R. Yarkony, *J. Chem. Phys.* **81**, 4549 (1984). (b) P. Saxe, B. H. Lengsfield, and D. R. Yarkony, *Chem. Phys. Lett.* **113**, 159 (1985).

81. N. C. Blais, D. G. Trublar, and B. C. Garrett, *J. Chem. Phys.* **78**, 2956 (1983).

82. P. Saxe and D. R. Yarkony, *J. Chem. Phys.* **86**, 321 (1985).

83. D. R. Yarkony, *J. Chem. Phys.* **84**, 3206 (1986).

84. D. R. Yarkony, *J. Chem. Phys.* **90**, 1657 (1989).

85. K. Yamashita, K. Morokuma, Y. Shiraishi, and I. Kusunoki, *J. Chem. Phys.* **92**, 2505 (1990).

86. J. K. Stine and J. T. Muckerman, *J. Chem. Phys.* **65**, 3975 (1976).

87. P. J. Kuntz, J. Kendrick, and W. N. Whitton, *Chem. Phys.* **38**, 147 (1979).

88. N. Blais and D. G. Truhlar, *J. Chem. Phys.* **79**, 1334 (1983).

89. C. W. Eaker, *J. Chem. Phys.* **87**, 4532 (1983).

90. B. Andresen and S. E. Nielsen, *Mol. Phys.* **21**, 523 (1971).

91. H. Gabriel and K. Taulbjerg, *Phys. Rev. A* **10**, 741 (1974).

92. M. Baer, *Mol. Phys.* **40**, 1011 (1980).

93. (a) Yu. N. Demkov, *Sov. Phys. JETP* **18**, 138 (1964). (b) E. E. Nikitin, in *Advances in Quantum Chemistry*, Academic, New York, 1970, Vol 5, p. 135. (c) B. M. Smirnov, *Asymptotic Methods in the Theory of Atomic Collisions*, Atomizdat, Moscow, 1973 (in Russian).

94. C. Kubach and V. Sidis, *Phys. Rev. A* **14**, 152 (1976).

95. (a) V. I. Bylkin, L. A. Pakina, and B. M. Smirnov, *Sov. Phys. JETP*, **32**, 540 (1971). (b) A. V. Evseev, A. A. Radtsig, and B. M. Smirnov, *Sov. Phys. JETP* **50**, 283 (1980). (c) A. V. Evseev, A. A. Radtsig, and B. M. Smirnov, *J. Phys. B* **15**, 4437 (1982).

96. V. Sidis and D. P. De Bruijn, *Chem. Phys.* **85**, 201 (1984).

97. D. Grimbert, B. Lassier-Govers, and V. Sidis, *Chem. Phys.* **124**, 187 (1988).

98. A. W. Kleyn, J. Los, and E. A. Gislason, *Phys. Rep.* **90**, 1 (1982).

99. H. C. Longuet-Higgins, *Adv. Spectrosc.* **2**, 429 (1961).

100. I. Ozkan and L. Goodman, *Chem. Revs* **79**, 275 (1979).

101. V. Sidis and H. Lefebvre-Brion, *J. Phys. B* **4**, 1040 (1971).

102. V. Sidis, *J. Phys. B* **6**, 1188 (1973).

103. V. Sidis, D. Dhuicq, and M. Barat, *J. Phys. B* **8**, 474 (1975).

104. V. Sidis, in *The Physics of Electronic and Atomic Collisons*, Invited Lectures and Progress Reports of IX ICPEAC, J. S. Risley and R. Geballe, eds., University of Washington, Seattle, 1976, p. 295.

105. V. Lopez, A. Macias, R. D. Piacentini, A. Riera, and M. Yanez, *J. Phys. B* **11**, 2889 (1978).

106. D. Dowek, D. Dhuicq, J. Pommier, Vu Ngoc Tuan, V. Sidis, and M. Barat, *Phys. Rev. A* **24**, 2445 (1981).

107. D. Dowek, D. Dhuicq, V. Sidis, and M. Barat, *Phys. Rev. A* **26**, 746 (1982).

108. V. Sidis and D. Dowek, in *Electronic and Atomic Collisions*, J. Eichler, I. V. Hertel, and M. Stolterofht, eds., North-Holland, Amsterdam, 1984, p. 403.

109. C. Kubach, C. Courbin-Gaussorgues, and V. Sidis, *Chem. Phys. Lett.* **119**, 523 (1985).

110. J. C. Brenot, D. Dhuicq, J. P. Gauyacq, J. Pommier, V. Sidis, M. Barat, and E. Pollack, *Phys. Rev. A* **11**, 1245 (1975).

111. A. Russek and R. J. Furlan, *Phys. Rev. A* **39**, 5034 (1989).

112. I. Shavitt, in *Methods of Electronic Structure Theory*, H. F. Schaeffer III ed., Plenum, New York, 1977, p. 189.

113. C. Courbin-Gaussorgues, V. Sidis, and J. Vaaben, *J. Phys. B.* **16**, 2817 (1983).

114. F. Spiegelmann and J. P. Malrieu, *J. Phys. B* **17**, 1235 (1984).

115. F. Spiegelmann and J. P. Malrieu, *J. Phys. B* **17**, 1259 (1984).

116. R. Cimiraglia, J. P. Malrieu, M. Persico, and F. Spiegelmann, *J. Phys. B* **18**, 3073 (1985).

117. J. C. Tully, *J. Chem. Phys.* **59**, 5122 (1973).

118. R. W. Numrich and D. G. Truhlar, *J. Phys. Chem.* **82**, 168 (1978).

119. V. Sidis, C. Kubach, and D. Fussen, *Phys. Rev. A* **27**, 2431 (1983).

120. D. R. Bates and R. McCarroll, *Adv. Phys.* **11**, 39 (1962).

122. W. Fritsch and C. D. Lin, in *Electronic and Atomic Collisions*, J. Eichler, I. V. Hertel, N. Stolterfoht, eds., North-Holland, Amsterdam, 1984, p. 33.

123. M. Kimura and N. F. Lane, *Adv. At. Mol. Opt. Phys.* **26**, 79 (1990).

124. S. B. Schneiderman and A. Russek, *Phys. Rev. A* **181**, 311 (1969).

125. M. E. Riley and T. A. Green, *Phys. Rev. A* **4**, 619 (1971).

126. D. S. F. Crothers and J. G. Hughes, *Proc. R. Soc. London Ser. A* **359**, 349 (1978).

127. V. H. Ponce, *J. Phys. B* **12**, 3731 (1979).

128. (a) W. R. Thorson and J. B. Delos, *Phys. Rev. A* **18**, 117 (1978). (b) W. R. thorson and J. B. Delos, *Phys. Rev. A* **18**, 135 (1978).

129. J. S. Briggs and K. Taulbjerg, *J. Phys. B* **8**, 1905 (1975).

130. T. G. Heil, S. E. Butler, and A. Dalgarno, *Phys. Rev. A* **23**, 100 (1981).

131. (a) L. R. Kahn, P. Baybutt, and D. G. Truhlar, *J. Chem. Phys.* **65**, 3876 (1976). (b) P. Valiron, R. Gayet, R. McCarroll, F. Masnou-Seeuws, and M. Philippe, *J. Phys. B* **12**, 53 (1979).

132. P. O. Löwdin, *Adv. Quant. Chem.* **5**, 185 (1970).

133. V. Sidis, C. Kubach, and J. Pommier, *Phys. Rev. A* **23**, 119 (1981).

134. F. X. Gadea, F. Spiegelmann, M. Pelissier, and J. P. Malrieu, *J. Chem. Phys.* **84**, 4872 (1986).

135. H. Heitler and F. London, *Z. Phys.* **44**, 455 (1927).

136. (a) M. Noll and J. P. Toennies, *J. Chem. Phys.* **85**, 3313 (1986). (b) V. Sidis, D. Grimbert, M. Sizun, and M. Baer, *Chem. Phys. Lett.* **163**, 19 (1989). (c) M. Sizun, D. Grimbert, and V. Sidis, *J. Phys. Chem.* **94**, 5674 (1990).

137. D. Dhuicq and V. Sidis, *J. Phys. B* **19**, 199 (1986).

138. (a) B. Levy, in *Spectral Line Shape*, B. Wende, ed., W. de Gruyter, Berlin, 1981, p. 615. (b) B. Levy, in *Current Aspects of Quantum Chemistry*, R. Carbo, ed., Elsevier, Amsterdam, 1981, p. 127.

139. F. X. Gadea, *Thèse de Doctorat d'Etat*, Univ. Paul Sabatier de Toulouse (1987).

140. J. des Cloizeaux, *Nucl. Phys.* **20**, 321 (1960).

141. J. P. Gauyacq, in *Electronic and Atomic Collisions*, G. Watel, ed., North-Holland, Amsterdam, 1978, p. 431.

142. V. Sidis, N. Stolterfoht, and M. Barat, *J. Phys. B* **10**, 2815 (1977).

143. M. Aubert and C. Le Sech, *Phys. Rev. A* **13**, 632 (1976).

144. M. J. Werner and W. Meyer, *J. Chem. Phys.* **74**, 5802 (1981).

145. S. Yu Ovchinnikov and E. A. Solov'ev, *Comments At. Mol. Phys.* **22**, 69 (1988).

146. J. M. Rost, J. S. Briggs, and P. T. Greenland, *J. Phys. B* **22**, L353 (1989).

147. B. Stern, J. P. Gauyacq, and V. Sidis, *J. Phys. B* **11**, 653 (1978).

148. (a) J. O. Olsen, T. Andersen, M. Barat, C. Courbin-Gaussorgues, V. Sidis, J. Pommier, J. Agusti, N. Andersen, and A. Russek, *Phys. Rev. A* **19**, 1457 (1979). (b) V. Sidis and J. P. Gauyacq, in *Electronic and Atomic Collisions*, Abstracts of Contributed Papers of the XI ICPEAC, K. Takayanagi and N. Oda, eds., The Society for Atomic Collision Research, Japan, 1979, p. 507. (c) U. Thielmann, J. Krutein, and M. Barat, *J. Phys. B* **13**, 4217 (1980).

149. (a) R. G. Gordon, *J. Chem. Phys.* **51**, 14 (1969). (b) R. G. Gordon, *Methods. Comput. Phys.* **10**, 81 (1969).

150. L. Eno and G. G. Balint-Kurti, *J. Chem. Phys.* **75**, 690 (1981).

151. M. Baer, G. dDolshagen, and J. P. Toennies, *J. Chem. Phys.* **73**, 1690 (1980).

152. E. E. Nikitin, M. Ya Ovchinnikova, and D. V. Shalashilin, *Chem. Phys.* **111**, 313 (1987).

153. C. Kubach, G. Vien, and M. Richard-Viard, *J. Chem. Phys.* **94**, 1929 (1991).

MODEL POTENTIAL ENERGY SURFACES FOR INELASTIC AND CHARGE-TRANSFER PROCESSES IN ION–MOLECULE COLLISIONS

F. A. GIANTURCO

Department of Chemistry, The University of Rome, Città Universitaria, Rome, Italy

F. SCHNEIDER

Central Institute for Physical Chemistry, Rudower Chaussee 5, Berlin, Germany

CONTENTS

State-Selected and State-to-State Ion–Molecule Reaction Dynamics, Part 2: Theory, Edited by Michael Baer and Cheuk-Yiu Ng. Advances in Chemical Physics Series, Vol. LXXXII. ISBN 0-471-53263-0   © 1992 John Wiley & Sons, Inc.

I. INTRODUCTION

Charge-transfer processes between subsystems A and B^+ may be written as

$$A + B^+ \rightarrow A^+ + B + IP(B) - IP(A), \tag{1.1}$$

where $IP(X)$ means the ionization potentials of the subsystems X. If A and B represent atoms, a model potential description leads to a potential curve (crossing) problem. For small differences $IP(B) - IP(A)$, often asymptotic interaction potential forms are acceptable descriptions:

$$V = -C_4 r^{-4} - C_6 r^{-6} \cdots, \tag{1.2}$$

where the coefficients C_i may be obtained from transport properties and crystal data. The first term in Eq. (1.2) describes the ion-induced dipole interaction, the following terms the van der Waals energy. Such a description is only valid for large distances of the constituent species, that is, atoms and ions. At smaller distances chemical forces come into play, and they have to be calculated by quantum-chemistry methods. Generally, quantum-chemical high-quality calculations of diatomic potential energy curves (PEC) are nowadays possible for all light constituent atoms or cations. A one-dimensional analytic representation of the energies as function of the inter-atomic distance r would then give the required model potential curve.

A similar but practically more difficult situation arises if the charged particle is an anion:

$$A + B^- \rightarrow A^- + B + EA(A) - EA(B). \tag{1.3}$$

Here, even in case of atoms $X = A, B$, not all electron affinities $EA(X)$ are accurately known. Furthermore, the long-range interactions are not as simply formulated as in the case of cations. The electronic states of some negative ions are not certain, since free-electron states, that is, continuum states, may play a role. And last but not least the quantum chemistry of negative ion molecules is still much more problematic than that of positively charged systems.

Charge-transfer processes as seen from the quantum chemist's view are nonadiabatic processes, since as reactant and product systems are differently charged, it is obvious that they are in different electronic states. The transition between different electronic states involves a passing of the system through a zone of strong electronic interaction (**nonadiabatic** interaction), where the processes do not follow the rules of the usual **adiabatic** dynamics, but are guided by the influence of at least two potentials (belonging to the different electronic states) and of the action of a nonadiabatic interaction operator.

In the case of very strong interaction in restricted regions of the nuclear coordinate space, a pure **diabatic** model may simplify the treatment. The same is true for very weak interaction of both states, where a true **adiabatic** description is applicable. Both diabatic and adiabatic descriptions of the dynamics may be treated as one-potential-surface problems, where the equations of motion may be formulated and solved in a standard way. Nonetheless, the general case for charge-transfer systems could be described by the **nonadiabetic** behavior of the system in limited regions of space and by **adiabatic** dynamics on different surfaces for most of the configuration space.

The task of constructing model potentials for charge-transfer systems thus consists in

1. modeling at least two potentials of the system in different electronic states (with differently charged subsystems), and
2. modeling the nonadiabatic couplings between them.

Taking into account that for A resp. B being molecules instead of atoms the number of degrees of freedom complicates the description, it must be confessed that in general systems quantum-chemical calculations of reliable accuracy are possible only for a limited number of spatial configurations of the nuclei.

Furthermore, even with a fairly complete grid of potential energy points that depend on nuclear coordinates, one needs a smooth (analytical) functional description of the energy values at all intermediate configurations between the grid points. That analytical description of potential energy surfaces (PES) is one of the major stumbling blocks of chemical-interaction theory.

Thus, in most practically interesting cases we depend on **semiempirical concepts** to describe the global topography of the PES. Moreover, when dealing with the calculation of nonadiabatic coupling matrices, it is by no means standard now to get them from accurate quantum-chemical methods and, therefore, for practical calculations we are forced to rely on **simplified model concepts**.

One useful method for calculating model potentials is the method of diatomics in molecules (DIM). It is structurally identical to *ab initio* VB calculations, in that it uses information from different asymptotic electronic states of atomic and diatomic fragments to construct a multiconfiguration basis set and applies the variation principle to obtain the wavefunctions, energies, and couplings of several electronic states.

The main differences with *ab initio* methods are, however, the following:

1. The approximate nature of hamiltonian matrix elements, which are mostly constructed from fragment data rather than by integral calculations.
2. The strong restrictions put on the viable number of configurations by the increasing complexity of the approach and by enormous increase of input data requirements with that number (kept usually lower than 30).

In practical terms the matrix diagonalization becomes the time-determining step once the molecular integral calculations can be avoided. With small secular matrix dimensions the whole calculation may be performed very rapidly during a dynamic calculation and subsequent fitting steps are not necessary.

As in every semiempirical method, experience is necessary in applying DIM and caution is needed when interpreting the results. But so far, there is simply no other method suited for effective calculations of the PES's for several electronic states and for nonadiabatic interactions that have correct spin and space symmetries, are asymptotically (for large nuclear distances) accurate, and reasonably resemble the physical interactions at short distances. DIM is not the only scheme for calculating model potentials, and it is not simply applicable to any system. Nonetheless, the DIM framework has some attractive features, for example, its physical well-founded background and the easy interpretability of its results. Furthermore, it allows one to include the preliminary knowledge on long-range forces between atoms and ions simply by designing the long-range part of the diatomic fragment interactions accordingly.

For the long-range part of the whole interaction the DIM potentials are automatically correct (that is, their inherent approximations become negligible), a fact that leads to reliable asymptotic interactions.

In this chapter the method of DIM and its application to several three-atomic ion–molecule systems are reviewed with respect to the description of charge-transfer processes. Future developments and applications are discussed in the final section.

II. THE DIM METHOD

The DIM method was originally proposed by Ellison,[1] who brought the atoms-in-molecule partition of the total hamiltonian of Moffit[2] to its logical consequence for polyatomic molecules. A number of applications to ionic systems were subsequently published by Ellison et al.,[3-8] which restricted themselves to linear or fixed-angle molecular geometries. Later, Kuntz et al.[9,10] and Tully[11,12] completed the theoretical framework for the description of arbitrary geometries, nonadiabatic and spin–orbit interactions, and used a non-hermitian matrix approach that involves less approximations than the usual hermitian one (Tully and Truesdale[13]). Faist and Muckerman[14,15], and still later Vojtik,[16,17] presented a comprehensive description of the whole formalism of the DIM method. Reviews were published, for instance by Kuntz[18] and by Tully.[19] Most common applications concerned the potential energy surfaces of three-atomic molecular systems.

A. General Formalism

The total nonrelativistic hamiltonian of a system made up of m electrons and N nuclei consists of an electronic hamiltonian $\hat{H}_{el}$, a nuclear kinetic energy operator $\hat{T}_k$, and the nuclear repulsion energy $\hat{V}_{nn}$. These operators all act on functions of the internuclear distances (collectively denoted by r), while the electronic hamiltonian also acts on functions of the electronic coordinates, here given as ξ.

Starting from the usual time-dependent Schrödinger equation in nonrelativistic form

$$(\hat{H}_{el}(\xi, r) + \hat{T}_k(r) + \hat{V}_{kk}(r))\,\Psi_{ke}(\xi, r, t) = i\frac{\partial}{\partial t}\Psi_{ke}(\xi, r, t) \tag{2.1}$$

and introducing the Born–Oppenheimer separation, that is, expressing Ψ_{ke} by the expansion

$$\Psi_{ke}(\xi, r, t) = \sum_n \Psi_n(\xi, r)\varphi(r, t), \tag{2.2}$$

where $\Psi_n(\xi, r)$ are the eigenfunctions of the stationary Schrödinger equation

$$(\hat{H}_{el}(\xi, r) + \hat{V}_{kk}(r))\,\Psi_n(\xi, r) = E_n(r)\Psi_n(\xi, r), \tag{2.3}$$

one arrives at the usual set of coupled differential equations describing the nonadiabatic dynamics of the system:

$$(\hat{T}_k(r) + E_n(r))\varphi_n(r, t) + \Sigma\,\hat{C}_{nm}(r)\varphi_m(r, t) = i\frac{\partial}{\partial t}\varphi_n(r, t). \tag{2.4}$$

Here $E_n(r)$ are the potential energy surfaces of the system, and $\hat{C}_{nm}(r)$ is the operator of the nonadiabatic coupling,

$$\hat{C}_{nm}(r) = \langle\Psi_n|\hat{T}_k(r)|\Psi_m\rangle - \sum_j M^{-1}\langle\Psi_n|\nabla_j(r)|\Psi_m\rangle\nabla_j(r). \tag{2.5}$$

Equation (2.3) is interpreted as an eigenvalue problem for the electronic functions $\Psi_n(\xi) = \Psi_n(\xi, \mathbf{r})|_{\mathbf{r}=\mathbf{r}_i}$, seen as a "clamped" nuclei problem depending only parametrically on the nuclear geometry $\mathbf{r}_i$. Similarly, the coupling operator, Eq. (2.5), is determined by calculating the brackets in Eq. (2.5) for fixed parameters $\mathbf{r}$. The DIM method differs from conventional *ab initio* methods by the way in which the functions $\Psi_n(\xi)$ are expanded as combinations of basis functions. It chooses a special form of representation that is similar to valence bond functions and that allows one to determine in an approximate way from diatomic data the matrix elements of the hamiltonian as well as the coupling matrix elements.

B. The Symmetry-Adapted Basis Functions

At variance with the *ab initio* MO methods, the DIM basis functions are built as m-electron functions. In particular, it means that these basis functions are not *orbitals* in the quantum-chemical sense. The second difference from *ab initio* basis functions is that the DIM basis is not used to calculate integrals, thus avoiding the difficulty to define explicitly functional forms of basis functions. Actually, the DIM basis functions are defined formally by specifying their space and spin symmetries and by demanding largely that symmetry-defined combinations of these functions be eigenfunctions of certain fragment hamiltonians.

The primitive functions *ansatz* of the DIM method was named "composite functions" by Moffit. It is a product of atomic eigenfunctions of atomic hamiltonians, atomic spin and space symmetry operators, and atomic anti-symmetrization operators, for example, in the case of three atoms A, B, and C:

$$\phi_k(1 \cdots \mathcal{N}) = \phi_\alpha(1 \cdots n_A)\phi_\beta(n_A + 1 \cdots n_A + n_B)\phi_\gamma(\mathcal{N} - n_C + 1 \cdots \mathcal{N}), \quad (2.6)$$

where $\mathcal{N} = n_A + n_B + n_C$ is the total number of electrons in the system. The distributions of the atoms over the atomic species (that accordingly may be ions as well) may differ between different basis structures k, that is, the atomic electron numbers n_P may depend on k; α, β, and γ are collective labels of atomic quantum numbers of the species of A, B, and C, respectively.

The primitive basis functions ϕ_k are the building blocks of basis functions Φ_k, which are antisymmetrized and symmetry consistent with respect to all electrons and all atoms of the system:

$$\Phi = A_{\mathcal{N}}\phi\Lambda = (\Phi_1 \cdots \Phi_m). \quad (2.7)$$

Here ϕ is the row matrix of the ϕ_k and Λ is a symmetry transformation matrix.

Now the eigenfunctions of the Schrödinger equation are expanded as linear combinations of the ϕ_k:

$$\Psi = (\Psi_1 \cdots \Psi_m) = \Phi\Gamma, \qquad (2.8)$$

where Γ is the matrix of the eigenvectors of the total Hamiltonian $\hat{H}$.

As is shown in the next section, it is necessary in the method of diatomics in molecules to define in addition to the N-atom basis functions [Eq. (2.7)], the basis functions of all possible diatomic fragment systems, and to establish well-defined relations between the diatomic basis function sets and the total basis on one hand and between the diatomic basis functions and the physical diatomic eigenstates on the other hand.

The diatomic basis functions of the fragment PQ are defined by following the philosophy of DIM as antisymmetrized composite functions from all atomic eigenstate functions of both the atoms P and Q, which participate in the building of the total basis functions. Thus, every primitive basis function from Eq. (2.6) corresponds to one primitive basis function of each fragment.

The relation between the diatomic basis functions and eigenfunctions of diatomic states is established by spin and space symmetry transformations similar to Eq. (2.7).

For the diatomic fragment AC in Eq. (2.6) that means, for instance,

$$\phi_K^{AC}(1 \cdots n_A + n_C) = p_K^{AC} \phi_\alpha^A(1 \cdots n_A) \phi_\gamma^C(n_A + 1 \cdots n_A + n_C), \qquad (2.9)$$

where p_K^{AC} is a phase factor ($+1$ or -1) accounting for the permutation symmetry of the number of electron permutations necessary for bringing the electrons of atom C in the desired order in the molecule AC, which here is different from their order in ABC.

$$\Phi^{PQ} = \hat{A}_{n_P + n_Q} \phi^{PQ} = (\Phi_1^{PQ} \cdots \Phi_m^{PQ}) \qquad (2.10)$$

are then the antisymmetrized product functions suited for describing the fragment PQ written as a row matrix.

They are subject to several matrix transformations, the first of which could be a spin transformation Λ_{PQ} accounting for the diatomic total spin, that is, expressing diatomic spin eigenfunctions by fixed linear combinations of products of atomic spin eigenfunctions. Second, it must be required that each diatomics is described by functions that have the correct angular momentum.

For that requirement the spin eigenfunctions are transformed by a rotation that brings the diatomic axis into the z-axis direction. The corresponding transformation matrix is dependent on the geometry of the total molecular system. It may be written as a product of two independent transformations of the atomic functions on P and Q, the matrices of which are R_P and R_Q, respectively. The diatomic spin and angular momentum eigenfunctions

$$\bar{\Psi}_{PQ} = (\bar{\Psi}_1^{PQ} \cdots \bar{\Psi}_m^{PQ}) = \Phi^{PQ} \Lambda_{PQ} R_P R_Q \tag{2.11}$$

often may already determine the diatomic eigenstates uniquely.

This is achieved by the particular way of defining DIM basis functions, as flexible as it is consistent with the most fundamental symmetry considerations, that is, nothing other than their symmetries was specified for the atomic functions from which the diatomic functions originate.

If several of the functions $\bar{\Psi}_1^{PQ}$ describe the same diatomic spin and angular momentum, and if they are constructed from different atomic-state functions, then several diatomic states of the same spin and angular momentum have to be taken into account, that is, a further transformation (Γ_{PQ}) combining the functions $\bar{\Psi}_{PQ}$ of the same momenta to different diatomic states must be applied:

$$\Psi_{PQ} = (\Psi_1^{PQ} \cdots \bar{\Psi}_m^{PQ}) = \bar{\Psi}^{PQ} \Gamma_{PQ}. \tag{2.12}$$

For homonuclear fragments PQ the elements of the matrix Γ_{PQ} may be constants (*gerade* and *ungerade* states) and, in general, they will depend on the interatomic distance of PQ. Γ_{PQ} plays the role of a matrix of eigenvectors of the diatomic hamiltonian matrix.

C. The DIM Hamiltonian

The equation for the m lowest approximate potential energy surfaces of the system reads, in matrix form, as follows:

$$\hat{H}\Psi = \Psi E, \tag{2.13}$$

with the diagonal matrix E of the m lowest PES. By multiplying Eq. (2.13) with the column matrix Ψ and integrating over the electron coordinates, it is found that

$$S^{-1}H\Gamma = \Gamma E, \tag{2.14}$$

where $H = \langle \Phi^T | \hat{H}\Phi \rangle$ and $S = \langle \Phi^T | \Phi \rangle$. Equation (2.14) is an eigenvalue problem of a nonsymmetric DIM matrix $S^{-1}H$ (see, for example, Tully and Truesdale[13]), but in most cases it can be made symmetrical without significant loss of accuracy.

The decisive point of the DIM method is the partition of the total hamiltonian into a number of atomic and diatomic hamiltonians, the matrix elements of which are easy to calculate from atomic and diatomic data. That partition was introduced by Ellison[1] and is accurate in the limit of a saturated (i.e., complete) basis set:

$$\hat{H} = \sum_{P<Q} \hat{H}_{PQ} - (N-2)\sum_{P} \hat{H}_P. \qquad (2.15)$$

The preceding partition, together with the special construction of the basis functions in Eq. (2.7), leads to matrix representations of the form

$$H = \langle \Phi^T | \hat{A}_{\mathcal{N}} \hat{H}\phi \rangle \Lambda \qquad (2.16)$$

that is also valid for the diatomic and atomic fragment hamiltonian matrices H_{PQ} and H_P, respectively. It should be mentioned that there is more than one correct partition (2.15), depending on the assignment of the $\mathcal{N}$ electrons to the N atomic species. In particular, it is possible to choose a hamiltonian partition that corresponds to the electron assignment suited to the basis function on which the hamiltonian acts in (2.16). The crucial problem consists then in expressing the functions ϕ through functions ϕ_{PQ}, which are appropriate for describing the diatomic fragment wavefunctions. By using the definition [for example, see Eq. (2.6)] of ϕ and defining different DIM partitions of the total hamiltonian for different electronic configurations (different ionicities), it may be shown that

$$S^{-1}H_{PQ} = \Lambda^T \Lambda_{PQ} R_P R_Q \Gamma_{PQ} E_{PQ} \Gamma^{-1} R^T R^T \Lambda_{PQ} \Lambda. \qquad (2.17)$$

Here Λ is the previously mentioned transformation matrix, which transforms the primitive basis functions ϕ into the spin and spatial symmetry of the whole system. Similarly, as mentioned in the preceding section, the matrices Λ_{PQ} are transformations of ϕ to spin eigenfunctions of the fragment PQ, the rotation matrices R_P transform the angular momentum functions of the atomic species P to an adequate representation of the diatomic PQ (that is, with an integer angular momentum projection on the diatomic axis). E_{PQ} and Γ_{PQ} are defined by the eigenvalue equation (2.18) of the diatomic fragment PQ:

$$\hat{H}_{PQ}\bar{\Psi}_{PQ}\Gamma_{PQ} = \bar{\Psi}_{PQ}\Gamma_{PQ}E_{PQ}. \qquad (2.18)$$

Most of the elements of the eigenvector matrices Γ_{PQ} are determined by symmetry, the others either have to be chosen semiempirically, possibly guided by diatomic VB expansion coefficients, or must be calculated *ab initio*, using orbital approximations for the formally defined DIM basis functions.

The diagonal eigenvalue matrix E_{PQ} is made up of the respective potential curves of different electronic states of the diatomic fragment species.

Approximate fragment calculations may also be necessary for unknown diatomic terms in E_{PQ}, which are commonly involved for highly excited diatomic fragment states.

For the atomic species fragment matrices it is found (owing to the way the basis was constructed) that

$$H_P = S\Lambda^T E_P \Lambda, \tag{2.19}$$

where E_P is the diagonal matrix of the electronic energy terms of the atomic species.

The construction of the transformation matrices is done by standard angular momentum coupling techniques, that is, the spin matrices consist of Clebsch–Gordan coefficients and the rotation matrices of elements of the Wigner rotation matrices.

D. The Nonadiabatic Coupling

The starting point for an approximate treatment of the nonadiabatic coupling operator is the adiabatic approximation, resulting in a perturbation of the adiabatic dynamics in the form of Eq. (2.5). Usually the first term, which depends on the kinetic energy, is neglected as small in comparison with the second, which is the only contribution to the nonadiabatic coupling needed in the usual semiclassical dynamic methods. This velocity-dependent term in Eq. (2.5) may be approximated by the DIM method as a matrix D with elements

$$d_{ij} = \tfrac{1}{2}(f_{ij} + f_{ji})/(E_i - E_j), \tag{2.20}$$

while the gradient of the energy needed for dynamical calculations is

$$\nabla E_i = -f_{ii}. \tag{2.21}$$

In the last two equations use was made of an auxiliary matrix F (with elements f_{ij}), which reads.

$$F = -\Gamma^{-1}(\nabla S^{-1}H)\Gamma, \tag{2.22}$$

that is, the gradient of the DIM matrix $S^{-1}H$, transformed by the eigenvector matrix of the DIM secular equation. As in Eq. (2.12), a symmetrized approximation for Eq. (2.22) turns out to be effective.

The treatment described here is approximate not only because of the neglect of the kinetic coupling terms but also because, in the course of obtaining Eq. (2.22), a part of the velocity-dependent contribution (the part that is related to the basis functions' dependence on the geometry) was neglected in accordance with the assumptions on the validity of the DIM method.[12]

Furthermore, the Coriolis (rotational) coupling was not taken into account. It could be obtained by describing the system not in the internal (body-fixed) coordinates, but in a space-fixed system, where the additional Coriolis coupling term arises from the overall rotation of the molecular frame during the process.

Accordingly, the preceding equations are adequate in systems, where (1) the assumption of the DIM method are fulfilled by the DIM model and (2) the "radial" part of the nonadiabatic coupling represents the largest contribution to the influence of other states on the processes under consideration.

Generally, it would be possible to formulate other contributions of the nonadiabaticity in the preceding formalism, too. As a drawback of such a project it must be mentioned, that some of the attractive features of DIM would be lost: one would need orbital approximations of the basis functions to calculate some types of integrals, which may not be derived from available diatomic data, while it would also be necessary to have the second derivatives of the DIM matrix.

E.   The Implementations of the Method

In an actual calculation there are different ways to implement the DIM method briefly described. One method would be to orient it to an *ab initio* way of calculation: define in terms of orbital approximations a many-electron basis function set (large enough to describe at least all interesting fragment states) and perform all necessary fragment calculations, including the integrals that would be restricted to at most two center types. Then the fragment DIM matrices could be determined *a priori*.

Another possibility of applying the DIM method (now by far the more common) is to restrict the formal basis set to the most relevant structures and to use empirical and semiempirical fragment information wherever it is available.

The disadvantage of the *ab initio* approach is clearly connected with the implicit effort to expand many-center integrals by two-center terms.[20] It is well known, in fact, that the convergence of such expansions is very poor, although some rare comparisons of the DIM *ab initio* approach and, for example, a VB *ab initio* approach with comparable basis sets, seem to point to the superiority of the DIM method.[21,22]

In the following sections DIM model calculations are described and use is made as much as possible of independently found, best available fragment potential data. This way of performing DIM calculations represents quite an effective method for obtaining model potentials. Its disadvantage stems from the need to include either many-center correction terms or empirical corrections to diatomic fragment terms and diatomic wavefunction expansion coefficients in order to account for basis set shortcomings. With small basis sets even the global features of the potential topology may be wrong, thus requiring predictive *ab initio* examination. On the other hand, much experience has been assembled with the semiempirical kind of DIM calculations for many three-atomic ion–molecule systems, generally showing the applicability of that approximate approach.

How important ionic structures are for *ab initio* VB wavefunction calculations is well known from H_2 Heitler–London calculations. It seems to be a reason for the success of the DIM method in several small basis set applications on charge-transfer systems that in the description of ion–molecule systems different ionic basis structures are naturally included in the DIM basis set, because the different asymptotic channels (reactant and product channels) have to be described by different ionic structures. Thus, a kind of *configuration-interaction* correction comes into the method, out of necessity, once it is applied to charge-transfer processes.

III. TRIATOMIC ION–MOLECULE SYSTEMS

The ionic systems that have been treated with the DIM method were mostly three- or four-atomic systems of the type $A + B_2^+$ or $A^+ + B_2$, where at least one of the species A and B was hydrogen or lithium. A simple DIM model for rare-gas charge-transfer processes in systems AB_2^+ has also been reported.[23]

We shall discuss in the following sections several three-atomic ionic systems for which DIM calculations have been performed mostly in our groups. To begin with, we discuss the less complicated systems for which part of the mathematical model of the DIM calculation is described, while in the cases with larger basis sets the details of the mathematical background are to be found in the respective original papers, as we prefer to concentrate here on the discussion of the final results.

A. The H_3^+ Case and Related Models

Some of the first applications of the DIM method were already dedicated to the hydrogen triatomic cation. Ellison, Huff, and Patel[3] calculated the ground H_3^+ PES using the simplest set of basis structures (only *s* functions). They found an equilibrium geometry at equilateral triangular configuration. Binding energy and vibrational frequencies were also determined. Their findings were confirmed by later papers, for examples, by Preston and Tully,[24] where the authors concentrated on the quantities of most interest to the dynamics of that system. They determined the ground- and lowest-excited-state surfaces as well as the radial nonadiabatic coupling in the DIM approximation. Further PES calculations with simple DIM models for excited states of H_3^+ were reported by Wu and Ellison.[7,8]

With the same basis set as used for the H_3^+ ground state the Li_3^+ system was calculated by Pfeiffer and Ellison.[5] DIM models of Li_3^+ and Li_4^+ were also calculated by Pickup[25] and optimized by Polak et al.[26] Vibrational frequencies of the complex and nonadiabatic couplings of Li_3^+ were obtained with the DIM method by Vojtik et al.[27]

For the ground-state and first-excited-state nonadiabatic dynamic processes in the H_3^+ or H_4^+ system, a number of comparisons between theory and experiment have been performed using the DIM model.[24,28-32]

In any application of the DIM method the starting point is the definition of the electronic states of the atomic species that have to be taken into account. For H_3^+ (indicating the three atoms by A, B, and C), these states (or DIM basis structures) are

$$1. \ H_A(1s) - H_B(1s) - H_C^+,$$

$$2. \ H_A(1s) - H_B^+ - H_C(1s),$$

$$3. \ H_A^+ - H_B(1s) - H_C(1s). \qquad (3.1)$$

Then the polyatomic basis functions (PBF) are defined by the (differing) quantum numbers of the atomic eigenfunctions, from which they are composed. All possible spin and angular momentum projections, which may lead to the desired overall symmetry, have to be included in the set of PBF. Here the quantum numbers sufficient to distinguish the PBF are the spin projections, which are given in the order of A–B–C for an overall projection of zero:

$$1. \ \alpha - \beta - 0,$$

$$2. \ \beta - \alpha - 0,$$

$$3. \ \alpha - 0 - \beta,$$

$$4. \ \beta - 0 - \alpha,$$

$$5. \ 0 - \alpha - \beta,$$

$$6. \ 0 - \beta - \alpha. \qquad (3.2)$$

It is easy to imagine how that table and the matrices belonging to it would extend when (in other systems) for every atomic function several angular projections have to be included. The number of PBF is the dimension of the matrices used in the following development and therefore we chose to examine here the construction of a DIM model only with the very case of the present system.

It is rather easy to determine in that simple case what the diatomic spin transformation matrices Λ_{PQ} [see Eq. (2.17)] look like (0 and 1 are submatrices of appropriate dimensions):

$$\Lambda_{AB} = \begin{bmatrix} -a & a & \vdots & \\ a & a & \vdots & 0 \\ \hdashline & 0 & \vdots & 1 \end{bmatrix}, \quad \Lambda_{BC} = \begin{bmatrix} 1 & \vdots & 0 \\ \hdashline & \vdots & -a & a \\ 0 & \vdots & a & a \end{bmatrix},$$

$$\Lambda_{AC} = \begin{bmatrix} -1 & \vdots & 0 & \vdots & 0 \\ \hdashline 0 & \vdots & -a & a & \vdots & 0 \\ & \vdots & a & a & \vdots & \\ \hdashline 0 & \vdots & 0 & \vdots & 1 \end{bmatrix}, \quad \text{with } a = 2^{-1/2}$$

(3.3)

where the 2×2 diagonal submatrix -1 in Λ_{AC} is negative because of the permutation phase factor mentioned for Eq. (2.9).

After the definitions of Eq. (3.3) the diatomic states for the six diatomic functions of AB are known to be

$$H_2(^1\Sigma_g^+), \quad H_2(^3\Sigma_u^+), \quad H_2(^2\Sigma^+), \quad H_2(^2\Sigma^+), \quad H_2(^2\Sigma^+), \quad H_2(^2\Sigma^+).$$

The H_2 functions must form *gerade* and *ungerade* combinations to define uniquely the final four energy terms in AB. An analogous situation is found in the other fragments BC and AC. The diatomic eigenvector matrices Γ_{PQ} are here uniquely defined by symmetry and thus are given by constant factors:

$$\Gamma_{AB} = \begin{bmatrix} 1 & \vdots & & & & 0 \\ \hdashline & \vdots & -a & 0 & 0 & a \\ 0 & \vdots & 0 & -a & a & 0 \\ & \vdots & 0 & a & a & 0 \\ & \vdots & a & 0 & 0 & a \end{bmatrix}, \quad \Gamma_{BC} = \begin{bmatrix} -a & 0 & 0 & a & \vdots & \\ 0 & -a & a & 0 & \vdots & \\ 0 & a & a & 0 & \vdots & 0 \\ a & 0 & 0 & a & \vdots & \\ \hdashline & 0 & & & \vdots & 1 \end{bmatrix},$$

$$\Gamma_{AC} = \begin{bmatrix} -a & 0 & \vdots & & \vdots & 0 & a \\ 0 & -a & \vdots & 0 & \vdots & a & 0 \\ \cdots & \cdots & \cdots & \cdots & \cdots & \cdots & \cdots \\ 0 & & \vdots & 1 & \vdots & 0 & \\ \cdots & \cdots & \cdots & \cdots & \cdots & \cdots & \cdots \\ 0 & a & \vdots & 0 & \vdots & a & 0 \\ a & 0 & \vdots & & \vdots & 0 & a \end{bmatrix}. \tag{3.4}$$

Now the parity of the diatomic states is obtained as

$$\begin{array}{cccccc}
g, & u, & u, & u, & g, & g & (AB), \\
u, & u, & g, & g, & g, & u & (BC), \\
u, & u, & g, & u, & g, & g & (CA),
\end{array} \tag{3.5}$$

that can be seen in connection with the previous definitions, where the ionicities of the H_2 species are obvious from the PBF and the spin quantum numbers follow from Eqs. (3.1).

Calculating the Clebsch–Gordan coefficients for the coupling of the spins of AB with C and multiplying the matrix Λ_{AB} with the matrix of these coefficients, we find for the total spin transformation matrix the following expression:

$$\Lambda = \begin{bmatrix} -a & 0 & 0 & 0 & a & 0 \\ a & 0 & 0 & 0 & a & 0 \\ 0 & 0 & a & a & 0 & 0 \\ 0 & 0 & -a & a & 0 & 0 \\ 0 & -a & 0 & 0 & 0 & a \\ 0 & a & 0 & 0 & 0 & a \end{bmatrix}, \tag{3.6}$$

which leads to a total spin of $(0, 0, 0, 1, 1, 1)$ for the six states that one may obtain from the simple H_3 model.

Applying Eq. (2.17) (where the rotation matrices R_P are unity because all atomic states are S states) and adding the three fragment contributions $S^{-1}H_{PQ}$ and the atomic terms $-S^{-1}H_P$, the matrix $S^{-1}H$ is constructed. The DIM matrix is symmetrical and block-diagonal in two 3×3 blocks respectively describing singlet and triplet states of the three-atom system.

Both blocks have the same analytical form:

$$S^{-1}H = \begin{bmatrix} V_{AB}(p) + Q_{BC} + Q_{CA} & pJ_A & J_{BC} \\ pJ_{CA} & Q_{AB} + V_{BC}(p) + Q_{CA} & J_{AB} \\ J_{BC} & J_{AB} & Q_{AB} + Q_{BC} + V_{CA}(p) \end{bmatrix},$$

(3.7)

where $p = 1$ for the singlet case and $p = -1$ for triplets. The symbols $V_{PQ}(p)$ mean H_2 diatomic terms (potential energy curves), while the H_2 terms enter the expressions J_{PQ} and Q_{PQ} as follows:

$$J_{PQ} = [V^+_{PQ}(g) - V^+_{PQ}(u)]/2, \quad Q_{PQ} = [V^+_{PQ}(g) + V^+_{PQ}(u)]/2.$$

(3.8)

The potential energy surfaces for the singlet states of H_3 in the DIM model are given by the eigenvalues of the matrix equation (3.4), with $p = 1$, that allows for the direct use of the DIM model potential in dynamic calculations.

In Fig. 1 we show a schematic view of the potential energy surfaces for an angle (BAC) of $\pi/3$. We recognize there the presence of the asymptotic region of the avoided-crossing seam as originating from the crossing of the H_2 ground-state PEC with the lowest H_2 PEC. For the interaction region of all three particles, the splitting of the two lowest potentials is the largest and thus that seam is not visible in the section through the potential surface diagonal, which is shown in Fig. 2 for D_{3h} and C_{2v} configurations. Here the D_{3h} degeneration leads to a true crossing of the two upper PES, which is resolved in C_{2v} symmetry.

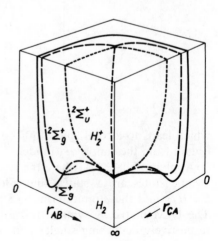

Figure 1. Schematic view on the topology of the H_3 singlet PES for an angle BAC of $60°$ according to a simple DIM model.

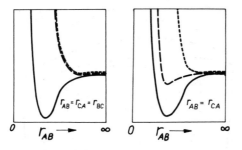

Figure 2. Cut through the diagonal of Fig. 1 (D_{3h}) (left part) and in C_{2v} symmetry (right part).

In the H_3^+ system the avoided crossing (roughly parallel to the entrance valley indicated at large r by a dashed line) is responsible for significant nonadiabatic coupling[24] and for charge-transfer processes with vibrationally excited reactants.[28,29]

While the form of Eq. (3.4) is a correct DIM model only for a homonuclear three-atom system of three S-state atoms and may also serve as a parametrized model for other ionic three-atom homonuclear molecules, it has to be modified for a three-atom heteronuclear system. Restricting ourselves to S-state atoms only, $Li^+ + H_2$ may be chosen to describe a simple DIM model for heteronuclear ion–molecule systems.

In that heteronuclear case the mixing matrices of the diatomic fragments are not orthogonal (as in the case of H_3^+) and the whole three-atom matrix becomes nonsymmetric. Either by symmetrizing or by diagonalizing the DIM matrix with a diagonalization procedure for nonsymmetric matrices one obtains three singlets and three triplets PES. For LiH_2 the topology of the singlet PES is illustrated in Fig. 3. For that system the differences of the ionization potentials lead to a clearly separated, lower PES for which nonadiabatic effects are unlikely to play a role. The corresponding DIM potential has ben studied by Wu and Ellison[6] and later on by us.[33]

The DIM matrix of this type of systems may be transformed to a nonsymmetric form that reflects the maximum symmetry $C_{2v}(D_{\infty h})$ of the molecular frame:

$$S^{-1}H = \begin{bmatrix} K+V+\bar{V}-Z & W+\bar{W} & W-\bar{W} \\ W'+\bar{W}' & G+X+\bar{X} & D-\bar{D} \\ W'-\bar{W}' & D-\bar{D} & U+X+\bar{X} \end{bmatrix}. \quad (3.9)$$

The symbols W, D, X, and V mean diatomic LiH- or LiH^+-term combinations that depend on the distance r_{BC} (and on r_{AC} in the symbols

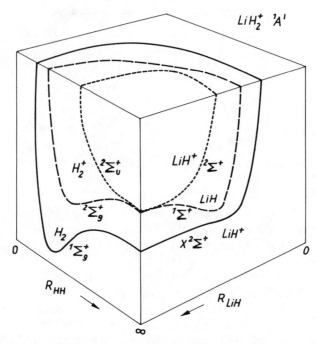

Figure 3. Schematic view on the topology of the LiH$_2$ singlet PES for C_s symmetry according to a simple DIM model.

with bar), while G, U, and K are H$_2^+$ or H$_2$ terms at r_{AB}. Z is the ionization potential of Li. The energy of the neutral atoms is taken to be zero. In particular, the abbreviations mean:

$$X = (E + V')/2, \quad D = (E - V')/2, \tag{3.10}$$

with $V' = (s_1 s_2 V_1 + c_1 c_2 V_2)/d$ and $d = c_1 c_2 + s_1 s_2$, taking

$$c_i = (1 - s_i^2)^{1/2},$$
$$V = (c_1 c_2 V_1 + s_1 s_2 V_2)/d,$$
$$w = c_1 s_2 (V_1 - V_2)/(2^{1/2} d),$$
$$W' = s_1 c_2 (V_1 - V_2)/(2^{1/2} d).$$

In these expressions the diatomic potential curves $E = V_{BC}(\text{LiH}, \ X\,^1\Sigma^+)$, $V_1 = V_{BC}(\text{LiH}^+, \ X\,^2\Sigma^+)$, $V_2 = V_{BC}(\text{LiH}^+, \ A\,^2\Sigma^+)$, $K = V_{AB}(\text{H}_2, \ X\,^1\Sigma_g)$, $G =$

$V_{AB}(H_2^+, 1s\sigma)$, and $U = V_{AB}(H_2^+, 2p\sigma)$ and the mixing parameters $s_1(r_{BC})$ and $s_2(r_{BC})$ (depending on r_{AC} when used in expressions with a bar) appear. In the diatomic PEC the correct asymptotic energies are included (those of neutral atoms assumed to be zero).

They are known from *ab initio* calculations and spectroscopic measurements. The mixing parameter functions were estimated in Ref. 33 from the results of *VB*-type atoms-in-molecules calculations.

In this example the specific origin of the nonhermiticity of the DIM matrix is obvious: it comes from the mixing parameters s_1 and s_2 (*resp.*, c_1 and c_2), the meaning of which is roughly that of configuration expansion coefficients for the ground the first-excited state of LiH$^+$ in terms of normalized composite basis functions. They are obviously different from each other. This result, on the other hand, follows from the fact that the matrices Γ_{PQ} are nonorthogonal and thus may be the reason for the nonhermiticity of the DIM method. The most natural way of symmetrizing the DIM matrices comes from a symmetrical orthogonalization of the diatomic fragment states and therefore of the diatomic fragment DIM matrices $S^{-1}\hat{H}_{PQ}$ (see, Ref. 20).

Results of the application of the symmetrized DIM model of Eq. (3.6) in comparison to *ab initio* (CEPA) results[34] (dashed curves) for the ground-state PES are shown in Fig. 4. The Pes was obtained without *a posteriori* fitting of input data. It is correct with respect to the global shape and topology of the PES, whence finer details of the potential (for example, location and

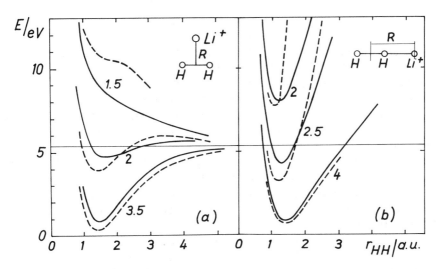

Figure 4. DIM potential (solid lines) and IEPA–PNO potentials[34] (dashed lines) for fixed $R(Li^+-H_2)$.

depth of the spurious ion-induced dipole minimum) differ significantly between various theoretical input data sources and with the experimental values.

The DIM potential models of Eq. (3.4) or (3.6) could be used for modeling the PES of similar systems, too; in that case some of the diatomic curves parameters (probably the mixing parameters) would have to be used as disposable parameters that are fitted to some specific property or to a more accurate calculation.

B.   The HeH$_2^+$ System

For the lower states of rare-gas/hydrogen molecular ion systems the difference in the ionization potentials of the rare-gas atom and a hydrogen atom often is large enough to allow for the neglect of DIM basis structures containing the rare-gas ionic species. In these cases (e.g., HeH$_2^+$,[35] NeH$_2^+$[36,37]) the most simple DIM model potential gives already a realistic ground, state PES. It is obtained as lower eigenvalue of a symmetric 2×2 matrix and thus may be given in closed form as follows (the index C means the rare-gas species):

$$E = Q_{AB} + Q_{AC} + Q_{BC} - [(J_{AC} - J_{BC})^2 + J_{AB}]^{1/2}. \tag{3.11}$$

Q_{AB} and J_{AB} are defined according to Eq. (3.5), while for the AC and BC fragments (rare-gas–hydrogen diatomic species) one gets:

$$\begin{aligned} J_{PQ} &= [V_{PQ}^+(X\,^1\Sigma^+) - V_{PQ}(X\,^2\Sigma^+)]/2, \\ Q_{PQ} &= [V_{PQ}^+(X\,^1\Sigma^+) + V_{PQ}(X\,^2\Sigma^+)]/2. \end{aligned} \tag{3.12}$$

Figure 5 shows the typical features of the topology of the PES which result from the simple DIM model. Here the upper surface is not realistic for any excited state of the system. Thus, if only the ground-state PES shall be modeled, it is possible to use the PEC asymptotically forming the upper surface (repulsive curves of H$_2^+$ and of HeH or NeH) to fit the ground-state PES to known potential data. This approach has been followed in a number of papers[35-37] with some success.

For a description of nonadiabatic charge-transfer processes the assumptions on which Eq. (3.4) is founded, that is, the neglect of excited-state basis functions in the DIM model, is no longer justified. The model must include at least the recharge structure He$^+$ + H$_2$ additional to He + H$_2^+$ (on which structure alone the above potential was built).

The potential topology of that model is indicated in Fig. 6. In contrast to the simple model of Fig. 5, here the topology of the entrance valley (R_{HH}) is for the lower excited states roughly correct, while in the exit valley (R_{HeH}) several lower excited states are missing. Taking into account that the only possible recharge processes on the first excited surface are He$^+$ + H$_2$ ⇒

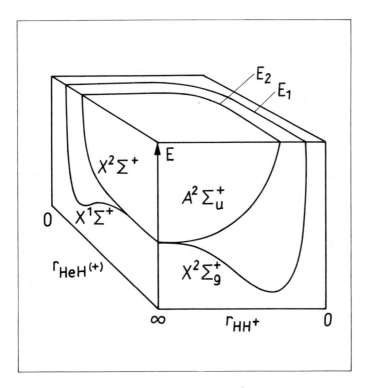

Figure 5. Schematic view on the topology of the HeH$_2$ $^2A'$ PES for C_s symmetry according to a simple DIM model.

He + H + H$^+$ in the lower eV range, that approximation is not too stringent, as long as the DIM model is used to analyze these processes only. The DIM matrix for the HeH$_2^+$ doublet states becomes a 4×4 nonsymmetric matrix with the following structure:

$$
S^{-1}\hat{H} =
\begin{bmatrix}
Q_{AB} + Z_1^{BC} + D_{CA} - I & J_{AB} & Z_3^{BC}/2 \\
J_{AB} & Q_{AB} + Z_1^{CA} + D_{BC} - I & Z_3^{CA}/2 \\
Z_4^{BC}/2 & Z_4^{CA}/2 & S_{AB} + 3(T_{BC} + T_{CA})/4 + (Z_2^{BC} + Z_2^{CA})/2 \\
\sqrt{3}Z_4^{BC}/2 & -\sqrt{3}Z_4^{CA}/2 & \sqrt{3}(T_{CA} - T_{BC} + Z_2^{BC} - Z_2^{CA})/4
\end{bmatrix}
$$

$$
\begin{matrix}
Z_3^{BC}/2 \\
-\sqrt{3}Z_3^{CA}/2 \\
\sqrt{3}(T_{CA} - T_{BC} + Z_2^{BC} - Z_2^{CA})/4 \\
T_{AB} + (T_{BC} + T_{CA})/4 + (Z_2^{BC} + Z_2^{CA})/2
\end{matrix}
$$

$$(3.13)$$

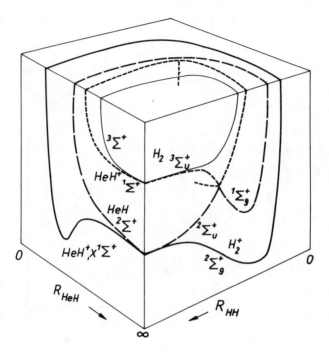

Figure 6. Schematic view on the topology of the HeH_2 $^2A'$ PES for C_s symmetry according to an extended DIM model.

The term combination expressions in Eq. (3.13) have the following meaning: J_{AB} and Q_{AB} are the same as in Eq. (3.8), D_{PQ} is the lowest $^2\Sigma^+$ state PEC of neutral HeH, I is the difference of the IP of H and He, S_{AB} and T_{AB} are the hydrogen PEC for $X\,^1\Sigma_g$ and $a\,^3\Sigma_u$, respectively, and T_{BC} and T_{CA} mean the triplet state PEC of the respective fragment diatomic HeH^+. The lowest two singlet states of HeH^+, X_{PQ} and Y_{PQ}, enter the matrix expressions through the terms Z^{PQ}, which read

$$Z_1^{PQ} = (c_1 c_2 X_{PQ} + s_1 s_2 Y_{PQ})/d, \qquad Z_2^{PQ} = (c_1 c_2 Y_{PQ} + s_1 s_2 X_{PQ})/d,$$
$$Z_3^{PQ} = c_1 s_2 (X_{PQ} - Y_{PQ})/d, \qquad Z_4^{PQ} = c_2 s_1 (X_{PQ} - Y_{PQ})/d. \tag{3.14}$$

Here s_i, c_i, and d are as in Eq. (3.10) mixing parameter functions, depending on the distance PQ.

For that DIM potential model, as formulated in Eq. (3.13), an analytic solution for the energy eigenvalues may still be found once the matrix is symmetrized. Every model for systems of that type using a more extended basis set will lead to a matrix representation that must be diagonalized using a numerical diagonalization routine. Approaches of that kind for HeH_2^+ have been reported with a DIM model including ionic structures, mainly for fitting purposes $(H^- H^+ He^+)$, that resulted in a 6×6 matrix[38] and with additional excited H atom states $(2s, 2p)$, where a 15×15 case is produced.[39] Both papers cited give the analytical forms of the DIM matrices as well as comparisons with *ab initio* PES results.

The main purpose of the 6×6 model[38] is the description of the avoided-crossing seam between the first- and second-excited-state PES in the entrance channel. It is marked with "S" in Fig. 7, where both surfaces are shown in $C_{\infty v}$ symmetry. In the collinear configurations the splitting of the PES is largest and the nonadiabatic coupling strength (Fig. 8) has a minimum. Only in that linear geometries a dissociation with charge transfer is probable.

In the DIM model calculations on HeH_2^+ reviewed here, the excited (charge-transfer) states of the system were not described *a priori* well enough by the DIM model; a fitting procedure had to be used to adjust some of the parameters of excited diatomic fragment states. The preceding seems to be a rather general finding from other DIM models of other systems: it is a necessity to plan a DIM calculation with some additional care in order to adjust the model to accurate data that are known only for a few points of the whole PES.

C. The FH_2^+ Interaction

To that charge-transfer system a DIM PES study of Kendrick et al.[40] was published, and corroborated by a more extended basis set DIM model of Schneider et al.[41] The channel $H^+ + HF$ and the probability of charge-transfer processes in that system was further studied by us[42] using the latter model.

The DIM model structure is based there on the atomic states $F^+(^3P_g)$, $F^+(^1D_g)$, $F(^2P_u)$, $F^-(^1S_g)$, H^+, and $H(^2S_g)$ which combine to five three-atom basis structures. This model leads to 17 triplet and to 15 singlet states; accordingly, the dimensions of the matrices are 17×17 and 15×15 as long as other symmetries are not taken into account. Transformation to the reflection symmetry at the plane of the three nuclei leads to separate problems for A' and A'' states of dimensions 9 and 8 for ^3A states and 9 and 6 for ^1A states. Fourteen different analytic representations (fits) of diatomic PEC and five fits of diatomic mixing coefficients are used for the construction of the matrix elements. The less accurately known fragment data were calculated by the semiempirical atoms-in-molecules method.[43] The channel for proton–HF collisions was compared with ground-state *ab initio* CEPA[44] results and analyzed with respect to the probability of nonadiabatic charge-transfer processes.[42]

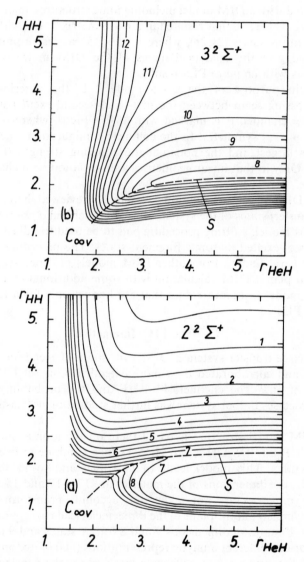

Figure 7. Collinear DIM PES (first and second excited state) for HeH$_2$ from an extended DIM model.[38]

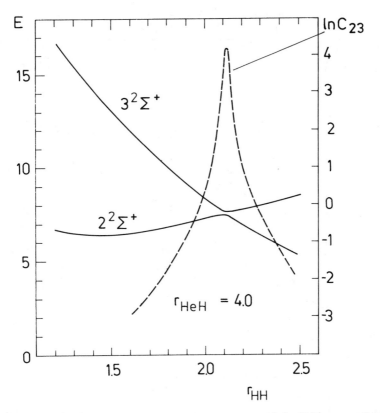

Figure 8. Potential terms and nonadiabatic coupling strength for HeH_2 perpendicular to the entrance valley in $C_{\infty v}$.[38]

The asymptotic interactions (diatomic PEC) taken into account for the DIM model are shown in Fig. 9. In Fig. 10, the resulting lowest and first-excited singlet surfaces show no evidence of a pronounced avoided crossing in $C_{\infty v}$ geometry, although a small barrier in the lower surface is visible. To visualize the possible orientation dependence of that barrier, Fig. 10 reports computed cuts through the full surfaces, with r_{HF} kept at a fixed value, as functions of orientation and relative distance R.

The quality of the DIM calculations can be assessed by comparing the DIM (solid) and CEPA (dashed) PEC; the overall shape is essentially the same with the CEPA results providing, as expected, deeper potential wells in the protonation region. It is important to note that, as the curves become increasingly repulsive with smaller γ values, an outer barrier appears in both the DIM and CEPA calculations, the latter being less pronounced than the

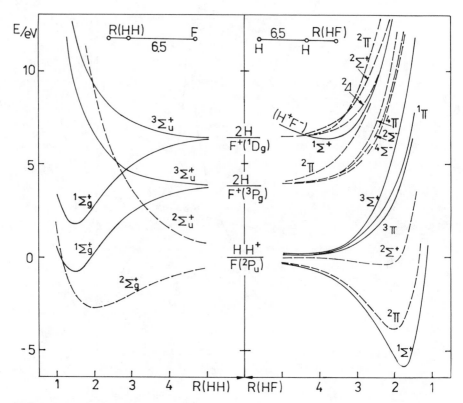

Figure 9. Asymptotic situation for the reactant and product valley of a DIM calculation of $(FHH)^{+}$[41] (unit of length a.u.).

former. Thus, the "cone" of approach defined by that range of γ values contains the only region where the lower singlet curves come close to the higher, charge-exchange curves [which asymptotically produce $H + HF^{+}(^{2}\Pi)$]. It seems, therefore, that only for protons approaching on the H side does the possibility exist for either radial coupling at a near-avoided crossing or for nonadiabatic Coriolis coupling. In this region, however, the energy gap is rather large and the differences in slope are highly localized, thus suggesting that both mechanisms would be inefficient in inducing surface-hopping processes.

Further examination of the energy as a function of the r_{HF} distance in the region of the smallest energy gap[42] indicates that even for large variations of the molecular coordinate the two surfaces remain apart, coming close to each other only for a very compressed H_2 geometry. For these r_{HH} distances

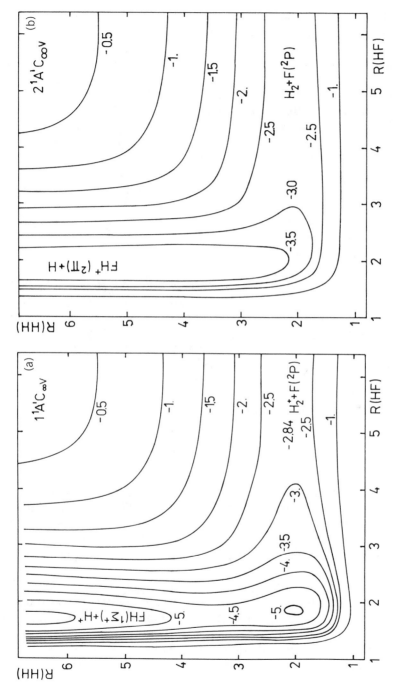

Figure 10. Collinear singlet DIM PES of $(FHH)^+$.[41] (energy in eV, length in a.u.): lower, upper surface.

of less than 1 a.u. a substantial amount of repulsion within the triatomic system is already produced, making it very unlikely that the internal energy available in the target HF would allow the interacting species to overcome such repulsion and to further apply the necessary torque needed for the nonadiabatic transition to occur.

In conclusion, the DIM model PES for $H^+ + HF$ collisions at relative energies up to about 100 eV indicate negligible coupling from the initial state to $HF^+ + H$, consistent with the experimental findings.[45] Furthermore, it allows one to explain the rather small vibrational energy transfer observed,[46] because the PES shows that the internal motion on the lowest surface is coupled with that of the incoming proton only over a very small range of values of the impact parameters and for relative distances that are not sampled at most collision energies.[47] It is worth noting here that a great deal of the physical details of the scattering, charge-exchange process could be gleaned by a simple analysis of the DIM calculations.

D. The BH_2^+ Surfaces

The BH_2^+ system was studied in two papers of Schneider et al.[48,49] by DIM potential energy calculations aimed at the interpretation of molecular beam data. The same DIM model was used by us in a study of the $H^+ + BH$ charge-transfer dynamics.[50] Furthermore, the DIM model was used for that system in an optimization of the stable complexes on the triplet PES and to characterize the spectroscopic constants.[51]

The basic structures that define the DIM model are

$$
\begin{aligned}
&1.\ B^+(^1S) + 2H(1s), \\
&2.\ B^+(^3P) + 2H(1s), \\
&3.\ B(^2P) + H(1s) + H^+.
\end{aligned}
\tag{3.15}
$$

The diatomic PEC used as input data for this model are shown in Fig. 11. Most of them are "rationalized" fits to *ab initio* calculated potential points (see Refs. 48 and 49). The resulting lower singlet and triplet PES seem quite reliable compared to existing experimental knowledge, thus some confidence in the model with respect to the charge-transfer channels appears to be justified.

The DIM model leads here to $7\,^1A'$ and $3\,^1A''$ states, among which the lowest three excited-state PES are most interesting for charge-transfer processes in the system $H^+ + BH(^1\Sigma^+)$. It is found in fact that symmetry considerations and different types of nonadiabatic couplings among these states markedly influence and control the relative probabilities of various elementary processes in such encounters.

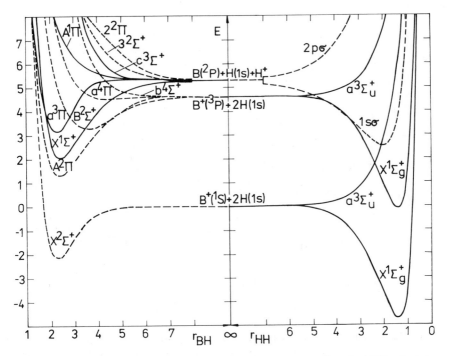

Figure 11. Asymptotic situation for the reactant and product valley of a DIM calculation of (BHH)$^{+49}$ (unit of length a.u., energy in eV).

Figures 12 and 13 report various cuts through the PES of the system at the BH equilibrium distance ($\approx 2.4\,$a.u.) that differ from each other by the value of the angle of approach, γ, which the proton forms with the bond at the BH center of mass. Here $\gamma = 0°$ is the linear approach from the H end of BH, while $\gamma = 180°$ is the approach from the B end, respectively.

The lowest potential curve, associated with the $^2\Sigma$ state of BH$^+$, is always too low in energy to be coupled dynamically to the entrance channel potential and therefore can be kept out of the present discussion by taking it to be unimportant for the dynamics. For the excited states, the two collinear configurations exhibit a lower surface that is of $^1\Pi$ symmetry and therefore it splits up once any bent configuration is reached and $^1A'$, $^1A''$ states are formed. Moreover, the $^1\Pi$ surface can cross the upper $^1\Sigma^+$ surface and transitions between these states can proceed only via Coriolis coupling; by the same token, in the bent configurations one can witness crossings between $^1A''$ and $^1A'$ surfaces, as it is seen for the values of γ up to 90° shown in the figures.

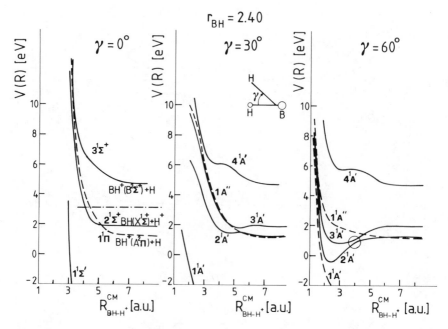

Figure 12. DIM PES sections for BH_2 showing the approach of H^+ from the H end of BH. The dashed line in the left panel is the height of the barrier to the exchange reaction leading to $B + H_2$. The encircled area in the right panel marks a region of enhanced nonadiabatic radial coupling.

The avoided-crossing regions that are exhibited by this system as γ increases (circled areas in Figs. 12 and 13) could be further characterized by evaluating the radial coupling strength,[12] which leads to electronically nonadiabatic collisions. In Figs. 13 and 14 we show the extension of the areas of strong coupling via the encircled regions, where one assumes collision energies up to $\approx 10\,eV$.

It is interesting to note here that, for smaller γ values (e.g., for $\gamma = 30°$), one already sees evidence for an avoided crossing between the second and third $^1A'$ states. The involved wavefunctions, however, still largely retain features of their symmetries at $\gamma = 0°$, that is, they are essentially of Σ and of Π character and therefore do not produce any significant values for the radial coupling matrix elements.

In case one expects the occurrence of rotational coupling, on the other hand, the presence of the $^1A''$ curves nearby and close to the $^{3\,1}A'$ makes the Coriolis coupling between the latter two states more probable than the

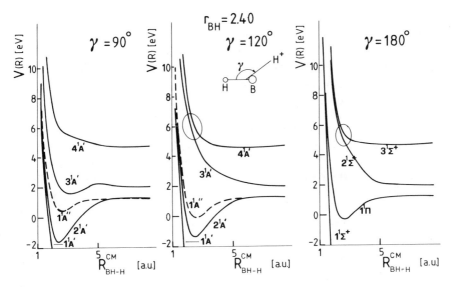

Figure 13. As Fig. 12 for the boron end of BH.

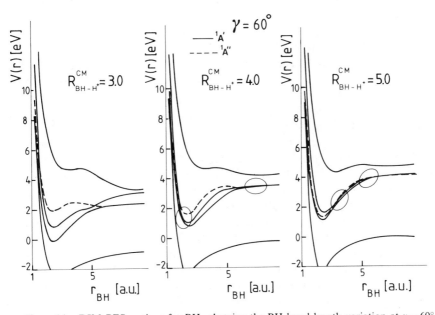

Figure 14. DIM PES sections for BH_2 showing the BH bond length variation at $\gamma = 60°$ for fixed proton distances to the center-of-mass of BH.

coupling between the two A' states and therefore suggests, at these γ values, possible charge-exchange processes into $BH^+(^2\Pi)$.

The possibility of exhibiting substantial Coriolis coupling through the appearance of a crossing seam is shown even more clearly for the $\gamma = 60°$ geometry (Fig. 12), at a relative distance of the colliding partners of about 4.5 a.u. Thus, surface hopping could occur to populate the $2\,^1A'$ state, and charge-exchange processes into $VH^+(^2\Pi)$ are possible.

Once we reach the rectangular geometry ($\gamma = 90°$), one sees from Fig. 13 that the situation has completely changed: the electronic state that describes the entrance channel (the $3\,^1A'$ term) is now fairly isolated, at nearly all distances, from all the other nearby states over the range of possible collision energies examined here. The $1\,^1A''$ curve has, in fact, become attractive as opposed to its repulsive behavior for smaller γ values and describes now (together with the $2\,^1A'$ state) an attractive interaction of the first electronically excited state of $BH^+(A\,^2\Pi)$ with an impinging H atom. One therefore expects that for the range of impact parameters which samples this orientation, no possibility of charge-transfer processes should exist.

By progressing to the B end of the system, that is, for γ values from about 100° through 180° as shown in Fig. 13, the $BH + H^+$ interaction along the $3\,^1A'$ curve has become repulsive. Its interaction with the first excited state of BH^+, the $A\,^2\Pi$ state, is now negligible at all distances, while its interaction with the next excited state, the $B\,^2\Sigma^+$ state, is instead becoming important. A sizable region over which large radial coupling appears for the avoided crossing between 2.0 and 3.0 a.u. is clearly shown. In going from the $\gamma = 120°$ orientation to the collinear geometry of H–B–H, one sees that the energy necessary to reach the crossing seam goes down to values slightly above 4.0 eV but can also increase up to and beyond 7.0 eV, provided we keep the molecular bond fixed at the $BH(X\,^1\Sigma^+)$ value. One knows, however, from Fig. 11, that the equilibrium bond length of $BH^+(B\,^2\Sigma^+)$ is 3.66 a.u. and therefore one should expect in this case the formation of strongly vibrationally excited BH^+, as opposed to what should have happened with the other orientations around the H end of the molecule, where only small vibrational excitation is probable during the charge-transfer process.

For the geometrical situation where the proton impinges on the H end of the target (Fig. 14), one sees that for distances inside the avoided crossings (left panel of the figure) the stretching of the bond causes increased coupling between states only when a large amount of vibrational energy already exists in the target molecule on the entrance channel ($3\,^1A'$); thus, for the usually "cold" molecular-beam measurements one expects very little coupling and also very little vibrational excitation as the forced-oscillator-model conditions for motion along one surface only are not satisfied in this system.[52] As the partners sample the avoided-crossing regions (center panel of Fig. 14), one sees that radial and angular couplings now occur not nly for a stretched BH bond but also for $BH(X\,^1\Sigma^+)$ around its equilibrium position, where hopping into the $BH^+(A\,^2\Pi)$ state can therefore take place. On the other hand, the ionic partner will not be formed in a vibrationally excited state and thus one should expect charge-transfer processes without substantial vibrational energy transfer.

For larger impact parameters (curves in the right panel of Fig. 14), the Coriolis coupling discussed above is now stronger and can induce the formation of vibrationally excited $BH^+(A\,^2\Pi)$.

In sum, encounters on the H end of BH are favoring the formation of $BH^+(A\,^2\Pi)$, either vibrationally excited or in its ground vibrational state, as a function of the collision energy (that is, of the dominant collision parameter values). they indicate a good probability of rotational coupling, hence suggesting possible formation of vibrationally cold $BH^+(A\,^2\Pi)$ if beam experiments are carried out. If the BH target molecule is prepared with a large content of vibrational energy, then one observes strong radial coupling, leading to larger quantities of electronically excited and vibrationally cold BH^+ in the $B\,^2\Sigma^+$ state.

E. The O_2H^+ Charge-Transfer Interactions

The application of the DIM method to systems containing several atoms with a complicated electronic structure as is the case in oxygen is either very difficult or restricted to simple model approaches that include an unreliably small basis set and try to account for its artifacts by semiempirical corrections. We have chosen this kind of approach for our description of the $O_2 + H^+$ potentials and the nonadiabatic couplings involved.[53,54] Preliminary (separate adiabatic) dynamic studies with the DIM potentials are reported in Ref. 55.

In particular, the DIM basis set should be chosen to be capable to describe excited states of the oxygen atom and its positive ion as well as at least the ground state of its negative ion. That last requirement was however ignored,[53] and only the minimal number of basis structures corresponding to the two energetically lowest channels,

$$(1)\quad 2O(3P) + H^+,$$
$$(2)\quad O^+(^4S) + O(^3P) + H(1s), \tag{3.16}$$

was taken into account. The reason for this choice relies on the experimental findings.[53] Starting from that basis all possible triplet state functions were constructed, already leading to $8\,A''$ and $13\,A'$ symmetry adapted functions. Most of the diatomic fragment information used was taken as "rationalized" analytic fit or spline interpolant to *ab initio* values (22 different PEC of O_2,

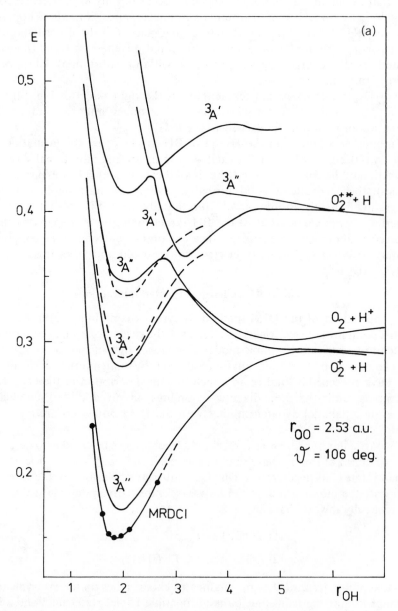

Figure 15. Comparison of DIM (solid lines) and CI (dashed lines) results for the lower triplet PES of O_2H^+ (energy and distances in a.u.).[53]

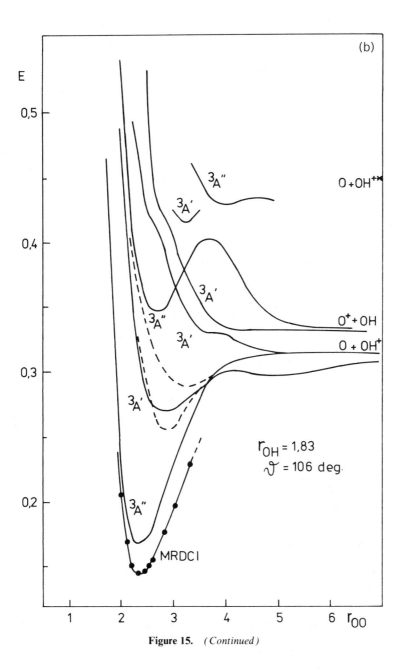

Figure 15. *(Continued)*

OH, O_2, and OH^+ states, 2 mixing parameters). Several excited-state curves (seven PEC, one mixing) were then freely varied to obtain the right energetic order and the right qualitative angular dependence of the lowest O_2H^+ states. The "trial-and-error" optimization of the free parameters was guided by the results of large-scale *ab initio* configuration interaction (CI) calculations. Figure 15 may provide a qualitative impression of the remaining differences between the DIM PES and the CI energies.

The DIM model[53] was used to explain the anomalous vibrational excitation of O_2 in proton encounters by analyzing the topography of the PES. In Ref. 55 the rotational inelasticity during inelastic and charge-transfer processes was studied by applying the above DIM model potentials. Nonadiabatic couplings between the two lowest[3]A'' PES were determined in Ref. 54.

Turning more in detail to the DIM potentials, Fig. 16 shows the collinear orientation as function of the interatomic distances d_{OH} and r_{OO}. When looking at the lower adiabatic surface (Fig. 16a), it is interesting to note the following points:

1. The charge-transfer channel, as d_{OH} increases, exhibits a small barrier at the seam between the $^3\Sigma^-$ and $^3\Pi$ configurations and a rather shallow valley as a function of the vibrational coordinate, r_{OO}. Thus, one should expect that reactions starting with "hot" molecules would more readily progress into charge-exchange products if the complex is formed in a Σ configuration.

2. The dissociation channel along the r_{OO} coordinate presents an even weaker energy dependence on the OH vibrational coordinate and a marked seam not far from the equilibrium geometry of the complex. The slightly higher barrier, therefore, requires more energy to be deposited into the d_{OH} coordinate.

The upper adiabatic surface (Fig. 16b) is rather instructive, on the other hand, in providing general, qualitative features of the other two possible channels for the break-up of the ionic complex:

1. In the upper surface, for the $O_2 + H^+$ entrance channel, the proton approach within the region of the O_2 equilibrium geometry corresponds always to a repulsive situation without complex formation, unless nonadiabatic coupling with the lower surface occurs on the way in, as marked by the seam in Fig. 16.

2. For the stretched values of the r_{OO} variable, the possibility exists of a shallow minimum (local complex formation) around the region of the crossing seam with the lower surface. Thus, the dissociative channel with O^+ formation can only occur if enough vibrational energy is deposited into the O_2 molecule as the proton approaches.

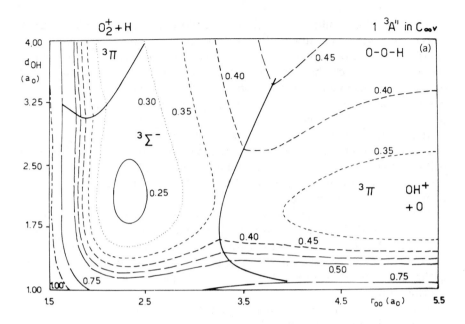

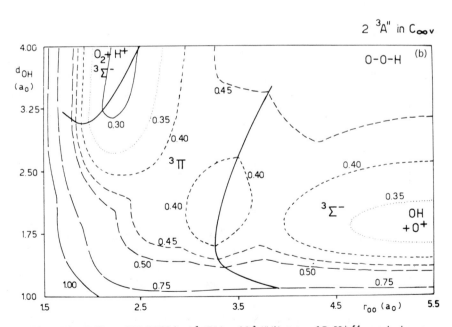

Figure16. Collinear DIM PES for $1\,^3A''$ (a) and $2\,^3A''$ (b) states of O_2H^+,[54] energies in a.u.

The preceding collinear geometries therefore suggest possible inelastic, direct processes in the upper PES with a small probability of dissociative charge-exchange occurrence once vibrational excited targets are formed in the former entrance channel. Moreover, charge-exchange mechanisms and dissociative attachment of the proton only appear to occur via nonadiabatic coupling mechanisms. For both surfaces, of course, Coriolis coupling is the physical origin of such nonadiabatic terms for linear orientations.

In an entirely similar manner one can now analyze the surface topography in the more general C_s symmetry. In Ref. 53 it was pointed out that most collision-partner approaches are indeed likely to occur for this type of relative orientation. It implies the passing of the system through avoided-crossing regions between adiabatic states.

Figure 17a and 17c present the results of calculations for the same lowest two $^3A''$ state PES as before but in a C_s configuration.

Figure 17b shows the general shape of an intermediate channel; of $^3A'$ symmetry, which is present for the same geometry. It stems from the splitting of the $^3\Pi$ surface parts of Fig. 16 ($C_{\infty v}$) into $^3A'$ and $^3A''$, when the symmetry is lowered (C_s).

Between the PES of Figs. 17a and 17c the avoided-crossing seams are not obviously visible in these plots. They are located for that angle of approach

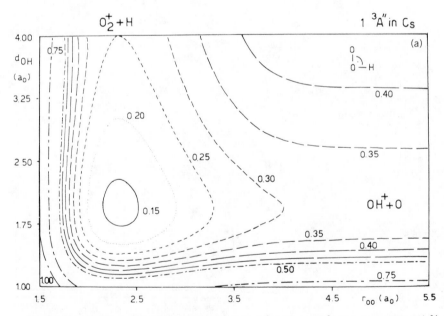

Figure 17. Nonlinear (C_s) DIM PES for $1\,^3A''$ (a), $1\,^3A'$ (b), and $2\,^3A''$ (c) states of O_2H^+,[54] energies in a.u.

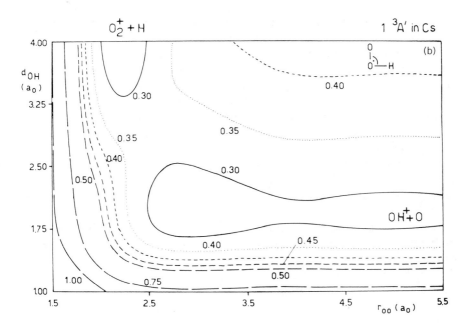

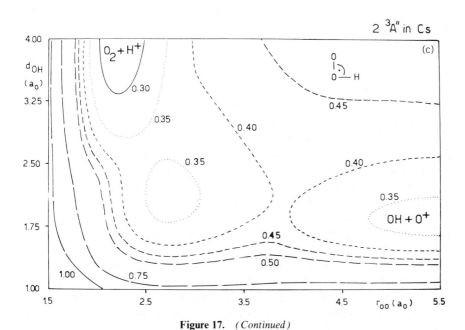

Figure 17. *(Continued)*

at d_{OH} values of about 3.8 a.u. for the entrance channel seam and at $r_{OO} \approx 3.0$ a.u. for the second pseudocrossing.

The lower of these two surfaces clearly shows the deep minimum region of the charge-exchange complex that dissociates into the charge-transfer channel (CT, $O_2^+ + H$) for O_2 targets not greatly vibrationally excited. The dissociative attachment channel (DA, $OH^+ + H$) evolves through a small barrier and a fairly steep repulsive region of the potential that does not easily allow for the vibrational excitation of the OH^+ system.

The corresponding upper surface for the same C_s symmetry is shown in Fig. 17c and indicates a more "floppy" structure for the $(O_2H)^+$ complexes, since a far more shallow minimum region exists for it and for a broader range of internal coordinates. Both the direct, inelastic channel shown in the upper left part of the figure and the CT + DA channel ($OH + O^+$) are reached via small repulsive barriers. Moreover, as the involved bond distances can markedly vary without much change of their internal-energy content, both exit channels allow for effective storage of energy into the vibrational mode during the encounters that lead to the two different asymptotic situations. This result seems to indicate that this type of behavior of the system is the most significant, since C_s collision approaches are most probable here.

Finally, Fig. 17b shows the presence of a sort of "intruder" state of $^3A'$ symmetry that appears in between the other two configurations discussed previously. Its asymptotic channels are the same as those for the lower surface of $^3A''$ symmetry, but its general shape is entirely different. If the system state changes in the entrance channel $O_2 + H^+$ to $^3A'$ by strong rotational coupling, then the DA channel can be reached here without a significant barrier across that channel.

Nonadiabatic radial coupling, which was also determined for that DIM model, is of interest here mostly for the transitions between the lower two $^3A''$ states. Founded on the coupling-strength function, an appropriately reduced "diabatic representation" of the PES allows one to treat the coupled nonadiabatic dynamics of the system.[56] The largest component of the coupling-strength vector from the described DIM model is the one dependent on r_{OO}. It is shown in Fig. 18 and reveals some features, which are general findings for all components: (1) sharply localized avoided-crossing seams, (2) enhancement of the coupling strength with O_2 vibrational excitation, (3) similar strong coupling for linear and T-shaped complexes, with a minimum coupling for $\gamma \approx 45°$.

It is interesting to note at this point that the rise of the DIM surfaces for this system within a coupled, quantum treatment of the CT dynamics has indeed provided very good agreement with experimental findings[56] and has confirmed the importance of nonadiabatic coupling between surfaces in order to transfer energy to the O_2 molecule.

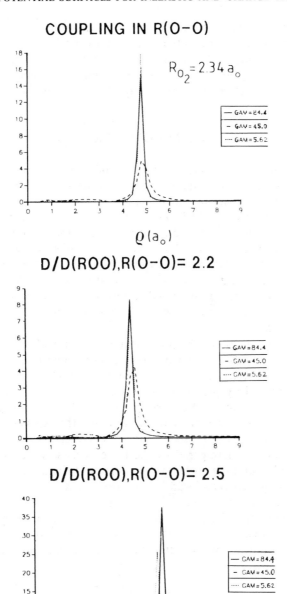

Figure 18. Nonadiabatic coupling elements of $O_2^+H^+$ from the DIM model[54] for fixed distances r_{oo}, units in a.u.

In other words, the scattering calculations have managed to show that, besides testing the general reliability of the DIM surfaces for such a system, they can specifically indicate whether or not the going through the avoided-crossing region constitutes the essential mechanism responsible for energy transfer processes in *both* asymptotic channels (i.e., $H^+ + O_2$ and $H + O_2^+$).

IV. MORE COMPLEX CASES

In principle, the specific model approach implied by the previous discussion should be well suited for extension to the more complex situations often encountered in ion–molecule chemical kinetics. We have tried to show in the previous analysis that the DIM description of polyatomic molecules does not really try to answer specific questions about the nature of a chemical bond, since the model starts with an assumed complete knowledge of all the diatomic fragments and component atoms. Thus, the diatomic bonds already exist at the outstart and the polyatomic situation is ultimately expressed in terms of diatomic behavior as the DIM approach takes account of bonding theory in an indirect way through the assumed electronic structure of the diatomic fragments.

The basic ingredients of the present model are therefore produced by a somewhat complicated interplay between spin-coupling, charge-transfer, and the rotational properties of the atomic functions. DIM is closely related to a valence bond (VB) description of a molecule but does not quite stand independently of it, since a VB description of the fragments is required as input data of the method. It therefore follows that the extension of the model PES calculations via the present method to larger systems and to more complex situations is hampered by the large basis sets required to handle the spin coupling and the rotational properties of the direct product basis.

The extensions have therefore followed very slowly the general applications to three-particle cases as those discussed in the previous section, and we will try to describe briefly some of them in the present section. It will be, out of necessity, a compact description of a few examples, but they should be sufficient to indicate what little has been done beyond the atom (ion)-diatomics case.

A. Rare-Gas Clusters

The general area of ionized clusters of simple rare gases is quite an interesting area of application of model interactions between ionic partners, especially since the presence of ions in the clusters strongly modifies both the bonding picture and the structure of equilibrium geometries.

In the case of the DIM approach the main problem is how to handle, on the one hand, the possibility of charge transfer and local strong binding and, on the other hand, the rest of the interaction, which is quite weak and dominated by dispersion terms. The final result may therefore turn out to be quite sensitive to the choice of input functions.

The simplest case is given by singly ionized rare-gas clusters, where applications to trimer ions have appeared in the recent literature.[57,58] In that instance one starts by admitting only one electronic state for each atomic center and neutral atoms are assumed to be in the ^1S_g state. The positive ions are instead taken to be in the ^2P_u state. The positive charge is also allowed to reside on any of the centers so that there are N state groups, which are possible for a singly ionized N-atom clusters:

$$
\begin{array}{lll}
1. & \mathrm{Rg}(^1S)\,\mathrm{Rg}(^1S) & \cdots \quad \mathrm{Rg}(^1S)\,\mathrm{Rg}^+(^2P) \\[4pt]
2. & \mathrm{Rg}(^1S)\,\mathrm{Rg}(^1S) & \cdots \quad \mathrm{Rg}^+(^2P)\,\mathrm{Rg}(^1S) \\[2pt]
\vdots & & \\[2pt]
N-1 & \mathrm{Rg}(^1S)\,\mathrm{Rg}^+(^2P) & \cdots \quad \mathrm{Rg}(^1S)\,\mathrm{Rg}(^1S) \\[4pt]
N & \mathrm{Rg}^+(^2P)\,\mathrm{Rg}(^1S) & \cdots \quad \mathrm{Rg}(^1S)\,\mathrm{Rg}(^1S)
\end{array}
\tag{4.1}
$$

here the Rg symbol stands for rare gas. Because of the three fold degeneracy of the ^2P state, there are three polyatomic basis functions corresponding to each of the state groups.

If such functions are denoted as P_x, P_y, and P_z and the $\mathrm{Rg}(^1S)$ function as s, t, then the three basis function for a particular group m are given by

$$
\begin{array}{llll}
mp_x & : & ss \quad \cdots \quad P_x \quad \cdots \quad ss \\[4pt]
mp_s & : & ss \quad \cdots \quad P_y \quad \cdots \quad ss \\[4pt]
mp_z & : & ss \quad \cdots \quad P_z \quad \cdots \quad ss
\end{array}
\tag{4.2}
$$

In total, there are $3N$ polyatomic basis functions for a singly ionized N-atom cluster. The corresponding DIM hamiltonian matrix $\mathbf{H}$ is then written as a sum of pair contributions, $\mathbf{H}^{ij}$ and of single-center contributions, $\mathbf{H}^i$, where i and j label the atomic centers. The atomic contributions are diagonal and the diagonal elements, H_{kk}, are either $E(\mathrm{Rg})$ or $E(\mathrm{Rg}^+)$, so that, if one adopts the convention that $E(\mathrm{Rg}) = 0$, their sum over the single-center matrices simply yields the ionization potential of Rg, $I\,(\mathrm{Rg})$. The atomic terms therefore all yield the same contribution, which is given by $(N-2)$ $I(\mathrm{Rg})$, to each diagonal element of the DIM hamiltonian matrix and therefore can be ignored.

The symmetric pair matrices $\mathbf{H}^{ij}$ have elements that are mostly zero, except for the elements between basis functions, which have a positive charge on either atom i or j. They also have nonvanishing values for the diagonal elements of all the remaining functions. The corresponding value of the elements is given in terms of the diatomic fragment energies[57]:

$$
\begin{aligned}
Rg_2(^1\Sigma_g) &= S, \\
Rg_2(^2\Sigma_g) &= G, \\
Rg_2(X\,^2\Sigma_u) &= U, \\
Rg_2(^2\Pi_g) &= G, \\
Rg_2(^2\Pi_u) &= U.
\end{aligned}
\tag{4.3}
$$

One can also define the following auxiliary quantities, which are given by $Q = \frac{1}{2}(U + G)$, $J = \frac{1}{2}(U - G)$, $\bar{Q} = \frac{1}{2}(\bar{U} + \bar{G})$, and $\bar{J} = \frac{1}{2}(\bar{U} - \bar{G})$. They remind us that the P state has u symmetry. If the two atoms i and j are along the z axis, then we get six functions from the possible groups of pairs: ip_y, ip_z, ip_x and jp_y, jp_x. The corresponding $\mathbf{H}$ matrix can then be written as follows:

$$
\mathbf{H}^{ij}(0,0) = \begin{array}{c} \\ ip_y \\ ip_z \\ ip_x \\ jp_y \\ jp_z \\ jp_x \end{array} \overset{\begin{array}{cccccc} ip_y & ip_z & ip_x & jp_y & jp_z & jp_x \end{array}}{\begin{bmatrix} \bar{Q}_{ij} & 0 & 0 & \bar{J}_{ij} & 0 & 0 \\ 0 & Q_{ij} & 0 & 0 & J_{ij} & 0 \\ 0 & 0 & \bar{Q}_{ij} & 0 & 0 & J_{ij} \\ \bar{J}_{ij} & 0 & 0 & \bar{Q}_{ij} & 0 & 0 \\ 0 & J_{ij} & 0 & 0 & Q_{ij} & 0 \\ 0 & 0 & \bar{J}_{ij} & 0 & 0 & \bar{Q}_{ij} \end{bmatrix}}
\tag{4.4}
$$

here the auxiliary quantities Q_{ij} and J_{ij} are obtained from the diatomic fragment PEC's at the internuclear distance given by r_{ij}. The index $(0,0)$ denotes the dependence of the DIM matrix on the rotational angles that are needed to produce total wavefunctions of the correct coupling symmetry[57]. All the remaining diagonal elements of the $\mathbf{H}$ matrix are given by the values of the S curve indicated in Eq. (4.3).

In a more general case, the two-body vector r_{ij} is not parallel to the z axis and therefore the preceding DIM matrix must be constructed from the rotated diatomic fragment matrices. This requires the transformation of $\mathbf{H}^{ij}(0,0)$ by a rotation matrix:

$$
\mathbf{H}^{ij}(\alpha, \beta) = \mathbf{R}(\alpha, \beta)\mathbf{H}^{ij}(0,0)\mathbf{R}^{-1}(\alpha, \beta),
\tag{4.5}
$$

where one defines the rotation matrix as follows:

$$\mathbf{R}(\alpha, \beta) = \mathbf{1}_{3N-6}\mathbf{R}^{(i)}(\alpha, \beta)\mathbf{R}^{(j)}(\alpha, \beta). \tag{4.6}$$

Here the $\mathbf{R}^{(i)}(\alpha, \beta)$ is a 3×3 matrix that connects with each other the three functions ^ip_y, ^ip_z, and ^ip_x:

$$\mathbf{R}^{(i)}(\alpha, \beta) = \begin{array}{c} \\ ip'_y \\ ip'_z \\ ip'_x \end{array} \overset{\begin{array}{ccc} ip_y & ip_z & ip_x \end{array}}{\begin{bmatrix} \cos \alpha & \sin \alpha \sin \beta & \sin \alpha \cos \beta \\ 0 & \cos \beta & -\sin \beta \\ -\sin \alpha & \cos \alpha \sin \beta & \cos \alpha \cos \beta \end{bmatrix}}. \tag{4.7}$$

The other matrix $\mathbf{R}^{(j)}(\alpha, \beta)$ is a similar 3×3 matrix that refers to the three functions from the center j. The result of multiplication (4.6) is a symmetric matrix that is diagonal everywhere except for the block of Eq. (4.4), where it has the following structure:

$$\mathbf{H}^{ij}(\alpha, \beta) = \begin{bmatrix} \mathbf{M} & \mathbf{N} \\ \mathbf{N} & \mathbf{M} \end{bmatrix}, \tag{4.8}$$

where $\mathbf{M}$ and $\mathbf{N}$ are both 3×3 symmetric matrices that can be written in terms of another matrix, $W(a, b)$, which is given by

$$\begin{aligned} W p_y p_y &= a(\cos^2 \alpha + \sin^2 \alpha \cos^2 \beta) + b \sin^2 \alpha \sin^2 \beta, \\ W p_z p_y &= (b - a) \sin \alpha \sin \beta \cos \beta, \\ W p_x p_y &= (b - a) \sin \alpha \cos \alpha \sin^2 \beta, \\ W p_z p_z &= a \sin^2 \beta + b \cos^2 \beta, \\ W p_x p_z &= (b - a) \cos \alpha \sin \beta \cos \beta, \\ W p_x p_x &= a(\sin^2 \alpha + \cos^2 \alpha \cos^2 \beta) + b \cos^2 \alpha \sin^2 \beta. \end{aligned} \tag{4.9}$$

The angles β and α, which define the transformation, are the angles of the polar coordinates of the relative position vector r_{ij}:

$$r_{ij} = (X_j - X_i)\hat{\imath} + (Y_j - Y_i)\hat{\jmath} + (Z_j - Z_i)\hat{k} \tag{4.10}$$

hence
$$\cos \beta = (Z_j - Z_i)/|r_{ij}|$$

$$\alpha = \tan^{-1}[(Y_j - Y_i)/(x_j - x_i)]. \tag{4.11}$$

In conclusion, once a specific general symmetry is chosen for the cluster of N atoms, ne could simplify the preceding construction of the DIM matrix by making use of the symmetry propeties of the corresponding point group representation of the cluster. In particular, the results for trimers of Ar, $(Ar)_3^+$, have been worked out in the C_s, C_{2v}, and D_{3h} symmetries.[57]

For more complicated cases with $N > 3$, the ionic clusters consist of a small ionic molecule (trimer or tetramer) surrounded by adjoining shells of increasingly more neutral atoms, which are weakly bound to the central core. The trimer ions appear to be particularly stable and form the cores of clusters up to size 13, when the first shell of neutrals is completed.[57] Once the second shell begins to be filled, there is a tendency for the central ion to change from a trimer to a tetramer and for a small fraction of the positive charge to spread out over the rest of the rare gases in the cluster. What happens because of that is that the overall structure remains very floppy and several configurations of the outer shells exist with an energy near the bottom of the potential well. The DIM model qualitatively reproduces such a behavior in the sense that the properties of the chosen diatomics, which are the building blocks of this model, turn out to be important in influencing the nature of bonding in the polyatomic structures.

B.   Interactions on Surfaces

A further development in complexity could naturally come from a situation in which the charge-exchange process takes place on a metal surface, for example, when impact ionization processes like

$$2A^* + S \rightarrow 2A^+ + S^- \rightarrow A_2 + S^-, \tag{4.12}$$

where S represents a metal surface and A^+ some metastable atomic projectile, are considered. Of course several different outcomes could also be taken into account when examining the final channels: the treatment of the relevant interactions would become accordingly more complicated with the consequence of needing even more extended basis functions to construct the DIM matrix.

This area of application of DIM models is relatively new and different approaches have been followed to deal with its difficulties. One possibility is to model the surface by a small cluster of atoms, so that the final treatment amounts to a computation for a large molecule. Examples are given by the study of H atoms interacting with Li_9 clusters[59] and by the analysis of the dissociative adsorption of H_2 on a Li_8 cluster.[60] The basic mechanism of a charge-transfer process on a surface, or in the vicinity of a surface, is perhaps even more amenable to the DIM method, since the process turns out to be very similar to the harpooning mechanism of gas-phase reactions,[61] whereby dynamical calculations could be carriedd out within a small diabatic basis, usually chosen to describe the system before and after the charge-transfer process.[62,63]

Another approach attempted to incorporate the entire physical surface into the DIM model by approximating its main physical features. The application was carried out for the case of the formation of Na^+ and Na_2 ions from $Na^+(^2P_u)$ atoms in the proximity of a tungsten surface partially covered with sodium.[64,65] If the covering could be considered of less than a monolayer, then the sodium could be treated as a layer of Na^+ on the surface. Since each ion is associated with an induced image charge on the metal, then the covered surface can be approximated by an nfinitely extended dipole layer, which in turn affects the shapes of the neutral and ionic PEC's from the surface itself. The DIM model could then be constructed by considering the entire tungsten surface to be an atom and then adding $(N - 1)$ Na^+ ions and one sodium atom. Additional contributions will also have to come from several Na_2 electronic states and the modeled Na^+–W potential curves.

A further extension of the modeling of ion–surface interaction within the DIM picture comes from a method that includes delocalized electronic interactions within the basic VB picture of the molecule–surface potentials.[66] Strickly speaking, such an approach is no longer a DIM procedure, but it may be a useful way to point at many-body improvements within the diatomics basis implied by the DIM modeling. For this reason, we are briefly describing the approach in what follows.

The essential approximation is based on an effective medium approach to represent the metal–metal bonding and the metal–nonmetal bonding outside the strictly covalent bonding described by DIM PEC's.[67,68]

The total system is thought of as made up of Ng nonmetal atoms interacting with a metal surface that is modeled by N_m metal atoms. The metal atoms are separated into a primary zone comprised of N_{mp} atoms that are allowed to move and a secondary zone in which the metal atoms are fixed at their lattice geometries.

The total number of atoms in the system is $N = Ng + N_m$ and the total number of atoms allowed to move is $Ng_{mp} = Ng + N_{mp}$. This separation therefore allows one to develop the interaction with the idea of the embedded atom method (EAM), the latter being the procedure through which the noncovalent part of the potentials is treated.[67] hence, the total potential is separated into two contributions: V_M, which includes metal–metal interactions within the solid and V_{DIM}, that accounts for the interactions between gas atoms and the metal and between all the gas atoms:

$$V(\mathbf{X}) = V_{DIM}[E_{Kl}(\mathbf{X}), E_{KM}(\mathbf{X})] + V_M(\mathbf{X}) + E_0, \qquad (4.13)$$

here $\mathbf{X}$ is a collective index for all the relative coordinates, E_0 is a reference scaling factor that can be chosen for convenience, and the V_{DIM} is a functional of the diatomic potential curves $E_{Kl}(X)$, for the pertinent electronic states of the nonmetal atoms K and l, and of the $E_{KM}(\mathbf{X})$ potential curves which describe the interactions between the relevant electronic state of the nonmetal atom K and the metal surface M. Thus, the indices K and l of the former curves run up to Ng, the number of nonmetal atoms, and describe two-body interactions. In the latter E_{KM} curves, the index K also runs up to Ng and the metal surface is described via the embedded atom potential as given in the V_M expression[67]:

$$V_M(\mathbf{X}) = \sum_{i=Ng+1}^{N} F_i[\rho_i(\mathbf{X})] + \sum_{i=Ng+1}^{N_P} \sum_{j>i} \phi_{ij}(R_{ij}), \qquad (4.14)$$

where the electron density at the location of the metal atom i is called $\rho_i(\mathbf{X})$ and can be written as a sum over atomic densities $\rho_j(R)$ of all the other atoms, metal and nonmetal, in the system under consideraton:

$$\rho_i(\mathbf{X}) = \sum_{i,j \neq i} \rho_j(R_{ij}). \qquad (4.15)$$

The new element in Eq. (4.14) is the embedding function $F_i[\rho_i(\mathbf{X})]$ and corresponds to the energy to embed the atom i into the homogeneous electron gas density used to describe the density of the host (that is, all the other atoms) at a particular location; it is an attractive function of the density. Finally, the ϕ_{ij} functions of Eq. (4.14) are short range, pairwise repulsive functions of the interatomic distances that are given by a simple Coulombic form[67]

$$\phi_{ij}(R) = CZ_i(R)Z_j(R)/R, \qquad (4.16)$$

where C is a constant and $Z_i(R)$ is the effective nuclear charge of atom i at a distance R from its nuclens. Both $F_i(\rho)$ and $Z_i(R)$ are empirical functions for a given atom.

One obviously sees that the DIM model provides here a way of selecting the dominating "curves" in the VB sense that play a major role in the charge-transfer processes on the surface or in the proximity of it. On the other hand, the EAM implied by Eq. (4.14) provides a description of the medium, that is, of the other metal atoms in the surface, which modifies the structure of the V_{DIM} curves. Instead of expanding the full PES itself in a many-body expansion,[69,70] one tries first to obtain a model hamiltonian for the full system and thus introduces the many-body effects into the correcting terms of the model potential.

As a matter of fact, it is not yet certain whether or not the DIM method is well suited to describe a surface structure itself. It is perhaps more likely to be of use in describing processes where covalent bonding is involved but is also perturbed by the vicinity of a surface or by a few neighboring atoms on a simple surface. In such instances, the choice of specific configurations via the DIM procedure may be ne of the few methods available to deal with the problems at hand.

V. FINAL CONSIDERATIONS

In the present chapter we have considered the possible way in which model potentials could be developed to treat inelastic collisions involving more than one PES during processes at thermal, or nearly thermal, energy conditions.

The main idea is to see if the use of diatomic bond formations, or the use of diatomic fragment excitations when describing the region of nonadiabatic couplings between surfaces, could provide us with a realistic description of the interactions that lead to the charge-transfer channels or that lead to internal excitation accompanying the charge-transfer process.

The concept of nonadiabatic transitions is a very general and interdisciplinary concept, which refers, as is well known, to a transition among the adiabatic states defined as the eigenstates of a system given at a fixed adiabatic parameter "R." The transition is indeed induced by a variation of that parameter R. The adiabatic states produced by the present DIM modeling of them therefore describe good basis states only when the separability of the parameter R from the other variables holds well, or when the variation of R is much slower than the motion with respect to the other variables. Since the adiabaticity breaks down, and therefore the required nonadiabatic transitions occur efficiently, in such regions of the parameter R where the adiabatic states come close together, then it becomes crucial for the present model to describe as correctly as possible the local occurrences of the avoided crossings where the coupling effects are most important.

In the examples discussed previously for three-particle systems, and in some of the further extensions outlined in Section IV, we have shown that the DIM model, being closely related to a VB description of the collision partners, appears as an efficient tool for selecting the dominant configurations for the nonadiabatic coupling regions. On the other hand, extension of such a method to larger molecules is hampered by the large basis sets required to handle the spincoupling and the rotational properties of the direct-product basis. Thus, one could profitably employ othe theoretical results to help out in the simplification. It is always possible, in fact, to omit many of the spin–orbit couplings at the design stage of the most important configurations, but the decision about which structures to leave out can only be made on the basis of some external experimental knowledge or of previous theoretical evaluations.

Whenever this is possible, however, one finds that the selection of even only a few, but judiciously chosen, excited VB configurations could guide one quite effectively in the design of approximate, model surfaces relevant to charge-transfer collisions.[56]

To move to larger cases, however, not only creates several complications in the handling of the dynamical process but also requires the introduction of external information, which can help us in gaging the weight of each selected configuration for a specific process. Typically, MRD-CI calculations could be employed for certain regions of a given surface or surfaces and the corresponding DIM calculations could be scaled accordingly and extrapolated outside the scaling region to save in computational efforts.[53]

Finally, the presence of strong, covalent bonding on a surface could be handled by extending to it the harpooning model of charge exchange already existing in the gas phase but also introducing the noncovalent interactions by some sort of effective potential that modifies the basic DIM approach to the metal–ion interactions. Here again, chemical intuition may help in selecting the relevant basis functions and external theories (for example, density functional approaches) could be profitably invoked to implement the basic modeling provided by the DIM approach.

References

1. F. O. Ellison, *J. Am. Chem. Soc.* **85**, 3540 (1963).
2. W. Moffitt, *Proc. R. Soc. (London)* A **210**, 245 (1951).
3. F. O. Ellison, N. T. Huff, and J. C. Patel, *J. Am. Chem. Soc.* **85**, 3544 (1963).
4. G. V. Pfeiffer, N. T. Huff, E. M. Greenawalt, and F. O. Ellison, *J. Chem. Phys.* **46**, 821 (1967).
5. G. V. Pfeiffer and F. O. Ellison, *J. Chem. Phys.* **43**, 3405 (1965).
6. A. A. Wu and F. O. Ellison, *J. Chem. Phys.* **47**, 1458 (1967).
7. A. A. Wu and F. O. Ellison, *J. Chem. Phys.* **48**, 1491 (1968).
8. A. A. Wu and F. O. Ellison, *J. Chem. Phys.* **48**, 5032 (1968).
9. P. J. Kuntz and A. C. Roach, *J. Chem. Soc. Faraday Trans. II* **68**, 259 (1972).
10. E. Steiner, P. R. Certain, and P. J. Kuntz, *J. Chem. Phys.* **59**, 47 (1973).
11. J. C. Tully, *J. Chem. Phys.* **58**, 1396 (1973).
12. J. C. Tully, *J. Chem. Phys.* **59**, 5122 (1973).
13. J. C. Tully and C. M. Truesdale, *J. Chem. Phys.* **65**, 1002 (1976).
14. M. B. Faist and J. T. Muckerman, *J. Chem. Phys.* **71**, 225 (1979).
15. M. B. Faist and J. T. Muckerman, *J. Chem. Phys.* **71**, 233 (1979).
16. J. Vojtik, *Int. J. Quantum. Chem.* **28**, 593 (1985).
17. J. Vojtik, *Int. J. Quantum Chem.* **28**, 943 (1985).

18. P. J. Kuntz, in *Atom-Molecule Collision Theory*, R. B. Bernstein, ed., Plenum, New York, 1979, p. 79.

19. J. C. Tully, in *Adv. Chem. Phys.* **42**, 63 (1980).

20. R. Polak, *Chem. Phys.* **103**, 277 (1986).

21. P. J. Kuntz and A. C. Roach, *J. Chem. Phys.* **74**, 3420 (1981).

22. A. C. Roach and P. J. Kuntz, *J. Chem. Phys.* **74**, 3435 (1981).

23. P. J. Kuntz and W. N. Whitton, *Chem. Phys.* **16**, 301 (1976).

24. R. K. Preston and J. C. Tully, *J. Chem. Phys.* **54**, 4297 (1971).

25. B. T. Pickup, *Proc. R. Soc. (London) A* **333**, 69 (1973).

26. R. Polak, J. Vojtik, and F. Schneider, *Chem. Phys. Lett.* **53**, 117 (1978).

27. J. Vojtik, A. Krtkova, and R. Polak, *Theoret. Chim. Acta (Berlin)* **63**, 235 (1983).

28. J. Krenos, R. Preston, and R. Wolfgang, *Chem. Phys Lett.* **10**, 17 (1971).

29. J. C. Tully and R. K. Preston, *J. Chem. Phys.* **55**, 526 (1971).

30. J. R. Krenos, K. K. Lehmann, J. C. Tully, P. M. Hierl, and G. P. Smith, *Chem. Phys.* **16**, 109 (1976).

31. C. G. Schlier and U. Vix, *Chem. Phys.* **113**, 211 (1987).

32. J. K. Badenhoop, G. C. Schatz, and C. W. Eaker, *J. Chem. Phys.* **87**, 5317 (1987).

33. F. Schneider, thesis, Academy of Sciences of the GDR, Berlin, 1981.

34. W. Kutzelnigg, V. Staemmler, and C. Hoheisel, *Chem. Phys.* **1**, 27 (1973).

35. P. J. Kuntz, *Chem. Phys. Lett.* **16**, 581 (1972).

36. E. F. Hayes, A. K. Q. Siu, F. M. Chapman Jr., and R. L. Matcha, *J. Chem. Phys.* **65**, 1901 (1976).

37. R. L. Matcha, B. M. Pettitt, P. F. Meier, and P. Pendergast, *J. Chem. Phys.* **69**, 2264 (1978).

38. F. Schneider and L. Zülicke, *Chem. Phys. Lett.* **67**, 491 (1979).

39. F. Schneider, L. Zülicke, R. Polak, and J. Vojtik, *Chem. Phys.* **76**, 259 (1983).

40. J. Kendrick, P. J. Kuntz, and I. H. Hillier, *J. Chem. Phys.* **68**, 2373 (1978).

41. F. Schneider, R. Polak, and J. Vojtik, *Chem. Phys.* **99**, 265 (1985).

42. F. A. Gianturco and F. Schneider, *Chem. Phys. Lett.* **129**, 481 (1986).

43. R. Polak, J. Vojtik, and F. Schneider, *Chem. Phys. Lett.* **85**, 107 (1982).

44. F. A. Gianturco, A. Palma, E. Semprini, F. Stefani, H. P. Diehl, and V. Staemmler, *Chem. Phys.* **107**, 293 (1986).

45. M. Noll, G. Niedner, and J. P. Toennies (unpublished results).

46. F. A. Gianturco, G. Niedner, M. Noll, and J. P. Toennies, *J. Phys. B* **20**, 3725 (1987).

47. F. A. Gianturco and V. Staemmlet, *Int. J. Quantum Chem.* **28**, 553 (1985).

48. F. Schneider, L. Zülicke, R. Polak, and J. Vojtik, *Chem. Phys.* **84**, 217 (1984).

49. F. Schneider, L. Zülicke, R. Polak, and J. Vojtik, *Chem. Phys. Lett.* **105**, 608 (1984).

50. F. A. Gianturco and F. Schneider, *Chem. Phys.* **111**, 113 (1987).

51. F. Schneider and A. Merkel, *Chem. Phys. Lett.* **161**, 527 (1989).

52. T. Ellenbroeck and J. P. Toennies, *Chem. Phys.* **71**, 309 (1982).

53. F. Schneider, L. Zülicke, F. Di Giacomo, F. A. Gianturco, I. Paidarova, and R. Polak, *Chem. Phys.* **128**, 311 (1988).

54. F. A. Gianturco, A. Palma, and F. Schneider, *Int. J. Quantum Chem.* **37**, 729 (1990).

55. F. A. Gianturco, A. Palma, and F. Schneider, *Chem. Phys.* **137**, 177 (1989).

56. F. A. Gianturco, A. Palma, M. Baer, E. Semprini, and F. Stefani, *Phys. Rev. A*, **42**, 3926 (1990).

57. P. J. Kuntz and J. Valldorf, *Z. Phys. D* **8**, 195 (1988).

58. R. Polák and P. J. Kuntz, *Mol. Phys.* **63**, 27 (1988).

59. J. Voitik and J. Fiser, *Chem. Phys. Lett.* **86**, 312 (1982).

60. J. Voitik, J. Savrda, and J. Fiser, *Chem. Phys. Lett.* **97**, 397 (1983).

61. J. C. Polanyi and J. L. Schreiber, in *Kinetics of gas Reactions*, W. Jost, ed., Academic Press, New York, 1974.

62. J. A. Olson and B. J. Garrison, *J. Vac. Sci. Technol. A* **4**, 1222 (1986).

63. J. A. Olson and B. J. Garrison, *J. Phys. Chem.* **91**, 1430 (1987).

64. P. J. Kuntz, *Int. J. Quantum Chem.* **29**, 1105 (1986).

65. P. J. Kuntz and K. Lacmann, *Sur. Sci.* **192**, 859 (1957).

66. T. N. Truang, D. G. Truhlar, and B. C. Garrett, *J. Phys. Chem.* **93**, 8227 (1989).

67. M. S. Daw and M. I. Baskes, *Phys. Rev. B* **29**, 6443 (1984).

68. J. K. Norskov and N. D. Lang, *Phys. Rev. B* **21**, 1131 (1980).

69. S. Carter, I. M. Mills, and J. N. Murrell, *Mol. Phys.* **37**, 1885 (1979).

70. A. J. C. Varandas and J. N. Murrell, *Faraday Disc. Chem. Soc.* **62**, 92 (1977).

QUANTUM-MECHANICAL TREATMENT FOR CHARGE-TRANSFER PROCESSES IN ION–MOLECULE COLLISIONS

MICHAEL BAER

Department of Physics and Applied Mathematics, Soreq Nuclear Research Center, Yavne, Israel

CONTENTS

State-Selected and State-to-State Ion–Molecule Reaction Dynamics, Part 2: Theory, Edited by Michael Baer and Cheuk-Yiu Ng. Advances in Chemical Physics Series, Vol. LXXXII. ISBN 0-471-53263-0 © 1992 John Wiley & Sons, Inc.

I. INTRODUCTION

Scattering processes taking place on more than one potential energy surface are the most common types of interactions between two (or more) molecular species. Among these processes the low-energy charge-transfer (CT) processes between ions and neutrals is of particular interest, because it is encountered in many systems of major importance such as plasmas, combustion, fusion, upper atmosphere chemistry, astrochemistry, and so on. In order to understand the behavior of these complicated systems under various conditions, very detailed information on the microscopic level of the CT process is required. This ranges from state-selected integral total cross sections through state-to-state integral cross sections down to state-to-state differential cross sections.

The experimental study of these processes started about two decades ago[1-12] but the first state-to-state cross sections were measured only a few years ago.[6,8-12] In this sense the most detailed measurements were carried out in Toennies' group[9,10] for state resolved differential cross sections for both the inelastic and CT processes and in Ng's group for absolute state-to-state integral cross sections.[11-12]

The theoretical treatment of CT processes in ion (atom)–molecule (ion) collisions started about two decades ago. The most important contributions at that time were the trajectory surface-hopping model (TSHM) first suggested by Bjerre and Nikitin[13] and then so skillfully applied by Tully and Preston[14] and the multivibronic (curve) crossing model due to Bauer, Fischer, and Gilmore.[15] These two models are based on the Landau–Zener formula,[16] which yields the transition probability from one adiabatic curve to another. It is important to mention that this formula is not always capable of yielding the correct values.[17] The TSHM and other related topics are discussed in Chap. 6 by Chapman.[18] Another approach mainly pursued by Gislason, Parlant, and Sizum[19] is based on the classical path technique (for single-surface studies see review by Billing[20]).

The first quantum-mechanical studies for CT processes were carried out within the group of the present author about 15 years ago.[21,22] Those were done for the collinear reactive $(H + H_2)^+$ and $(Ar + H_2)^+$ systems. About 10 years later these studies were extended to three dimensions employing the infinite-order sudden approximation (IOSA).[23,24]

The basic theory for electronic transition in heavy-particle collisions (atoms and molecules) was given by Born and Oppenheimer.[25] The electronic nonadiabatic transitions occur due to the breakdown of the Born–Oppenheimer approximation. In general, one distinguishes between two types of nonadiabatic transitions: (1) those originating from the various relative motions of the heavy particles[16,26-30] and (2) those originating from the rotation of the body frame of the heavy particles with respect to a system of coordinates fixed in space.[31,32] In this review our main concern is with the first type of transition and therefore the theoretical part is devoted to this topic.

The subject of nonadiabatic transitions in atom–molecule collisions has been reviewed several times in the past,[33,34] but the quantum-mechanical aspect of it was mainly discussed by the present author[34a] and recently also by Sidis.[34b] The foremost difference between this and the previous review[34a] is that here we emphasize the (approximate) three-dimensional treatment, which also finally leads to numerical results that can be directly compared with experiment.

II. THEORY

A. The Schrödinger Equation

To treat electronic transitions in heavy-particle (atoms, molecules) collisions, we consider the Schrödinger equation, which describes the motion of the electrons and the nuclei, namely,

$$(T_n + H_e)\Psi = E\Psi. \tag{1}$$

Here H_e is the electronic hamiltonian of the electrons assuming the nuclei are fixed, T_n is the nuclear kinetic energy, and $\Psi(e, n)$ is the total wavefunction of both the electrons and the nuclei ('e' stands for the electronic coordinates and 'n' for the nuclear coordinates). The function $\Psi(e, n)$ can be expanded in the following form[25]:

$$\Psi = \sum_{i=1}^{N} \psi_i(n)\zeta_i(e; n), \tag{2}$$

where $\psi_i(n)$ are the nuclear wavefunctions and $\zeta_i(e; n)$, $i = 1 \dots N$, are a complete set of electronic wavefunctions that are solutions of the eigenvalue problem:

$$[H_e - V_i(n)]\zeta_i(e; n) = 0. \tag{3}$$

Here $V_i(n)$ are the corresponding electronic eigenvalues that are calculated for the given nuclear configuration n. Substituting Eq. (2) and (3) into Eq. (1), multiplying it from the left by an electronic basis function $\zeta_j(e; n)$, and integrating over the electronic coordinate e yields the Born–Oppenheimer set of coupled equations:

$$\left\langle \zeta_j \middle| T_n \sum_i \zeta_i \psi_i \right\rangle + (V_j - E)\psi_j = 0, \quad j = 1, \ldots, \tag{4}$$

where the bra and the ket notation is used for integration over electronic coordinates. To continue we have to be more specific. In the next section is presented the theory for a three nuclei system composed of two coordinates: the translational coordinate R and the (internal) vibrational coordinate r. For this sake the three particles are assumed to move along a straight line (the collinear system). The extension to a more complicated case (either more atoms and/or more internal coordinates) is usually straightforward (for the three-atom three-dimensional case see Ref. 34a). In Section II C we again discuss a two-coordinate system, but this time it will be for a three-dimensional system treated within the IOSA. This exceptional extension is done because, so far, all three-dimensional treatments were carried out within this framework.

B. The Collinear Case

1. *The Schrödinger Equation*

For the three-atom collinear case the kinetic energy operator can be presented as

$$T_n = -\frac{\hbar^2}{2} \nabla \cdot \nabla_m, \tag{5}$$

where

$$\nabla_m = \left(\frac{1}{m_R} \frac{\partial}{\partial R}, \frac{1}{m_r} \frac{\partial}{\partial r} \right) \tag{6a}$$

and

$$\nabla = \left(\frac{\partial}{\partial R}, \frac{\partial}{\partial r} \right), \tag{6b}$$

and the dot means scalar product. Here m_R and m_r are the translational and internal reduced masses of the interacting particles. Substituting Eq. (5) into Eq. (4) yields

$$\left(-\frac{\hbar^2}{2} \nabla_m \cdot \nabla + V_i(R, r) - E \right) \psi_i(R, r) - \hbar^2 \sum_{j=1}^{N} \tau_{ij}^{(1)} \cdot \nabla_m \psi_j - \frac{\hbar^2}{2} \sum_{j=1}^{N} \tau_{ij}^{(2)} \psi_j = 0, \tag{7}$$

where

$$\tau_{ij}^{(1)} = \langle \zeta_i | \nabla \zeta_j \rangle; \qquad \tau_{ij}^{(2)} = \langle \zeta_i | \nabla \cdot \nabla_{\widetilde{m}} \zeta_j \rangle. \tag{8}$$

Equation (7) can also be written in a form of a matrix equation:

$$\left[\nabla \cdot \nabla_m - \frac{2}{\hbar^2}(\mathbf{V} - E\mathbf{I}) \right] \mathbf{\Psi} + 2\tau \mathbf{\Psi} + \tau^{(2)} \mathbf{\Psi} = 0. \tag{7'}$$

Here $\mathbf{V}$ is a diagonal matrix that contains the (adiabatic) potential energy surfaces and $\tau^{(1)}$ is the (antisymmetric) matrix that contains the nonadiabatic coupling terms which are responsible for the transitions from one adiabatic potential surface to another. The matrix $\tau^{(2)}$ is also a (potential kind) matrix that couples the various adiabatic potentials, but is of much smaller elements compared to those of $\tau^{(1)}$.

2. The Adiabatic–Diabatic Transformation

The adiabatic equation as presented in Eq. (7) is probably the more efficient one for treating the heavy-particle motion. However, from various numerical studies it is known that numerical instabilities are encountered owing to the very abrupt behavior of the $\tau^{(1)}$ matrix elements. Therefore, it is most advisable before starting any numerical treatment to eliminate these troublesome elements. This can be done by the following transformation:

$$\mathbf{\Psi} = \mathbf{A}\mathbf{\eta} \tag{9}$$

Substituting Eq. (9) into Eq. (7) leads to the following expression:

$$\mathbf{A}(\nabla \cdot \nabla_m)\mathbf{\eta} + 2(\nabla \mathbf{A} + \tau^{(1)}\mathbf{A}) \cdot \nabla_m \mathbf{\eta} + (\tau^{(2)} + \nabla \cdot \nabla_m + 2\tau^{(1)} \cdot \nabla_m)\mathbf{A}\mathbf{\eta} - \frac{2}{\hbar^2}(\mathbf{V} - E\mathbf{I})\mathbf{A}\mathbf{\eta} = 0. \tag{10}$$

So far the matrix $\mathbf{A}$ is undetermined and therefore it will be chosen in such a way that the coefficients of $\nabla_m \mathbf{\eta}$ will vanish, namely, $\mathbf{A}$ will be assumed to fulfill the first-order differential equation

$$\nabla \mathbf{A} + \tau^{(1)}\mathbf{A} = 0. \tag{11}$$

Equation (11) and the ability to solve it is the bottleneck through which we have to pass if electronic transitions (including charge transfer) in heavy-particle collisions are to be studied.

In order to find the condition for having a solution, we write Eq. (11)

explicitly

$$\frac{\partial \mathbf{A}}{\partial R} + \tau_R^{(1)} \mathbf{A} = 0, \tag{12a}$$

$$\frac{\partial \mathbf{A}}{\partial r} + \tau_r^{(1)} \mathbf{A} = 0. \tag{12b}$$

Differentiating Eq. (12a) with respect to r and Eq. (12b) with respect to R and subtracting the two resulting equations leads to an important requirement to be fulfilled by $\tau_R^{(1)}$ and $\tau_r^{(1)}$, namely,

$$\frac{\partial \tau_R^{(1)}}{\partial r} - \frac{\partial \tau_r^{(1)}}{\partial R} = [\tau_R^{(1)}, \tau_r^{(1)}], \tag{13}$$

which can also be written as

$$\text{rot } \tau^{(1)} + [\tau^{(1)} \times \tau^{(1)}] = 0, \tag{14}$$

or

$$[\nabla \times \tau^{(1)}] + [\tau^{(1)} \times \tau^{(1)}] = 0, \tag{14'}$$

where the $\times$ sign stands for the vectorial product. (Here the vector product between two identical vector matrices is not necessarily zero.) Equation (13) is a *necessary* condition to be fulfilled by $\tau_R^{(1)}$ and $\tau_r^{(1)}$ in order for having a unique solution to Eq. (12).

The next question to be asked is whether $\tau_R^{(1)}$ and $\tau_r^{(1)}$ as defined in Eq. (8) really fulfill Eq. (13). The treatment of this problem will not be given here (see Ref. 35), but it can be shown that in order for Eq. (13) to be fulfilled, the size N of the electronic manifold has to be large enough so that the condition

$$\sum_{j=1}^{N} |\zeta_j\rangle\langle\zeta_j| = I \tag{15}$$

is satisfied.

In case N is large enough so that Eq. (15) is fulfilled, it can be shown[35] that $\mathbf{A}$ also satisfies the equation

$$(\tau^{(2)} + \nabla \cdot \nabla_m + 2\tau^{(1)} \cdot \nabla_m)\mathbf{A} = 0. \tag{16}$$

Considering, again, Eq. (10) and recalling Eqs. (11) and (16), we obtain that the equation for η reduces to

$$\mathbf{A}\nabla\cdot\nabla_m\eta - \frac{2}{\hbar^2}(\mathbf{V} - E\mathbf{I})\mathbf{A}\eta = 0. \tag{17}$$

Next we prove that $\mathbf{A}$ which fulfills Eq. (11) is a unitary matrix. The proof is based on the fact that $\tau^{(1)}$ is an antisymmetric matrix, namely,

$$\tau^{(1)\dagger} = -\tau^{(1)}. \tag{18}$$

Consequently, taking the transpose of Eq. (11) we get

$$\nabla\mathbf{A}^\dagger - \mathbf{A}^\dagger\tau^{(1)} = 0. \tag{11'}$$

Multiplying Eq. (11) from the left by $\mathbf{A}^\dagger$ and Eq. (11') from the right by $\mathbf{A}$ and adding the results, we obtain

$$\mathbf{A}^\dagger\nabla\mathbf{A} + (\nabla\mathbf{A}^\dagger)\mathbf{A} = 0 \tag{19}$$

or

$$\nabla(\mathbf{A}^\dagger\mathbf{A}) = 0 \Rightarrow \mathbf{A}^\dagger\mathbf{A} = \mathbf{C}, \tag{20}$$

where $\mathbf{C}$ is a constant matrix independent of R and r. Without losing generality, one may choose $\mathbf{A}$ at a given point (r_0, R_0) to be equal to unity, namely,

$$\mathbf{A}(r_0, R_0) = \mathbf{I}. \tag{21}$$

Consequently, $\mathbf{C}$ is equal to $\mathbf{I}$ and therefore, for every r and R, we have

$$\mathbf{A}^\dagger\mathbf{A} = \mathbf{A}\mathbf{A}^\dagger = \mathbf{I} \tag{22}$$

or in other words $\mathbf{A}$ is a unitary matrix at every point.

Returning now to Eq. (17) we multiply this equation from the left by $\mathbf{A}^\dagger$ and obtain the equation

$$\nabla_m\cdot\nabla\eta - \frac{2}{\hbar^2}(\mathbf{W} - E\mathbf{I})\eta = 0, \tag{23}$$

where $\mathbf{W}$ is an ordinary potential matrix

$$\mathbf{W} = \mathbf{A}^\dagger\mathbf{V}\mathbf{A}. \tag{24}$$

Thus one may conclude that the adiabatic–diabatic transformation shifts the nonadiabatic coupling terms into the potential matrix.

To solve Eqs. (12) we have to convert them into an integral equation. However, there are infinite numbers of these equations and we will mention two of them (here and in what follows we replace $\tau^{(1)}$ by τ):

$$\mathbf{A}(r, R) = \mathbf{A}(r_0, R_0) + \int_r^{r_0} \tau_r(r, R)\mathbf{A}(r, R)dr + \int_R^{R_0} \tau_R(r_0, R)\mathbf{A}(r_0, R)dR$$

(25a)

or

$$\mathbf{A}(r, R) = \mathbf{A}(r_0, R_0) + \int_R^{R_0} \tau_R(r, R)\mathbf{A}(r, R)dR + \int_r^{r_0} \tau_r(r, R_0)\mathbf{A}(r, R_0)dr.$$

(25b)

In order to show that Eqs. (25) are equivalent to Eq. (12) we consider, as an example, Eq. (25a).

The differentiation of Eq. (25a) with respect to r furnishes immediately Eq. (12b). The derivation of (12a) is somewhat more complicated. Differentiating Eq. (25a) with respect to R yields

$$\frac{\partial \mathbf{A}}{\partial R} = \int_r^{r_0} \frac{\partial}{\partial R}[\tau_r(R, r)\mathbf{A}(r, R)]dr - \tau_R(r_0, R)\mathbf{A}(r, R).$$

(26)

Recalling Eqs. (12) (and assuming $\mathbf{A}$ to be analytic), it can be shown that

$$\frac{\partial}{\partial R}[\tau_r(r, R)\mathbf{A}(r, R)] = \frac{\partial}{\partial r}[\tau_R(R, R)\mathbf{A}(R, R)].$$

(27)

Substituting Eq. (27) into Eq. (26) produces

$$\frac{\partial \mathbf{A}}{\partial R} = \tau_R(r, R)\mathbf{A}(r, R)|_r^{r_0} - \tau_R(r_0, R)\mathbf{A}(r_0, R)$$

(28)

or

$$\frac{\partial \mathbf{A}}{\partial R} + \tau_R \mathbf{A} = 0.$$

(12a')

A similar proof can be given for Eq. (25b).

Equation (25a) or (25b) are particularly convenient for the computation of $\mathbf{A}(r, R)$. Equation (25a), for instance, furnishes the value of $\mathbf{A}$ at $Q(r_1, R_1)$

once its value at $P(r_0, R_0)$ is known in two following steps (see Fig. 1a):

1. The derivation of an intermediate matrix $\mathbf{A}(r_0, R_1)$ is carried out by solving the integral equation.

$$\mathbf{A}(r_0, R_1) = \mathbf{A}(r_0, R_1) + \int_{R_1}^{R_0} \tau_R(r_0, R)\mathbf{A}(r_0, R)dR. \qquad (29a)$$

2. The derivation of $\mathbf{A}(r_1, R_1)$ is obtained by solving the equation:

$$A(r_1, R_1) = A(r_0, R_1) + \int_{r_1}^{r_0} \tau_r(r, R_1)A(r, R_1)dr \qquad (29b)$$

It is noted that the derivation of $\mathbf{A}(r_1, R_1)$ involves two line integrals (see Fig. 1a): one along the straight line $r = r_0$ from $R = R_0$ to $R = R_1$ and the second along the line $R = R_1$ from $r = r_0$ to $r = r_1$. Similarly, in case of Eq. (25b), we first integrate along the line $R = R_0$ from $r = r_0$ to $r = r_1$ and then along the line $r = r_1$ from $R = R_0$ to $R = R_1$.

This result can be generalized to any contour line that connects the two points $P(r_0, R_0)$ and $Q(r_1, R_1)$ (see Fig. 1a). If $\mathbf{ds}$ is a (directed) length element along a given contour line $\Gamma(s)$, then Eq. (11) can be written in the form

$$d\mathbf{A} = -\tau \cdot \mathbf{ds}\,\mathbf{A}. \qquad (30)$$

Integrating both sides produces the result

$$\mathbf{A}(s_1) = \mathbf{A}(s_0) + \int_{s_1}^{s_0} \tau \cdot \mathbf{ds}\,\mathbf{A}(s), \qquad (31)$$

where s_i, $i = 0, 1$, stand for (r_i, R_i) respectively. It is well noted from Fig. 1a that each two paths form a contour. The fact that the result, at the end point of each path, does not depend on the path implies that any integral along a complete contour is zero.

The theory presented so far was based on the assumption that τ_R and τ_r fulfill Eq. (13). The fulfillment of Eq. (13) is guaranteed as long as the electronic basis set comprises all the electronic wavefunctions. In what follows we discuss the situation when this requirement is not fulfilled.

3.   The Quasi-Adiabatic–Diabatic Transformation

It is not realistic to expect that in a numerical treatment a complete adiabatic manifold will be used. In fact, so far, in all applications only two (adiabatic) states were included. In such cases Eq. (13) is not always fulfilled and therefore

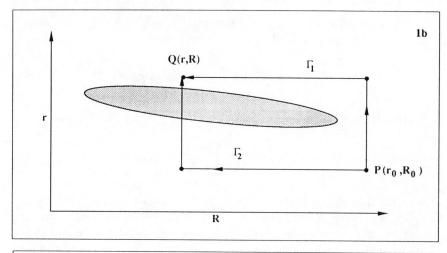

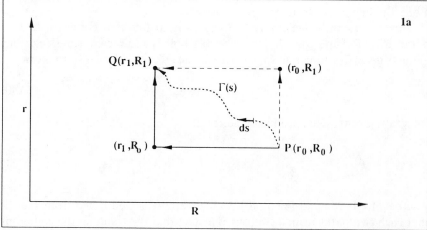

Figure 1. (a) Three different paths that combine the two points P and Q; two of them are composed for straight lines parallel to the R and r axes. Along the third, which is an arbitrary path, we designated the length element **ds**. (b) Two different paths which combine the points P and Q; one avoiding the shadowed area and the other crossing it. The shadowed area designates the region for which the completeness requirement is not fulfilled.

the solutions as obtained in Eq. (25) or (31) may not be unique. This problem was discussed in a series of papers[36,37] and in a few of them also recipies were given how, partially, to resolve this difficulty. In what follows we first discuss an interesting idea by Pacher et al.[37] and then present our approach.

The basic observation by Pacher et al.[37] is that in the ideal case when Eq. (13) is fulfilled the transformed nonadiabatic coupling terms, $\tilde{\tau}$ given in the form

$$\tilde{\tau} = \mathbf{A}^\dagger \tau \mathbf{A} + \mathbf{A}^\dagger \nabla \mathbf{A} \tag{32}$$

are identically zero. This requirement is not always fulfilled in the reduced space. Consequently, they propose to define quasidiabatic states by demanding that the integral of the Euclidean norm for matrices

$$\| \tilde{\tau} \|^2 = \mathrm{Tr}(\tilde{\tau}^\dagger \cdot \tilde{\tau}) \tag{33}$$

over the nuclear coordinates

$$Q = \int dV \| \tilde{\tau} \|^2 \tag{34}$$

is minimal. This requirement led to an interesting equation for $\mathbf{A}$:

$$\nabla \cdot \nabla \mathbf{A} + \mathbf{A} \nabla \mathbf{A}^\dagger \cdot \nabla \mathbf{A} + \mathbf{A} \nabla \cdot \mathbf{A}^\dagger \tau \mathbf{A} = 0. \tag{35}$$

This equation, as they put it, is too complicated for practical use. Moreover, since no analysis is given, the nature of the solution of such an equation is not clear. Some insight can be gained by assuming $\mathbf{A}$ to be *unitary*; then it can be shown, following some simple algebra, that is equation reduces to

$$(\nabla + \mathbf{A} \nabla \mathbf{A}^\dagger) \cdot (\nabla \mathbf{A} + \tau \mathbf{A}) = 0. \tag{36}$$

From this representation it is well noted that every solution of Eq. (11) is also a solution of Eq. (36). Thus, it is not obvious that the uniqueness of the quasidiabatic representation is guaranteed by this equation.

Here we would like to suggest a different approach, namely, to try and modify the τ matrices so that a unique solution is obtained. As an example we consider the two-surface case for which the condition for a unique solution is

$$\frac{\partial \tau_r}{\partial R} - \frac{\partial \tau_R}{\partial r} = 0. \tag{37}$$

Since the reduced representation of τ_r and τ_R does not satisfy this relation, we may modify either τ_r or τ_R (or even both) so that Eq. (37) is fulfilled. For instance, we introduce $\tilde{\tau}_R$:

$$\tilde{\tau}_R = \tau_R + \varepsilon \qquad (38)$$

so that

$$\frac{\partial \tau_r}{\partial R} - \frac{\partial \tilde{\tau}_R}{\partial r} = 0. \qquad (39)$$

Substituting Eq. (38) into Eq. (39) yields an equation for $\varepsilon(r, R)$:

$$\frac{\partial \varepsilon}{\partial r} = \frac{\partial \tau_r}{\partial R} - \frac{\partial \tau_R}{\partial r}, \qquad (39')$$

which can be solved for each value of R

$$\varepsilon(r, R) = \varepsilon(r_0, R) - [\tau_R(r, R) - \tau_R(r_0, R)] + \int_r^{r_0} dr \frac{\partial \tau_r}{\partial R}. \qquad (40)$$

Now, if along $r = r_0$ Eq. (37) is satisfied, this implies that

$$\varepsilon(r_0, R) = 0, \qquad (41)$$

and, consequently, the modified value of $\tau_R(r, R)$, that is, $\tilde{\tau}_R(r, R)$, is given in the form

$$\tilde{\tau}_R(r, R) = \tau_R(r_0, R) + \int_{r_0}^r dr \frac{\partial \tau_r(r, R)}{\partial R}. \qquad (42)$$

If Eq. (37) is satisfied along the line $R = R_0$ (instead of along the line $r = r_0$), then $\tau_r(r, R)$ will be modified so that:

$$\tilde{\tau}_r(r, R) = \tau_r(r, R_0) + \int_{R_0}^R dR \frac{\partial \tau_R(r, R)}{\partial r}. \qquad (42')$$

In many cases we may not need to do any corrections. The reason being that Eq. (31) is usually fulfilled except at regions where the nonadiabatic coupling is large. This happens in small regions of space configurations where the adiabatic surfaces come close to each other. If that is the case, then while applying the line integral to calculate **A** [eqs. (25) or (31)] one may distinguish between "safe" paths and "unsafe" paths. The "safe" paths are those along which Eq. (37) is fulfilled and the "unsafe" ones are those along which Eq. (37) is not always satisfied. Two such paths are shown in Fig. 1b, where the shadowed area is the region of the strong nonadiabatic couplings [where Eq. (37) is not expected to be satisfied]. Thus, it is noted that in order to derive the value of $A(r, R)$ at $Q(r, R)$ it is best to follow path Γ_1, whereas complications are expected when following path Γ_2.

In summary it seems rather difficult to give a general solution for the nonuniqueness of A. Each system has to be analyzed separately and, accordingly, the path for solving A has to be chosen. In most cases the calculation of A will be carried out using τ_R and τ_r as such. In a few others they may have to be modified as suggested in Eq. (42) or (42').

4. The Two-Surface Case

The two-surface case is of particular interest because the representation of A becomes very simple. Since A is unitary, it can be written as

$$A = \begin{pmatrix} \cos \alpha & -\sin \alpha \\ \sin \alpha & \cos \alpha \end{pmatrix}, \tag{43}$$

where $\alpha = \alpha(r, R)$. For this case τ_R and τ_r have the form

$$\tau_x = \begin{pmatrix} 0 & \tau_x \\ -\tau_x & 0 \end{pmatrix}, \quad x = R, r, \tag{44}$$

and it can be shown that in this case τ_R and τ_r commute. Substituting Eqs. (43) and (44) into Eqs. (12) yields two equations for α:

$$\frac{\partial \alpha}{\partial r} = \tau_r, \quad \frac{\partial \alpha}{\partial R} = \tau_R, \tag{45}$$

and the solution of these equations can be presented as line integrals along various paths. For instance,

$$\alpha(r, R) = \alpha(r_0, R_0) + \int_{R_0}^{R} \tau_R(r_0, R)dR + \int_{r_0}^{r} \tau_r(r, R)dr. \tag{46}$$

Thus the main simplification in the two-state case is that instead of solving integral equations we compute integrals.

Having calculated α we can obtain the diabatic potential matrix elements

W [see Eq. (24)]:

$$W_{11} = V_1 \cos^2 \alpha + V_2 \sin^2 \alpha, \tag{47a}$$

$$W_{22} = V_1 \sin^2 \alpha + V_2 \cos^2 \alpha, \tag{47b}$$

$$W_{12} = W_{21} = \tfrac{1}{2}(V_2 - V_1)\sin 2\alpha. \tag{47c}$$

The resulting diabatic surfaces are expected to cross each other along a line (called *seam*), which can be represented as

$$r = r(R). \tag{48}$$

The equation of the *seam* can be calculated from the equation

$$w(r(R), r) = 0, \tag{49}$$

where

$$w(r, R) = W_{22}(r, R) - W_{11}(r, R). \tag{50}$$

Next we briefly study the various functions in the vicinity of the *seam*. To do that we express the angle α in terms of $w(r, R)$. Substracting Eq. (47a) from (47b) yields

$$w = (V_2 - V_1)\cos 2\alpha, \tag{51}$$

and taking the ratio between Eqs. (47c) and (51) leads to

$$\alpha = \tfrac{1}{4}\pi + \tfrac{1}{2}\tan^{-1}(w(r, R)/2W_{12}(r, R)). \tag{52}$$

Since at $r = r(R)$ the value of $w(r, R)$ is zero, we obtain the value of $w(r, R)$ at the vicinity of $r(R)$ as

$$w'(r, R) = [r - r(R)]\frac{\partial w}{\partial r}\Bigg|_{r = r(R)}. \tag{53}$$

Defining now $\lambda(R)$ as

$$\lambda(R) = 2W_{12}(r(R), R)/w'_r(r(R), r), \tag{54}$$

where $w'_r = (\partial w/\partial r)$ and assuming that $W_{12}(r(R), R)$ in the vicinity of $r(R)$ is at most, weakly dependent on r, we obtain for $\alpha(r, R)$ [see Eq. (52)] at the vicinity of the *seam* that

$$\alpha(r, R) = \tfrac{1}{4}\pi + \tfrac{1}{2}\tan^{-1}(r - r(R))/\lambda(R)). \tag{55}$$

Consequently, the corresponding expression for τ_r is

$$\tau_r = \frac{\partial \alpha}{\partial r} = \frac{1}{2} \frac{\lambda(R)}{\lambda(R)^2 + (r - r(R))^2}. \tag{56}$$

A similar treatment can be carried out if the *seam* is presented as

$$R = R(r). \tag{48'}$$

However, in this case we obtain the translational nonadiabatic coupling term τ_R:

$$\tau_R = \frac{\partial \alpha}{\partial R} = \frac{v(r)}{v(r)^2 + (R - R(r))^2}, \tag{56'}$$

where

$$v(r) = 2W_{12}(r, R(r))/w'_R(r, R(r)) \tag{54'}$$

and $w'_R(r, R) = (\partial w/\partial R)$.

Thus, in general, if $\kappa(x)$ is defined as either $\lambda(R)$ or $v(r)$ and (x, y) are either (R, r) or (r, R), then

$$\tau_y = \frac{1}{2} \frac{\kappa(x)}{\kappa(x)^2 + (y - y(x))^2}. \tag{57}$$

It happens that at $x = x_s$ the value of $W_{12}(x, y(x))$ becomes zero, which implies that $\kappa(x)$ also becomes zero. Then in the vicinity of $x = x_s$ we have

$$\lim_{x \to x_s} \tau_y(x, y) = \frac{1}{2} \lim_{x \to x_s} \frac{\kappa}{\kappa^2 + (y - y(x))^2} \tag{58}$$

or applying a definition of the Dirac δ function leads to

$$\lim_{x \to x_s} \tau_y(x, y) = \tfrac{1}{2}\pi\delta(y - y(x_s)). \tag{59}$$

This result implies that at every point along the *seam* where the diabatic coupling term becomes zero at least one nonadiabatic coupling term becomes a Dirac δ function.

A very familiar case of a nonadiabatic coupling term becoming a Dirac δ function is encountered in treating charge-transfer processes in atom–diatom collisions. Here the reagents are either (A^+, BC) or (A, BC^+) and therefore in many cases a well-defined *seam* is encountered in the asymptotic and near-asymptotic region. The *seam*, in this case, is a straight line along the R axis. Consequently, the vibrational nondadiabatic coupling term $\tau_r(r, R)$ takes the form

$$\lim_{R \to \infty} \tau_r(r, R) = \tfrac{1}{2}\pi\delta(r - r_s), \tag{56''}$$

where $r = r_s$ is the location of the seam.

As for $\tau_R(r, R)$ it can be shown that

$$\lim_{R \to \infty} \tau_R(r, R) = 0. \tag{60}$$

This behavior of the two nonadiabatic coupling terms as $R \to \infty$ yields the following values for $\alpha(r, R)$:

$$\lim_{R \to \infty} \alpha(r, R) = \begin{cases} 0, & r < r_s, \\[2mm] \dfrac{\pi}{4}, & r = r_s, \\[2mm] \dfrac{\pi}{2}, & r > r_s. \end{cases} \tag{61}$$

The asymptotic behavior of $\alpha(r, R)$ guarantees the correct asymptotic limit of the potential, namely, it leads to the correct diatomic potentials:

$$\lim_{R \to \infty} W_1(r, R) = \begin{cases} V_1(r), & r < r_s, \\ V_2(r), & r > r_s, \end{cases} \tag{62}$$

$$\lim_{R \to \infty} W_2(r, R) = \begin{cases} V_2(r), & r > r_s, \\ V_2, & r < r_s, \end{cases} \tag{63}$$

and $W_{12} \equiv 0$.

It is noted that in this case W_1 and W_2 are the potentials of BC and BC^+, respectively.

C. The Diatomic-in-Molecules Potential

In contrast to *ab initio* treatments that always yield adiabatic information (adiabatic potential energy surfaces or adiabatic electronic wavefunctions),[38] the semiempirical methods usually deliver diabatic information, in particular, the diabatic potential energy matrix. Here the diagonal elements serve as the diabatic surfaces and the off-diagonal elements are the diabatic coupling terms. The most common method in this sense is the diatomics-in-molecules (DIM) method, which is now frequently used.[39] Here one constructs a potential matrix with dimensions equal to the number of electronic states that are assumed to be relevant. Usually the dimensionality of this matrix is too large to be applied directly in a scattering treatment, and the question is whether the dimensionality can be reduced in an inherent consistent way. It turns out that this can be achieved by transforming to the corresponding adiabatic representation. Since the adiabatic surfaces can be ordered according to their energy values, the reduction within this framework is straightforward. Whereas the derivation of the adiabatic potential energy surfaces is achieved by diagonalizing the diabatic potential matrix, the derivation of the nonadiabatic coupling term is some what more complicated. The latter can be obtained either by employing the Hellman–Feymnan theorem[14a,34a] or by other methods. Let us consider the first method.

If $\mathbf{W}$ and $\mathbf{V}$ are the diabatic and the adiabatic potential matrices, then the two are related by [see Eq. (24)]

$$\mathbf{W} = \mathbf{A}^\dagger \mathbf{V} \mathbf{A}, \tag{24'}$$

where $\mathbf{A}$ is the unitary transformation matrix. Our aim is now to calculate τ employing Eq. (11). Differentiating Eq. (24') and applying Eq. (11) we obtain

$$\mathbf{A}(\nabla \mathbf{W})\mathbf{A}^\dagger = \nabla \mathbf{V} + \mathbf{V}\tau - \tau \mathbf{V}. \tag{64}$$

If $\mathbf{S}$ is defined as the sum of the three products on the r.h.s. of Eq. (64), then it can be shown that the elements of $\mathbf{s}$, that is, s_{lm}, are

$$S_{lm} = \begin{cases} \nabla V_{ll} & l = m, \\ (V_l - V_m)\tau_{lm}, & l \neq m. \end{cases} \tag{65}$$

In the same way, if $\bar{\mathbf{S}}$ is defined as the product appearing on the l.h.s. of Eq. (64), then

$$\bar{S}_{lm} = \mathbf{A}_l \nabla \mathbf{W} \mathbf{A}_m^\dagger, \tag{66}$$

where A_l and $A_m^\dagger$ are the lth and the mth row of $\mathbf{A}$. Equating the two expressions, namely, Eq. (65) and (66), one finds

$$\tau_{lm} = \frac{A_l(\nabla W)A_m^\dagger}{V_l - V_m}. \tag{67}$$

Here τ_{lm} is a vector with the components

$$\tau_{plm} = \frac{A_l(\partial W/\partial p)A_m^\dagger}{V_l - V_m}, \quad p = r, R, \ldots. \tag{68}$$

A simpler expression can be obtained by making use of Eq. (11), namely,[34a]

$$\tau = -(\nabla A)A^\dagger, \tag{69}$$

or if a moresymmetrical representation is desired, τ may also be written

$$\tau = \tfrac{1}{2}(A\nabla A^\dagger - (\nabla A)A^\dagger). \tag{70}$$

Once the nonadiabatic coupling terms, in addition to the adiabatic surfaces, are also known, one is able to transform to a diabatic representation of a *reduced* dimensionality. The back transformation is achieved by calculating

$$W_{re} = A_{re}V_{re}A_{re}^\dagger \tag{71}$$

where A_{re} is the solution of the equation

$$\nabla A_{re} + \tau_{re}A_{re}^\dagger = 0. \tag{11'}$$

Here, both V_{re} and τ_{re} have dimensions that are smaller than those of V and τ but contain the same corresponding elements.

4.   *The Treatment of Reactive Systems*

In this chapter we concentrate on two types of systems: reactive and nonreactive. To treat a reactive system in its full dimensionality (including charge transfer) is a very time-consuming process. For this purpose we recently introduced a new method to treat reactive systems that is based on employing short-range negative imaginary potentials [which are termed as negative imaginary arrangement decoupling potentials (NIADP)] to convert a multiarrangement system into a quasiinelastic single-arrangement system.[40] It is found that by doing so one obtains not only the correct inelastic transition probabilities but also the correct total (and sometimes even state-to-state) transition probabilities for the various ignored product-arrangement channels. This conversion is done by substituting these potentials at the entrances of all arrangement channels. However, they must be located deep enough into each arrangement channel so that they absorb outgoing fluxes only. A valuable feature of the NIADPs is that they do not affect the wavefunction in the regions external to them, but once the wavefunction passes through them, it decays to zero within a short distance. The fact that the product-arrangement channels can be so harmless eliminated enables a reactive system to be treated as an inelastic one and, consequently, the various methods and approximations developed for the purpose of treating inelastic systems can now be used on reactive systems without any difficulties.

To show how this conversion is done we consider a collinear reactive system (see Fig. 2) where the NIADP $u_I(r, R)$ is located at the entrance to the reactive arrangement—the shadowed area in the figure. We found that a convenient form for such a potential is

$$u_I(r, R) = \begin{cases} -iu_{I0}\dfrac{r - r_{1I}}{r_{2I} - r_{1I}}, & r_{1I} \leqslant r \leqslant r_{2I}, \\ 0, & \text{otherwise,} \end{cases} \tag{72}$$

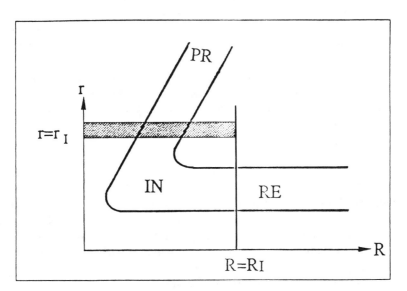

Figure 2. The collinear-type mass-scaled potential energy surface. The negative imaginary arrangement decoupling potential $u_I(r, R)$ is located in the shadowed area along the line $r = r_I$.

where the range for R is essentially unlimited and u_{I0} must fulfill the two inequalities[40]

$$\hbar E_{CM}^{1/2}/(\Delta r_I \sqrt{8m}) \ll u_{I0} \ll \Delta r_I \sqrt{8m} E_{CM}^{3/2}/\hbar. \tag{73}$$

Here E_{CM} is the translational energy of the particle interacting with the NIADP, m is its mass and Δr_I is the NIADPs nonzero range:

$$\Delta r_I = r_{2I} - r_{1I}. \tag{74}$$

In Eq. (73) the left-hand inequality guarantees that all the flux passing through this potential is absorbed and the right-hand inequality guarantees that no flux is reflected while passing through it. Once the parameters of this potential are determined (they can be determined *a priori*), the NIADP is added to the ordinary potential, which governs the motion of the interacting particles and considers the sum as an ordinary potential for an inelastic system.

E. The Three-Dimensional System—The Introduction of the Infinite-Order Sudden Approximation

In this chapter we do not intend to analyze the general three-dimensional system (this was done by us on several other occasions,[34a,41]) because no numerical applications are available for this case. To our knowledge most of the three-dimensional numerical studies were carried out within the framework of the IOSA[12,23,24,42–46] and therefore we will concentrate on this approximation.

The general conditions for the validity of the IOSA were presented and discussed extensively elsewhere[47] and will not be repeated here. The ion–molecule systems usually differ significantly from the atom–molecule systems, because the potential is much less dependent on the orientation angle γ[23] defined as

$$\gamma = \cos^{-1}(\hat{R} \cdot \hat{r}). \tag{75}$$

Thus, the IOSA, which becomes more relevant the weaker the dependence of the potential on γ, is expected to yield more reliable results for charged systems than for neutral ones. It is important to emphasize that the IOSA becomes exact for isotropic potentials being either rigid-rotor-type or breathing-sphere-type potentials.

As was shown in Section II B 4 the most important function for the adiabatic–diabatic transformation for a Born–Oppenheimer two-surface system is the angle α given in Eq. (46). One way of extending the two-dimensional expression to the three-dimensional case is

$$\alpha(r, R, \gamma) = \alpha_{00}(r_0, R_0, \gamma_0) + \int_{r_0}^{r} \tau_r(r, R, \gamma) dr$$

$$+ \int_{R_0}^{R} \tau_R(r_0, R, \gamma) dR + \int_{\gamma_0}^{\gamma} \tau_\gamma(r_0, R_0, \gamma) d\gamma, \qquad (76)$$

where $\alpha_{00}(r_0, R_0, \gamma_0)$ is some initial α value and τ_γ is

$$\tau_\gamma = \left\langle \zeta_1 \left| \frac{\partial}{\partial \gamma} \right| \zeta_2 \right\rangle. \qquad (77)$$

Since the IOSA is essentially a "frozen" γ dynamics, it will be convenient for our purposes to define for each γ an intermediate angle $\alpha_0(r_0, R_0, \gamma)$:

$$\alpha_0(r_0, R_0, \gamma) = \alpha_{00}(r_0, R_0, \gamma_0) + \int_{\gamma_0}^{\gamma} \tau_\gamma(r_0, R_0, \gamma) d\gamma, \qquad (78)$$

so that $\alpha(r, R, \gamma)$ is given in form

$$\alpha(r, R, \gamma) = \alpha_0(r_0, R_0, \gamma) + \int_{r_0}^{r} \tau_r(r, R, \gamma) dr + \int_{R_0}^{R} \tau_R(r_0, R, \gamma) dR. \qquad (79)$$

Equation (79) is the one to be employed within the charge-transfer IOSA (CT-IOSA). The corresponding CT-IOSA equation to Eq. (23) is given in the form[23,24]

$$\left[-\frac{\hbar^2}{2m_R} \frac{\partial^2}{\partial R^2} - \frac{\hbar^2}{2m_r} \frac{\partial^2}{\partial r^2} + \frac{\hbar^2}{2m_R} \frac{l(l+1)}{R^2} + \frac{\hbar^2}{2m_r} \frac{j(j+1)}{r^2} + W_{11} - E \right] \eta_1 + W_{12}\eta_2 = 0,$$

$$(80)$$

$$\left[-\frac{\hbar^2}{2m_R} \frac{\partial^2}{\partial R^2} - \frac{\hbar^2}{2m_r} \frac{\partial^2}{\partial r^2} + \frac{\hbar^2}{2m_R} \frac{l(l+1)}{R^2} + \frac{\hbar^2}{2m_r} \frac{j(j+1)}{r^2} + W_{22} - E \right] \eta_2 + W_{12}\eta_1 = 0,$$

where l and j are the orbital and the internal angular momentum quantum numbers assumed to be conserved during the collision (the actual numerical treatment is carried out for $j = 0$). In Eq. (80) all three matrix elements W_{11}, W_{22}, and W_{12} are functions of r, R, and γ, but whereas r and R are variables, γ is a parameter. Consequently, the η_i functions are also parametrically dependent on γ. The final physical magnitudes (as will be shown) are obtained following an integration over the angle γ.

To treat a reactive system all we have to do (as was discussed in the previous section) is to add NIADP's to the ordinary potential. In case the two surfaces are reactive, the W_{ii}, $i = 1, 2$, in Eq. (80) are replaced by[42,43,45]

$$W_{iic} = W_{ii} + u_{Ii} \tag{81}$$

where

$$u_{Ii} = u_{Ii}(r, R; \gamma). \tag{82}$$

However, if only the lower surface is reactive, then W_{22} is left unchanged and only W_{11} is modified accordingly. Consequently, Eq. (80) becomes (assuming also that $j = 0$)

$$\left[-\frac{\hbar^2}{2m_R} \frac{\partial^2}{\partial R^2} - \frac{\hbar^2}{2m_r} \frac{\partial^2}{\partial r^2} + \frac{\hbar^2}{2m_R} \frac{l(l+1)}{R^2} + W_{11c} - E \right] \eta_1 + W_{12} \eta_2 = 0,$$

$$\left[-\frac{\hbar^2}{2m_R} \frac{\partial^2}{\partial R^2} - \frac{\hbar^2}{2m_r} \frac{\partial^2}{\partial r^2} + \frac{\hbar^2}{2m_R} \frac{l(l+1)}{R^2} + W_{22} - E \right] \eta_2 + W_{12} \eta_1 = 0. \tag{83}$$

Equations (83) were used (with some modifications that will be discussed) to calculate reactive cross sections for the $(Ar + H_2)^+$ system.[42]

The solution of Eqs. (83) yields S matrix elements of the form $S(E, \gamma, l | q_i, v_i, q_f, v_f)$, where E, γ, and l were introduced earlier, q_i and q_f stand for the initial and final electronic states, respectively, and v_i and v_f are the corresponding initial and final vibrational states. The fact that the interaction potential contains an imaginary component does not affect the hermitian property of the S matrix, but it may destroy its unitarity. The nonunitarity is related to the fact that the interaction may lead to exchange. Consequently, $P_r(E, \gamma, l | q_i, v_i)$, the reaction probability for a given initial state (q_i, v_i), is calculated from the expression

$$P_r(E, \gamma, l | q_i, v_i) = 1 - \sum_{q_f, v_f} |S(E, \gamma, l | q_i, v_i, q_f, v_f)|^2, \tag{84}$$

where the summation is carried out with respect to all (asymptotic) open states.

In this work we consider four types of cross sections.

1. Differential total (nonreactive) cross sections:[23,24]

$$\frac{d\sigma(E | q_i, v_i, q_f)}{d\Omega} = \frac{1}{8k_{q_i v_i}^2} \sum_{l=0}^{l_m} \sum_{l'=0}^{l_m} (2l+1)(2l'+1) P_l(\cos\theta) P_{l'}(\cos\theta),$$

$$\sum_{v_f} \int_{-1}^{+1} d(\cos\gamma) S(E, \gamma, l | q_i, v_i, q_f, v_f) S^*(E, \gamma, l' | q_i, v_i, q_f, v_f). \tag{85}$$

Here $P_l(\cos\theta)$ is the lth Legendre polynomial, θ is the scattering angle, and $k_{q_i v_i}$ is the wave vector defined as

$$k_{q_i v_i} = \sqrt{2\mu(E - E_{q_i v_i})}/\hbar, \tag{86}$$

where $E_{q_i v_i}$ is the initial vib-electronic eigenvalue.

2. State-to-state (nonreactive) integral cross sections:

$$\sigma(E|q_i, v_i, q_f, v_f) = \frac{\pi}{2k_{q_i v_i}^2} \sum_{l=0}^{l_m} (2l+1) \int_{-1}^{+1} d(\cos\gamma) |S(E, \gamma, l|q_i, v_i, q_f, v_f)|^2 \tag{87}$$

and integral total reactive cross sections[42,43]

$$\sigma_r(E|q_i, v_i) = \frac{\pi}{k_{q_i v_i}^2} (l_m + 1)^2 - \sum_{q_f v_f} \sigma(E|q_i, v_i, q_f, v_f), \tag{88}$$

where $\sigma(E|q_i, v_i, q_f, v_f)$ is defined in Eq. (87)

3. Initial state (nonreactive) opacity functions:

$$P(E, l|q_i, v_i, q_f) = \sum_{v_f} \frac{1}{2} \int_{-1}^{+1} d(\cos\gamma) |S(E, \gamma, l|q_i, v_i, q_f, v_f)|^2 \tag{89}$$

and the corresponding reactive functions[42,43]

$$P_r(E, l|q_i, v_i) = 1 - \sum_{q_f} P(E_t, l|q_i, v_i, q_f), \tag{90}$$

where $P(E_f, l|q_i, v_i, q_f)$ is defined in Eq. (89).

4. γ-dependent initial state (nonreactive) cross sections:

$$\sigma(E, \gamma|q_i, v_i, q_f) = \frac{\pi}{k_{q_i v_i}^2} \sum_{l=0}^{l_m} (2l+1) \sum_{v_f} |S(E, \gamma, l|q_i, v_i, q_f, v_f)|^2 \tag{91}$$

and the corresponding reactive cross sections:

$$\sigma_r(E, \gamma|q_i, v_i) = \frac{\pi}{k_{q_i v_i}^2} (l_m + 1)^2 - \sum_{q_f} \sigma(E, \gamma|q_i, v_i, q_f), \tag{92}$$

where $\sigma(E, \gamma|q_i, v_i, q_f)$ is defined in Eq. (91).

This completes our theoretical section.

III. STUDIES OF SPECIFIC SYSTEMS

A. Background

The first quantum-mechanical studies of CT processes between an atom (ion) and a diatomic molecule were carried out for the two reactive systems $(H_2 + H)^+$ and $(Ar + H_2)^+$ assuming the three atoms to be aligned, that is, the system to be collinear.[21,22] The extension to three dimensions was done about a decade later employing the IOSA.[23,24] Here we distinguish between reactive (exchange) and inelastic treatments. The inelastic studies were carried out for relative high energies ($\geqslant 20$ eV), and the systems that were considered were $(H_2 + H)^+$,[23] $(O_2 + H)^+$,[44] and $(Ar + H_2)^+$.[12] In all three cases the reactive channels were ignored and the justification for this is that at such high energies exchange processes usually do not take place. The only low-energy study of this kind was carried out for the reactive $(Ar + H_2)^+$ and here we distinguish between two studies: (1) the earlier one[23] for which the ordinary reactive IOSA was employed[48] but without satisfactory results and (2) the latter[42] in which the new IOSA with NIADP[45] is used and for which much better results were obtained. The latter treatment, in contrast to the previous one, also utilizes three surfaces, which is the minimal number of surfaces required to obtain meaningful results that can be compared with experiment.

To our knowledge there exists only one additional quantum-mechanical CT treatment that is based on the coupled-states-distorted-wave Born approximation.[49] As the results of this study are presented elsewhere in this volume,[19] they will not be discussed here.

In what follows we mainly present results for two systems, the $H_2 + H^+$ system and its isotopic analogs and the $(Ar + H_2)^+$ system. A few results only will be shown for $O_2 + H^+$ as this system is discussed extensively in other chapters (see Chapters 3 , 4, and 5).

B. The $H_2(D_2, HD) + H^+$ Systems

The study of the $H_2 + H^+$ system is motivated by the experimental available vibrational resolved differential cross sections[9] for the inelastic and charge transfer channels for the collision processes:

$$H_2(v = 0) + H^+ \rightarrow \begin{cases} H_2(v') + H^+, & \Delta E_0 = 0, & \text{(93a)} \\ H + H_2^+(v^+), & \Delta E_0 = 1.83 \text{ eV}. & \text{(93b)} \end{cases}$$

The experiments were performed at an energy of E_{Lab} ($\equiv \frac{3}{2} E_{CM}$) = 30 eV and encompassed scattering angles up to $20°$. At these high collision energies the reactive (exchange) process has a very low probability[9] and can be completely neglected.

With only two electrons, this is the simplest of all ion–molecule collision systems. The potential hypersurface has extensively been studied, and it has been found that the computationally convenient DIM potential is reasonably accurate.[14a,39] This potential was used by Niedner et al.[9] within the framework of the TSHM to calculate differential and total cross sections for direct comparison with the preceding experiment.

1. *Differential Cross Sections*

Differential cross sections for the *vibrationally inelastic* channel were calculated for the ground potential surface and compared with the experiment. The results are shown in Fig. 3a for $v' = 0, 2, 4$, and 6 in Fig. 3b for $v' = 1, 3, 5$ and for the "total." Since the experiment only provides relative cross sections, they were arbitrarily matched to the theoretical total cross section by setting the experimental "total" cross section equal to the theoretical cross section at $\theta_{CM} = 11°$. The agreement between the theoretical calculations and the experiments is quite good. The IOSA treatment predicts the correct relative magnitude of the inelastic cross section all the way up to $v' = 6$. If shifted by 2.5° to smaller angles, the IOSA also nicely reproduces the overall shape of the cross section at angles smaller than the rainbow and even suggests that some of the predicted system undulation may also be present in the experimental distributions. The main observed discrepancy between theory and experiment is with respect to the position of the rainbow angle. It is noted that the rainbow angles calculated with IOSA are all shifted by about 2.5° toward larger angles.

Differential quantum-mechanical cross sections for vibrational resolved CT processes are presented in Fig. 3c, where they are compared with the experiment. It is seen that the relative cross sections are once more nicely reproduced by the IOSA calculations. However, at small angles $\theta \leqslant 3°$, there is a serious disagreement between theory and experiment, where the present IOSA predicts an increase in cross sections whereas the experiment shows a fall off. As for the CT rainbow, it is well reproduced by the IOSA treatment. However, as in the inelastic case, it is shifted by 2.5° to larger angles.

Total differential cross sections for CT are presented in Fig. 3d, where the IOSA results are compared with both experiment and TSH results. It is well noted that whereas the IOSA produces the expected rainbow structure the TSH fails to do so. For the comparison with the experimental cross sections, which are only relative, an arbitrary normalization had to be introduced. In this case the magnitude of the experimental cross section was matched to the IOSA cross section at the rainbow angle. Since both theories provide

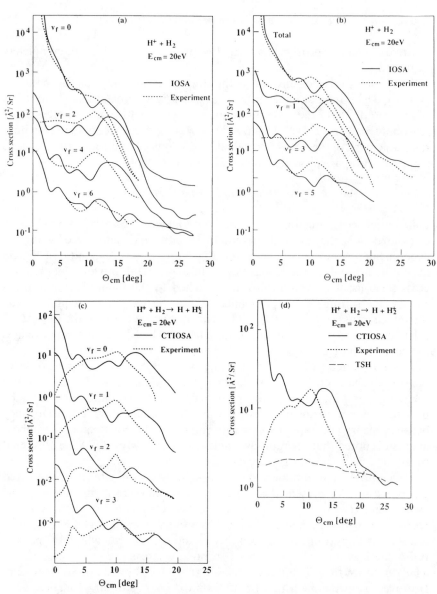

Figure 3. Vibrational inelastic (elastic) and CT differential cross sections; comparison between theory and experiment: ———, theory, ⋯, experiment. The normalization at each case is done by making the experimental "total" curve equal to the corresponding theoretical curve at its rainbow angle (see Fig. 3b for the inelastic case and Fig. 3d for the CT case). (a) Vibrational state-to-state inelastic (elastic) and total differential cross sections; even transitions. (b) Like (Fig. 3a), but for the odd transitions. (c) State-to-state CT differential cross sections. (d) Total CT differential cross sections. A comparison between theoretical (IOSA and TSH) and experimental results. The experimental curve is shifted in such a way that its maximum coincides with the IOSA maximum at their corresponding rainbow angles. The TSH curve is scaled correctly with the IOSA curve.

absolute cross section, the magnitudes can be compared directly, revealing that the TSH cross sections are smaller by nearly an order of magnitude.

2. *Integral Cross Sections*

The experimental (partial) integral cross sections as a function of v' and v^+ are shown in Figs. 4a and 4b, respectively, where they are compared with the IOSA and TSH results. For the inelastic case the experiments were adjusted to match the theoretical absolute values for $v' = 0$, where both theories predict very similar values. It is seen that the IOSA calculation agree almost perfectly with the observed rapid decrease in cross section with v', whereas the TSH calculations are not nearly as good. For the CT case the experiments were adjusted to match the IOSA $v^+ = 0$ result and an excellent agreement in the dependence of v^+ is found. It is important to mention that neither the two theories nor the experiment come close to the distributions expected from the commonly used Franck–Condon factors.

State-to-state integral cross sections, although not directly measured in the apparatus, can be derived from the differential cross sections. The

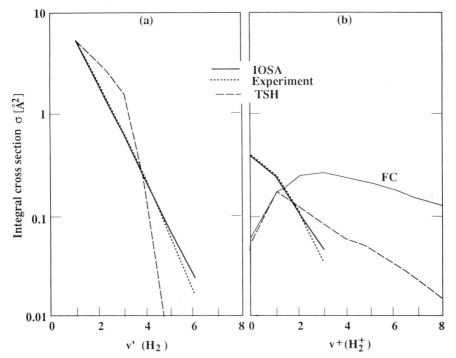

Figure 4. Experimental and theoretical (incomplete) final vibrational state distribution (see also Table I): (a) inelastic process and (b) CT process. (FC stands for Frank–Condon-type transitions.)

procedure requires an integration of the angular distributions out to the largest angle. In the present study the detector probed only a limited angular range, that is, $0° \leqslant \theta_{CM} \leqslant 18°$ and, consequently, only partial integral cross sections could be obtained. However, since this range includes the rainbow, most of the contributions to the inelastic and CT processes are expected to be from this region. This assumption was checked by integrating the IOSA differential cross sections over the entire angular range and comparing with the cross sections obtained by integrating just beyond the rainbow region. The results are compared in Table I, and it is seen that whereas most processes indeed take place in the rainbow region, contributions from the larger angular region cannot be ignored. In particular, it is noted that 40% of the CT processes occur in the angular region beyond the rainbow.

Finally in Table I, the absolute cross section calculated within the IOSA are also listed. The total integral cross section for CT is found to be 1.1 Å^2. The closest available experimental result is $\sigma = 0.226 \text{ Å}^2$ at $E_{CM} = 50 \text{ eV}$.[50] Assuming that the CT cross section decreases as E_{CM}^{-1}, it is estimated to be 0.57 Å^2 at $E_{CM} = 20 \text{ eV}$, which is about one-half of the IOSA value. The agreement is considered satisfactory in view of the approximate nature of the potential surface. At the same energy (i.e., $E_{CM} = 20 \text{ eV}$) the TSH theory predicts a value of 0.3 Å^2.

TABLE I

Quantum-mechanical Integral Cross sections for the processes

$$H_2(v = 0) + H^+ \rightarrow \begin{cases} H_2(v') + H^+ \\ H_2^+(v^+) + H \end{cases}$$

Final Vibrational State (v', v^+)	Cross Sections (Å^2)	
	Inelastic	Charge Transfer
0	52.4 (52.4)[a]	0.53 (0.36)[a]
1	5.3 (5.3)	0.33 (0.23)
2	2.2 (2.06)	0.16 (0.10)
3	0.77 (0.68)	0.088 (0.044)
4	0.27 (0.209)	—
5	0.13 (0.076)	—
6	0.065 (0.026)	—
Total	61.1 (60.75)	1.10 (0.73)

[a]The numbers in parentheses are the corresponding *partial* cross sections calculated for the (rainbow) angular range $0 < \theta_{CM} = 18°$.

2.   On The Mechanism of Charge Transfer in The $H_2(v=0) + H^+$ System

The (nonreactive) vibronic states of the two coupled systems: $(H_2 + H^+)$ and $(H_2^+ + H)$ are presented in Fig. 5 for the orientation angle $\gamma = 60°$. A similar situation is encountered for all other angles. In order for the CT process

$$H_2(v=0) + H^+ \rightarrow H_2^+(v^+) + H \tag{94}$$

to take place the system has to be able to transform translational energy into internal energy (most likely vibrational energy) so that the transition from one set of vibronic states to the other becomes a resonantic-type transition. Consequently, the CT process can be visualized as a two-step CT

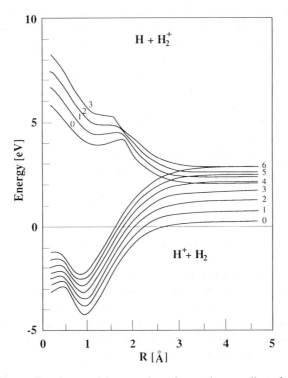

Figure 5.   Electrovibronic potential curves along the reaction coordinate for the two lowest quasiadiabatic surfaces. (The calculations are done for $\gamma = 60°$ and for $r_{min} = 0.3$ Å and $r_{max} = 2.2$ Å.)

process[21c]:

$$H_2(v = 0) + H^+ \rightarrow H_2(v' \geq 4) + H^+, \tag{95a}$$

$$H_2^+(v' \geq 4) + H^+ \rightarrow H_2^+(v^+) + H, \tag{95b}$$

where the most crucial step is the inelastic vibrational excitation process. Thus, the next question to be asked is how does the $H_2 + H^+$ system get so strongly vibrationally excited.

The $(H_2 + H)^+$ is characterized by a relatively deep potential well (~ 4.8 eV) located at the strong interaction region. As can be seen from Fig. 5, this well may contain a large number of vibrational states all of them strongly coupled and, consequently, partially populated while the three particles are in close proximity to each other. These states are usually expected to become depopulated while the particles are receding from the potential well, but, since the higher vibrational states of the $H_2 + H^+$ system are in a resonance condition with the lower vibrational states of the (H_2^+, H) system, the CT process may successfully compete with the depopulation process. In order to support this model, we consider the CT orientational-dependent integral cross sections (hence called CT steric factor) for the $H_2 + H^+$ system and its two isotopic analogs $D_2 + H^+$ and $HD + H^+$,[46] all shown in Fig. 6. At the bottom of the figure the meaning of the orientation angle γ for HD is explicitly presented; thus, the proton is approaching the HD molecule from the H side for $0 \leq \gamma \leq \pi/2$ and from the D side for $\pi/2 \leq \gamma \leq \pi$.

It is well noted that, whereas H_2 and D_2 possess only weak steric factors, a relatively strong steric factor is obtained for HD. In Ref.[46] we also showed that a similar situation is encountered for the *vibrational* excitation steric factors. These findings indicate that the two processes, namely, vibrational excitation and CT transfer, as suggested by Eqs. (95), are strongly coupled and the second takes place following the first.

C. The $(Ar + H_2)^+$ System

The $(Ar + H_2)^+$ is unique in that, on the one hand, it contains all possible competing processes that may occur in an ion (atom)–diatom (ion) collision ranging from vibrational (inelastic) transitions through CT and spin-flip transitions to chemical reactions and dissociations and, on the other hand, it is still simple enough to be treated quantum mechanically.

The processes that will be treated here are

$$Ar + H_2^+(v^+) \rightarrow \begin{cases} Ar^+(^2P_j) + H_2, & j = \tfrac{1}{2}, \tfrac{3}{2}, \tag{96a} \\ ArH^+ + H, \tag{96b} \end{cases}$$

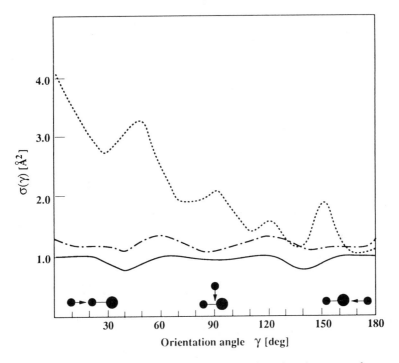

Figure 6. Orientational-dependent integral cross sections for charge transfer: ———, $H^+ + H_2$; ---, $H^+ + HD$, ----, $H^+ + D_2$.

and

$$Ar^+(^2P_j) + H_2(v = 0) \rightarrow \begin{cases} Ar + H_2^+(v^+), & \text{(97a)} \\ ArH^+ + H, & j = \tfrac{1}{2}, \tfrac{3}{2}, & \text{(97b)} \\ Ar^+(^2P_{j'}) + H_2(v'), & j'(\neq j) = \tfrac{1}{2}, \tfrac{3}{2}. & \text{(97c)} \end{cases}$$

The relative positions of the various vibrational levels are shown in Fig. 7.

This section is organized in the following way: we start by discussing the potential and its extension to include the spin-flip transitions, next we consider high-energy results ($\geqslant 20$ eV) for which the reactive channel is ignored, and finally we discuss the low-energy region for which reactive cross sections are also calculated.

1. The Three-Surface System

The potential energy surface assumed to govern the three-particle system is a DIM surface as derived by Kuntz and Roach.[51] However, in this potential

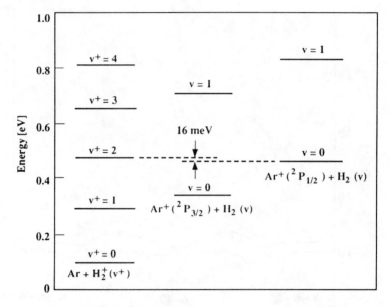

Figure 7. The eigenstate diagram of the $(Ar + H_2)^+$ system. Note the resonance condition between $Ar + H_2^+(v^+ = 2)$ and $Ar^+(^2P_{1/2}) + H_2(v = 0)$.

no spin–orbit coupling is built in and, therefore, a brief description of how this coupling is incorporated is given.

The original DIM matrix is an 8×8 diabatic potential matrix, which is reduced to a 2×2 diabatic potential following diagonatization of the original matrix and employing the adiabatic–diabatic transformation for the 2×2 adiabatic representation. Thus, we obtain

$$\mathbf{W}^{(2)} = \begin{pmatrix} W_1 & W_{12} \\ W_{21} & W_2 \end{pmatrix}, \tag{98}$$

where

$$W_1 = V_1 \cos^2 \alpha + V_2 \sin^2 \alpha,$$

$$W_2 = V_1 \sin^2 \alpha + V_2 \cos^2 \alpha,$$

$$W_{12} = W_{21} = \tfrac{1}{2}(V_2 - V_1)\sin^2 \alpha, \tag{47'}$$

Here the meaning of V_i, $i = 1, 2$, and α was discussed in Section II B 4. It is important to emphasize that once the translational coordinate R becomes sufficiently large, $W_1(R, r)$ and $W_2(R, r)$, which depend on R and the vibrational coordinate r of the diatom, can be identified as the diatomic potentials of H_2^+ and H_2, respectively. The energy difference of $W_1(R,r)$ and $W_2(R,r)$ at their asymptotic limits are equal to the difference of the ionization energies for H_2 and Ar.

Next we incorporate the spin–orbit coupling to distinguish between the the two states of Ar^+, namely, $Ar^+(^2P_{3/2})$ and $Ar^+(^2P_{1/2})$. Consequently, the $\mathbf{W}^{(3)}$ matrix is constructed:

$$\mathbf{W}^{(3)} = \begin{pmatrix} \bar{W}_1 & \bar{W}_{12} & \bar{W}_{13} \\ \bar{W}_{21} & \bar{W}_2 & \bar{W}_{23} \\ \bar{W}_{31} & \bar{W}_{32} & \bar{W}_3 \end{pmatrix}, \tag{99}$$

where $\bar{W}_i$, $i = 1\text{–}3$, are the new diabatic surfaces that correlate asymptotically with the $Ar + H_2^+$, $Ar^+(^2P_{3/2}) + H_2$, and $Ar^+(^2P_{1/2}) + H_2$ states, respectively. The $\bar{W}_{ij}$, $i,j = 1\text{–}3$, are the diabatic couplings. To achieve this goal, we have made the following assumptions:

1. The lowest diabatic surface within the new scheme remains unchanged, namely,

$$\bar{W}_1 \equiv W_1. \tag{100}$$

2. The two higher surfaces $\bar{W}_2$ and $\bar{W}_3$ are chosen to be identical to W_2, but shifted one with respect to the other:

$$\bar{W}_2 = W_2 + \delta_2,$$
$$\bar{W}_3 = W_3 + \delta_3, \tag{101}$$

where $\delta_3 - \delta_2 = 0.178\,\text{eV}$, the energy difference between the two spin–orbit states, $Ar^+(^2P_{3/2})$ and $Ar^+(^2P_{1/2})$. The value of δ_3 is determined in such a way that, asymptotically, the difference between the eigenvalue of the $H_2^+(v^+ = 2) + Ar$ state differs by $0.016\,\text{eV}$ from the eigenvalue of the $H_2(v = 0) + Ar^+(^2P_{1/2})$ ground state (see Fig. 7). Thus, δ_3 is assumed to be $\delta_3 = 0.233\,\text{eV}$. Having these values, δ_2 becomes $\delta_2 = 0.055\,\text{eV}$.

3. Next we consider the two diabatic off-diagonal terms $\bar{W}_{12}$ and $\bar{W}_{31}$, which couple the ground-state surface with the two corresponding excited surfaces associated with the spin–orbit states of Ar^+. Here, we follow a procedure devised by Tanaka et al.[6a] and implement their zeroth-order case. According to their analysis, $\bar{W}_{12}$ and $\bar{W}_{13}$ are related to W_{12} as

$$\bar{W}_{12} = (\sqrt{6}/3)W_{12}, \quad \bar{W}_{13} = -(\sqrt{3}/3)W_{12}. \tag{102}$$

4. The spin–orbit coupling energy term is known to be in general much weaker than the regular electronic terms. Consequently, the coupling energy term between the two surfaces due to the spin–orbit states, that is, $\bar{W}_{23}$ and $\bar{W}_{32}$ are assumed to be zero throughout the numerical treatment:

$$\bar{W}_{23} = \bar{W}_{32} = 0. \tag{103}$$

Thus, the final 3×3 potential matrix $\mathbf{W}^{(3)}$, which follows from $\mathbf{W}^{(2)}$, is

$$\mathbf{W}^{(3)} = \begin{pmatrix} W_1 & (\sqrt{6}/3)W_{12} & -(\sqrt{3}/3)W_{12} \\ (\sqrt{6}/3)W_{12} & W_2 + \delta_2 & 0 \\ -(\sqrt{3}/3)W_{12} & 0 & W_2 + \delta_3 \end{pmatrix}. \tag{104}$$

This completes our discussion of the potential energy matrix employed in the calculations, except for one additional important comment: Whereas the $Ar^+(^2P_{1/2})$ ion may always give rise to CT while colliding with H_2, only half of the $Ar^+(^2P_{3/2})$ ions (those with $m_j = \pm\frac{1}{2}$) are capable of doing so. Consequently, all calculated cross sections for $Ar^+(^2P_{3/2})$ are divided by two before the comparison with experiment is made.

2.   The Study of CT in The High-Energy Region

The calculation for CT were carried out at two energies: $E_{CM} = 19.3\,\text{eV}$ and $E_{CM} = 47.6\,\text{eV}$ for which experimental results are available.

The CT cross section for reaction (96a) at the preceding two energies are presented in Figs. 8(a)–8(d), where they are compared with the experimental results. Figures 8(a) and 8(c) show the state-selected cross sections, $\sigma_{v^+}(Ar^+)$, and Figs. 8(b) and 8(d) show the partial state-to-state cross sections, $\sigma_{v^+ \to 3/2,1/2}$, $v^+ = 0, \ldots, 4$.

The main observations for Figs. 8(a) and 8(c) are as follows:

1. The overall shapes of the experimental and calculated curves are very similar; the lowest cross section is observed to be for $v^+ = 0$, whereas the largest cross section is found to be for $v^+ = 2$.

2. With the exception of the theoretical cross sections for $v^+ = 0$, which seem to be too low compared to the experimental values, most of the theoretical cross sections are in fair agreement with the experimental results.

3. It is expected that once the collision energy becomes sufficiently high, the importance of the *vibronic* coupling will disappear, yielding similar cross sections for all the lower vibrational states. This tendency is clearly seen in the theoretical results, but not to the same extent as in the experimental ones.

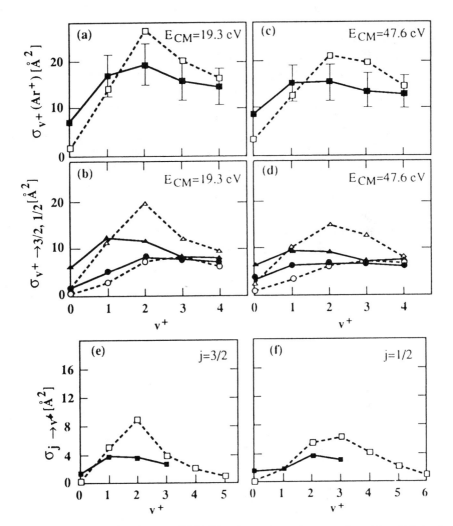

Figure 8. Comparison between CT-IOSA and experimental cross sections; (*a*) and (*c*) stand for the process $Ar + H_2^+(v^+) \rightarrow Ar^+ + H_2$ and (*b*) and (*d*) for the state-to-state processes $Ar + H_2^+(v^+) \rightarrow Ar^+(^2P_j) + H_2, j = \frac{1}{2}, \frac{3}{2}$. (*e*) and (*f*) are for the state-to-state process $Ar^+(^2P_j) + H_2(v = 0) \rightarrow Ar + H_2^+(v^+)$. (*a*) and (*b*) are for $E_{CM} = 19.3$ eV; (*c*) and (*d*) are for $E_{CM} = 47.6$ eV; and (*e*) and (*f*) are for $E_{CM} = 19.3$ eV. Solid symbols are for experimental results and open ones are for theoretical results. In Figs. 7*b* and 7*d*: $\bigcirc, \bullet - j = \frac{1}{2}; \triangle, \blacktriangle - j = \frac{3}{2}$.

The comparison of partial state-to-state cross sections in Figs. 8(b) and 8(d) gives more detailed information about where the source for the difference between the experimental and theoretical results lies. It is seen that the calculated and measured curves for $\sigma_{v^+ \to 1/2}$ overlap reasonably well. Therefore, the calculated values for $\sigma_{v^+ \to 3/2}$ are the main cause for the observed discrepancies.

The CT cross sections for reaction (97a) at the energy $E_{CM} = 19.3\,\text{eV}$ is presented in Figs. 8(e) and 8(f), where they are compared with the corresponding experimental results. The fit between the two kinds of results are reasonable except for the $\sigma_{3/2 \to 2}$ and $\sigma_{1/2 \to 3}$ values shown in Figs. 8(e) and 8(f), respectively. The discrepancy for $\sigma_{3/2 \to 2}$ is reminiscent of a similar discrepancy found between the experimental and theoretical values for $\sigma_{2 \to 3/2}$ [see Fig. 8(b)]. In contrast to the discrepancy for $\sigma_{3/2 \to 2}$, the discrepancy for $\sigma_{1/2 \to 3}$ is somewhat unexpected because good agreement between the experimental and theoretical values for $\sigma_{3 \to 1/2}$ was observed [see Fig. 8(b)].

TABLE II

Exchange Cross Sections (Å^2) for the reactions $Ar + H_2^+(v^+) \to ArH^+ + H$

Translational energy (eV)	v^+	v^+ 0	1	2	3	4
0.22	T1[a]	37.06	46.2	2.9	17.2	24.8
	T2	—	20.0	—	—	—
	T3	—	37.0	—	—	—
0.48	T1	24.5	32.8	12.9	27.4	28.0
	T2	8.8	—	—	—	—
	T3	31.0	—	—	—	—
	E1	27.0 ± 1.4	31.0 ± 1.6	22.5 ± 1.5	26.5 ± 1.8	28.0 ± 3.2
	E2	27.4	34.0	21.3	32.3	36.4
0.75	T1	17.8	26.4	19.9	29.7	—
	E2	22.6 ± 4.7	24.5 ± 4.7	28.0 ± 4.7	26.6 ± 4.7	32.8 ± 4.7
1.00	T1	17.8	23.1	22.0	—	—
	T3	15.0 ± 1.3	28.6 ± 3.1	26.2 ± 2.9	—	22.5 ± 2.6
	E1	25.6 ± 1.3	28.0 ± 1.4	22.0 ± 1.5	23.3 ± 1.6	25.0 ± 3.7
	E3	51.8	66.1	69.1	67.7	73.4
1.30	T1	16.4	21.7	—	—	—
	E2	21.5 ± 5.4	23.6 ± 5.4	26.0 ± 5.4	31.8 ± 5.4	25.4 ± 5.4
1.50	T1	15.3	—	—	—	—

[a]T1, complex IOSA (Ref. 42); T2, previous RIOSA (Ref. 23); T3, TSH (Ref. 52); E1, Liao et al. (Ref. 12); E2, Tanaka et al. (Ref. 6); E3, Houle et al. (Ref. 5).

The fact that microscopic reversibility is fulfilled theoretically[23] seems to point to the conclusion that the experimental value for $\sigma_{3 \to 1/2}$ is too small.

3.   The Study of Charge Transfer, Exchange, and Spin Transition in The Low-Energy Region

In the low-energy region, $0.22 \leqslant E_{CM} \leqslant 1.5\,\mathrm{eV}$, all processes listed in Eqs. (96) and (97) are encountered and were studied. The calculated cross sections were compared with experimental results obtained in three different laboratories. The comparison is presented in several tables and figures.

a.   *Results for $Ar + H_2^+(v^+) \to ArH^+ + H$.* The initial state-selected total cross sections for these processes are presented in Table II and in Figs. 9

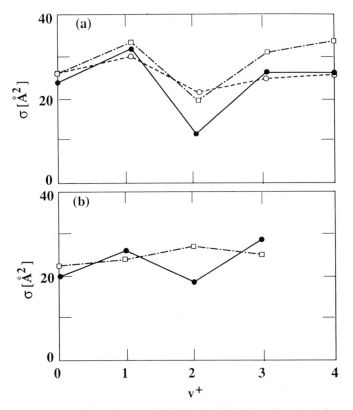

Figure 9.   Initial state-dependent total (absolute) cross sections for the exchange process $Ar + H_2^+(v^+) \to ArH^+ + H$. ●————●, IOSA results (Ref. 42); ○—○- -○, experimental results (Ref. 12); □·, ·—□ ·—, experimental results (Ref. 6). (a) Results for $E_{CM} = 0.48\,\mathrm{eV}$; (b) results for $E_{CM} = 0.75\,\mathrm{eV}$.

and 10. The dependence on v^+ for several translational energies E_{CM} is shown in Fig. 9, and the dependence on E_{CM} for various v^+ states in Fig. 10. In Fig. 9(a) the IOSA cross sections for $E_{CM} = 0.48$ eV are compared with experimental results obtained by Liao et al.[12] and by Tanaka et al.[6] It is seen that the features found in the experiment are nicely reproduced by the IOSA calculation, in particular, the strong dip at $v^+ = 2$ and the shoulder at $v^+ = 1$. It is important to emphasize that here and elsewhere, unless otherwise specified, the comparison is between absolute cross sections for both theory and experiment, and therefore no normalization factors have been used. The theoretical IOSA cross sections are compared with Tanaka

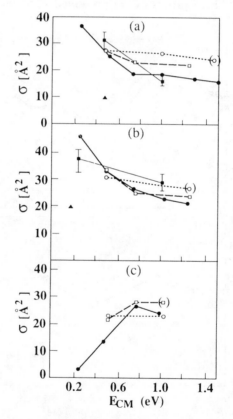

Figure 10. Translational energy-dependent (absolute) cross sections for the exchange process $Ar + H_2^+(v^+) \rightarrow ArH^+ + H$. ●——●, complex IOSA results (Ref. 42); ▲ (previous) RIOSA results (Ref. 23); ○-○--○ experimental results (Ref. 12); □--□--□ experimental results (Ref. 6). (a) Results for $v^+ = 0$; (b) results for $v^+ = 1$; (c) results for $v^+ = 2$. Symbols between parantheses stand for linearly interpolated values with respect to the next measured cross section.

et al.'s[6] results for $E_{CM} = 0.75$ eV in Fig. 9(b). Again, a reasonably good fit is obtained.

Energy-dependent cross sections for $v^+ = 0, 1, 2$ are presented in Figs. 10(a), 10(b) and 10(c), respectively. It is seen that for $v^+ = 0, 1$ the fit among the different kinds of results is reasonably good. Note that, as far as the present version of the IOSA is concerned,[42] not only does it reproduce the experimental results reasonably well, but it also gives values very close to the TSH ones.[52] In Figs. 10(a) and 10(b), we give two results due to the previous IOSA version.[23] It is seen how they differ significantly from all other results.

In contrast to the results in Figs. 10(a) and 10(b), the fit between theory and experiment presented in Fig. 10(c) is somewhat less encouraging. The theoretical curve shows a stronger dependence on the energy than the experimental curves do, although all results seem to decrease as the energy decreases. This behavior is unexpected, because it hints at the existence of a potential barrier, which does not seem to exist for the initial $v^+ = 0, 1$ states. We discuss this finding in greater detail in Section III C 4.

b.   *Results for* $Ar^+(^2P_j) + H_2 \rightarrow ArH^+ + H$. Energy-dependent cross sections for $Ar^+(^2P_{3/2})$ and $Ar^+(^2P_{1/2})$ are shown in Figs. 11(a) and 11(b), respectively. Again, the theoretical results are compared with the experimental. In general, Liao et al.'s experimental results fit the theoretical results much better than do those of Tanaka et al.[17,18] For both spin states, Tanaka et al.'s cross sections are not only larger than the others, but they also show a much stronger energy dependence. The more encouraging outcome of this comparison is the nice fit between Liao et al.'s experimental and the theoretical results for energies higher than 0.4 eV. Still, the different behavior at low energies is puzzling.

c.   *Results for* $Ar + H_2^+(v^+) \rightarrow Ar^+(^2P_j) + H_2$. The cross sections for these CT processes are presented in Tables III and IV and in Figs. 12–14.

Initially we discuss the CT process in general, not distinguishing between the two spin states of Ar^+. The energy-dependent cross sections for $v^+ = 1$ and $v^+ = 2$ are in Figs. 12(a) and 12(b), respectively. In general, a reasonable fit between theory and experiment is obtained. This means that for $v^+ = 1$ the cross sections are small (2–6 Å^2), whereas for $v^+ = 2$ they are much larger (20–40 Å^2). However, whereas for $v^+ = 1$ we obtain the correct energy dependence, we fail to do so in the case of $v^+ = 2$. The reason may be associated with the fact that ($v^+ = 2$) is in resonance with ($v = 0, j = \frac{1}{2}$) state, and therefore the numerical results are much more sensitive to the details of the potential (which is semiempirical, as was mentioned). In Fig. 13 we present the dependence of the CT cross sections on the initial vibrational

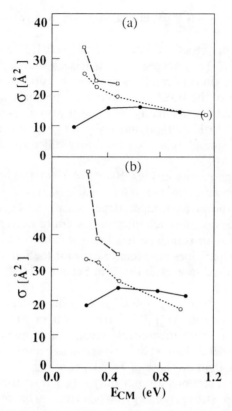

Figure 11. Translational energy-dependent (absolute) cross sections for the exchange process $Ar^+(^2P_j) + H_2(v = 0) \rightarrow ArH^+ + H$. Symbols as in Fig. 10. (a) Results for $j = \frac{3}{2}$; (b) results for $j = \frac{1}{2}$.

states of H_2^+. The results for $E_{CM} = 0.48\,eV$ are presented in Fig. 13(a) and those for $E_{CM} = 0.75\,eV$ in Fig. 13(b). In Fig. 13(a), the IOSA results are compared with Liao et al. experimental results[12] as well as with those of Tanaka et al.,[6] while in Fig. 13(b), they are compared only with those of Tanaka et al.[6] A nice agreement is shown in Fig. 7(a) (note that we show absolute, not relative, cross sections). A less encouraging fit is shown in Fig. 13(b), mainly due to the large cross section at $v^+ = 2$. More detailed results are shown in Fig. 14, where we distinguish between the two final $Ar^+(^2P_j)$ states. The theoretical results are compared with Liao et al. experimental values (at $E_{CM} = 0.48\,eV$), which are the only ones available. Again, a nice fit is obtained.

TABLE III

Charge-Transfer Cross Sections ($Å^2$) for the Process $Ar + H_2^+(v^+) \rightarrow Ar^+ + H_2$

Trans-lational energy (eV)	v^+	v^+ 0	1	2	3	4
0.22	T1[a]	Closed	1.7	17.2	7.5	10.8
	T2	Closed	24.0	—	—	—
	T3	Closed	31.0	—	—	—
0.48	T1	0.2	3.3	31.2	7.7	12.0
	T2	2.6	—	—	—	—
	T3	0.8	—	—	—	—
	E1	2.0 ± 0.1	5.3 ± 0.3	27.5 ± 2.0	14.6 ± 0.9	13.6 ± 2.1
	E2	0.0 ± 2.8	3.3 ± 2.8	25.9 ± 2.8	10.8 ± 2.8	6.9 ± 2.8
0.75	T1	0.7	4.2	40.5	8.3	—
	E2	0.3 ± 2.6	3.6 ± 2.3	25.8 ± 2.6	14.2 ± 2.6	9.6 ± 2.6
1.00	T1	0.6	6.6	44.0	—	—
	T3	1.9 ± 0.5	26.6 ± 3.1	22.4 ± 3.6	—	21.1 ± 3.6
	E1	2.0 ± 0.1	5.9 ± 0.3	27.0 ± 1.9	15.5 ± 1.1	11.7 ± 1.8
	E3	3.1	17.1	47.2	28.4	24.4
1.25	T1	0.8	7.1	—	—	—
	E2	1.0 ± 3.4	6.4 ± 3.4	28.4 ± 3.4	16.9 ± 3.4	9.4 ± 3.4
1.5	T1	0.9	—	—	—	—

[a]T1, Complex IOSA (Ref. 42); T2, previous RIOSA (Ref. 23); T3, TSH (Ref. 52); E1, Liao et al. (Ref. 12); E2, Tanaka et al. (Ref. 6); E3, Houle et al. (Ref. 5).

d. *Results for* $Ar^+(^2P_j) + H_2(v=0) \rightarrow Ar + H_2^+(v^+)$. Results for these CT processes are presented in Figs. 15 and 16. In Fig. 15 we show the energy-dependent cross sections for the two initial spin states of Ar^+. The theoretical cross sections are compared with Liao et al.[12] experimental values and with those of Henri et al.[8] In general, the agreement is good, namely, for $j = \frac{3}{2}$ both theory and experiment yield small cross sections (less than $6 Å^2$) and for $j = \frac{1}{2}$ they yield larger cross sections ($> 15 Å^2$). However, whereas for $j = \frac{3}{2}$ the IOSA correctly reproduces the energy dependence of these cross sections, it fails to do so for the $j = \frac{1}{2}$ case, where the steep increase of the cross sections as a function of the energy is not observed in the experiments.

The final vibrational distribution for the product ion H_2^+ for $E_{CM} = 1.0 \, eV$ is presented for $j = \frac{3}{2}$ in Fig. 16(a) and that for $j = \frac{1}{2}$ in Fig. 16(b). Whereas the fit between experiment (Liao et al.)[12] and theory is not satisfactory for $Ar^+(^2P_{3/2})$, a nice agreement is obtained for $Ar^+(^2P_{1/2})$. One possible reason

TABLE IV
State-to-State Charge-Transfer Cross Sections (Å^2) for $Ar + H_2^+(v^+) \rightarrow Ar^+(^2P_j) + H_2$

Translational energy (eV)		j	v^+				
			0	1	2	3	4
0.22	T^a	$\frac{3}{2}$	Closed	1.7	0.5	1.1	5.8
	E	$\frac{3}{2}$	Closed	—	—	—	—
	T	$\frac{1}{2}$	Closed	Closed	17.1	6.4	5.0
	E	$\frac{1}{2}$	Closed	Closed	—	—	—
0.48	T	$\frac{3}{2}$	0.2	3.1	1.0	1.6	5.4
	E	$\frac{3}{2}$	2.0 ± 0.1	5.3 ± 0.4	6.0 ± 0.5	4 ± 3	4 ± 3
	T	$\frac{1}{2}$	Closed	0.2	30.6	6.1	5.0
	E	$\frac{1}{2}$	Closed	0.0	21.5 ± 1.5	11 ± 3	10 ± 3
0.75	T	$\frac{3}{2}$	0.6	3.8	1.8	0.9	—
	E	$\frac{3}{2}$	—	—	—	—	—
	T	$\frac{1}{2}$	0.03	0.4	38.7	6.3	—
	E	$\frac{1}{2}$	—	—	—	—	—
1.00	T	$\frac{3}{2}$	0.6	6.2	3.0	—	—
	E	$\frac{3}{2}$	2.0 ± 0.1	5.9 ± 0.4	6.3 ± 0.6	4 ± 1.5	8.2 ± 3.5
	T	$\frac{1}{2}$	0.07	0.4	40.8	—	—
	E	$\frac{1}{2}$	0.0	0.0	21.3 ± 1.5	11.5 ± 1.5	3.5 ± 3.5
1.25	T	$\frac{3}{2}$	0.6	6.8	—	—	—
	T	$\frac{1}{2}$	0.1	0.5	—	—	—
1.50	T	$\frac{3}{2}$	0.8	—	—	—	—
	T	$\frac{1}{2}$	0.1	—	—	—	—

^{a}T, Complex IOSA; E, Liao et al. (Ref. 12).

for the poor fit in the case of $j = \frac{3}{2}$ could be the overall *small* cross sections in this case, which may lead to larger *experimental* errors.

e. *Results for* $Ar^+(^2P_j) + H_2(v = 0) \rightarrow Ar^+(^2P_{j'}) + H_2$. Cross sections for spin transitions as a function of energy are presented in Fig. 17. Except for $E_{CM} = 1.2\,\text{eV}$, experimental values are unavailable. Note that at this energy a good agreement is obtained with the experimental results of Liao et al.[12] for the endothermic processes cross section $\frac{3}{2} \rightarrow \frac{1}{2}$, but the agreement is less than satisfactory for the reverse process. As for the energy dependence, measurements at high energies indicate that both cross sections increase. This finding is confirmed in the calculations.

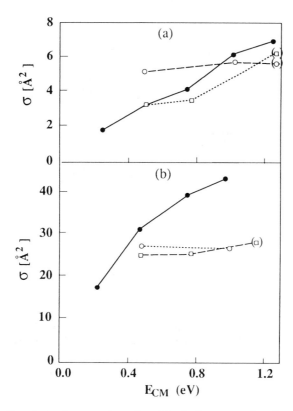

Figure 12. Translational energy-dependent (absolute) cross sections for the CT process $Ar + H_2^+(v^+) \rightarrow Ar^+ + H_2$. Symbols as in Fig. 10. (*a*) Results for $v^+ = 1$; (*b*) results for $v^+ = 2$.

4. *Discussion*

a.   The Resonances.   The dominant feature of the $(Ar + H_2)^+$ ionic system is the existence of vibronic resonance in the entrance (reagents) channel between $v = 0$ of the $Ar^+(^2P_{1/2}) + H_2$ system and the $v^+ = 2$ of the $Ar + H_2^+$ system. This situation is best described in Fig. 18 (but see Fig. 6), where we present vibronic curves, as a function of the translational coordinate R for the two $Ar^+(^2P_{:}) + H_2$ systems and the $Ar + H_2^+$ system. Note that, not only does the previously-mentioned resonance [designated as $(\frac{1}{2}, 2)$] exist in the asymptotic region, but it continues to smaller R values, where the diabatic coupling terms between the lower and the two upper surfaces become strong enough. Somewhat less pronounced is the resonance between the $Ar^+(^2P_{3/2}) + H_2(v = 0)$ and the $Ar + H_2(v^+ = 1)$ [designated as $(\frac{3}{2}, 1)$]. The

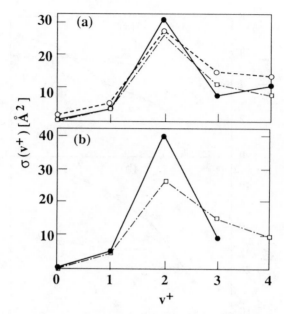

Figure 13. Initial state-selected total (absolute) cross sections for the CT process $Ar + H_2^+(v^+) \to Ar^+ + H_2$. Symbols as in Fig. 9. (a) Results for $E_{CM} = 0.48$ eV; (b) results for $E_{CM} = 0.75$ eV.

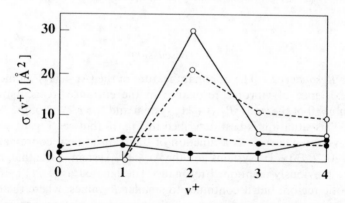

Figure 14. State-to-state integral (absolute) cross section for the CT process $Ar + H_2^+(v^+) \to Ar^+(^2P_j) + H_2$ as obtained for $E_{CM} = 0.48$ eV. ———, IOSA results (Ref. 42); ---, experimental results (Ref. 12); $\bigcirc$, results for $j = \frac{1}{2}$; $\bullet$, results for $j = \frac{3}{2}$.

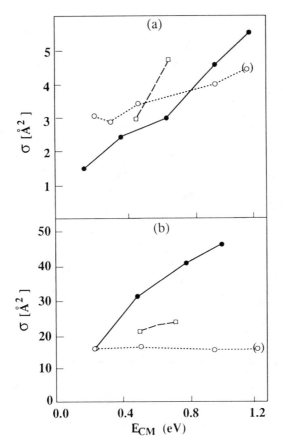

Figure 15. Translational energy-dependent (absolute) cross sections for the CT process $Ar^+(^2P_j) + H_2(v = 0) \rightarrow Ar + H_2^+$. ●—●—●, IOSA results (Ref. 42); ○--○--○, experimental results (Ref. 12); --□--, experimental results (Ref. 8). (a) $j = \frac{3}{2}$, (b) $j = \frac{1}{2}$.

$(\frac{1}{2}, 2)$ resonance is much stronger than the $(\frac{3}{2}, 1)$ not only due to the differences in the (asymptotic) energy gaps (0.019 eV vs 0.063 eV), but also because this smaller energy gap continues to smaller R distances. Many of the results shown in the previous section, both in the figures and in the tables, support this observation.

At this point we examine the existence of the resonance from a different perspective. In order to do so, we calculated differential cross sections, which are presented for $E_{tot} = 1.435$ eV in Fig. 19. In Fig. 19(a) we show differential cross sections for the two previously mentioned processes. For both we obtain strong forward scattering, but two main differences are apparent: (1) whereas

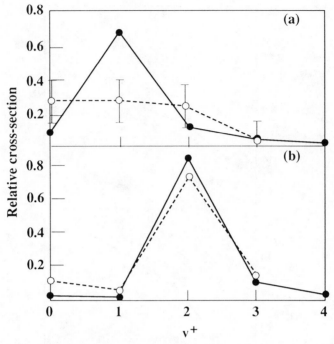

Figure 16. Final vibrational distribution. (Relative) cross-sections $(\sigma(j \to v^+))/\sigma(j \to \Sigma(v^+))$ for the charge-transfer process $Ar^+(^2P_j) + H_2(v = 0) \to Ar + H_2^+(v^+)$, as calculated at $E_{CM} = 1.0\,eV$. ●—●—●, IOSA results (Ref. 42); ○-○--○, experimental results (Ref. 12). (a) Results for $j = \frac{3}{2}$; (b) results for $j = \frac{1}{2}$.

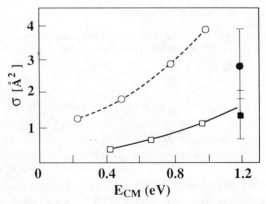

Figure 17. Translational energy-dependent (absolute) cross section for the spin-transition process $Ar^+(^2P_j) + H_2(v = 0) \to Ar^+(^2P_{j'}) + H_2; j(\neq j') = \frac{1}{2}, \frac{3}{2}$. ————, IOSA results for the $(\frac{3}{2} \to \frac{1}{2})$ transition; ---, IOSA results for the $(\frac{1}{2} \to \frac{3}{2})$ transition; ■, experimental result for the $(\frac{3}{2} \to \frac{1}{2})$ transition: ●, experimental result for the $(\frac{1}{2} \to \frac{3}{2})$ transition.

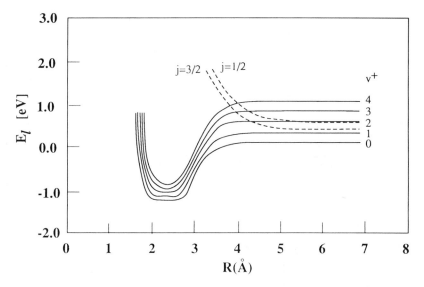

Figure 18. Vibronic potential curves along the translational coordinate R for the three quasiadiabatic potential energy surfaces. The calculations were carried out for $\gamma = 30°$, $r_{min} = 0.3$ Å, and $r_{max} = 4$ Å.

a very clear rainbow scattering is seen in the $(\frac{1}{2}, 2)$ case, it seems not to be present in the $(\frac{3}{2}, 1)$ case. (2) The relative contribution of the forward angular range to the integral cross section is much larger for the $(\frac{1}{2}, 2)$ case. For instance, the contribution of the $(0, 50°)$ angular range in the $(\frac{1}{2}, 2)$ case is about 70%, whereas in the $(\frac{3}{2}, 1)$ case it is only 43%.

The importance of the resonance becomes even more apparent in Fig. 19(b), where the differential cross sections for the reactions $Ar + H_2^+(v^+ = 2) \rightarrow Ar^+(^2P_j) + H_2$ [designated as $(2, j)$] are presented. For all practical purposes the $(2, \frac{1}{2})$ curve is identical to the $(\frac{1}{2}, 2)$ curve, which is another strong indication that these two states are strongly coupled. As an example of a nonresonance case, we show in the same figure the differential cross sections for the $(2, \frac{3}{2})$ case. Note that the strong forward region is nearly missing and also that the rainbow shows up at smaller angles, indicating a shallower potential well.

Experimental differential cross sections for the charge-transfer process are hardly available and those that are available are not state-to-state cross sections. Still the cross sections relevant to ours seem to fit rather well, qualitatively, with our calculated cross sections:

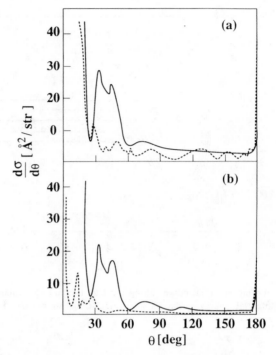

Figure 19. State-to-state differential cross sections for the charge-transfer process as calculated for $E_{tot} = 1.435\,\text{eV}$. ———, Results for $j = \frac{1}{2}$; ---, results for $j = \frac{3}{2}$. (a) $Ar^+(^2P_j) + H_2(v = 0) \rightarrow Ar + H_2^+(v^+ = 2)$; (b) $Ar + H_2^+(v^+ = 2) \rightarrow Ar^+(^2P_j) + H_2$.

1. Hierl et al.[3] measured the differential cross section for the process

$$Ar^+ + H_2(v = 0) \rightarrow Ar + H_2^+ \qquad (105)$$

and found, for the energy value $E_{CM} = 0.45\,\text{eV}$ (which is that closest to ours) essentially a similar distribution; namely, a strong forward peak in the angular range $0 \leqslant \theta_{CM} \leqslant 60$ and only minor contributions from the rest of the angular range.

2. Bilotta et al.[4] measured the differential cross section for the reversed reaction

$$H_2^+ + Ar \rightarrow H_2 + Ar^+, \qquad (106)$$

where the H_2^+ beam possesses, owing to electron impact of H_2, the corresponding vibrational Frank–Condon distribution. For the translational energy $E_{CM} = 0.45\,\text{eV}$, a dominant forward distribution is obtained (the authors labeled it "backward"), which again qualitatively fits ours.

However, whereas the calculated differential cross sections show a clear rainbow structure, this structure is lacking in the experimental cross sections. It is quite possible that since neither the initial nor the final states in the experiment are well selected, this structure, which is typical only for the resonantic transitions, is smeared out.

In Fig. 20 we show l-weighted opacity functions $(2l + 1)P(E_t, l | \cdots)$, where $P(E_t, l | \cdots)$ is defined in Eq. (89). In Fig. 20(a) are presented the opacity functions for $(\text{Ar}^+(^2P_{1/2}), \text{H}_2)$ and in Fig. 20(b) those for $(\text{Ar}^+(^2P_{3/2}), \text{H}_2)$. In each figure we show curves, one for the exchange and the other for the CT process. The most interesting pattern is that seen in Fig. 20(a) for the CT process. The curve has at least three maxima: the largest maximum stands for the forward scattering, the next can be identified as responsible for the rainbow scattering, and the third, and so on, are responsible for the supernumerary rainbows. A similar structure is obtained in Fig. 20(b), but its size is much smaller.

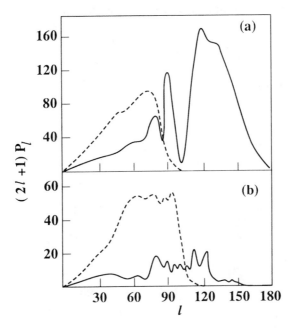

Figure 20. l-weighted opacity functions $((2l + 1)P_l)$ for charge-transfer and exchange processes, as calculated for $E_{tot} = 1.435\,\text{eV}$. (a) Results for $j = \frac{1}{2}$; (b) results for $j = \frac{3}{2}$. ———, $\text{Ar}^+(^2P_j) + \text{H}_2(v = 0) \rightarrow \text{Ar} + \text{H}_2^+(v^+ = 2)$; ---, $\text{Ar}^+(^2P_j) + \text{H}_2(v = 0) \rightarrow \text{ArH}^+ + \text{H}$.

b.   *Exchange Versus Charge Transfer.* The second feature that characterizes the $(Ar + H_2)^+$ system is the competition between CT and exchange processes. This is best presented in terms of the *l*-weighted opacity functions shown in Fig. 20. Note that for the more resonant case, $Ar^+(^2P_{1/2})$, the CT process is not much affected by the exchange, because the CT takes place mainly at the large-*l* region, whereas the exchange occurs at the lower-*l* region. However, since the CT process is also relatively strong in the low-*l* region, it competes directly with the reaction process and probably decreases the reactive cross sections. In the less resonant case, $Ar^+(^2P_{3/2})$, the competition between the two processes is direct, both are affected, and probably this results in smaller cross sections compared to those without competition. This is expected to occur in all other cases, as is noted from the reactive opactiy functions presented in Fig. 21 for $E_{tot} = 1.435\,eV$. All of these functions have a similar pattern with similar numerical values. The probabilities for $Ar^+(^2P_{3/2}) + H_2$ are smaller because the calculated values are divided by 2 for symmetry considerations.

D.   The $(H + O_2)^+$ System

The theoretical investigation of the $(H + O_2)^+$ was inspired by the experimental study of Noll and Toennies[10] of the process:

$$H^+ + O_2(v = 0) \rightarrow \begin{cases} H^+ + O_2(v') \\ H + O_2^+(v^+) \end{cases} \tag{107}$$

carried out at $E_{CM} = 23.00\,eV$.

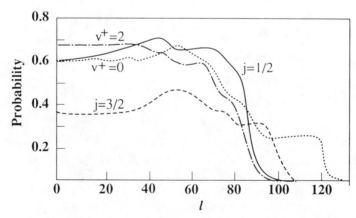

Figure 21. Opacity functions for exchange processes as calculated at $E_{tot} = 1.435\,eV$. v^+ stands for the process $Ar + H_2^+(v^+) \rightarrow ArH^+ + H$. j stands for the process $Ar^+(^2P_j) + H_2(v = 0) \rightarrow ArH^+ + H$.

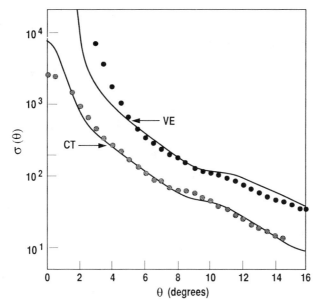

Figure 22. A comparison between experimental (Ref. 10) and theoretical (Ref. 44a) cross sections at $E_{CM} = 23.0\,eV$. The theoretical results were obtained using the Effective Model Potential approach; •, experimental results. ———, theoretical results. VE, pure elastic and inelastic cross sections. CT, charge-transfer cross sections.

At this energy the reactive (exchange) processes are not likely to take place and therefore the reactive arrangement channel can be ignored in the treatment. Since this system is also reviewed in other chapters (Chapters 3–5) we will mention here only a few results.

The $(H + O_2)^+$ system was treated by two groups; both groups used the same method, but each employed a different potential energy surface. In Fig. 22 we compare the experimental and quantum mechanical cross sections for the (pure) inelastic-elastic process and the CT process, obtained using the effective model potential approach.[44a] A very nice fit is noted. In Fig. 23 three sets of vibrational-resolved angular-dependent CT probabilities are shown: experimental results and theoretical results based on the DIM potential approach[44b] and the effective model potential approach.[44a] In general the results produced with the DIM potential seem to fit experiment better.

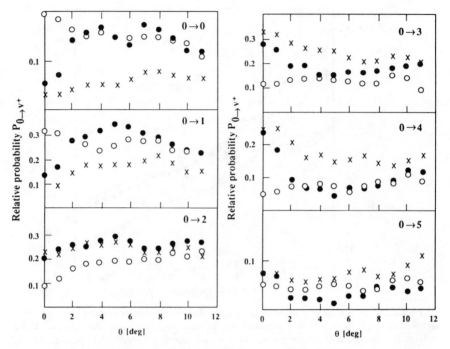

Figure 23. A comparison between experimental and theoretical results for the CT process $H^+ + O_2(v=0) \rightarrow H + O_2^+(v^+)$ as carried out at $E_{CM} = 23.0 \, eV$. Relative CT probabilities (for each of the final vibrational states) as a function of the center of mass scattering angle θ. ●, experimental results (Ref. 10). +, theoretical results—effective potential (Ref. 44a). ○, theoretical results—DIM potential (Ref. 44b).

IV. SUMMARY

In this chapter we concentrated on ion–molecule collisions accompanied by charge transfer. The emphasis was on a rigorous quantum-mechanical approach (rather than models), which is composed of three parts:

1. A consistent treatment that enables the reduction of a large-dimensional DIM matrix (such as 8×8 or 16×16) to a small-dimensional (such as 2×2) diabatic matrix.

2. The incorporation of the negative imaginary arrangement decoupling potentials, which enables the treatment of a reactive system as if it were nonreactive and still obtaining the correct reactive (exchange) and CT cross sections.

3. The introduction of the infinite-order sudden approximation, which enables the treatment of these processes in three dimensions.

In this review were presented CT (and other) cross sections for three different system. Part of the cases were treated in the high-energy region ($E_{CM} \geqslant 20\,eV$), but we also reported on many results for the low-energy region ($E_{CM} \leqslant 1.5\,eV$). In the high-energy region the exchange channel was ignored but in the low-energy region it was incorporated employing the NIADP. All cross sections were compared with experimental results. In many cases the comparison was carried out with respect to absolute values [this accounts mainly for the $(Ar + H_2)^+$ system]. The comparison was more than encouraging. This applies for total integral cross sections as well as the vibrational resolved differential cross sections.

The fact that we have a reliable numerical method available for treating quantum-mechanical CT processes for nonreactive and reactive system calls for extension of this approach to larger (and therefore to more interesting) systems. In fact the first steps in this direction were already initiated. Recently, we completed a preliminary study[43] of the di-ion–diatom system, that is, $H_2 + H_2^+$ and the results that followed were most interesting. The fit with experiment was reasonable but still more calculations are necessary.

References

1. Z. Herman, K. Kerstetter, T. Rose, and R. Wolfgang, *Disc. Farad. Soc.* **44**, 123 (1967). J. R. Krenos and R. Wolfgang, *J. Chem. Phys.* **52**, 5961 (1970); J. R. Krenos, R. K. Preston, R. Wolfgang and J. C. Tully, *J. Chem. Phys.* **60**, 1634 (1971).

2. W. A. Chupka and M. E. Russel, *J. Chem. Phys.* **49**, 5426 (1968).

3. P. M. Hierl, V. Pacak, and Z. Herman, *J. Chem. Phys.* **67**, 2678 (1977).

4. M. Billota, F. N. Preuminger, and J. M. Farrar, *Chem. Phys. Lett.* **74**, 95 (1980); *J. Chem. Phys.* **73**, 1637 (1980); **74**, 1699 (1981).

5. F. A. Houle, S. L. Anderson, D. Gerlich, T. Turner, and Y. T. Lee, *Chem. Phys. Lett.* **82**, 392 (1981); *J. Chem. Phys.* **77**, 748 (1982).

6. (a) K. Tanaka, J. Durup, T. Koto, and I. Koyano, *J. Chem. Phys.* **74**, 5561 (1981). (b) K. Tanaka, T. Kato, and I. Koyano, *J. Chem. Phys.* **75**, 4941 (1981).

7. K. M. Ervin and P. B. Armentrout, *J. Chem. Phys.* **83**, 166 (1985); P. Tosi, F. Boldo, F. Eccher, M. Fillipi, and D. Bassi, *Chem. Phys. Lett.* **164**, 471 (1989).

8. G. Henri, M. Lavallee, O. Dutuit, J. B. Ozenne, P. M. Guyon, and E. A. Gislason, *J. Chem. Phys.* **88**, 6381 (1988).

9. G. Niedner, M. Noll, J. P. Toennies, and Ch. Schlier, *J. Chem. Phys.* **87**, 2685 (1987).

10. M. Noll and J. P. Toennies, *J. Chem. Phys.* **85**, 3313 (1986).

11. C. L. Liao, C. Y. Liao, and C. Y. Ng, *Chem. Phys. Lett.* **81**, 5672 (1984); **82**, 5489 (1985). J. D. Shao and C. Y. Ng, *Chem. Phys. Lett.* **118**, 481 (1985); *J. Chem. Phys.* **84**, 4317 (1986). C. L. Liao, R. Xu, G. D. Flesch, Y. G. Li, and C. Y. Ng, *J. Chem. Phys.* **85**, 3874 (1986); J. D. Shao, Y. G. Li, G. D. Flesch, and C. Y. Ng, *J. Chem. Phys.* **86**, 170 (1987).

12. (a) C. L. Liao, R. Xu, G. D. Flesch, M. Baer and C. Y. Ng, *J. Chem. Phys.* **93**, 4818 (1990). (b) C. L. Liao, R. Xu, S. Nourbaksh, G. D. Flesch, M. Baer, and C. Y. Ng, *J. Chem. Phys.* **93**, 4832 (1990).

13. A. Bjerre and E. E. Nikitin, *Chem. Phys. Lett.* **1**, 179 (1967).

14. (a) R. K. Preston and J. C. Tully, *J. Chem. Phys.* **54**, 4297 (1971); (b) J. C. Tully and R. K. Preston, *J. Chem. Phys.* **55**, 562 (1971).

15. E. Bauer, E. R. Fisher, and F. R. Gilmore, *J. Chem. Phys.* **51**, 4173 (1969). A. detailed application of this model is given by M. S. Child and M. Baer, *J. Chem. Phys.* **74**, 2832 (1981).

16. L. D. Landau, *J. Chem. Phys.* **82**, 4033 (1985); C. Zener, *Proc. R. Soc. London Ser. A* **137**, 696 (1932).

17. I. Last and M. Baer, *Molec. Phys.* **54**, 265 (1985).

18. S. Chapman (Chap. 6, this volume).

19. E. Gislason, G. Parlant, and M. Sizum (Chap. 5, this volume).

20. G. D. Billing, *Comput. Phys. Dept.* **1**, 237 (1984).

21. (a) Z. H. Top and M. Baer, *J. Chem. Phys.* **64**, 3078 (1976); (b) *J. Chem. Phys.* **66**, 1363 (1977); (c) *Chem. Phys.* **25**, 1, 1977.

22. M. Baer and J. A. Beswick, *Chem. Phys. Lett.* **51**, 360 (1977); M. Baer, *Molec. Phys.* **35**, 1637 (1978); M. Baer and J. A. Beswick, *Phys. Rev. A* **19**, 1559 (1979).

23. M. Baer and H. Nakamura, *J. Chem. Phys.* **87**, 4651 (1987); M. Baer, H. Nakamura, and A. Ohsaki, *Chem. Phys. Lett.* **131**, 468 (1986); M. Baer and H. Nakamura, *J. Phys. Chem.* **91**, 5503 (1987).

24. M. Baer, G. Niedner-Schatteburg, and J. P. Toennies, *J. Chem. Phys.* **88**, 1461 (1988); **91**, 4169 (1989).

25. M. Born and J. R. Oppenheimer, *Ann. Phys. (Leipzig)* **84**, 457 (1927).

26. E. C. G. Stuckelberg, *Helv. Phys. Acta* **5**, 89 (1932).

27. W. Lichten, *Phys. Rev.* **131**, 229 (1963).

28. W. D. Hobie and A. D. Mclachlan, *J. Chem. Phys.* **33**, 1695 (1960).

29. M. S. Child, *Molec. Phys.* **20**, 171 (1971).

30. F. T. Smith, *Phys. Rev.* **179**, 111 (1969).

31. R. de L. Kronig, *Band Spectra and Molecular Spectra*, Cambridge University Press, New York, 1930, Chap. 6.

32. D. R. Bates, *Proc. R. Soc. London Ser. A* **240**, 22 (1952); **257**, 22 (1960).

33. M. S. Child, in *Atom-Molecule Collision Theory*, R. B. Bernstein, Ed., Plenum Press, New York, 1979, p. 427; J. C. Tully, in *Dynamics of Molecule Collisions, Part B*. W. H. Miller, Ed., Plenum Press, New York 1976, p. 217.

34. (a) M. Baer in *The Theory of Chemical Reaction Dynamics*, M. Baer, Ed. CRC, Boca-Raton, FL, 1985, Vol. II, Chap. IV. (b) V. Sidis, in *Advances in Atomic and Molecular Physics*, D. R. Bates and B. Baderson, Eds., Academic Press, New York, 1989.

35. M. Baer, *Chem. Phys. Lett.* **35**, 112 (1975).

36. H. Koppel, W. Domke, and L. S. Cederbaum, *Adv. Chem. Phys.* **57**, 59 (1984); C. Petrongolo, R. J. Buenker, and S. D. Peyerimhoff, *J. Chem. Phys.* **78**, 7284 (1983); T. Pacher, L. S. Cederbaum, and H. Koppel, *J. Chem. Phys.* **89**, 7367 (1988).

37. T. Pacher, C. A. Mead, L. S. Cederbaum, and H. Koppel, *J. Chem., Phys.* **91**, 7057 (1989).

38. D. Yarkony (this volume, Chap. 1).

39. F. O. Ellison, *J. Am. Chem. Soc.* **85**, 3540, 3544 (1963); P. J. Kuntz, in *The Theory of Chemical Reaction Dynamics*, M. Baer, Ed., CRC, Boca Raton, FL, 1985, Vol. I, Chap. 2. See also F. A. Gianturco (this volume, Chap. 4).

40. D. Neuhauser and M. Baer, *J. Chem. Phys.* **90**, 4351 (1989); *J. Phys. Chem.* **93**, 2862 (1989); *J. Chem. Phys.* **91**, 4651 (1989); *J. Phys. Chem.* **94**, 185 (1990); *J. Chem. Phys.* **92**, 3419 (1990); M. Baer, D. Neuhauser, and Y. Oreg. *J. Chem. Phys. Faraday Transactions* **86**, 1721 (1990); D. Neuhauser, M. Baer and D. J. Kouri, *J. Chem. Phys.* **93**, 2499 (1990); M. S. Child, *Molec. Phys.* **72**, 89 (1991).

41. M. Baer, *Chem. Phys.* **15**, 49 (1976).

42. M. Baer, C-L Liao, R. Xu, S. Naurbahksh, G. D. Flesch, C. Y. Ng, and D. Neuhauser, *J. Chem. Phys.* **93**, 4845 (1990).

43. M. Baer and C. Y. Ng, *J. Chem. Phys.* **93**, 4845 (1990); **93**, 7787 (1990).

44. (*a*) V. Sidis, D. Grimbert, M. Sizum, and M. Baer, *Chem. Phys. Lett.* **163**, 19 (1989) M. Sizum, D. Grimbert, V. Sidis, and M. Baer, *J. Chem. Phys.* (in press); (*b*) F. A. Gianturco, A. Palma, E. Semprimi, F. Stefani, and M. Baer, *Phys. Rev. A.* **42**, 3926 (1990).

45. M. Baer, C. Y. Ng, and D. Neuhauser, *Chem. Phys. Lett.* **169**, 534 (1990).

46. M. Baer, G. Niedner, and J. P. Toennies, *Chem. Phys. Lett.* **167**, 269 (1990).

47. L. Munchick and E. A. Mason, *J. Chem. Phys.* **35**, 1671 (1961); K. Takayanaki *Progr. Theor. Phys. (Kyoto) Suppl.* **25**, 40 (1963); K. Kramer and R. B. Bernstein, *J. Chem. Phys.* **44**, 4473 (1964); C. F. Curtiss, *J. Chem. Phys.* **49**, 1952 (1968); T. P. Tsien, G. P. Parker, and R. T. Pack, *J. Chem. Phys.* **59**, 5373 (1973); D. Secrest, *J. Chem. Phys.* **62**, 710 (1975); V. Khare, D. J. Kouri, and D. K. Hoffman, *J. Chem. Phys.* **74**, 2275 (1981); M. A. Wartel and R. J. Cross, *J. Chem. Phys.* **55**, 4983 (1971); J. M. Bowman and J. Arruda, *Chem. Phys. Lett.* **41**, 43 (1976).

48. V. Khare, D. J. Kouri, and M. Baer, *J. Chem. Phys.* **71**, 1188 (1979); J. Jellinek and M. Baer, *J. Chem. Phys.* **76**, 4883 (1982); H. Nakamura, A. Ohaski, and M. Baer, *J. Phys. Chem.* **90**, 6176 (1976); J. M. Bowman and K. T. Lee, *J. Chem. Phys.* **72**, 5071 (1980); G. D. Barg and G. Drolshagen, *Chem. Phys.* **47**, 209 (1980); D. C. Clary and G. Drolshagen, *J. Chem. Phys.* **76**, 5027 (1982); D. C. Clary, *Chem. Phys.* **71**, 117 (1982); **81**, 379 (1983); G. Grossi, *J. Chem. Phys.* **81**, 3355 (1984); B. M. D. D. Jensen-op-de-Haar and G. G. Balint-Kurti, *J. Chem. Phys.* **85**, 329 (1987); **90**, 888 (1989); H. Nakamura, *Phys. Reps.* **187**, 1 (1990); M. Nakamura and H. Nakamura, *J. Chem. Phys.* **90**, 4835 (1989); T. Takayanagi, S. Tsunashima, and S. Sato, *J. Chem. Phys.* **93**, 2487 (1990).

49. D. C. Clary and D. M. Sonnenfroh, *J. Chem. Phys.* **90**, 1686 (1989).

50. W. H. Cremer, *J. Chem. Phys.* **35**, 836 (1961).

51. P. J. Kuntz and A. C. Roach, *J. Chem. Soc., Faraday Trans.* **268**, 259 (1972).

52. S. Chapman, *J. Chem. Phys.* **82**, 4033 (1985). A few of the lower-energy cross sections were obtained by private communication.

SEMICLASSICAL APPROACH TO CHARGE-TRANSFER PROCESSES IN ION–MOLECULE COLLISIONS

HIROKI NAKAMURA

Division of Theoretical Studies, Institute for Molecular Science, Myodaiji, Okazaki, Japan

CONTENTS

State-Selected and State-to-State Ion–Molecule Reaction Dynamics, Part 2: Theory, Edited by Michael Baer and Cheuk-Yiu Ng. Advances in Chemical Physics Series, Vol. LXXXII. ISBN 0-471-53263-0 © 1992 John Wiley & Sons, Inc.

I. INTRODUCTION

The terminology of "semiclassical theory" is used in a variety of ways. It is quite often used to mean generally the case that the electronic degrees of freedom are treated quantum mechanically but the nuclear degrees of freedom are approximated by the pure classical mechanics. In the impact parameter method, for instance, the nuclear trajectory (relative motion of atoms and molecules) is given *a priori* as a function of time and is treated as a kind of external parameter for the electronic degrees of freedom. More appropriate usage of the terminology from the author's view point is to mean that the nuclear degrees of freedom themselves are treated semiclassically. Even in this category, however, there exist many versions of semiclassical theory, depending on how we use semiclassical approximation (certain kind of expansion with respect to $\hbar$). The most standard one is the JWKB theory and its extension.[1-4] The other examples are the wave-packet approach in the coordinate space and the phase-space distribution function theory.[5-8] Our stand point in this paper is basically the first one. Since our main concern here is charge transfer, we present the semiclassical theories for the electronically nonadiabatic transitions between adiabatic electronic states at low collision energies.

Semiclassical theory is, of course, an approximation to quantum mechanics, but has a great significance in the sence that it provides a useful analytical theory, using the feasibility of classical mechanics and yet taking into account the correct concepts of quantum mechanics. As is well known, the JWKB theory can provide an analytical expression of wavefunction and can describe quite accurately the quantum-mechanical effects such as tunneling, nonadiabatic transition, interference, and resonance. These various effects can be treated in a nonperturbative way: the Landau–Zener type of theory for nonadiabatic transition, for instance, is *not* a simple perturbation theory. More importantly, with these capabilities of the semiclassical theory, this enables us to grasp the underlying physics of the phenomena. This is valuable even when accurate quantum-mechanical calculations become possible, since the latter results are often obtained by the heavy numerical computations. One-dimensional semiclassical theory has been well developed for describing the previously mentioned various effects, and presents a powerful method to investigate the dynamics of diatomic systems such as ion–atom inelastic collisions.[4,9,10] Unfortunately, however, the multi-dimensional theory has not been well developed yet compared to the one-dimensional case. In spite of the efforts made by Maslov and many other authors, it is still in a primitive stage,[2,3,11,12] and no such theory exists that can be directly and effectively applied to ion–molecule collisions.

There have been published quite a few review articles on ion–molecule collisions.[13-18] Most of them are concerned about the simple orbiting collision model of Langevin. Some others are basically based on the impact parameter method. Nice reviews are available by Kleyn and Los[17] and by Sidis.[18] This chapter is written from a different view point, since the author understands that an important mission of this chapter is to review and introduce the basic semiclassical theories, especially those for nonadiabatic transition.[9,10,19-21] Even these one-dimensional theories can be useful to comprehend the various phenomena and processes qualitatively or semiquantitatively. Since electronically nonadiabatic transition and particle rearrangement (reaction) are the two fundamental mechanisms of the charge-transfer ion–molecule collisions and reactions, we focus our attention on these basic subjects from the view point of semiclassical theory.

This chapter is organized as follows: In Section II basic mechanisms of nonadiabatic electronic transitions associated with charge transfer are summarized and their qualitative characteristics are discussed. These are the Landau–Zener and the Rosen–Zener type of nonadiabatic transitions, and the transitions induced by Coriolis (rotational) coupling, spin–orbit interaction, and the coupling due to the electron momentum transfer or the ETF (electron-translation factor) in charge transfer. It is shown that the semiclassical theory can be utilized to analyze all these transitions uniformly by introducing the new (dynamical-state) representation. Qualitative discussions are also presented for reactive transition, or particle rearrangement by emphasizing the role of the potential ridge, the watershed dividing the reactant and product valleys. Section III summarizes the basic formulas for the two-state nonadiabatic electronic transitions and explains the dynamical-state representation mathematically in more detail. One of the important generalizations of the one-dimensional two-state semiclassical theories is the multichannel curve-crossing problem, which is discussed in Section IV. With use of the two state theories a general formalism is described for the multilevel system involving closed channels. The simplified treatment, that is, the BFG (Bauer–Fisher–Gilmore) model, for the vibronic transitions in ion–molecule collisions is also briefly touched upon. Some of the numerical applications reported so far are presented together. Section V is devoted to reaction. After briefly summarizing the historical orbiting model of ion–polar-molecule collisions and the classical S-matrix theory, the hyperspherical coordinate approach is explained and is demonstrated to be powerful for grasping the reaction mechanisms. In Section VI a simple semiclassical generalization of the trajectory-surface-hopping method is proposed. Essential idea is to use the well-developed one-dimensional semiclassical theories on the curvilinear one-dimensional classical trajectories. Other desirable challenging problems to be developed in future are also briefly discussed. Section VII summarizes this chapter.

II. BASIC MECHANISMS AND THEIR CHARACTERISTICS

Qualitative explanations of the two basic mechanisms, that is, nonadiabatic electronic transition and particle rearrangement, are presented here before giving their detailed mathematical expressions. Since the region of potential curve (or surface) crossing where electronic transition predominantly occurs is generally located away from the so-called reaction zone where particle rearrangement occurs, we can consider and treat these two mechanisms separately.[22]

Let us first consider the one-dimensional case, namely, a diatomic system such as ion–atom collision process. At relatively low collision energies, it is generally better to use the Born–Oppenheimer adiabatic electronic states as the basis states to treat the dynamics. Then the best known nonadiabatic transition is the Landau–Zener type of transition at an avoided-crossing point of Born–Oppenheimer states induced by the non-Born–Oppenheimer radial derivative operator, or the relative radial motion of the colliding particles. Since the nature of the electronic states of the two Born–Oppenheimer states changes at the avoided-crossing point, the nonadiabatic transition is induced by the variation of the adiabatic parameter R (internuclear distance). The transition is very much spatially localized at the avoided-crossing point. This is because the energy transfer between different kinds of degrees of freedom is generally most effective when the amount of transferred energy is minimum, and also because the nonadiabatic (non-Born–Oppenheimer) coupling term has a maximum at the avoided-crossing point. As is well known, the original Landau–Zener formula for the nonadiabatic transition probability between two adiabatic states is given by[19-21,23,24]

$$p_{LZ}^{(0)} = \exp[-2\delta_{LZ}^{(0)}] = \exp\left[-2\frac{\pi V_x^2}{\hbar v_x \Delta F}\right], \qquad (2.1)$$

where V_x is the *diabatic* coupling strength, ΔF is the difference between the slopes of the two *diabatic* potential curves, and v_x is the velocity of the relative motion at the crossing point R_x. This is a useful formula to comprehend the qualitative features of the transition. We have to be careful, however, about its quantitative accuracy in application, since several conditions are assumed in the derivation of this formula. The details are discussed in the next section.

There is another kind of radially induced nonadiabatic transition called Rosen–Zener (or Demkov) type.[9,10,19-21,25,26] This is a noncrossing near-resonant type in which two adiabatic states are parallel asymptotically. Although there is no rapid change in the nature of electronic states and no conspicuous avoided crossing, the nonadiabatic transition occurs in a spatially localized region at the point where the two adiabatic potential curves start to diverge. The original Rosen–Zener formula for the transition probability is

$$p_{RZ}^{(0)} = [1 + \exp(2\delta_{RZ}^{(0)})]^{-1} = [1 + \exp(\pi\Delta/\hbar\alpha v_x)]^{-1}, \qquad (2.2)$$

where Δ is the asymptotic energy difference of the adiabatic states and α is the exponent of the exponential diabatic coupling ($= Ae^{-\alpha R}$). The limit of $\Delta \to 0$ is called the exact resonance case. In the adiabatic limit, that is, when the exponent $\delta_{RZ}^{(0)}$ in Eq. (2.2) is large, the probability $p_{RZ}^{(0)}$ behaves similarly as $p_{LZ}^{(0)}$, but in the other limit ($\delta \to 0$) they behave quite differently. The similar care should be taken for the quantitative applications of this primitive formula (2.2). The more sophisticated formula is given in the next section. Both Landau–Zener and Rosen–Zener type of transitions are induced by the motion with respect to the internuclear distance R, and thus occur only between the states of the same electronic symmetry, namely, the selection rules are $\Delta\Lambda = 0$, $\pm \leftrightarrow \pm$ and $g(u) \leftrightarrow g(u)$, where Λ is the ordinary quantum number of the electronic angular momentum.

Coriolis coupling, on the other hand, causes a transition between the states of different electronic symmetry with the selection rules $|\Delta\Lambda| = 1$, $\pm \leftrightarrow \pm$ and $g(u) \leftrightarrow g(u)$. This is a nonadiabatic transition induced by the relative angular (rotational) motion of the colliding particles. This transition is of quite different nature from the radially induced one. First of all, the Born–Oppenheimer electronic states can cross with each other, since the symmetries are different. The transition is not spatially localized at the crossing point, even if a real crossing occurs and this position is most favorable for the transition from the energetics point of view as mentioned before. This is because the Coriolis (or rotational) coupling possesses very different nature from the radial coupling. The nonadiabatic coupling between Σ and Π states, for instance, is given by[27,28]

$$T_{rot} = -\frac{\hbar}{\mu R^2}[(K - \Lambda)(K + \Lambda + 1)]^{1/2}V_0|_{\Lambda=0} \qquad (2.3a)$$

with

$$V_0 = \frac{\hbar}{2}\langle \phi_e(\Pi)|L_+|\phi_e(\Sigma)\rangle, \qquad (2.3b)$$

where μ is the reduced mass of the system, K is the total (electronic L plus rotational L_R) angular momentum quantum number, and V_0 represents the electronic matrix element with L_+ the raising angular momentum operator. As is seen from Eqs. (2.3), the coupling has basically the R^{-2} dependence on

R and does not show any peak at the crossing point, but becomes strongest at the closest approach. There can be the following three typical cases:[29] (1) real curve crossing of the Born–Oppenheimer adiabatic states, (2) Rosen–Zener-type parallel states, and (3) crossing or degeneracy at united atom limit. Because of the peculiar properties mentioned above, the rotationally induced transitions cannot, in general, be treated by the semiclassical theories available for the radially induced transitions as far as we employ the Born–Oppenheimer state representation. This is an important thing to be kept in mind. The third case mentioned above seems to be all right for the application of the Landau–Zener formula, since the crossing point and the pole of the coupling coincide at $R = 0$. Even this is not true, however, because the analytical properties of the energies and the coupling as a function of R are different. The detailed discussions are given in the next section. Another important feature of the Coriolis coupling is the fact that this can lead to such states that cannot be reached by the radial coupling because of the different selection rule, although the coupling itself is not so strong compared to the latter especially at low collision energies.

The next coupling to be considered here is the spin–orbit interaction. This coupling causes a transition among the states of different Λ and Σ (molecular axis component of the spin angular momentum S) within the manifold of the same $\Omega = \Lambda + \Sigma$. The more convenient representation is to define the Born–Oppenheimer states $(\Omega_{g,u}^{\pm})$ of the electronic hamiltonian including the spin–orbit interaction. Then there remain again the two kinds of nonadiabatic (radially induced and rotationally induced) transitions. The selection rules are $\Delta\Omega = 0$, $\pm \leftrightarrow \pm$ and $g(u) \leftrightarrow g(u)$ for the radially induced transitions and $|\Delta\Omega| = 1$, $\pm \leftrightarrow \pm$ and $g(u) \leftrightarrow g(u)$ for the transitions induced by Coriolis coupling. Namely, all we have to do is just to replace the angular momenta L and K in the $L \cdot S$ coupling scheme by $J = L + S$ and $J_T = J + L_R$, respectively.

Finally, the effective coupling peculiarly associated with charge transfer is discussed. This originates from the fact that momentum transfer of an electron inevitably accompanies charge transfer, namely, the electron changes its momentum abruptly at the moment of transfer. In other words, this comes from the intrinsic nature of the three-body rearrangement collision problem. The ordinary close-coupling-type expansion in the Jacobi coordinates cannot satisfy the correct boundary condition, that is, the traveling electron boundary condition. In order to take this effect into account and to recover the Galilean invariance, the so-called electron translation factor (ETF) is incorporated into the expansion of the total wavefunction.[30] This effect of ETF was shown to be expressed as an additional coupling matrix and to give a correction to the original nonadiabatic couplings, both radial and rotational.[30,31] This effect corrects the origin dependence and the fictitious long-range coupling, which sometimes occur and give serious errors in the rotational coupling matrix element. Apart from these the ETF effect generally becomes important at high collision energies.

It is very important to note that the various couplings mentioned previously, which have different origins and characteristics, can be uniformly handled by the semiclassical theories developed for the radially induced transitions, if we introduce the new representation called "dynamical-state representation",[21,28][29,32] the basis states of which are the eigenstates of the hamiltonian matrix composed not only of the original electronic hamiltonian matrix, but also of all the other coupling matrices except for the radial coupling matrix. That is to say, if we diagonalize the whole hamiltonian of the system except for the *radial* kinetic energy operator, then all kinds of transitions associated with this original hamiltonian can be dealt with by the semiclassical theories for the radially induced transitions. The more substantial mathematical explanations are given in the next section. So far discussion was made only to the nonadiabatic transition probabilities, but in the actual collision processes the various phases as well play an important role. This is discussed also in the next section.

Let us next look into a multi-dimensional case. In general, a real surface intersection of dimension $N-2$ is possible, when the potential energy surfaces of the same electronic symmetry are the functions of N independent variables. The cases of $N = 2$ can be classified into the following three cases:[33] (1) avoided crossing, (2) conical intersection, and (3) glancing interaction. A typical example of the case (1) is shown in Fig. 1. The diabatic surfaces cross with each other along a line (this is called "seam"), and the adiabatic surfaces never cross and have the shape of hyperbolic cylinder oriented along the seam.[34] Thus, in the direction perpendicular to the seam, we have the same situation as the one-dimensional avoided-crossing case and the nonadiabatic coupling has a sharp peak at the avoided crossing. In the direction along the seam, however, there is no nonadiabatic coupling. The conical intersection is schematically shown in Fig. 2. The adiabatic surfaces are discontinuous at the crossing point. The Jahn–Teller intersection is one example of this. In all directions passing through the intersection point, real curve crossing occurs and the nonadiabatic coupling has a delta function peak at the crossing point. Along the lines passing nearby the crossing point, the ordinary one-dimensional avoided crossing scheme holds, namely, the nonadiabatic coupling has a Lorentzian shape with a peak at the minimum gap. The nonadiabatic transition probability along the trajectory passing through the apex of the cone is unity, although the measure of these trajectories is zero. The third case is shown in Fig. 3, where the two adiabatic surfaces are continuous everywhere. This corresponds to the Renner–Teller interaction. Along the trajectories passing through the point of contact, the nonadiabatic

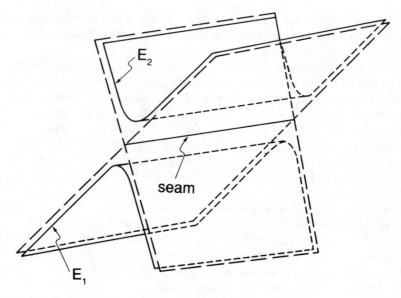

Figure 1.   Schematic view of an avoided crossing of the two-dimensional potential energy surfaces.

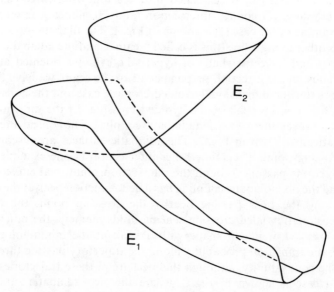

Figure 2.   Schematic view of a conical intersection of the two-dimensional potential energy surfaces

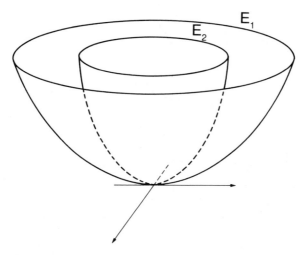

Figure 3. Schematic view of a glancing intersection of the two-dimensional potential energy surfaces.

coupling vanishes. Since the hamiltonian matrix contains quadratic terms only, the Landau–Zener type of analysis is not adequate and a separate careful consideration is required.

Coriolis coupling connects the states of different electronic symmetry also in the multidimensional case. In the case of triatomic system, for instance, the A' and A'' states of the C_s symmetry are coupled rotationally. A' (A'') is symmetric (antisymmetric) in reflection with respect to the molecular plane. The idea of the unified treatment of radial and rotational couplings mentioned previously also holds true here. The angular parts can be diagonalized first to define a new set of effective potential energy surfaces.[32]

Let us finally consider briefly the mechanism of particle rearrangement (reaction). Reaction occurs in the so-called reaction zone where the particles get close together. Generally, the (conventional) transition state, or the saddle point of potential energy surface, is considered to be the most important representative point for reaction to occur. This point gives a potential barrier for reaction and is considered to represent its bottleneck. The potential ridge dividing the reactant and the product valleys, however, gives more general concept to represent the position for reaction to occur. In most cases a saddle point is actually contained within the ridge line as a limiting point. Sometimes the saddle point is located quite far away from the central reaction zone around the ridge line. In this case the saddle point represents, of course, the reaction barrier, but cannnot represent the position for rearrangement to

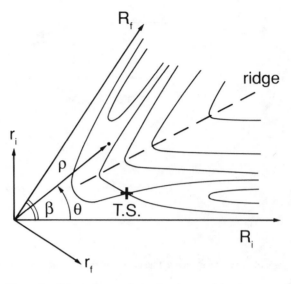

Figure 4.   Shcematic two-dimensional potential energy contour.

occur. In the terminology of the classical trajectory method, the "recrossing" correction to the conventional transition-state theory becomes important in this case. This has actually motivated the construction of the variational transition-state theory.[35] The effect and the role of potential ridge in reaction dynamics can be well visualized and incorporated into the theoretical framework by using the hyperspherical coordinates.[36] Figure 4 schematically shows a potential energy contour for a two-dimensional (collinear reaction) system. The hyperspherical coordinates in this case are nothing but the polar coordinates (ρ, θ). The cross sections of potential energy surface at constant hyperradius ρ are schematically shown in Fig. 5. At large ρ (Fig. 5a) the reactant and the product valleys are well separated by a high potential barrier in between (fragmentation region). The vibrational states localized in each valley can hardly interact with each other, even though they can be accidentally degenerate. At small ρ (Fig. 5c), on the other hand, there is no potential barrier and all particles move collectively in a single well (condensation region). At intermediate ρ (Fig. 5b) the vibrational states, asymptotically corresponding to either reactant or product vibrational state, interact strongly through the potential barrier. The trace of this barrier constitutes the potential ridge, which provides the boundary of a transition from the condensation region to the fragmentation region.[37] This transition

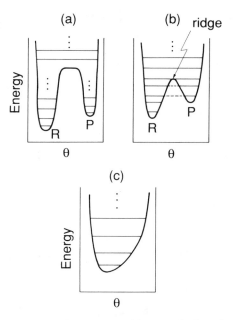

Figure 5. Potential energies as a function of hyperangle θ at fixed ρ: (*a*) large ρ, (*b*) intermediate ρ, (*c*) small ρ.

represents nothing but "reactive transition." This feature can be better visualized by drawing the vibrational eigenvalues as a function of ρ. Figure 6 shows such an example. This is the case for the collinear Cl + HBr → HCl + Br reaction.[36] It should be noted, however, that the LEPS potential energy surface employed is not designed to accurately mimic the real one for this system. This presents a rather peculiar system that the two levels HBr(v) and HCl($v + 2$) are in near resonant and are well separated from the others, and the vibrational adiabaticity holds well. Reactions occur predominantly between these pairs of states. Besides, the reactive transition is very much spatially localized near the potential ridge. This is explicitly demonstrated in Fig. 7. This shows the numerically calculated accumulated reaction probability as a function of ρ. Although Fig. 7. is the result of the 3-D reaction for (relative angular momentum) = 0 in the adiabatic-bend approximation explained in Section V C, the essential feature mentioned below is the same as in the collinear case. The transition clearly occurs in a very narrow region near the potential ridge. As is seen from Fig. 6, the two near-resonant levels start to diverge near the potential ridge because of this tunneling interaction.

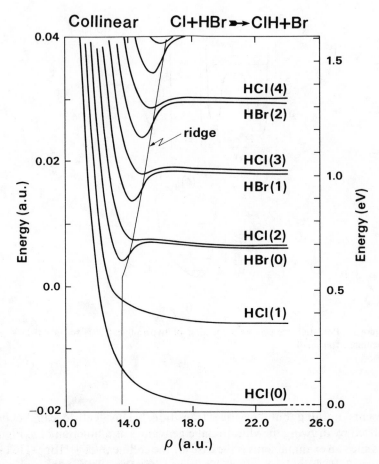

Figure 6. Adiabatic vibrational potential energy curves of the collinear ClHBr system (from Ref. 36).

This level diagram reminds us of the Rosen–Zener model. This reactive transition can actually be treated analytically by the semiclassical theory of nonadiabatic transition. More detailed discussions are given in Section V. Reaction can thus be viewed as a vibrationally nonadiabatic transition. Potential energy diagram such as Fig. 6 changes very much, of course, depending on potential energy surface topography and mass combination. In other words, this indicates that the hyperspherical coordinate approach is convenient to investigate these effects. Some more detailed explanations and examples are given in Section V.

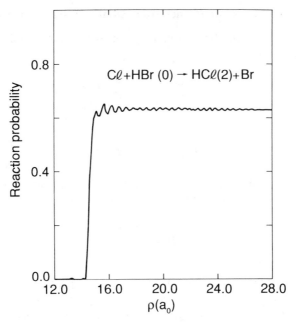

Figure 7. Reaction probability as a function of ρ at $E_{coll} = 0.77\,\text{eV}$ for $Cl + HBr(0) \rightarrow HCl(2) +$ Br (from Ref. 36).

III. SEMICLASSICAL THEORY OF NONADIABATIC TRANSITION

Since the pioneering works done by Landau,[23] Zener,[24] Stueckelberg,[38] and Rosen and Zener,[25] a lot of effort has been paid for developing better semiclassical theories of nonadiabatic transition. Many review articles have been written.[4,9,10,19–21] Because of the interdisciplinarity of the concept as a basic mechanism of state or phase change, the significance of nonadiabatic transition has recently been more recognized in various fields.[36,39–47] Here, the presently available best formulas for the most basic two-state problems are presented and compared with the original ones. First, the dynamical state (or generalized adiabatic state) representation is introduced.[28,29,32] As was briefly explained in the previous section, this representation enables us to treat all kinds of coupling by the semiclassical theories for radially induced transition. This is because the representation transforms the analytical property of the energies and couplings as a function of complex-R into the same one as that of radial coupling problem. This is explained in this section in more detail.

A. Dynamical State (Generalized Adiabatic State) Representation

The total hamiltonian of a diatomic system with the reduced mass μ can be generally written as

$$H = -\frac{\hbar^2}{2\mu}\frac{1}{R^2}\left(\frac{\partial}{\partial R}R^2\frac{\partial}{\partial R}\right) + H_{\text{dyn}} \tag{3.1a}$$

and

$$H_{\text{dyn}} = H_{\text{el}} + H_{\text{cor}} + H_{\text{rot}} + H', \tag{3.1b}$$

where H_{el}, H_{cor}, and H_{rot} represent the electronic hamiltonian, the Coriolis interaction, and the rotational motion, respectively, and

$$H' = \frac{\hbar^2}{2\mu R^2}L^2 \tag{3.2}$$

with L being the electronic angular momentum operator. The nonadiabatic radial and rotational (Coriolis) couplings originate from the first term of Eq. (3.1a) and H_{cor}, respectively. The spin–orbit coupling can be included in H_{el}. H' couples the states of the same electronic symmetry, but its contribution is small and is usually neglected.

The dynamical states (DS) are defined as the eigenstates of H_{dyn},[28,29,32]

$$H_{\text{dyn}}\Psi_n^K(\mathbf{r}, \hat{R}:R) = E_n^K(R)\Psi_n^K(\mathbf{r}, \hat{R}:R), \tag{3.3}$$

where $\mathbf{r}$ represents the electron coordinates collectively. Since H_{dyn} is hermitian, and depends on R only parametrically, the eigenvalues $E_n^K(R)$ are real functions of R. In this representation the ordinary quantum number Λ is not a good quantum number any more, and the eigenstates depend on the total angular momentum K $(= L + L_R)$, since the Coriolis coupling is diagonalized. That is to say, the dynamical states are the eigenstates of the rotating complex of the whole system. As is clear from the definition, transitions among the dynamical states are exclusively induced by the radial coupling,

$$T_{\text{rad}} = \left\langle \Psi_1 \left| \frac{\partial}{\partial R} \right| \Psi_2 \right\rangle_r. \tag{3.4}$$

The reason why the DS representation is convenient is explained subsequently. Let us introduce first the original treatment of Landau.[23,48] By avoiding the very oscillatory integral on the real R axis and going into the complex R plane, the nonadiabatic transition probability is given as

$$p \simeq \left| \exp\left\{ i \int_{Re(R_*)}^{R_*} [k_1(R) - k_2(R)] dR \right\} \right| \tag{3.5}$$

with

$$k_n(R) = \left(\frac{2\mu}{\hbar^2} (E - E_n(R)) \right)^{1/2}, \tag{3.6}$$

where $E_n(R)$ is the ordinary adiabatic (i.e., Born–Oppenheimer) state, and R_* is the complex crossing point,

$$E_1(R_*) = E_2(R_*). \tag{3.7}$$

If we introduce the effective relative velocity

$$v(R) = \sqrt{\frac{1}{2\mu}} [\sqrt{E - E_1(R)} + \sqrt{E - E_2(R)}], \tag{3.8}$$

then Eq. (3.5) can be rewritten as

$$p \simeq \left| \exp\left\{ \frac{i}{\hbar} \int_{Re(R_*)}^{R_*} \frac{\Delta E(R)}{v(R)} dR \right\} \right|, \tag{3.9}$$

with $\Delta E(R) = E_2(R) - E_1(R) > 0$ for real R. Here the following natural question arises: Why is the pre-exponential factor in Eq. (3.5) unity? If we employ the first-order perturbation theory in the adiabatic-state representation, then there should naturally appear some kind of scar of the nonadiabatic radial coupling T_{rad}, which does not seem to be unity at all. The "unity" is actually the result of renormalization of the higher-order terms of T_{rad}. If we use the Hellmann–Feynman theorem, we obtain

$$T_{rad} = \left\langle \Psi_1 \left| \frac{\partial H_{el}}{\partial R} \right| \Psi_2 \right\rangle \Big/ [\Delta E(R)]^2. \tag{3.10}$$

With use of the diabatic potentials $V_n(R)$ and coupling $V(R)$, we have as usual

$$\Delta E(R) = \{ [V_1(R) - V_2(R)]^2 + 4V^2(R) \}^{1/2}. \tag{3.11}$$

This implies that the complex zero R_* of $\Delta E(R)$ is of order one-half and that T_{rad} has a pole of order unity at the same position. This "analytical property"

TABLE I

Analytical Properties of the Various Nonadiabatic Coupling Schemes

Coupling Scheme	Potential Energy Difference $\Delta E \propto$	Coupling $T \propto$
Adiabatic-State Representation		
Radial	$(R - R_*)^{1/2}$	$(R - R_*)^{-1}$
(R_*: complex)		
Rotational		
(a) Degeneracy at $R = 0$	R^2	
(b) Crossing at finite $R = R_x$	$R - R_x$	R^{-2}
(c) No crossing	Constant	
Dynamical-State Representation		
Any transition	$(R - R_*)^{1/2}$	$(R - R_*)^{-1}$
(R_*: complex)		

underlies the Landau–Zener problem and is actually the reason why the preexponential factor is exactly unity. It is clear now why the Landau–Zener formula cannot be applied to the Coriolis coupling problem. The Coriolis coupling given by Eqs. (2.3) has no pole at crossing point, but has a pole of order 2 at $R = 0$. The DS representation can transform any coupling scheme into the same analytical structure as that of the ordinary radial coupling problem. Whatever the coupling $V(R)$ is, $\Delta E(R)$ and $T_{\rm rad}$ in the new representation have a complex zero of order one-half and a pole of order unity at the same position, respectively. The various cases mentioned in the previous section are summarized in Table I.[29] Thus, once we move into the DS representation, the semiclassical theories well developed for the radial coupling problems can be utilized uniformly for the various coupling cases, no matter what the original coupling scheme is. It should also be noted that the nonadiabatic transition occurs in a spatially localized region at the *new* avoided-crossing point. This is a useful property for generalizing the two-state theory to a multilevel problem, as will be discussed in Section IV. The various coupling cases can be basically classified into the following two: Landau-Zener type (Fig. 8a) and Rosen–Zener (or Demkov) type (Fig. 8b). The best available formulas for these cases are given in the following subsections.

The idea of the DS representation can be generalized, at least conceptually, to a N-particle multidimensional system with use of the hyperspherical

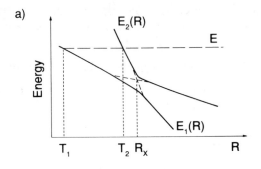

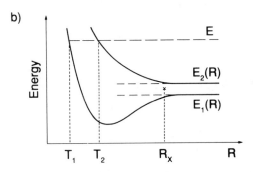

Figure 8. Schematic potential energy curves. (*a*) Landau–Zener type, (*b*) Rosen–Zener (Demkov) type.

coordinate system (ρ, Ω_H). The total hamiltonian can be expressed as

$$H = -\frac{\hbar^2}{2\mu_N} \frac{1}{\rho^{3N-4}} \frac{\partial}{\partial \rho} \rho^{3N-4} \frac{\partial}{\partial \rho} + H_{dyn}, \qquad (3.12)$$

with

$$H_{dyn} = \frac{\hbar^2}{2\mu_N} \frac{\Lambda^2(\Omega_H)}{\rho^2} + H_{el}, \qquad (3.13)$$

where $\Lambda(\Omega_H)$ represents the grand angular momentum operator with respect to the hyperangle variables Ω_H, and μ_N is the reduced mass of the system. The hyperradius ρ and the eigenstates of H_{dyn} represent, respectively, the size of the system and the rotating collision complex at fixed ρ. The eigenstates may be called "generalized adiabatic states."

In the following discussions, DS representation or the generalized adiabatic-state representation is tacitly assumed whenever it is appropriate, and the superfix K to energies is omitted.

B.   Landau–Zener-Type Nonadiabatic Transition

This is an avoided-crossing case. The nonadiabatic transition probability amplitudes are given as follows:

$$I_{1 \leftarrow 1}^{\mathrm{LZ}} = \sqrt{1 - p_{\mathrm{LZ}}} \exp(i\phi_s), \tag{3.14a}$$

$$I_{2 \leftarrow 2}^{\mathrm{LZ}} = \sqrt{1 - p_{\mathrm{LZ}}} \exp(-i\phi_s), \tag{3.14b}$$

$$I_{2 \leftarrow 1}^{\mathrm{LZ}} = -\sqrt{p_{\mathrm{LZ}}} \exp(-i\sigma_0), \tag{3.14c}$$

and

$$I_{1 \leftarrow 2}^{\mathrm{LZ}} = \sqrt{p_{\mathrm{LZ}}} \exp(i\sigma_0), \tag{3.14d}$$

where

$$p_{\mathrm{LZ}} = \exp(-2\delta) \tag{3.15}$$

$$\sigma_0 + i\delta = \int_{R_X}^{R_*} [k_1(R) - k_2(R)] dR, \tag{3.16}$$

$$\phi_s = (\delta/\pi) \log(\delta/\pi) - \delta/\pi + \pi/4 - \arg \Gamma(1 + i\delta/\pi), \tag{3.17}$$

and

$$R_x = \mathrm{Re}(R_*). \tag{3.18}$$

The derivation of these formulas is based on the comparison equation method or the phase integral method.[3,9,49,50] Since it is given elsewhere,[21a,c] it is not repeated here. Unfortunately, Eq. (3.16) involves a complex integral in the complex R plane and presents an obstacle to understanding by the experimentalists. It should be kept in mind, however, that the preceding formulas are quite accurate.

In the rest of this subsection let us focus our attention on the characteristics of the preceding formulas especially in comparison with the original Landau–Zener formula given by Eq. (2.1). First of all, it should be noted that the formulas are not based on the perturbation theory and interestingly depend only on the (generalized) adiabatic potentials $E_n(R)$, apparently not including any scars of T_{rad}. This is actually one of the big advantages of going into the complex R plane. The effects of T_{rad} are incorporated effectively to an infinite order. The second thing to be noted is that Eqs. (3.14) present

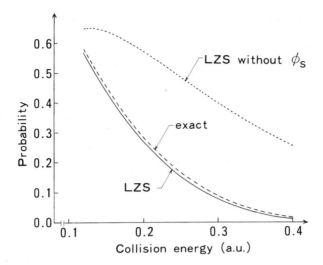

Figure 9. Transition probabilities as a function of collision energy for a model of the exponentially coupled two attractive Morse potentials (from Ref. 32). ---: exact numerical calculation; ———: two-state Landau–Zener–Stückelberg formula; ⋯: two-state Landau–Zener–Stückelberg formula without ϕ_s.

not the probabilities, but the more fundamental transition probability amplitudes. The phases σ_0 and ϕ_s are the phases due to the nonadiabatic transition. These are totally missing in the original Landau–Zener formula. Sometimes these give a nonnegligible contribution. Figure 9 gives an example. The phase ϕ_s is usually called Stokes phase correction, since this originates in the Stokes phenomenon of asymptotic function in the complex R plane.[51-53] The exponent δ defined by Eq. (3.16), which looks totally different from $\delta_{LZ}^{(0)}$ given in Eq. (2.1), can actually be reduced to $\delta_{LZ}^{(0)}$ under the following assumptions: (1) linear diabatic potentials $V_n(R) = F_n(R - R_x) + V_0$, (2) constant diabatic coupling $V(R) = V_x$, and (3) constant relative velocity $v(R) = v_x$ (i.e., straight line trajectory of relative motion). Using these simplifications and the transformation from Eq. (3.5) to Eq. (3.9), we have

$$\sigma_0 = 0, \tag{3.19}$$

and

$$\delta = \frac{1}{\hbar} \mathrm{Im} \int_{R_x}^{R_*} \frac{\Delta E(R)}{v(R)} \, dR$$

$$= \frac{\Delta F}{\hbar v_x} \int_0^{y_0} \sqrt{y_0^2 - y^2} \, dy = \frac{\pi V_x^2}{\hbar v_x \Delta F} = \delta_{LZ}^{(0)}, \tag{3.20}$$

where $y_0 = \mathrm{Im}(R_*) = 2V_x/\Delta F$. The preceding analysis clearly tells the limits of the original Landau–Zener formula (2.1). Namely, in the quantitative applications of this original formula (2.1), we have to pay attention to the validity of the preceding assumptions and the neglect of the various phases, not only σ_0 and ϕ_S but also other phases along the potentials $E_n(R)$ (see next section). For instance, it is clear that the formula is not applicable when the turning point coincides with or becomes larger than the crossing point R_x (classically forbidden case). This case gives a difficulty also to the accurate treatment described previously, but can be managed reasonably well by modifying the exponent δ. For instance, when both turning points T_1 and T_2 of the two potentials are larger than R_x, then δ may be modified to[54]

$$\delta = \mathrm{Im} \int_{R_x}^{R_*} [k_1(R) - k_2(R)] dR + \int_{R_x}^{T_2} |k_2(R)| dR - \int_{R_x}^{T_1} |k_1(R)| dR. \quad (3.21)$$

When the diabatic potentials have different signs of the slope *and* the energy is lower than the crossing point (nonadiabatic tunneling), even the accurate formulas given by Eqs. (3.14) are not applicable. The presently available semiclassical theory for this case is less accurate, unfortunately.[21,55] We do not go into the details here.

As was mentioned before, the Landau–Zener type of nonadiabatic transition is very much spatially localized around the avoided crossing point. This is demonstrated in Fig. 10. This is the case of rotationally coupled

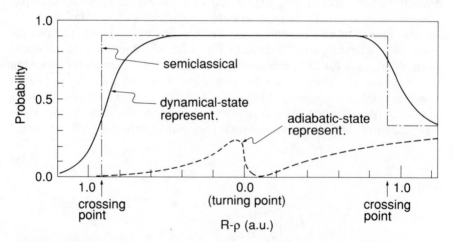

Figure 10. Localization of rotationally induced transition in the dynamical-state representation (from Ref. 21c).

two-state problem.[21c,29b] Transitions occur very locally at the crossing points both on the way in and the way out in the DS representation. The step function is the semiclassical approximation in the DS representation. Interestingly, the transitions in the adiabatic (Born–Oppenheimer) state representation are not localized at all. This clearly demonstrates the advantage of the DS representation.

C. Rosen–Zener (Demkov)-Type Nonadiabatic Transition

This is a noncrossing case (Fig. 8b). In contrast to the Landau–Zener type, the potential energy difference does not diverge as a function of R, or not strongly dependent on R, but the coupling depends strongly on R.[25,26] Analytical properties of the (generalized) adiabatic potential energy difference and the nonadiabatic coupling are essentially the same as those in the Landau–Zener case (complex zero of order one-half and pole of order unity). The derivation of the formula is, however, different from the Landau–Zener case because of the previously mentioned differences in potential coupling scheme.[56] The nonadiabatic transition probability amplitudes are given as follows:[21]

$$I^{RZ}_{1 \leftarrow 1} = I^{RZ}_{2 \leftarrow 2} = \sqrt{1 - p_{RZ}}, \tag{3.22a}$$

$$I^{RZ}_{2 \leftarrow 1} = -\sqrt{p_{RZ}} \exp(-i\sigma_0), \tag{3.22b}$$

and

$$I^{RZ}_{1 \leftarrow 2} = \sqrt{p_{RZ}} \exp(i\sigma_0), \tag{3.22c}$$

with

$$p_{RZ} = [1 + \exp(2\delta)]^{-1}, \tag{3.23}$$

where σ_0 and δ are the same as those defined by Eq. (3.16). It should be noted that the Stokes phase correction ϕ_S does not appear in this case. Exactly speaking, what was derived mathematically is the total transition probability $\text{sech}^2(\delta)/2$ corresponding to a double (in and out) passage of the transition point.[56] The expression (3.23) is just obtained by the physical interpretation of the double passage, that is, $2p_{RZ}(1 - p_{RZ}) = \text{sech}^2(\delta)/2$.

The exponent δ can be reduced to $\delta^{(0)}_{RZ}$ of Eq. (2.2) under the following assumptions: (1) constant potential energy difference Δ, (2) exponential dependence on R of the coupling potential ($= Ae^{-\alpha R}$), (3) constant relative velocity $v(R) = v_x$, and (4) neglect of the contributions from the complex crossing points except for the one closest to the real R axis. The complex crossing points R_* in the upper-half plane of R are given by

$$R_* = R_x + i\left(\frac{\pi}{2\alpha} + \frac{2n\pi}{2\alpha}\right) \quad (n = 0, 1, 2, \ldots), \tag{3.24}$$

where

$$R_x = \frac{1}{\alpha}\log(2A/\Delta). \tag{3.25}$$

This indicates that R_x corresponds to the position where $\Delta = 2V(R_x) = 2Ae^{-\alpha R_x}$ is satisfied. $\delta_{RZ}^{(0)}$ is obtained as follows:

$$\delta = \mathrm{Im}\int_{R_x}^{R_*(n=0)} [k_1(R) - k_2(R)]dR$$

$$\simeq \frac{\Delta}{\hbar v_x}\int_0^{y_0} \mathrm{Re}[1 + e^{-2\alpha iy}]^{1/2}dy = \frac{\pi\Delta}{2\hbar v_x\alpha} \equiv \delta_{RZ}^{(0)}, \tag{3.26}$$

here $y = \mathrm{Im}(R)$ and $y_0 = \pi/2\alpha$. It should be noted that the phase σ_0 does not vanish in this case.

As in the Landau–Zener case, in the quantitative applications we have to pay attention to the justification of the previously mentioned assumptions, although the fourth assumption is generally not bad. When the collision energy is low and the turning points become larger than R_x, the modification of δ should be made in the same way as in Eq. (3.21).

IV.   MULTICHANNEL CURVE CROSSING PROBLEM

A.   General Formulation of Electronic Transitions
in Diatomic Systems

Since the DS representation makes all transitions spatially localized at R_x, a multichannel curve crossing problem can be decomposed into a series of two-state problems. This is not possible, however, if we employ the ordinary Born–Oppenheimer state representation for a collision system involving Coriolis couplings. Since a transition induced by Coriolis coupling is delocalized in between turning point and crossing point in the adiabatic state representation as was demonstrated in Fig. 10, other crossings located in that region can not be separated out and treated by the two-state theories. This is a big advantage of the DS representation. With use of this representation a multichannel curve crossing problem is formulated here.

1. *The Case Without Closed Channel*

In order to facilitate understanding of the formulation, we employ an example shown in Fig. 11, which is a three-open-channel problem.[54] This corresponds to the $(Li + Na)^+$ collision system and represents the three dynamical states at the total angular momentum quantum number $K = 1700$. The avoided crossing at R_{c_2} between $E_2(R)$ and $E_3(R)$ is originally a real crossing between Σ and Π states coupled by Coriolis interaction. The noncrossing situation at R_{c_1} corresponds not only originally, but also in the DS representation to the Rosen–Zener type radial coupling problem. The potential $E_2(R)$ is,

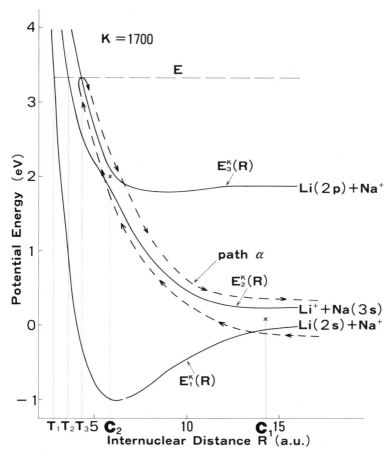

Figure 11. Lowest three dynamical-state potential energy curves for $(Li + Na)^+$ with $K = 1700$. Dashed line represents a possible path for the transition $Li(2s) + Na^+ \rightarrow Li^+ + Na(3s)$.

however, modified, very slightly though, from the original adiabatic Σ state by the Coriolis coupling with the above Π state at R_{c_2}. This presents another nice thing about the DS representation: a third state effect can be incorporated. An interesting example is shown in Section IV C.

Since R_{c_1} and R_{c_2} are well separated, the nonadiabatic transitions there can be described by the semiclassical two-state theories given in the previous section. Thus the present collision system can be expressed by a diagram shown in Fig. 12. The solid lines with arrow, the circles with T, the rectangular with $I(O)$ represent, respectively, wave propagation along dynamical state without any transition, wave reflection at turning point, and nonadiabatic transition in the inward (outward) relative motion. The wave propagation means the phase accumulation along the specified distance. $I(O)$ represents a matrix connecting the waves on the right and the left side of the transition point in the incoming (outgoing) segment.

This physical picture of the scattering process can be embodied into the following explicit expression of S matrix[28]:

$$S = P_{C_1 \infty} O_{C_1} P_{C_1 C_2} O_{C_2} P_{C_2 T C_2} I_{C_2} P_{C_1 C_2} I_{C_1} P_{C_1 \infty}, \qquad (4.1)$$

where P is a diagonal matrix, representing wave propagation in the region specified by the suffices. Equation (4.1) can be read from the right to the left. Explicit expressions of each matrix (3×3) are as follows:

$$(P_{A\infty})_{nm} = \delta_{nm} \exp[i\eta_n^{(0)}(A)], \qquad (4.2)$$

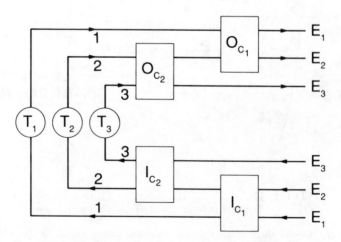

Figure 12. Diagram corresponding to Fig. 11.

$$(P_{AB})_{nm} = \delta_{nm} \exp\left(i \int_A^B k_n(R)dR \right), \tag{4.3}$$

$$(P_{ATA})_{nm} = \delta_{nm} \exp\left(2i \int_{T_n}^A k_n(R)dR + \frac{\pi}{2}i \right), \tag{4.4}$$

$$(I_{C_1})_{nm} = \begin{cases} I_{n \leftarrow m}^{RZ}, & \text{for } n, m = 1, 2, \\ \delta_{nm}, & \text{otherwise,} \end{cases} \tag{4.5}$$

$$(I_{C_2})_{nm} = \begin{cases} I_{n \leftarrow m}^{LZ}, & \text{for } n, m = 2, 3, \\ \delta_{nm}, & \text{otherwise,} \end{cases} \tag{4.6}$$

and

$$O_A = \tilde{I}_A \text{ (transposed)}, \tag{4.7}$$

where

$$\eta_n^{(0)}(A) = \int_A^\infty [k_n(R) - k_n(\infty)]dR - ik_n(\infty)A + \frac{i}{2}K\pi, \tag{4.8}$$

and $I_{n \leftarrow m}^{RZ}(I_{n \leftarrow m}^{LZ})$ is the Rosen–Zener (Landau–Zener) type nonadiabatic transition amplitude for $m \rightarrow n$ given by Eqs. (3.22) [Eqs. (3.14)]. It should be noted that $k_n(R)$ is defined by Eq. (3.6) with the dynamical state potential energy $E_n^K(R)$. The matrices I and O are defined as follows: When we write the semiclassical wave functions on the right and left sides of the transition point, R_{C_1} for instance, as

$$\Psi_n^{\text{right}} \cong - V_n' \varphi_n^+ (R:R_{C_1}) + V_n'' \varphi_n^- (R:R_{C_1}) \tag{4.9a}$$

and

$$\Psi_n^{\text{left}} \cong - U_n' \varphi_n^+ (R:R_{C_1}) + U_n'' \varphi_n^- (R:R_{C_1}), \tag{4.9b}$$

I and O connect the coefficient vectors as

$$\mathbf{U}'' = I\mathbf{V}'' \quad \text{and} \quad \mathbf{V}' = O\mathbf{U}', \tag{4.10}$$

where $\mathbf{V}'$ is a column vector with component V_n', and so on and

$$\varphi_n^\pm (R:R_{C_1}) = \sqrt{\frac{\mu}{2\pi k_n(R)}} \exp\left(\pm i \int_{R_{C_1}}^R k_n(R)dR \right). \tag{4.11}$$

2.   *General Case Involving Closed Channels*

In this subsection we consider a general case, which involves closed channels. The diagrammatic technique[57] explained in the previous subsection is still useful, of course, but unfortunately not straightforward to derive explicit expression of S matrix. This is especially so when the system contains three or more channels. Take, for instance, the system shown in Fig. 13. This is an elastic scattering problem with two closed channels. The corresponding diagram is given by Fig. 14. The total elastic scattering phase shift η is

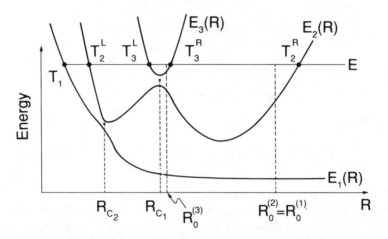

Figure 13.   Model three-state potential energy diagram.

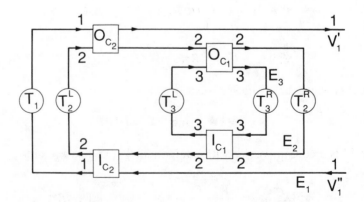

Figure 14.   Diagram corresponding to Fig. 13.

expressed as

$$\eta = \eta^{(0)}(R_{C_2}) + \frac{1}{2i} \log(V'_1/V''_1), \tag{4.12}$$

where V'_1 and V''_1 are the coefficients of the outgoing and incoming waves at $R = R_{C_2} + 0$ along the potential $E_1(R)$, respectively. The derivation of V'_1/V''_1 from the diagram is not straightforward, as can be easily conjectured. A much easier and more direct general procedure is explained here. The idea is not to try to obtain S matrix directly, but to construct a χ matrix, which spans not only open but also closed channels. This matrix is equivalent to the χ matrix introduced by Seaton in his formulation of the multichannel quantum defect theory[58], and is related to the S matrix by

$$S = \chi_{OO} - \chi_{OC}(\chi_{CC} - 1)^{-1}\chi_{CO}. \tag{4.13}$$

The suffix $O(C)$ means open (closed), and χ_{OO} represents the open channel–open channel block of the whole χ matrix ($N \times N$), and so on. The number of open (closed) channels is assumed to be $N_O(N_C)$ with $N = N_O + N_C$.

A general way of constructing the χ matrix is explained and the more convenient relation with S matrix than Eq. (4.13) is derived below. First, the χ matrix can be expressed as

$$\chi = X\bar{\chi}X, \tag{4.14}$$

where

$$(X_{OO})_{nm} = \delta_{nm} \exp[i\eta_n^{(0)}(R_0^{(n)})] \quad (n, m = 1, 2, \dots, N_O), \tag{4.15a}$$

$$X_{OC} = X_{CO} = 0, \tag{4.15b}$$

$$(X_{CC})_{nm} = \delta_{nm} \exp[\pi i v_n(R_0^{(n)})] \quad (n, m = N_O + 1, \dots, N_O + N_C = N), \tag{4.15c}$$

and

$$v_n(R_0^{(n)}) = \frac{1}{\pi}\left(\int_{R_0^{(n)}}^{T_n^R} k_n(R)dR + \frac{\pi}{4}\right). \tag{4.16}$$

T_n^R is the right-side turning point of the potential $E_n(R)$, and $R_0^{(n)}$ is an appropriately chosen position larger than the right-most basic element along the state n in the diagram and is smaller than T_n^R if the latter exists (see Fig. 14 in the present example). The thus defined $\bar{\chi}$ matrix describes the dynamics at $R \leqslant R_0$, and does not distinguish open and closed channels. Therefore, this matrix can be directly constucted in the same way as in the previous subsection. In the case of the example of Fig. 13, the $\bar{\chi}$ matrix is given explicitly as follows:

$$\bar{\chi} = P_{C_1 R_0} O_{C_1} P'_{C_1 C_2} O_{C_2} P'_{C_2 T C_2} I_{C_2} P'_{C_1 C_2} I_{C_1} P_{C_1 R_0}, \tag{4.17}$$

where the matrices P_{AB}, I_A, and O_A are the same as before,

$$P'_{C_1 C_2} = \begin{cases} \delta_{nm} \exp\left(i \int_{R_{C_1}}^{R_{C_2}} k_n(R) dR \right), & \text{for } n, m = 1, 2, \\ \delta_{nm}, & \text{otherwise,} \end{cases} \tag{4.18}$$

and

$$P'_{C_2 T C_2} = \begin{cases} \delta_{nm} \exp\left(2i \int_{T_n^L}^{R_{C_2}} k_n(R) dR + \frac{\pi}{2} i \right), & \text{for } n, m = 1, 2 \\ \delta_{nm} \exp\left(2i \int_{T_3^L}^{R_{C_1}} k_3(R) dR + \frac{\pi}{2} i \right), & \text{otherwise.} \end{cases} \tag{4.19}$$

In this case R_0 can be taken to be independent of the channel number n as $R_{C_2} < R_0 < T_3^R$. It should be noted that the (3.3)—elements of $P'_{C_1 C_2}$ and $P'_{C_2 T C_2}$ are different from the other diagonal elements. Even though T_3^L is larger than R_{C_2}, the adiabatic propagation on $E_3(R)$ from R_{C_1} to T_3^L and back to R_{C_1} is put in $P'_{C_2 T C_2}$ for convenience. This element is, of course, interchangable with the (3.3) element (unity) of $P'_{C_1 C_2}$.

Inserting Eq. (4.14) into Eq. (4.13) and using the fact that $X_{OC} = X_{CO} = 0$, then we obtain

$$S = X_{OO} \bar{S} X_{OO}, \tag{4.20}$$

with

$$\begin{aligned} \bar{S} &= \bar{\chi}_{OO} - \bar{\chi}_{OC} (\bar{\chi}_{CC} - X_{CC}^{-2})^{-1} \bar{\chi}_{CO} \\ &= \bar{\chi}_{OO} - \bar{\chi}_{OC} (\bar{\chi}_{CC} - e^{-2\pi i v})^{-1} \bar{\chi}_{CO}, \end{aligned} \tag{4.21}$$

where

$$(e^{-2\pi i v})_{nm} = \delta_{nm} \exp(-2\pi i v_n). \tag{4.22}$$

It is clear now that by the formulation described previously the effects of closed channels can be straightforwardly incorporated into the S matrix without difficulty. Resonances in the scattering correspond to the complex energy solutions of

$$\det[\bar{\chi}_{CC} - e^{-2\pi i v}] = 0. \tag{4.23}$$

The real and imaginary parts of the solutions give the resonance position and width, as usual.

The preceding method can further be generalized to the case that involves tunneling and nonadiabatic tunneling. We do not go into the details here. Furthermore, the present formulation can be shown to have a close connection with the multichannel quantum defect theory and to be suitable for uniformly describing the various processes: not only scattering, but also bound state and radiative (or half-collision) processes. More detailed discussions about the present formulation and the close connections to the quantum defect theory are presented elsewhere.[59]

Once the scattering matrix is obtained in the DS representation, the calculation of not only total but also differential cross sections can be formulated without any ambiguity.[54] The Coriolis coupling effect can be incorporated naturally. The scattering amplitude for the transition $n \to m$ can be generally given by

$$f_{nm}(\theta) = \frac{1}{2ik_n} \sum_K (2K + 1) S_{mn}^K P_K(\cos \theta), \qquad (4.24)$$

where S_{mn}^K is the S matrix element, $P_K(X)$ is the Legendre polynomial, and θ is the scattering angle. The differential cross section $\sigma_{nm}(\theta)$ is given by

$$\sigma_{nm}(\theta) = |f_{nm}(\theta)|^2. \qquad (4.25)$$

For convenience we introduce the impact parameter b by $b = (K + \frac{1}{2})/k = (K + \frac{1}{2})/\sqrt{k_n k_m}$ and the conventional approximation

$$P_K(\cos \theta) \simeq \left[\frac{2}{\pi (K + \frac{1}{2}) \sin \theta} \right]^{1/2} \cos \left[(K + \frac{1}{2})\theta - \pi/4 \right], \qquad (4.26)$$

and express S_{mn}^K as

$$S_{mn}^K = \sum_\alpha [P_{mn}^\alpha(b)]^{1/2} \exp[i\phi_{mn}^\alpha(b)], \qquad (4.27)$$

where the suffix α distinguishes the possible paths which enter along the potential $E_n(R)$ and exit along $E_m(R)$, and $P_{mn}^\alpha(b)$ and $\phi_{mn}^\alpha(b)$ represent the overall transition probability and the total phase along the path α. Then from Eq. (4.24) we have

$$f_{nm}(\theta) \simeq \left[\frac{k_n^2}{2\pi k \sin \theta} \right]^{1/2} \sum_\alpha \int db [b P_{mn}^\alpha(b)] [\exp(i\varphi_-) - \exp(i\varphi_+)], \qquad (4.28)$$

where

$$\varphi_\pm = \phi^\alpha_{mn} \pm (kb\theta + \pi/4). \tag{4.29}$$

Here we employ the stationary phase approximation to evaluate integrals over b. The phase stationary points are given by the solutions of

$$\theta = \mp \Theta^\alpha_{nm}(b) = \mp \frac{\partial \phi^\alpha_{nm}}{k\partial b}. \tag{4.30}$$

There is only one phase stationary point b_3 in the φ_- branch (see Fig. 15), and the corresponding integral $f^{(-)}_{nm}(\theta)$ is given by

$$f^{(-)}_{nm}(\theta) \simeq \sum_\alpha [\sigma^{(3)}_{\alpha,nm}(\theta)]^{1/2} \exp[i\beta^{(-)}_{\alpha,nm} - (\pi/2)i], \tag{4.31}$$

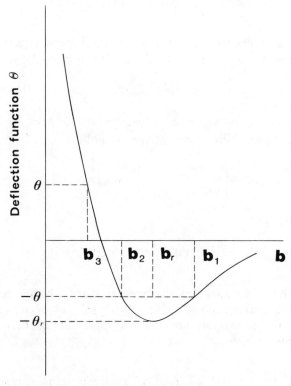

Figure 15. Schematic diagram of deflection function Θ as a function of impact parameter b.

where $\sigma_{\alpha,nm}^{(3)}$ is the classical cross section defined by

$$\sigma_{\alpha,nm}^{(3)}(\theta) = \frac{b_3}{\sin\theta\left[\dfrac{d\Theta_{nm}^\alpha}{db}\right]_{b_3}} \tag{4.32}$$

and

$$\beta_{\alpha,nm}^{(-)} = \phi_{nm}^\alpha(b_3) - kb_3\theta. \tag{4.33}$$

Since there exist two phase stationary points b_1 and b_2 (see Fig. 15) in the φ_+ branch, the corresponding integral $f^{(+)}(\theta)$ is evaluated by the uniform semiclassical approximation. The final result is

$$f_{nm}^{(+)}(\theta) \simeq \sum_\alpha [A_{nm}^\alpha A_i(-Z_\alpha) + iB_{nm}^\alpha A_i'(-Z_\alpha)]\exp[i\beta_{\alpha,nm}^{(+)} + \tfrac{5}{4}\pi i], \tag{4.34}$$

where $A_i(-Z)$ and $A_i'(-Z)$ are the Airy function and its derivative, and

$$A_{nm}^\alpha = \sqrt{\pi}Z_\alpha^{1/4}\{[\sigma_{\alpha,nm}^{(1)}(\theta)]^{1/2} + [\sigma_{\alpha,nm}^{(2)}(\theta)]^{1/2}\}, \tag{4.35a}$$

$$B_{nm}^\alpha = \sqrt{\pi}Z_\alpha^{-1/4}\{[\sigma_{\alpha,nm}^{(1)}(\theta)]^{1/2} - [\sigma_{\alpha,nm}^{(2)}(\theta)]^{1/2}\}, \tag{4.35b}$$

$$Z_\alpha = \{\tfrac{3}{4}[\phi_{mn}^\alpha(b_2) - \phi_{mn}^\alpha(b_1) + (b_2 - b_1)\theta]\}^{2/3}, \tag{4.35c}$$

and

$$\beta_{\alpha,nm}^{(+)} = \tfrac{1}{2}[\phi_{nm}^\alpha(b_2) + \phi_{nm}^\alpha(b_1) + (b_1 + b_2)\theta]. \tag{4.35d}$$

Thus the differential cross section is finally given by

$$\sigma_{nm}(\theta) \simeq |f_{nm}^{(+)}(\theta) + f_{nm}^{(-)}(\theta)|^2. \tag{4.36}$$

B.   Vibronic Transition in Ion–Molecule Collisions

In the case of ion–molecule charge-transfer collisions, we have to deal with the multiple curve crossings between the rovibrational level manifolds of the initial and the final electronic state. Since it is practically impossible to take into account all the rotational and vibrational levels, some kind of approximate reduction of the level numbers is inevitable. What is usually employed to this end is either to utilize the IOS (infinite-order-sudden) approximation with respect to rotational degree of freedom[18,60,61] or to simply neglect that degree of freedom. In either case, however, we still have multiple crossings between the two vibrational manifolds. Each crossing represents a coupling between a certain vibrational state of the initial channel and one of the final channel. The simplest model to this problem is the BFG (Bauer–Fischer–Gilmore) model,[62] in which the overall transition probability is expressed in terms of a product of the Landau–Zener transition probabilities. If there are more than one possible paths connecting initial state and final state, then the overall probability is given by a sum over the paths. The exponent of the Landau–Zener probability is usually approximated by a product of the electronic part and the Franck–Condon factor.[63] In the case of the IOS approximation, a certain function of molecular orientation angle is assumed in this exponent and an average over this angle is carried out finally.[63]

This BFG model has been applied to various systems so far and seems to work reasonably well.[17–19,63–66] As is clearly understood from the discussions made so far in this chapter, however, there arise the following questions: (1) Is it all right to use the original Landau–Zener formula for nonadiabatic transition probability [Eq. (2.1)]? (2) All phases are completely neglected, and the interferences among possible paths are totally disregarded. (3) Vibrational transitions within the manifold of the same electronic channel are ignored. None of these questions has been seriously and satisfactorily investigated yet. The favorable situations, however, might fortunately hold in many cases. For instance, the linear crossing of potential curves is expected to hold well between covalent and ionic levels. When the colliding particles are heavy, the various phases accumulate easily and the random phase

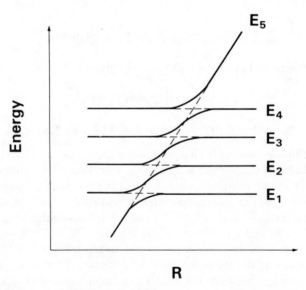

Figure 16.   Potential energy level scheme studied by Demkov and Osherov[67].

approximation might hold well. It is interesting to note that, in such a special level scheme as the one shown in Fig. 16, the overall transition probabilities can be simply expressed as a product of the transition probabilities at each crossing point without any phases, no matter how close to each other the levels are.[67] In any case, this is a very special case, and more careful investigation is still required in general about the questions raised.

C.   Numerical Examples

Some of the numerical examples relevant to the present section are presented here.

1.   Vacancy Migration in the $Ne^+ + Ne$ Collision—Catalysis Effect

The collision system considered here imitates a vacancy migration from the $2p$ shell to the $1s$ or the $2s$ shell in the Ne^+–Ne system.[29b] There are two Coriolis couplings: one between $1\pi_u(\varepsilon_1)$ and $2\sigma_u(\varepsilon_2)$, and the other between $1\pi_u$ and $1\sigma_u(\varepsilon_3)$ (see Fig. 17). The model potentials employed are as follows (in atomic units):

$$\Delta\varepsilon_{12} = \varepsilon_1 - \varepsilon_2 = 2.71(1/1.5 - 1/R),$$

$$\Delta\varepsilon_{13} = \varepsilon_1 - \varepsilon_3 = 536.9R^2,$$

$$V_{12} = -0.71vb/R^2, \tag{4.37}$$

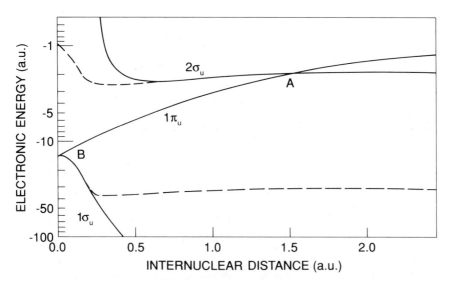

Figure 17.   Electronic energy diagram of the $1\sigma_u$, $1\pi_u$, and $2\sigma_u$ states of the Ne^+–Ne system (from Ref. 29b). ---: variable screening model; ———: model potential used.

and

$$V_{13} = - vb/R^2,$$

where ε_n's represent the adiabatic (Born–Oppenheimer) electronic energies without the nuclear Coulomb repulsion potentials. Since the collision velocity (v) considered here is quite high, the impact-parameter (b) method with the straight line trajectory approximation is employed and the factor $\hbar[(K - \Lambda)(K + \Lambda + 1)]^{1/2}$ μ in Eq. (2.3a) is replaced by vb. The coupling between $1\pi_u$ and $1\sigma_u$ corresponds to the Coriolis coupling at united atom limit shown in Table I. In order to apply the semiclassical theory the model potentials are transformed into the dynamical states that can be obtained analytically by using the method of Cardano. The numerical results are presented in Figs. 18–20. Solid (dashed) lines are the results of the semiclassical approximation (numerical solution of coupled equations). These figures clearly show the effectiveness of the semiclassical theory in the DS representation. Without introducing the DS representation this problem can not be treated by the conventional semiclassical theories. Figure 20 demonstrates another interesting phenomenon. The dotted line in this figure is the result of numerical solution of the two-state ($1\sigma_u$ and $1\pi_u$) coupled equations. The big difference between the solid line and the dotted line indicates interestingly an effect of the $2\sigma_u$ state on the transition between $1\pi_u$ and $1\sigma_u$. This is called the catalysis effect of the $2\sigma_u$ state. The $1\sigma_u$ state was

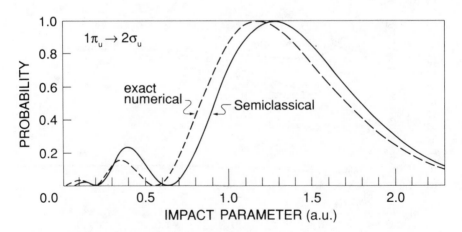

Figure 18. Impact-parameter dependence of the transition probabilities for $1\pi_u \rightarrow 2\sigma_u$ at $v = 0.9$ a.u. (from Ref. 29b). These are the three-state results, but the two-state results are practically the same as these. ———: semiclassical approximation; ---: exact numerical calculation.

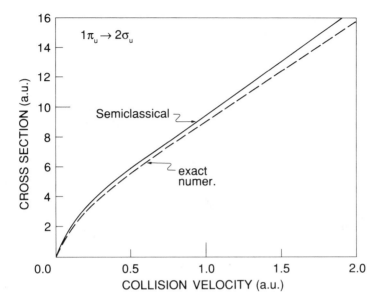

Figure 19. Total cross sections for $1\pi_u \rightarrow 2\sigma_u$ as a function of collision velocity (see the caption of Fig. 18) (from Ref. 29b).

confirmed not to affect the transition between $1\pi_u$ and $2\sigma_u$. This catalysis effect is due to the fact that the Coriolis coupling between $1\pi_u$ and $2\sigma_u$ deforms the $1\pi_u$ state strongly near $R = 0$, since the Coriolis coupling diverges as R^{-2}. This interesting phenomenon leads to the following conclusion: Rotational couplings not directly associated with the transitions concerned may affect these transitions if the avoided-crossing points corresponding to the transitions are located at small R. This phenomenon can be reproduced by the semiclassical theory based on the DS representation. It is true that the constant $(= 0.71)$ electronic angular momentum coupling matrix element is an approximation and probably exaggerates the catalysis effect, but this phenomenon can be expected to occur in such a situation mentioned above.

2. $Li^+ + Na$ and $Li + Na^+$ Collisions

Collision processes considered here are

$$Li(2s) + Na^+ \rightarrow Li^+ + Na(3s),$$

and

$$Li^+ + Na(3s) \rightarrow Li(2s) + Na^+, Li(2p) + Na^+. \qquad (4.38)$$

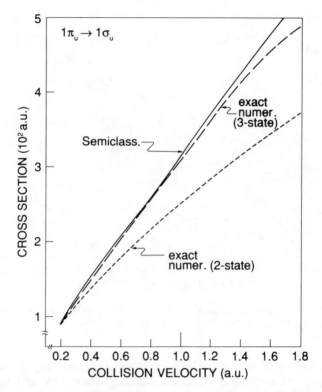

Figure 20. Total cross sections for $1\pi_u \rightarrow 1\sigma_u$ as a function of collision velocity (from Ref. 29b). ———: semiclassical approximation; ---: exact numerical calculation (three-state); ⋯: exact numerical calculation (two-state).

The adiabatic potential energy curves are shown in Fig. 21. The lowest three states are enough to be considered for treating the processes (4.38). There is a Coriolis coupling between 2Σ and 1Π, and the coupling between 1Σ and 2Σ represents the Rosen–Zener type. This problem can also be well treated by the semiclassical theory in the DS representation.[54] The coupling between 2Σ and 1Π presents a Landau–Zener-type problem in this representation. The dynamical states are shown in Fig. 11 for $K = 1700$. Some of the numerical results are shown in Figs. 22, 23, and 24. Figure 22 shows the validity of the semiclassical theory in the DS representation. These are the results just within the two-state (2Σ and 1Π) approximation to test the semiclassical theory. Figures 23 and 24 show the differential cross sections for the $1\Sigma \rightarrow 2\Sigma$ transition (Fig. 23) and for the $2\Sigma \rightarrow 1\Pi$ transition and the total

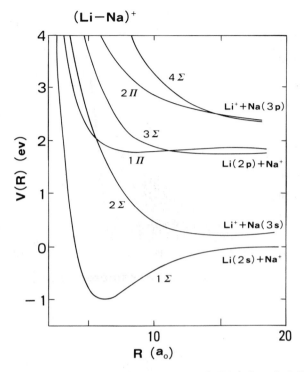

Figure 21. Adiabatic potential energy curves of LiNa$^+$ (from Ref. 54).

charge transfer from 2Σ ($2\Sigma \rightarrow 1\Sigma + 1\Pi$) (Fig. 24), in comparison with experiments. The calculations followed the method described in Section IV A 2. Since the experimentally observed differential cross sections are, however, the low-resolution cross sections, we have used the expression

$$\sigma_{nm}(\theta) \simeq |f_{nm}^{(-)}(\theta)|^2 + |f_{nm}^{(+)}(\theta)|^2 \qquad (4.39)$$

instead of Eq. (4.36). Considering the accuracy of the adiabatic potential curves employed in the calculations and the low resolution of the experiments, the agreement between theory and experiment is fairly good. Apart from the comparison with experiment, the more important thing is that the semiclassical theory proves to become an effective powerful method within the DS representation and that understanding of the mechanisms of rotationally induced transitions can be cast into the same level as that of the ordinary radially induced transitions. For instance, the DS representation

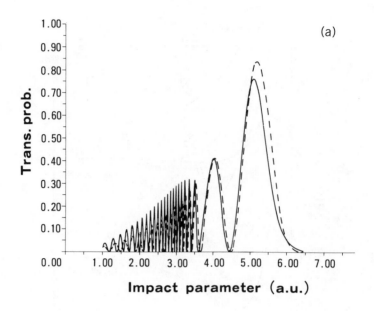

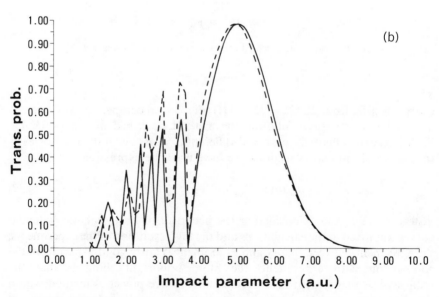

Figure 22. Transition probabilities as a function of impact parameter for $2\Sigma \rightarrow 1\Pi$ (two-state approximation) (from Ref. 54). ————: exact numerical calculation; ---: semiclassical theory. (a) $E_{coll} = 30\,eV$; (b) $E_{coll} = 750\,eV$.

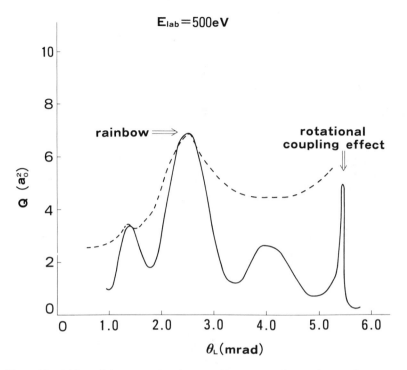

Figure 23. Differential cross section for $1\Sigma \to 2\Sigma$ at $E_{lab} = 500\,eV$ (from Ref. 54). ———: semiclassical theory; ---: experiment (see Ref. 54).

enables us to incorporate the multitrajectory effect naturally, and thus to uniquely define the deflection functions for any transitions.

3. Three- and Four-Level Model Systems

The semiclassical theory presented in Section IV A 1 has been applied to the model three- and four-state systems including the cases that the avoided crossings can not be regarded to be well separated from each other.[55] The model diabatic potentials employed are (in atomic units)

$$V_1(R) = V_0 e^{-R/c},$$

$$V_2(R) = V_1(R) + V_{20},$$

$$V_3(R) = d e^{-a(R - R_e)}[e^{-a(R - R_e)} - 2] + V_{30}, \qquad (4.40a)$$

and

$$V_4(R) = V_3(R) + V_{40},$$

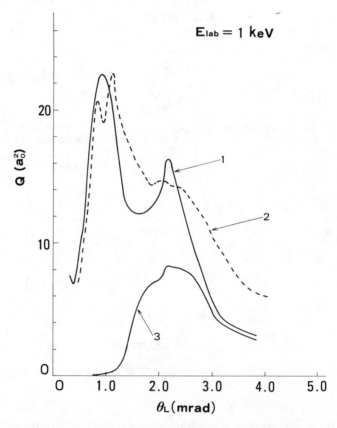

Figure 24. Differential cross sections for $2\Sigma \rightarrow 1\Pi$ and for the total charge transfer $(2\Sigma \rightarrow 1\Sigma + 1\Pi)$ at $E_{lab} = 1$ keV (from Ref. 54). 1: semiclassical theory for $2\Sigma \rightarrow 1\Sigma + 1\Pi$; 2: experiment for $2\Sigma \rightarrow 1\Sigma + 1\Pi$ (see Ref. 54); 3: semiclassical theory for $2\Sigma \rightarrow 1\Pi$.

where $V_0 = 129.62$, $V_{30} = 0.08$ (three-state case), 0.06 (four-state case), $V_{20} = 0.005$–0.03, $V_{40} = 0.004$–0.02, $a = 0.3, c = 0.5915, d = 0.16$, and $R_e = 6.0$. The diabatic couplings are taken as follows:

$$V_{13} = V_{14} = V_{23} = V_{24} = 0.002. \qquad (4.40b)$$

The potential curves are shown in Figs. 25 and 26. Numerical results for the transition probabilities are shown in Figs. 27 for the three-state models $(V_{20} = 0.02$ and $0.005)$ and in Figs. 28 and 29 for the four-state models $[(V_{20}, V_{40}) = (0.03, 0.004)$ and $(0.01, 0.004)]$. Although the complex integrals defined by Eq. (3.16) were performed for the three-state case, the approximations $\sigma_0 \simeq 0$ and $\delta \simeq \delta_{LZ}^{(0)}$ were used in the four-state case for simplicity. These approximations are not bad at all, because the model potentials are so designed. Exact numerical solutions are the results of close coupled equations. As can be seen from these figures, the semiclassical theory works very well, although it is slightly worse in the four-state case. It is not very surprising that the semiclassical theory works all right for the cases that the avoided crossings are well separated. However, it works surprisingly well even in the cases that the adjacent avoided crossings cannot be regarded to

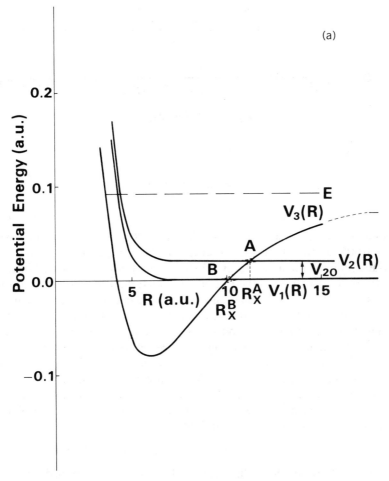

Figure 25. Three-state model potentials (from Ref. 55). (a) $V_{20} = 0.02$; (b) $V_{20} = 0.005$.

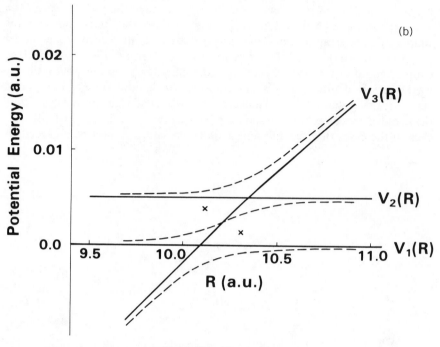

Figure 25. *(Continued)*

be well separated as shown in Figs. 25b and 26b. Actually, in the cases of $V_{20} = 0.005$ of the three-state model and $(V_{20}, V_{40}) = (0.01, 0.004)$ of the four-state model, the adjacent avoided-crossing points are located inside the half-maximum position of the Lorentzian function of nonadiabatic coupling. This surprisingly good behavior of the semiclassical approximation is probably due to the fact that the fundamental analytical properties of adiabatic potential energies and nonadiabatic couplings in the complex R plane are correctly taken into account in the basic two-state theory. However, this is still just a kind of case study and more extensive work is required. Finally, it is interesting to note the role of phase. If all the phases are neglected, no oscillation of transition probability as a function of collision energy appears as is seen in Figs. 28. The significant effect of the Stokes phase ϕ_S can also be seen from Figs. 27b and 28a.

4. Application of the BFG Model to Vibronic Transitions

The BFG (Bauer–Fisher–Gilmore) model[62] has been applied to various practical processes, since this is one of the simplest generalization of the

Landau–Zener theory to vibronic transition. This model seems generally to work all right at relatively low collision velocities and for the systems with widely spaced crossings. The transition between covalent and ionic states presents a good example, since the original Landau–Zener situation holds well in this case. Child and Baer introduced a parameter γ as a measure of the validity of the BFG approximation.[64] This parameter γ is defined as

$$\gamma = \bar{v}_R \Delta G_R / \bar{v}_r \Delta G_r, \tag{4.41}$$

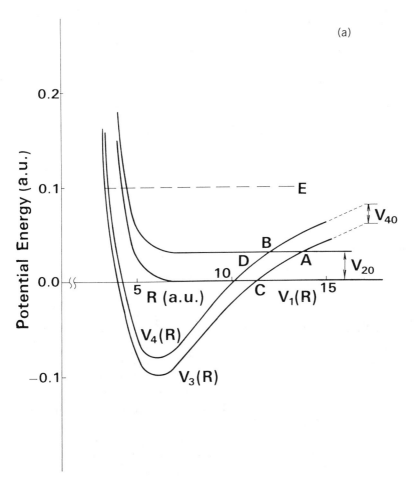

Figure 26. Four-state model potentials (from Ref. 55): (a) $V_{20} = 0.03$ and $V_{40} = 0.02$; (b) $V_{20} = 0.01$ and $V_{20} = 0.04$.

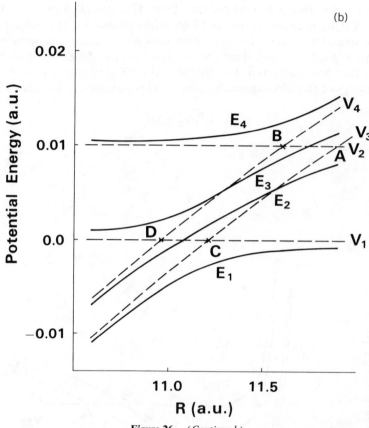

Figure 26. *(Continued)*

where $\bar{v}_s$ is the average translational ($s = R$) or the maximal vibrational ($s = r$) velocity, and $\Delta G_s(s = r, R)$ is the derivative with respect to s of the potential energy difference at the crossing seam. The BFG approximation works all right at $\gamma \ll 1$. Since there have been published quite a few articles on this approximation,[17-19] we do not go into the details here and just give two recent examples. Application of the Rosen–Zener formula to multiple crossing problem has also been considered.[68,69]

Using the BFG, or the multiple curve crossing model and the quasifree electron model, Desfrancois et al.[66] discussed the electronic transitions of alkali atoms (A) by collisions with diatomic molecules,

$$A(nl) + M \rightarrow A(n'l') + M. \tag{4.42}$$

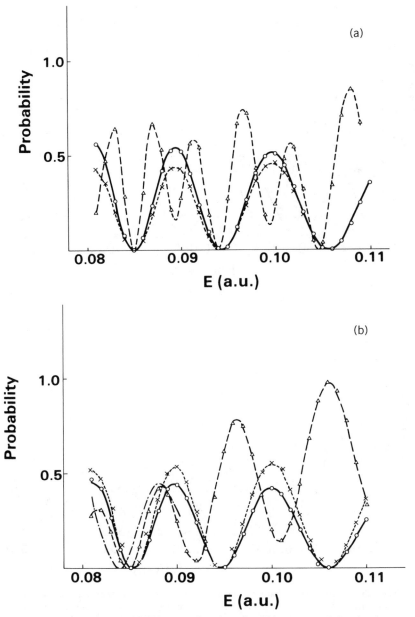

Figure 27. Transition probabilities as a function of collision energy E for the three-state model (from Ref. 55). (a) $V_{20} = 0.02$, (b) $V_{20} = 0.005$. P_{12}: ———, semiclassical, $\bigcirc$ exact; P_{23}: ---, semiclassical, $\triangle$ exact; P_{13}: $\cdots$ semiclassical, $\times$ exact. The dash-dot line is the semiclassical result for P_{12} with ϕ_s and σ_0 neglected.

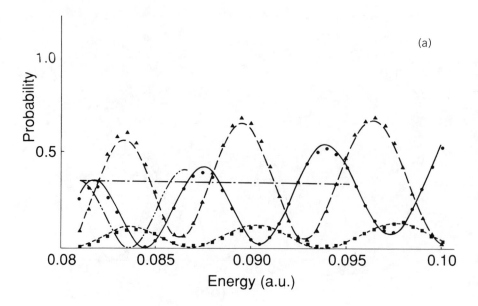

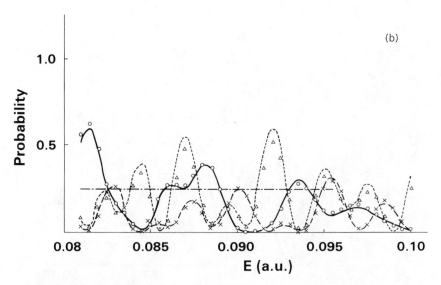

Figure 28. Transition probabilities as a function of collision energy E for the four-state model with $V_{20} = 0.03$ and $V_{40} = 0.004$ (from Ref. 21c): (a) P_{12}: ———, semiclassical, ● exact; —·—, semiclassical with phases totally neglected; —··—, semiclassical with ϕ_S neglected; P_{13}: ——— semiclassical, ▲ exact, P_{14}: ··· semiclassical, ■ exact. (b) P_{23}: ———, semiclassical, ○ exact, —·—, semiclassical with phases totally neglected; P_{34}: --- semiclassical, △ exact; P_{24}: ··· semiclassical, × exact.

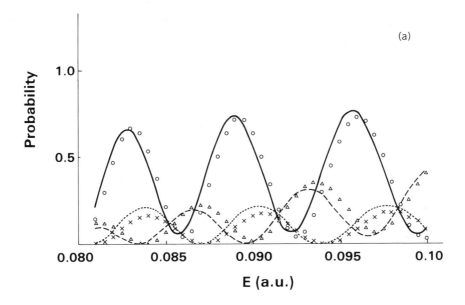

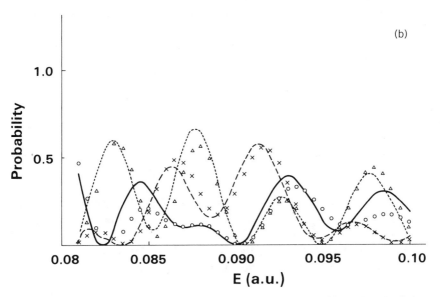

Figure 29. The same as Figs. 28 for $V_{20} = 0.01$ and $V_{40} = 0.004$. (a) $\bigcirc$: P_{12}, $\triangle$: P_{13}, $\times$: P_{14}; (b) the same as Fig. 28b.

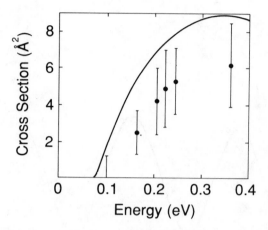

Figure 30.   Cross sections for $Rb(5d) + O_2 \rightarrow Rb(7s) + O_2$ (from Refs. 66 and 70). ——— BFG model, ● experiment.

In this case the ionic states $(A^+ - M^-)$ are assumed to play an important role of intermediate state, namely, the process (4.42) is mediated by the curve crossings with the ionic states. The authors found that the quasifree electron model appears as a first-order treatment of the multiple curve crossing model and their main difference arises from the determination of the ionic–covalent couplings. One of their results in comparison with experiment is shown in Fig. 30. This is the result for

$$Rb(5d) + O_2(v = 0) \rightarrow Rb(7s) + O_2.$$

The BFG result is the one obtained by Paillard et al.[70]
The second example is the charge-transfer collision,[71]

$$Ne^{2+}(^1S) + H_2 \rightarrow Ne^+(^2S) + H_2^+(v).$$

The diabatic model potentials are shown in Fig. 31. Only the attractive polarization potential and the repulsive Coulomb potential are assumed for the initial and final channels, respectively. The diabatic coupling strength (electronic part) was estimated by the formula of Olson et al.[72] Figure 32 gives the relative distribution of the final vibrational states. As is clearly seen, the multichannel Landau–Zener model (BFG model) reproduces the experimental result very well, while the simple Franck–Condon factor fails to do so.[73]

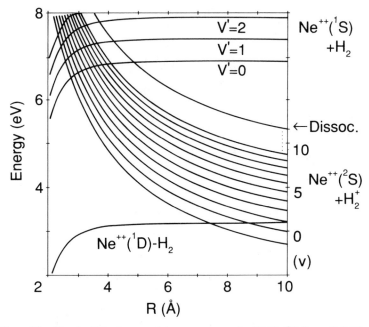

Figure 31. Simple diabatic potential energy curves for $(NeH_2)^{2+}$ (from Ref. 71).

In spite of the various questions raised in Section IV B, the simple multiple crossing model seems to be working generally quite well. This is probably because in many circumstances the various conditions are satisfied to a good extent. For instance, as mentioned previously, the conditions for the validity of the original Landau–Zener formula are probably well satisfied in the ionic–covalent intersections, and the phase randomization occurs effectively in the case that heavy atoms and molecules are involved. Besides, the detailed structures in the transition probabilities are usually smeared out in the total cross section. In any case, however, more detailed careful studies are required on the applicability of the various versions of semiclassical theory.

V. CHEMICAL REACTION—PARTICLE REARRANGEMENT

Potential energy surface (avoided) crossing is usually located in either reactant or product region distant from the reaction zone where particle rearrangement occurs. In such a case, electronic transition and particle rearrangement can be separately treated; and the latter can be discussed on a single potential energy surface.[22] Some semiclassical methods for chemical

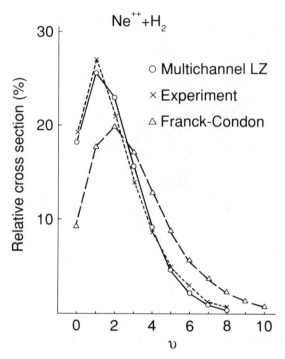

Figure 32. Relative final vibrational state distribution in $Ne^{2+} + H_2^+(v'=0) \rightarrow Ne^+ + H_2^+(v)$ (from Ref. 71). $\times$: experiment; $\bigcirc$: multichannel Landau–Zener model; $\triangle$: Franck–Condon factor.

reaction are reviewed and explained here. First, the simplest orbiting (Langevin) model is briefly discussed.[13–16] The PRS (perturbed rotational state) theory proposed by Takayanagi and Sakimoto[16,74–77] is introduced, which takes into account the discreteness of the rotational states. Next, the classical S matrix theory is briefly explained for later convenience.[4,78] Finally, the hyperspherical coordinate approach is introduced and its power is demonstrated.[36]

A. Orbiting (Langevin) Model for Ion–Molecule Collision

If a molecule is nonpolar and does not possess a dipole moment, the long-range interaction potential between an ion and the molecule is given by a sum of the polarization potential and the centrifugal potential,

$$V(R) = -\frac{\alpha e^2}{2R^4} + \frac{Eb^2}{R^2}, \qquad (5.1)$$

where α is the polarizability, e is the ion charge, b is the impact parameter, and E represents the relative kinetic energy. The orbiting model[13-15] simply assumes that the capture ($\sim$ reaction) probability $P_c(b, E)$ is given by

$$P_c(b, E) = \begin{cases} 1, & b \leqslant b_c, \\ 0, & b > b_c, \end{cases} \tag{5.2}$$

where b_c is such an impact parameter that

$$\max_{R} [V(R)] = E \tag{5.3}$$

is satisfied. From Eqs. (5.1) and (5.3), we have

$$b_c = (2\alpha e^2/E)^{1/4}, \tag{5.4}$$

and thus the capture cross section $\sigma_c(E)$ is given by

$$\sigma_c(E) = \pi b_c^2 = \pi e(2\alpha/E)^{1/2}. \tag{5.5}$$

If the molecule has a dipole moment D, then the interaction potential becomes

$$V(R) = -\frac{\alpha e^2}{2R^4} - \frac{De}{R^2}\cos\theta + \frac{Eb^2}{R^2}, \tag{5.6}$$

where θ is the angle that the dipole makes with the line of centers of the collision. Since the interaction potential is dependent on θ, it is not straightforward to calculate the capture cross section. If we lock the molecule at $\theta = 0$, then we have

$$b_c^2 = e(2\alpha/E)^{1/2} + \frac{De}{E} = \frac{De}{E}\left(1 + \sqrt{\frac{E}{E_c}}\right) \tag{5.7}$$

with $E_c = D^2/(2\alpha)$. If we take an average over θ, then b_c is given by[79]

$$b_c^2 = \frac{De}{4E}\left(1 + \sqrt{\frac{E}{E_c}}\right)^2. \tag{5.8}$$

These approximations generally give larger cross sections than experiment. If we want to take into account the discreteness of the molecular rotational states at low energies, it is convenient to employ the PRS representation.[16,74-77]

This is the representation of the eigenstates $\{\chi_{jm}\}$ of the rotating collision complex,

$$\left(B\mathbf{j}^2 + \frac{De\cos\theta}{R^2}\right)\chi_{jm} = \varepsilon_{jm}(R)\chi_{jm}, \tag{5.9}$$

where B is the rotational constant of the molecule, $\mathbf{j}$ is the rotational angular momentum operator and $\varepsilon_{jm}(R)$ is the adiabatic potential. In order to obtain $\varepsilon_{jm}(R)$ semiclassically, Sakimoto[77] used the following adiabatic invariant:

$$(n + \tfrac{1}{2})\pi = \oint p_\theta d\theta = \int_{\theta_1}^{\theta_2}\left[\frac{\varepsilon}{B} - \frac{De\cos\theta}{BR^2} - \frac{m}{\sin^2\theta}\right]^{1/2} d\theta. \tag{5.10}$$

Here p_θ is the momentum conjugate to the polar coordinate θ of the molecular orientation with the z axis along the intermolecular direction, and is equal to

$$p_\theta^2 = \mathbf{j}^2 - \frac{m^2}{\sin^2\theta}. \tag{5.11}$$

θ_1 and θ_2 are the turning points of the θ motion. The adiabatic potential energies $u_{jm} = \varepsilon/B$ are numerically obtained from Eq. (5.10) as a function of R. Then the effective potential in this PRS representation is given by

$$V(x) = \frac{\bar{E}b^2}{x^2} - \frac{\kappa}{2x^4} + u_{jm}(x) - (j + \tfrac{1}{2})^2, \tag{5.12}$$

where $\bar{E} = E/B$, $x = R(B/De)^{1/2}$, $\bar{b} = b(B/De)^{1/2}$, and $\kappa = \alpha B/D^2$. The critical (orbiting) impact parameter $\bar{b}_0$ is determined from the conditions

$$\bar{E} = V(x) \quad \text{and} \quad dV(x)/dx = 0. \tag{5.13}$$

Then the jm-dependent cross sections can be obtained. This method has been proved to be very useful.

B. Classical S-Matrix Theory

The classical S-matrix theory is conceptually beautiful semiclassical theory, which can describe chemical reactions with the various quantum effects incorporated by using the feasible classical trajectories.[4,78] The quantum-mechanical effects are (1) quantization of internal states, (2) resonance, (3) interference effect, (4) tunneling effect, and (5) electronically nonadiabatic transition. The outline of this theory is very briefly described here, especially for the convenience in the discussions in the next section. The theory is well grounded by the Feynman's path integral formulation of quantum mechanics.[80] From all the conceivable paths the classical trajectories are selected by using the stationary phase approximation. The final expression for the S matrix is given as

$$S_{n_2,n_1}(E) = \sum_{\alpha} S^{\alpha}_{n_2,n_1}(E) \tag{5.14a}$$

with

$$S^{\alpha}_{n_2,n_1}(E) = i \frac{1}{(-2\pi i\hbar)^{N-1}} \left[\frac{\partial^2 \Phi_{\alpha}(n_2,n_1:E)}{\partial n_2 \partial n_1} \right]^{1/2} \exp\left(\frac{i}{\hbar} \Phi_{\alpha}(n_2,n_1:E) \right), \tag{5.14b}$$

where $n_1(n_2)$ represents collectively the initial (final) internal state quantum numbers (action variable), N is the total number of the degrees of freedom (one translational $+ N - 1$ internal), and $\Phi_{\alpha}(n_2,n_1:E)$ is the phase (action integral) along the classical trajectory α defined by

$$\Phi_{\alpha}(n_2,n_1:E) = \int_{-\infty}^{\infty} dt [-R(t)\dot{P}(t) - \mathbf{q}(t)\cdot\dot{\mathbf{n}}(t)]_{\alpha}. \tag{5.15}$$

$R(t)$ and $P(t)$ are the ordinary translational coordinate and momentum, and $\mathbf{q}(t)$ are the angle variables conjugate to the action variables $\mathbf{n}(t)$. For simplicity, the ordinary nonreactive scattering process is assumed, namely, the translational coordinate remains the same during the scattering process. In the case of reaction, additional labels to distinguish arrangement channels are necessary. The generalization is, however, straightforward and is omitted here. One of the important features about this theory is that the classical trajectories chosen here satisfy the quantization of internal states at both ends (double-ended boundary conditions). The classical trajectories starting with the quantized condition $n_1 = $ integer generally lead to the non integer final actions $n_2(n_1,q_1^0)$ as a function of the initial angle variable $q_1^{(0)}$. By varying $q_1^{(0)}$, the trajectories which satisfy

$$n_2(n_1,\bar{q}_1^{(0)}) = n_2(\text{integer}) \tag{5.16}$$

are searched. The function $\bar{q}_1^{(0)}(n_1,n_2)$ is a multivalued function and these selected trajectories are distinguished by the level α. Since we are not dealing with probability, but with probability amplitude, the quantum-mechanical interference effect is taken into account as the interference between the selected trajectories. Quantum-mechanical tunneling effect is incorporated into the formalism by considering the complex-valued trajectories. The phase

Φ possesses an imaginary part corresponding to barrier penetration and the transition probability becomes an exponentially decaying function of the imaginary phase. Electronic transitions can also be incorporated either by continuing the classical trajectories analytically into the complex coordinate space with branch cuts or by borrowing the method of semiclassical theory for nonadiabatic transition. This is discussed in the next section in a bit more detail.

In addition to the conceptual beauty of the formalism, this theory had been very successful in explaining various phenomena such as threshold behavior (tunneling effect), interference effect, and resonance. Unfortunately, however, there have arisen some serious difficulties in the practical applications. These are the difficulties of finding the double-ended trajectories (especially in the case of complex-valued trajectories) and of the complex calculus with branch cuts, breakdown of the stationary phase approximation, and appearance of the chattering phenomenon. Because of these difficulties the practical applications have virtually disappeared. The chattering, first noted in collinear reactive scattering, is the phenomenon that a final quantity such as product vibrational action becomes randomly dependent on an initial quantity such as initial vibrational angle variable.[81,82] In this situation classical trajectories are very unstable against initial conditions: neighboring trajectories diverge exponentially in a finite time period. This phenomenon is closely related to the so-called classical chaos and irregular scattering, and is associated with a complex formation.[83-88] Instability of classical trajectory is caused by other reasons also. Recently, we have devised a classical-mechanical method for decoupling collisional and vibrational variables (in phase space) in scattering process, in attempt to find a way to classify variables into relevant and irrelevant ones and also to elucidate the reaction dynamics[89]. This method can be used to diagnose collision system and to find out whether it contains instabilities or not, and if so, what kind. However, as far as we use classical trajectories, there is no treatment to avoid the chattering.

C. Hyperspherical Coordinate Approach

The hyperspherical coordinate system is a unique system in the sense that there is only one radial coordinate ρ (hyperradius) and all the others are angle variables (collectively denoted as Ω_H), three of which are usually taken as the Euler angles. The hyperradius ρ represents the size of the whole system, and the angle variables distinguish the various arrangement channels inside the sphere. This coordinate system is convenient to describe the reaction zone in which all particles get close together, and has recently attracted much attention in the study of reaction dynamics.[36,90-92] The hyperradius ρ in a triatomic system is defined as

$$\rho = [R_\lambda^2 + r_\lambda^2]^{1/2}, \tag{5.17}$$

where λ represents the arrangement channel $(\lambda + \kappa v)$. R_λ and r_λ are the mass scaled Jacobi coordinates defined by

$$R_\lambda = a_\lambda^{-1} R_\lambda' \quad \text{and} \quad r_\lambda = a_\lambda r_\lambda' \tag{5.18}$$

with

$$a_\lambda = \left[\frac{m_k m_v (m_\lambda + m_k + m_v)}{m_\lambda (m_k + m_v)^2} \right]^{1/4}. \tag{5.19}$$

Here R_λ' and r_λ' are the original translational and vibrational coordinates, and m_y is the mass of atom γ. There is no unique definition of hyperangles. Johnson's angle variables, for instance, are given by

$$\cos \vartheta = \frac{2}{\rho^2} r_\lambda R_\lambda \sin \gamma_\lambda, \tag{5.20a}$$

and

$$\cot \phi = \frac{2R_\lambda r_\lambda}{R_\lambda^2 - r_\lambda^2} \cos \gamma_\lambda, \tag{5.20b}$$

where γ_λ is the angle between the vibrational coordinate vector $\mathbf{r}_\lambda$ and the translational coordinate vector $\mathbf{R}_\lambda$.

The power of the hyperspherical coordinate approach may be summarized as follows:

1. A unified treatment of various arrangement channels is possible, since there is only one coordinate (ρ) that can be infinity.
2. The approach is numerically efficient.
3. Reactive transition is localized in a region along the potential ridge.
4. Even analytical treatment becomes possible in some cases.
5. Various effects such as those of potential energy surface topography, mass combination of reactants (isotope effect), and internal energy of reactants, and steric effects can be visualized and comprehended much better by drawing a potential curve diagram.
6. Thus the reaction mechanisms, in general, can be made more transparent.

Although accurate numerical solutions of some 3-D reactions are becoming possible,[90-93] we consider here rather a collinear reaction and the approximate treatments of 3-D reactions with an emphasis on facilitating a conceptual and qualitative understanding of the mechanisms. Most of the contents here can be found in the recent review article.[36] Some basic concepts of the hyperspherical coordinate approach are already explained in Section II. So by presenting some examples, these concepts and the power of the approach mentioned above are exemplified here.

The first example is the collinear reaction $Cl + HBr(v=0) \rightarrow HCl(v=2) + Br$, which was already briefly discussed in Section II. Figure 33 shows the reaction probability as a function of collision energy. This clearly indicates that not only the two-state close coupling but even the two-state Rosen–Zener formula gives good results. The resonance near the threshold is not reproduced by the latter, but this is simply because the resonance effect is not built into the formula. This can also be treated analytically by the semiclassical theory as will be described. The vibrational adiabaticity well visualized by the potential curve diagram of Fig. 6 is thus substantiated by Fig. 33. It should be noted that the analytical fitting was made directly to the vibrationally adiabatic potential energies (actually to the sum and the square of the difference) in

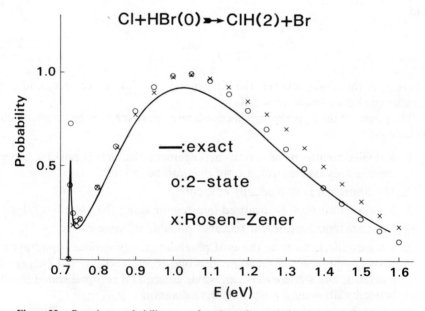

Figure 33. Reaction probability as a function of translational energy for the collinear reaction $Cl + HBr(v=0) \rightarrow HCl(v=2) + Br$ (from Ref. 107).

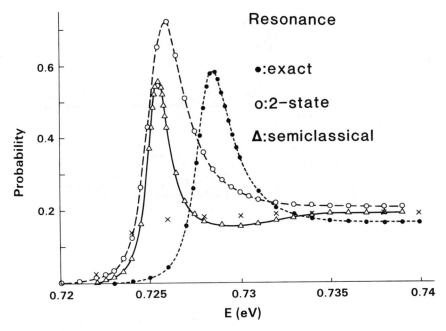

Figure 34. Resonance peak in the collinear reaction $Cl + HBr(v = 0) \to HCl(v = 2) +$ Br (from Ref. 94). $\times$: Rosen–Zener formula without resonance.

order to estimate the parameters σ_0, δ, and phases. That is, the nonuniqueness problem of "diabatization" can be avoided. Localization of reactive transition in the region of potential ridge as demonstrated by Fig. 7 is confirmed here by the fact that $\rho_x \equiv \text{Re}(\rho_*)$ is equal to $\sim 14.64a_0$. The resonance seen in Fig. 33 is an interesting one caused by a combination of the tunneling in the upper potential curve and the Rosen–Zener-type nonadiabatic transition between the two states. Neither one of these can produce resonance independently. Using the semiclassical theories mentioned in Section III and IV, this resonance effect can be incorporated into the Rosen–Zener formula[94]. The procedure is not repeated here, and only numerical results are presented in Fig. 34. The agreement with the exact result is not perfect, but this clearly demonstrates that the semiclassical theory can elucidate the mechanism. Similar resonances were found and analyzed in other systems[95].

Next, we demonstrate how a difference of potential energy surface topography is visualized in the form of potential energy diagram. Potential energy contours shown in Fig. 35 are the model LEPS potentials used by Polanyi and Sathyamurthy.[96] Both potentials have a saddle point of the same height at the same position, but differs in the slope up to the saddle

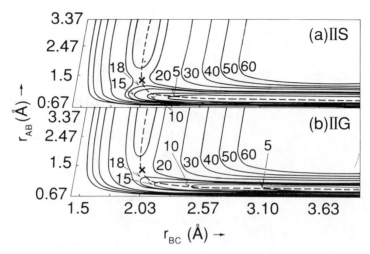

Figure 35. Model LEPS potentials (from Ref. 96). (*a*) LEPS-IIS; (*b*) LEPS-IIG.

point in the $AB + C$ channel (S stands for sudden rise and G for gradual). Figure 36 is the corresponding potential curve diagrams for the mass combination $m_A = m_H$ (hydrogen), $m_B = 19m_H$ and $m_C = 23m_H$.[97] The dashed lines represent the ridge and the bottoms of the potential energy surface. The difference in topography, which can not easily be detected in Fig. 35, can be more clearly seen in Figs. 36. The AB curves are attractive in the case of II S, while the corresponding curves in the case of II G are repulsive before reaching the potential ridge line. This means that the kinetic energy is inevitable for the reaction from the AB channel to occur in the latter case. Some other qualitative features of this system can be understood from these diagrams. For instance, vibrational excitation of AB is necessary for the reaction from the AB channel to occur effectively, and the product BC is expected to be vibrationally excited. The reaction from the BC channel is generally not probable, because the AB curves are much more sparse and there is no appropriate adjacent AB level to most of the BC levels. This is called heavy-particle anomaly. If we change the mass combination with the potential energy surface kept the same, the potential curve diagrams change drastically and interestingly.[36] Figure 37 is the case of the heavy–light–heavy combination ($m_A = m_C = 20m_H$ and $m_B = m_H$), and Fig. 38 for the light–heavy–light combination ($m_A = m_C = m_H$ and $m_B = 20m_H$). Vibrational adiabaticity holds well, as is expected, in the heavy–light–heavy system. At collision energies $\leqslant 1.0\,\mathrm{eV}$ only these transitions marked by rectangulars in Fig. 37 are prominent because of the favorable level arrangement in the vicinity of the potential ridge. In the light–heavy–light system, vibrational adiabaticity does not hold well and the various transitions become possible with more complicated dynamics. Hyperspherical mode is another example to demonstrate the power of this coordinate system.[98-100] This mode is a vibrational motion along or approximately along the hyperangle θ, namely, at constant or approximately constant hyperradius ρ. This corresponds to the antisymmetric stretching mode. One eminent feature about this mode is that this is expected to exist in a real system such as highly excited linear dihydride. Some resonances in collinear reactions are also assigned to the hyperspherical modes.

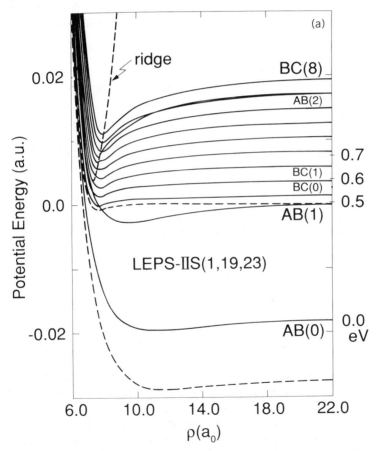

Figure 36. Vibrationally adiabatic potential energy curves of the LEPS potentials of Fig. 35 for the mass combination $m_A = m_H$, $m_B = 19m_H$ and $m_C = 23m_H$ (from Ref. 97). (a) LEPS-IIS; (b) LEPS-IIG.

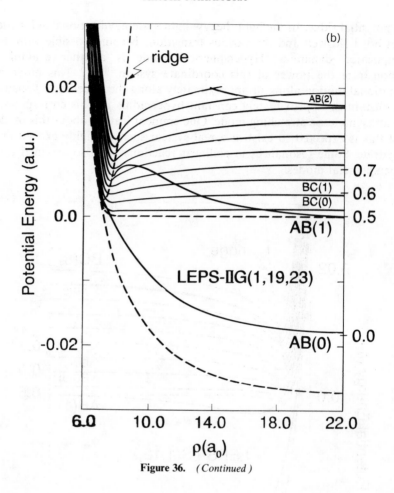

Figure 36. *(Continued)*

Finally, we give some examples of the approximate treatments of 3-D reactions. In the case of 3-D reaction, the potential energy surface at fixed hyperradius ρ is mapped onto the two dimensional space. If we use the variables $x = \sin \vartheta \sin \phi$ and $y = \sin \vartheta \cos\phi$, for instance, then this mapping can be done onto the unit circle. There are generally three ridge lines, each of which represents the region for one of the three possible reactions to occur.[36] See also the representation discussed by Aquilanti et al.[101] In the approximate treatments we concentrate our attention only on one of the reactions. Approximations usually employed are the IOS (infinite order sudden)[102] and the adiabatic-bend (or reduced dimensionality)[103,104] approximation. These treat the rotational degrees of freedom in the sudden and the adiabatic way, respectively. That is to say, the molecular rotational motion (γ motion) is frozen in the sudden approximation and is solved first at fixed translational and vibrational coordinates in the adiabatic approximation. In either case, the dimensionality of the dynamics is reduced to two, and the dynamics can be analyzed in the same way as in the collinear case. In the IOSA, the dynamics is solved at each γ_i, and γ_i-dependent S matrix and reaction probability (P^{IOS}) are obtained. The rotationally averaged and summed cross section for the vibrational transition $v_i \to v_f$ is given by

$$\sigma^{IOS}(v_i \to v_f : E) = \frac{\pi}{2k_{v_i}^2} \sum_l (2l + 1) \int_0^\pi P_l^{IOS}(v_i \to v_f : \gamma_i, E) \sin \gamma_i d\gamma_i. \quad (5.21)$$

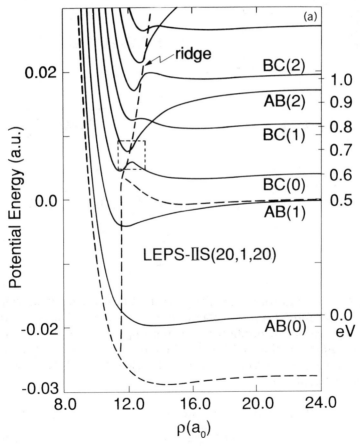

Figure 37.   The same as Fig. 36 for $m_A = m_C = 20 m_H$ and $m_B = m_H$.

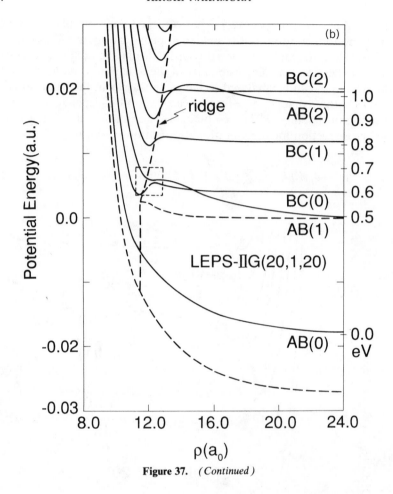

Figure 37. *(Continued)*

Since a reactive transition is spatially localized near the potential ridge, the centrifugal potential may be replaced by the constant value estimated at a certain representiative position (†) on the ridge line (constant centrifugal potential approximation: CCPA).[105] In this approximation, the reaction probability $P_l(v_i \rightarrow v_f : \gamma, E)$ for nonzero l can be estimated from the s-wave reaction probability by

$$P_l(v_i \rightarrow v_f : \gamma, E) \cong P_{l=0}(v_i \rightarrow v_f : \gamma, E - B^\dagger l(l+1)), \qquad (5.22)$$

where

$$B^\dagger = \hbar^2 / 2\mu(R^\dagger)^2. \qquad (5.23)$$

When we do not have to pay much attention to the details of the dynamics such as a resonance at particular l, this CCPA approximation provides us with a practically useful method for not only estimating cross section and rate constant, but also grasping the overall reaction mechanisms. The cross section and the rate constant in this approximation are given as follows:

$$\sigma_{CCPA}^{IOS}(v_i \to v_f : E) = \frac{\pi}{2k_{v_i}^2} \int_0^\pi d\gamma_i \frac{\sin \gamma_i}{B^\dagger(\gamma_i)} \int_0^E d\varepsilon P_{l_i=0}^{IOS}(v_i \to v_f : \gamma_i, \varepsilon), \qquad (5.24)$$

and

$$\begin{aligned}
k_{CCPA}^{IOS}&(v_i \to v_f : T) \\
&= \frac{h^2}{2(2\pi\mu_{A,BC}\kappa T)^{3/2}} \int_0^\pi d\gamma_i \sin \gamma_i \\
&\quad \cdot \left(\sum_l (2l+1) e^{-B^\dagger(\gamma_i)l(l+1)/\kappa T} \right) \int_0^\infty d\varepsilon\varepsilon^{-\varepsilon/\kappa T} P_{l=0}^{IOS}(v_i - v_f : \gamma_i, \varepsilon), \qquad (5.25)
\end{aligned}$$

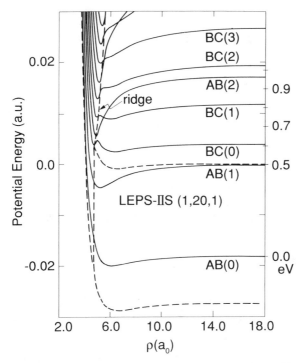

Figure 38. The same as Fig. 36 for $m_A = m_C = m_H$ and $m_B = 20m_H$.

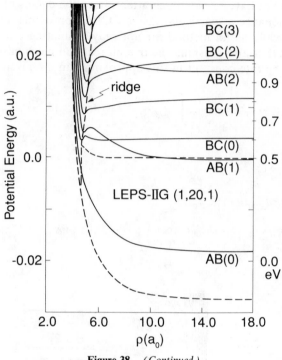

Figure 38. *(Continued)*

where $\mu_{A,BC}$ is the reduced mass and κ is Boltzmann constant. In the adiabatic-bend approximation, the hindered rotational eigenvalues estimated at each grid point (R_λ, r_λ) are added to the original potential in collinear arrangement and then the dynamics is solved for each hindered rotational state n. The cross section for the transition $v_i \to v_f$ is given by

$$\sigma^{\mathrm{AD}}(v_i \to v_f : E) = \frac{\pi}{(\bar{k}_{v_i})^2} \sum_l (2l+1) \sum_n P_l^{\mathrm{AD}}(v_i \to v_f : n, E) \qquad (5.26)$$

with

$$(\bar{k}_{v_i})^2 = \sum_{j_i = \mathrm{open}} (2j_i + 1) k_{v_i j_i}^2, \qquad (5.27)$$

where $k_{v_i j_i}$ is the wave number of the relative motion in the initial channel. If we use the CCPA, then the cross section and the rate constant are reduced to[103-105]

$$\sigma^{AD}_{CCPA}(v_i \rightarrow v_f:E) = \frac{\pi}{(\bar{k}_{v_i})^2} \sum_n \frac{1}{B^\dagger(n)} \int_0^E d\varepsilon P^{AD}_{l=0}(v_i \rightarrow v_f:n, \varepsilon), \qquad (5.28)$$

and

$$k^{AD}_{CCPA}(v_i \rightarrow v_f:T) = \frac{h^2}{(2\pi\mu_{A,BC}\kappa T)^{3/2}} \frac{1}{f_{rot}} \sum_n \left(\sum_l (2l+1)e^{-B^\dagger(n)l(l+1)/\kappa T} \right)$$
$$\cdot \int_0^\infty d\varepsilon e^{-\varepsilon/\kappa T} P^{AD}_{l=0}(v_i \rightarrow v_f:n, \varepsilon), \qquad (5.29)$$

where f_{rot} is the rotational partition function. Both IOS and adiabatic-bend approximations are big approximations, but are still quite useful for understanding the mechanisms. The IOS approximation is considered to be a good approximation for light–heavy–light system[106], while the adiabatic-bend approximation should work better for heavy–light–heavy system at low energies.[107]

Here we show an example of application of the adiabatic-bend approximation to the reaction $O + HCl \rightarrow OH + Cl$.[108] Figure 39 shows the potential curve diagram obtained with use of the model potential LEPS-I of Persky and Broida.[109] The dashed lines represent the bottoms of the potential energy surface. From this figure we can see easily that the reactions $O + HCl(v = 0) \leftrightarrow OH(v = 0) + Cl$ and $O + HCl(v = 4) \leftrightarrow OH(v = 3) + Cl$ are favorable, and that the kinetic energy more than $\sim 0.2\,eV$ is necessary for the former reaction to occur effectively. Figure 40 demonstrates again the importance of the potential ridge in reaction. The sharply rising position $(\rho \cong 9.6a_0)$ roughly corresponds to the position where the two potential curves cross the potential ridge line. More accurately, the position is slightly shifted to the right of the ridge, reflecting the fact that the hydrogen atom tunnels through the barrier. Figure 41 shows the l-dependence of the reaction probability at several kinetic energies. At higher energies the probability oscillates. This is one of the characteristics of a heavy–light–heavy system. The reaction probabilities shown here are calculated quantum mechanically by solving the close coupled equations. Finally, the kinetic isotope effect, k_{O+HCl}/k_{O+DCl}, is plotted in Fig. 42 as a function of the inverse temperature. The CCPA approximation works fairly well in spite of the fact that the reaction probability shows an oscillatory behavior as a function of l. The present quantum-mechanical results (both l-sum and CCPA) are in fairly good agreement with the experiment by Brown and Smith.[110] The experimental result by Wong and Bell[111] shows a temperature dependence

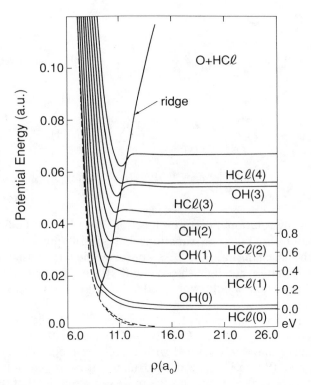

Figure 39. Vibrationally adiabatic potential energy curves for the O + HCl system for $l = 0$ and $n = 0$ (ground hindered rotational state) (from Ref. 108).

opposite to the others and seems to be contradictory to the normal behavior. Since the potential energy surface employed here is a model potential, the comparison between the theoretical results and the experiment can not have a great significance. The big difference between the present results and the quasiclassical trajectory results, however, indicates clearly the quantum-mechanical tunneling effect, because this reaction system has the potential barrier height of 8.12 kcal/mol. Thus the reaction system chosen here demonstrates the importance of the potential ridge, the significance of tunneling effect and the usefulness of the CCPA approximation. As is conceivable from what has been mentioned here, it would be significant and useful to develop the multidimensional semiclassical theory which can describe wave propagation in the region of (or along) the potential ridge.

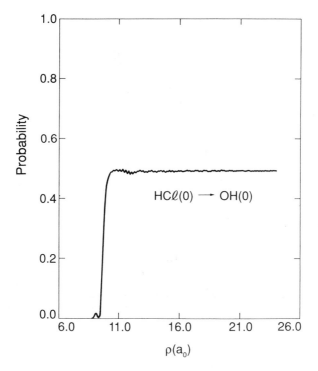

Figure 40.   Reaction probability at $l = 0$ as a function of hyperradius ρ at $E = 0.4\,\text{eV}$ for $O + HCl(v = 0) \rightarrow OH(v = 0) + Cl$ (from Ref. 108).

VI.   SEMICLASSICAL TREATMENT OF ELECTRONICALLY NONADIABATIC CHEMICAL REACTION—FUTURE DEVELOPMENTS

As was mentioned in the Introduction there is no such multidimensional semiclassical theory, unfortunately, that can be straightforwardly and effectively applied to electronically nonadiabatic ion-molecule reactions. In this section, we first briefly review the semiclassical surface hopping method based on the classical S-matrix theory,[78,112] and then try to have a perspective view of possible future developments by taking what have been reviewed and discussed in the previous sections into consideration.

The classical trajectory surface hopping (TSH) method, a review of which is given in this volume by Chapman,[113] is the most convenient and actually the most commonly used procedure to investigate the practical ion–molecule

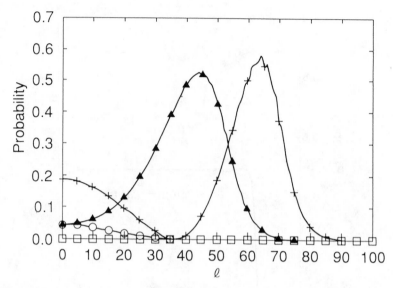

Figure 41. l-dependence of reaction probability for $O + HCl(v = 0) \to OH(v = 0) + Cl$ at various collision energies (from Ref. 108). $\square$: $E = 0.1\,eV$, $\bigcirc$: $E = 0.2\,eV$; $\blacktriangle$: $E = 0.3\,eV$; $+$: $E = 0.36\,eV$.

reaction processes. This method incorporates the effects of nonadiabatic transition into the ordinary classical trajectory procedure by using either the original Landau–Zener formula or the local numerical solution of time-dependent coupled equations.[114] This is useful but possesses some drawbacks: all phases are completely ignored and only the probabilities (not the probability amplitudes) are handled, the detailed balance is not necessarily satisfied, and nonadiabatic tunneling is usually neglected. A more sophisticated one is the semiclassical surface hopping theory, which is a generalization of the classical S-matrix theory made by Miller and George.[112] The double-ended trajectories are analytically continued into the complex R plane in the (avoided) crossing region to go around the branch cut between the two electronic potential energy surfaces. This detour into the complex plane gives a complex action to the S matrix. The imaginary part of the action provides the corresponding nonadiabatic transition probability. Inspite of the conceptual beauty, however, this theory turned out to be very cumbersome and time consuming because of the necessity of handling the full complex trajectories and their complicated branching patterns. In order to remedy this drawback, Kormonicki et al.[115] proposed a decoupling scheme in which the evaluation of electronic transition is decoupled from

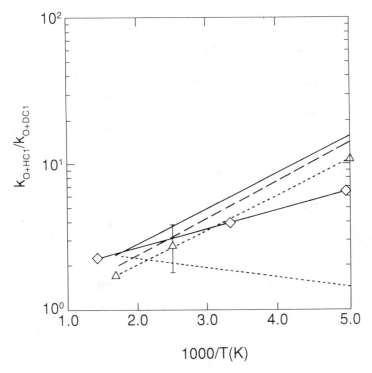

Figure 42. Kinetic isotope effect as a function of $1000/T$ (from Ref. 108). ———: present l-sum result; ---: present CCPA result; $\cdot\cdot\blacktriangle\cdot\cdot$: experiment (Ref. 110); $\cdots$: experiment (Ref. 111); — ◇ —: classical trajectory result (Ref. 109).

the trajectory claculations. The electronic transition calculation is carried out by the local complex integral *at* the avoided crossing point as is given by Eq. (3.16). They have neglected the phases σ_0 and ϕ_S, however, and employed Eq. (3.15). Not many applications have been made since then except for some which were performed within the framework of TSH.[116,117] In the author's opinion, it is now worthwhile to revive the method with some modifications to be mentioned. Classical trajectory constitutes a curvilinear one-dimensional space, and thus the one-dimensional semiclassical theories described in Section III can be utilized. The transition probability amplitudes I given by Eqs. (3.14) or (3.22) accompanied by phases can be incorporated into the formalism. Probably in many occasions we do not have to rely on the complex integral of Eq. (3.16), but can simply use the original Landau–Zener formula Eq. (2.1). Before doing actual dynamics calculations we can analyze the geometry of potential energy surface crossing and know in what situations of classical trajectories encountering the crossing region we can use Eq. (2.1) or we have to rely on Eq. (3.16). The square roots of the nonadiabatic transition probabilities $\sqrt{p}$ or $\sqrt{1-p}$ are multiplied to $S^{\alpha}_{n_2,n_1}$ of Eq. (5.14b) and the phases associated with nonadiabatic transition are added to the phase Φ_{α} along the relevant classical trajectory α. It may be true that in many cases the effects of phases are smeared out and the random-phase approximation seems to hold well. But the problem is that we can not know in advance whether this is really true or not. In any case, it is always valuable to carry out better calculations as far as they are feasible. With this semiclassical surface hopping method, a choice of initial condition for the trajectory which has hopped a surface becomes somewhat problematic. The conventional choice is, however, considered to be reasonable and acceptable.[113] That is to say, only the momentum component perpendicular to crossing seam is changed by the amount corresponding to the adiabatic potential energy difference. As is conceivable from the discussions in Sections III and IV, we have now larger flexibility and versatility. Depending on the schemes of avoided crossing, a proper use can be made for the probability amplitude. For instance, not Eq. (3.15) but Eq. (3.23) should be used in the near-resonant case. Besides, with use of the dynamical-state (DS) representation, any kind of coupling can be equally handled in a unified way. This is important in the case of Coriolis coupling, for instance, because the nonadiabatic transition is delocalized and surface hopping can not be well defined in the Born–Oppenheimer representation. Once we move into the DS representation, all transitions are localized at new avoided crossings and the surface hopping procedure can be straightforwardly applied. As was mentioned in Section III B, the semiclassical theory can deal with the nonclassical case that a crossing point is located in the classically forbidden region. For instance, the modification of δ by Eq. (3.21) is usable for the case that both turning points are larger than the crossing point. This kind of modification is not very accurate, of course, but is much better than simply neglecting the contribution of this case, since this case usually occurs at large impact parameters (or angular momenta of relative motion) and gives a significant contribution to the total transition probability. Furthermore, the nonadiabatic tunneling (tunneling through an adiabatic potential barrier created by the coupling with the other state) can also be treated by the semiclassical theory, again not very accurately though. As far as the author knows, this case has been totally neglected so far in the practical calculations. The one-dimensional semiclassical working equations for this case are not given in this chapter and the readers who are interested in these should refer to the other references.[55] There is one difficult problem, however, in the application to a multidimensional system. That is, how to determine a tunneling trajectory in a multidimensional potential barrier. May be, we can employ the idea from the recent development of the WKB method in multidimensional tunneling.[118] Thanks to the decoupling of electronic transition from classical trajectory, the latter can be run only on the real coordinate space. A search for the double-ended trajectories can be troublesome, however, as in the case of the original classical S-matrix theory. In this case we may relax even the double-end boundary conditions. Reaction probability can be estimated by[78]

$$P_{n_2 n_1} = \int_{n_2 - 1/2}^{n_2 + 1/2} d\bar{n}_2 |S_{\bar{n}_2, n_1}|^2 \qquad (6.1a)$$

or

$$P_{n_2 n_1} = \int_{n_1 - 1/2}^{n_1 + 1/2} d\bar{n}_2 |S_{\bar{n}_2, \bar{n}_1}|^2, \qquad (6.1b)$$

where S_{n_2, n_1} is a scattering matrix with $\bar{n}_j$ ($j = 1, 2$) continuous action variable. Equation (6.1b) can preserve the detailed balance. The last difficulty associated with the usage of classical trajectory is the chattering explained in Section V B. This has no direct connection with nonadiabatic transition, but is an inherent phenomenon in classical mechanics originated from its nonlinearity and is considered to be one possible manifestation of quantum-mechanical overlapping resonances. The semiclassical surface hopping method described above shares the similarity in basic idea with the generalized surface hopping procedure discussed by Herman[119,120] and the semiclassical theory in phase space of Takatsuka and Nakamura.[121] The latter is a theory of phase-space distribution function into which the semiclassical theories given in Section III can be directly incorporated. The theory is essentially based on the wave packet expansion and its practical feasibility is still an open question. Herman's idea is to propagate a multidimensional wavefunction, which is expanded with respect to non-adiabatic coupling. This perturbation expansion can be avoided and improved by using the semiclassical theories of Section III as was discussed previously. If a multidimensional wavefunction can be constructed effectively on a single potential energy surface, then we can pursue the same idea as that described in Section IV A and patch the basic matrices to build up the scattering matrix. Namely, the overall reaction S-matrix element is reduced to a sum of the products of a transition amplitude for inelastic (vibronic) transition in initial channel, that for reactive transition, and that for inelastic (vibronic) transition in final channel.[22] This method can be, in principle, free from the chattering phenomenon. However, the above mentioned "if" is still a big "if." Multidimensional semiclassical mechanics has been discussed by many authors,[2,11,12,118] but still a lot of effort should be paid for its development especially with an emphasis on its practical usuability. This is one of the most challenging subjects. Much simpler treatment would be to approximate the transition amplitudes for the inelastic transitions by these discussed in Section IV and the one for reactive transition by one of those obtained from the methods of Section V. Further simplification is to approximate the total reaction probability in this way.

VII. SUMMARY

Basic mechanisms of charge transfer and particle rearrangement are summarized and clarified. They are radially induced nonadiabatic transition, rotationally induced nonadiabatic transition, spin–orbit interaction, electron momentum transfer effect, and vibrationally nonadiabatic transition around potential ridge. Fundamental semiclassical theories for the two-state radially induced nonadiabatic transitions are reviewed. It is shown that with use of the dynamical-state representation all the transitions mentioned above associated with charge transfer can be basically classified into either Landau–Zener type or Rosen–Zener type and can be treated in a unified way by the sophisticated semiclassical theories. Transitions in this representation are all well localized at the new avoided-crossing points and thus the basic two-state theory can be easily extended to a general multilevel curve-crossing problem. The latter problem involving closed channels can be conveniently formulated by introducing an χ matrix, which spans not only open but also closed channels. This χ matrix can be constructed by patching the element matrices that represent the basic phenomena such as wave propagation, nonadiabatic transition, and wave reflection. The most simplified version of this multilevel curve crossing model is the BFG (Bauer–Fisher–Gilmore) model for the vibronic transitions in ion–molecule collisions. Some numerical applications are presented for ion–atom and ion–molecule collisions.

As for the particle rearrangement (chemical reaction), the simplest Langevin model, the classical S-matrix theory, and the hyperspherical coordinate approach are briefly reviewed. The PRS (perturbed rotational state) representation proposed by Takayanagi and Sakimoto is explained to be useful to estimate the rotational-state-dependent capture cross sections in low-energy ion–molecule collisions. The hyperspherical coordinate approach is shown to be powerful to clarify the reaction mechanisms. Particle rearrangement, especially light-atom transfer, occurs in a spatially localized region around the ridge of potential energy surface. Reaction can be clearly viewed as a vibrationally nonadiabatic transition. Some numerical examples are presented.

The basic theories explained in Sections III–V cannot be directly applied to the practical 3-D ion–molecule reactions. These theories must be useful, however, to grasp the physics of the reaction mechanisms at least qualitatively. Moreover, it seems to be worthwhile to formulate a kind of modified semiclassical surface hopping method by combining the sophisticated semiclassical theories given in Section III and the basic idea of the classical S-matrix theory, as is discussed in Section VI. This would be a realizable nice step forward to tackle a 3-D ion–molecule reaction. Semiclassical theory is always useful to comprehend the mechanisms, even when quantum-mechanical accurate calculations become feasible, which actually does not seem to be the case in near future for the 3-D electronically nonadiabatic reactions. On the other hand, it is very important to direct much effort to further develop and generalize the basic multidimensional semiclassical theories so that they can deal with tunneling and be practically useful in chemical reaction dynamics.

Acknowledgments

The author would like to thank Mr. M. Namiki, Dr. A. Ohsaki, and Dr. S. C. Park for the collaborations in some of the works reported in this article. This work was supported in part by a Grant in Aid from the Ministry of Education, Science and Culture of Japan. The numerical computations were carried out at the computer center of Institute for Molecular Science.

References

1. N. Fröman and P. O. Fröman, *JWKB Approximation. Contributions to the Theory*, North-Holland, Amsterdam, 1965.
2. V. P. Maslov and M. V. Fedoriuk, *Semi-classical Approximation in Quantum Mechanics*, D. Reidel, Boston, 1981.
3. M. V. Berry and K. E. Mount, *Rep. Prog. Phys.* **35**, 315 (1972).
4. M. S. Child, ed., *Semiclassical Methods in Molecular Scattering and Spectroscopy*, D. Reidel, Boston, 1980; *Molecular Collision Theory*, Academic, New York, 1974.
5. D. Huber and E. J. Heller, *J. Chem. Phys.* **87**, 5302 (1987), and references therein.
6. M. Hillery, R. F. O'Connell, M. O. Scully, and E. P. Wigner, *Phys. Rep.* **106**, 121 (1984).
7. P. Carruthers and F. Zachariasen, *Rev. Mod. Phys.* **55**, 245 (1983).
8. K. Takatsuka and H. Nakamura, *J. Chem. Phys.* **82**, 2573 (1985); **83**, 3491 (1985).
9. D. S. F. Crothers, *Adv. Phys.* **20**, 405 (1971); *Adv. Atom. Mol. Phys.* **17**, 55 (1981).
10. E. E. Nikitin and S. Ya. Umanskii, *Theory of Slow Atomic Collisions*, Springer, Berlin, 1984.
11. J. B. Delos, *Adv. Chem. Phys.* **65**, 161 (1986).
12. J. R. Klauder, *Phys. Rev. Lett.* **56**, 897 (1986); in *Random Media*, The IMA Volume in Mathematics and Its Application, G. Papanicolauou, ed., Springer, New York, 1987, Vol. 7, p. 163.
13. M. T. Bowers and T. Su, in *Interaction Between Ions and Molecules*, P. Ausloos, ed., Plenum, New York, 1975, p. 163.

14. T. Su and M. T. Bowers, in *Gas Phase Ion Chemistry*, M. T. Bowers, ed., Academic Press, New York, 1979, Vol. 1, Chap. 3.

15. A. B. Weiglein and D. Rapp, in *Gas Phase Ion Chemistry*, M. T. Bowers, ed., Academic Press, New York, 1979, Vol. 2, Chap. 16.

16. K. Takayanagi, in *Physics of Electronic and Atomic Collisions*, Invited Papers of the XII International Conference on the Physics of Electronic and Atomic Collisions, S. Datz, ed., North-Holland, Amsterdam, 1982, p. 343.

17. A. W. Kleyn and J. Los, *Phys. Rep.* **90**, 1 (1982).

18. V. Sidis, *Adv. Atom. Mol. Opt. Phys.* **26**, 161 (1990).

19. M. S. Child, in *Atom-Molecule Collision Theory*, R. B. Bernstein, ed., Plenum, New York, 1979, Chap. 13.

20. K. S. Lam and T. F. George, in Ref. 4, p. 179.

21. H. Nakamura, (a) *Butsuri (in Japanese)* **41**, 413 (1986); (b) in *Electronic and Atomic Collisions*, Invited Papers of the XV International Conference on the Physics of Electronic and Atomic Collisions, H. B. Gilbody et al., eds., North-Holland, Amsterdam, 1988, p. 413; (c) *Int. Rev. Phys. Chem.* **10**, 123 (1991).

22. See, for instance, H. Nakamura, *Mol. Phys.* **26**, 673 (1973).

23. L. D. Landau, *Phys. Zeit. Sowjet.* **2**, 46 (1932).

24. C. Zener, *Proc. Roy. Soc. A* **137**, 696 (1932).

25. N. Rosen and C. Zener, *Phys. Rev.* **40**, 502 (1932).

26. Y. N. Demkov, *Sov. Phys. JETP* **18**, 138 (1984).

27. W. R. Thorson, *J. Chem. Phys.* **34**, 1744 (1961).

28. H. Nakamura, *Phys. Rev. A* **26**, 3125 (1982); *Chem. Phys.* **78**, 235 (1983).

29. H. Nakamura and M. Namiki, (a) *J. Phys. Soc. Japan* **49**, L843 (1980); (b) *Phys. Rev. A* **24**, 2963 (1981).

30. M. Kimura and N. F. Lane, *Adv. Atom. Mol. Opt. Phys.* **26**, 80 (1990).

31. M. Kimura and W. R. Thorson, *Phys. Rev. A* **24**, 1780 (1981).

32. H. Nakamura, in *Electronic and Atomic Collisions*, Invited Papers of the XIII International Conference on the Physics of Electronic and Atomic Collisions, J. Eichler et al., eds., North-Holland, Amsterdam, 1984, p. 661; *J. Phys. Chem.* **88**, 4812 (1984).

33. M. Desouter-Lecomte, D. Dehareng, B. Leyh-Nihant, M. Th. Praet, A. J. Lorquet, and J. C. Lorquet, *J. Phys. Chem.* **89**, 214 (1985).

34. M. Desouter-Lecomte, B. Leyh-Nihant, M. T. Praet, and J. C. Lorquet, *J. Chem. Phys.* **86**, 7025 (1987).

35. D. G. Truhlar, A. D. Isaacson, and B. C. Garrett, in *Theory of Chemical Reaction Dynamics*, M. Baer, ed., CRC, BocaRaton, Fl., 1985, Vol. 4, Chap. 2.

36. A. Ohsaki and H. Nakamura, *Phys. Rep.* **187**, 1 (1990), and references therein.

37. U. Fano, *Phys. Rev. A* **24**, 2402 (1981).

38. E. C. G. Stückelberg, *Helv. Phys. Acta* **5**, 369 (1932).

39. K. Nasu and Y. Kayanuma, *Butsuri (in Japanese)* **35**, 226 (1980).

40. Y. Kayanuma, *J. Phys. Soc. Japan* **51**, 3526 (1984); **53**, 108 (1984); **53**, 118 (1984); **54**, 2037 (1985).

41. A. Yoshimori and M. Tsukada, eds., *Dynamical Processes and Ordering on Solid Surfaces*, Springer, Berlin, 1985.

42. D. DeVault, *Quantum Mechanical Tunneling in Biological Systems*, Cambridge Univ. Press, Cambridge, 1984.

43. P. Wolynes, *J. Chem. Phys.* **86**, 1957 (1987).

44. Y. Abe and J. Y. Park, *Phys. Rev. C* **28**, 2316 (1983).

45. B. Imanishi and W. von Oertzen, *Phys. Rev.* **155**, 29 (1987).

46. R. Ramaswamy and R. A. Marcus, *J. Chem. Phys.* **74**, 1385 (1981).

47. P. Gaspard, S. A. Rice, and K. Nakamura, *Phys. Rev. Lett.* **63**, 930 (1989).

48. L. D. Landau and E. M. Lifshitz, *Quantum Mechanics Nonrelativistic Theory*, Pergamon Press, Oxford, 1965.

49. S. C. Miller and R. H. Good, *Phys. Rev.* **91**, 174 (1953).

50. G. V. Dubrovskii, *Sov. Phys. JETP* **19**, 591 (1964).

51. J. Heading, *An Introduction to Phase-Integral Methods*, Methuen, London, 1962.

52. M. A. Evgrafov and M. V. Feroryuk, *Russian Mathematical Surveys* **21**, 1 (1966).

53. Y. Shibuya, *Global Theory of a Second Order Linear Ordinary Differential Equation with a Polynominal Coefficient*, North-Holland, Amsterdam, 1975.

54. R. Suzuki, H. Nakamura, and E. Ishiguro, *Phys. Rev. A* **29**, 3060 (1984).

55. H. Nakamura, *J. Chem. Phys.* **87**, 4031 (1987).

56. D. S. F. Crothers, *J. Phys. B* **9**, 635 (1976).

57. M. S. Child, *J. Mol. Spect.* **53**, 280 (1974).

58. M. J. Seaton, *Rep. Prog. Phys.* **46**, 167 (1983).

59. H. Nakamura and F. H. Mies, *VI-th Discussions on Chemical Reaction* (in Japanese), Tokyo, June, 1990, p. 195.

60. D. H. Kouri, in *Atom-Molecule Collision Theory, A Guide for the Experimentalist*, R. B. Bernstein, ed., Plenum, New York, 1979, Chap. 9.

61. R. Schinke and J. M. Bowman, in *Molecular Collision Dynamics*, J. M. Bowman, ed., Springer, Berlin, 1983, p. 61.

62. E. Bauer, E. R. Fisher, and F. R. Gilmore, *J. Chem. Phys.* **51**, 4173 (1969).

63. E. A. Gislason and J. G. Sachs, *J. Chem. Phys.* **62**, 2678 (1975).

64. M. S. Child and M. Baer, *J. Chem. Phys.* **74**, 2832 (1981).

65. U. C. Klomp, M. R. Spalburg, and J. Los, *Chem. Phys.* **83**, 33 (1984).

66. C. Desfrancois, J. P. Astuc, R. Barbe, and J. P. Schermann, *J. Chem. Phys.* **88**, 3037 (1988).

67. Y. N. Demkov and V. I. Osherov, *Sov. Phys. JETP* **26**, 916 (1968).

68. V. Aquilanti, S. Cavalli, and G. Grossi, *Chem. Phys. Lett.* **133**, 531 (1987).

69. V. Aquilanti and S. Cavalli, *Chem. Phys.* **133**, 538 (1987).

70. D. Paillard, J. M. Mestdagh, J. Cuvelier, P. de Pujo, and J. Berlande, *J. Chem. Phys.* **87**, 2084 (1987).

71. A. Fukuroda, N. Kobayashi, and Y. Kaneko, *J. Phys. B* **22**, 3457 (1989).

72. R. E. Olson, F. T. Smith, and E. Bauer, *Appl. Opt.* **10**, 1848 (1971).

73. M. Halmann and I. Laulicht, *J. Chem. Phys.* **43**, 438 (1965); **43**, 1503 (1965).

74. K. Takayanagi, *Comments on Atom. Mol. Phys.* **9**, 143 (1980).

75. K. Sakimoto and K. Takayanagi, *J. Phys. Soc. Japan* **48**, 2076 (1980).

76. K. Sakimoto, *J. Phys. Soc. Japan* **48**, 1683 (1980); *Chem. Phys.* **63**, 419 (1981); *Chem. Phys.* **68**, 155 (1982).

77. K. Sakimoto, *Chem. Phys.* **85**, 273 (1984).

78. W. H. Miller, *Adv. Chem. Phys.* **25**, 69 (1974); **30**, 77 (1975).

79. J. V. Dugan Jr., *Chem. Phys. Lett.* **21**, 476 (1973).

80. R. P. Feynman and A. R. Hibbs, *Quantum Mechanics and Path Integrals*, McGraw Hill, New York, 1965.

81. C. C. Rankin and W. H. Miller, *J. Chem. Phys.* **55**, 3150 (1971).

82. J. R. Stine and R. A. Marcus, *Chem. Phys. Lett.* **29**, 575 (1974).

83. L. Gottdiener, *Mol. Phys.* **29**, 1585 (1975).

84. D. W. Noid, S. K. Gray, and S. A. Rice, *J. Chem. Phys.* **84**, 2649 (1986).

85. P. Gaspard and S. A. Rice, *J. Chem. Phys.* **90**, 2242 (1989); **90**, 2255 (1989).

86. B. Eckhardt, *Physica D* **33**, 89 (1988).

87. R. Blümel and U. Smilansky, *Phys. Rev. Lett.* **60**, 477 (1988); *Physica D* **36**, 111 (1989); *Phys. Rev. Lett.* **64**, 241 (1990).

88. S. Bleher, E. Ott, and C. Grebogi, *Phys. Rev. Lett.* **63**, 919 (1989).

89. K. Someda and H. Nakamura, *J. Chem. Phys.* **94**, 4260 (1991).

90. B. LePetit and J. M. Launay, *Chem. Phys. Lett.* **151**, 287 (1988).

91. J. M. Launay and M. Le Dourneuf, *Chem. Phys. Lett.* **163**, 178 (1989).

92. G. A. Parker, R. T. Pack, A. Lagana, B. J. Archer, J. D. Kress, and Z. Bacic, in *Supercomputer Algorithms for Reactivity, Dynamics and Kinetics of Small Molecules*, A. Lagana, ed., Kluwer Academic, Dordrecht, 1989, p. 105.

93. J. Z. H. Zhang and W. H. Miller, *J. Chem. Phys.* **91**, 1528 (1989).

94. H. Nakamura, *Chem. Phys. Lett.* **141**, 77 (1987).

95. R. A. Fischer, P. L. Gertitschke, J. Manz, and H. H. R. Schor, *Chem. Phys. Lett.* **156**, 100 (1989).

96. J. C. Polanyi and N. Sathyamurthy, *Chem. Phys.* **33**, 287 (1978); **37**, 259 (1979).

97. H. Nakamura and A. Ohsaki, *J. Chem. Phys.* **83**, 1599 (1985).

98. J. Manz and H. H. Schor, *Chem. Phys. Lett.* **107**, 542 (1984).

99. K. C. Kulander, J. Manz, and H. H. Schor, *J. Chem. Phys.* **82**, 3088 (1985).

100. P. L. Gertischke, P. Kiprof, and J. Manz, *J. Chem. Phys.* **87**, 941 (1987).

101. V. Aquilanti, A. Lagana, and R. D. Levine, *Chem. Phys. Lett.* **158**, 87 (1989).

102. J. Jellinek and D. J. Kouri, in *Theory of Chemical Reaction Dynamics*, M. Baer, ed., CRC, BocaRaton, Fl., 1985, Vol. II, Chap. 1.

103. J. M. Bowman, *Adv. Chem. Phys.* **61**, 115 (1985).

104. J. W. Bowman and A. F. Wagner, in *The Theory of Chemical Reaction Dynamics*, D. C. Clary, ed., Reidel, Dordrecht, 1986, p. 47.

105. A. Ohsaki, H. Nakamura, and S. C. Park, *Comput. Phys. Commun.* **52**, 291 (1989).

106. D. C. Clary, *Mol. Phys.* **44**, 1083 (1981); in *The Theory of Chemical Reaction Dynamics*, D. C. Clary, ed., Reidel, Dordrecht, 1986, p. 331.

107. A. Ohsaki and H. Nakamura, *Chem. Phys. Lett.* **142**, 37 (1987).

108. S. C. Park, H. Nakamura, and A. Ohsaki, *J. Chem. Phys.* **92**, 6538 (1990).

109. A. Persky and M. Broida, *J. Chem. Phys.* **81**, 4352 (1984).

110. R. D. H. Brown and I. W. Smith, *Int. J. Chem. Kinet.* **10**, 1 (1978).

111. E. L. Wong and F. E. Bell, *Chem. Abstr.* **76**, 1832 (1972).

112. W. H. Miller and T. F. George, *J. Chem. Phys.* **56**, 5637 (1972).
113. S. Chapman, in this volume.
114. G. Parlant and E. A. Gislason, *J. Chem. Phys.* **91**, 4416 (1989).
115. A. Kormonicki, T. F. George, and K. Morokuma, *J. Chem. Phys.* **65**, 48 (1976).
116. J-H. Yuan and T. F. George, *J. Chem. Phys.* **70**, 990 (1979).
117. H-W. Lee, K. S. Lam, P. L. DeVries, and T. F. George, *J. Chem. Phys.* **73**, 206 (1980).
118. Z. H. Huang, T. E. Feuchtwang, P. H. Cutler, and E. Kazes, *Phys. Rev. A* **41**, 32 (1990).
119. M. F. Herman, *J. Chem. Phys.* **81**, 754 (1984); **81**, 764 (1984); **82**, 3666 (1985).
120. R. Currier and M. F. Herman, *J. Chem. Phys.* **82**, 4509 (1985).
121. K. Takatsuka and H. Nakamura, *J. Chem. Phys.* **85**, 5779 (1986).

THE SEMICLASSICAL TIME-DEPENDENT APPROACH TO CHARGE-TRANSFER PROCESSES

ERIC A. GISLASON

Department of Chemistry, University of Illinois at Chicago, Chicago, Illinois

GÉRARD PARLANT[a]

Department of Chemistry, The Johns Hopkins University, Baltimore, Maryland

MURIEL SIZUN

Laboratoire des Collisions Atomiques et Moléculaires,[b] Université de Paris-Sud, Orsay, France

CONTENTS

[a] Permanent address: Physico-Chimie des Rayonnements (UA 75, Associé au CNRS), Bâtiment 350, Université de Paris-Sud, 91405 Orsay Cedex, France.
[b] Unité associé au CNRS-No. 281.

State-Selected and State-to-State Ion–Molecule Reaction Dynamics, Part 2: Theory, Edited by Michael Baer and Cheuk-Yiu Ng. Advances in Chemical Physics Series, Vol. LXXXII. ISBN 0-471-53263-0 © 1992 John Wiley & Sons, Inc.

I. INTRODUCTION

A number of new experimental techniques have made the field of ion–molecule charge transfer (CT) reactions very exciting in recent years. Several reviews[1–6] have summarized much of this work. Particularly important are the new techniques that allow the state selection of reactants and the state analysis of products. For example, photoionization techniques including the threshold photoelectron–photoion coincidence method[6–8] have made it possible to measure absolute total CT cross sections as a function of the initial internal state of the reactants. In addition, laser-induced fluorescence has been used to analyze the rotational and vibrational product state distributions from CT reactions.[1] Crossed-molecular-beam apparatuses are now able to measure high-resolution angular and translational distributions for CT processes. A number of laboratories have succeeded in resolving individual vibrational levels of molecular CT products.[2] Perhaps the most remarkable experiments done to date are the measurements of state-to-state cross sections carried out in Ng's laboratory.[6]

The wide range of state-specific experimental data now available for CT processes presents many challenges and opportunities for theoreticians. Since these reactions always involve two or more potential energy surfaces, the theory of nonadiabatic transitions[9-12] must be used to treat them. The wide variety of theoretical techniques that have been used to study CT processes has been reviewed by several authors.[12-17]

The prototypical CT system is the reaction

$$Ar^+(^2P_{3/2}, {}^2P_{1/2}) + N_2(v = 0) \rightarrow Ar + N_2^+(v') \tag{1}$$

and its reverse. A wide variety of experimental results have been obtained for reaction (1).[1,2,6,8,18] It would be desirable to calculate exact state-to-state total and differential cross sections for this process. This is not possible at the present time. Nevertheless, a set of potential energy surfaces has been computed for this system.[19] In addition, considerable progress has been made in the dynamical calculations.[8,14,16-18] The results to date indicate that accurate results can only be obtained if the vibrational motion, as well as the electronic degrees of freedom, are treated quantum mechanically. Thus, until now, only two theoretical procedures have worked well for processes such as reaction (1).

The first of these involves a quantum-mechanical treatment of all degrees of freedom of the system.[20] In principle, this gives exact results, but in practice it is necessary to make approximations, such as the infinite-order-sudden (IOS) approximation for molecular rotation. Even then, the calculations are time consuming and difficult to perform at high collision energies, owing to the large number of open channels. In addition, the IOS approximation is not expected to work well at high energies.[17] In spite of these problems, important contributions have been made using this procedure.[21-23] Quantum-mechanical calculations of CT reactions are reviewed in Chapter 2.

The other very useful procedure for studying processes such as reaction (1) is the classical path technique. As the name suggests, the translational and rotational motions are treated classically, while the vibrational and electronic motions are treated quantum mechanically. Although this method is not an exact procedure, it has the advantage of being easier to carry out than quantum-mechanical calculations, while still treating the vibrational and electronic degrees of freedom on an equal footing. This technique is the subject of the present chapter. We restrict the discussion to positive ion CT involving at least one molecule; thus, atom–atom collisions are excluded.[24] In general, we consider only nonreactive systems such as reaction (1). The classical path technique has been widely used to study energy transfer on a single potential energy surface.[25] The first application of this method to study ion–molecule CT processes was the paper by Bates and Reid, which appeared in 1969[26]. Since that time many other papers have appeared that use the classical path technique, and this work has been reviewed a number of times.[12,13,16–18,27]

Theoretical calculations based on the classical path procedure for molecular CT processes have been carried out by a limited number of groups. After the work of Bates and Reid,[26] Moran, Flannery, and co-workers carried out in the 1970s theoretical and experimental studies of a number of symmetric CT reactions[28–38] such as

$$O_2^+ + O_2 \rightarrow O_2 + O_2^+. \tag{2}$$

They also studied a few atom–molecule systems.[39] In the early 1980s, more sophisticated calculations were carried out on symmetric systems by DePristo and coworkers,[40–47] using more accurate potential energy surfaces.[48] During this period the "semiclassical energy-conserving trajectory technique" was applied for the first time to CT reactions[38,42]. In 1985 and 1986 Kimura and co-workers reported calculations in the keV energy range for $H^+ + H_2$[49] and $Ar^+ + H_2$.[50] At about the same time Spalburg and Gislason used the classical path technique to study reaction (1) and its reverse.[14] They established the importance of fine-structure transitions in the $Ar^+(^2P_J)$ ion in $Ar^+–N_2$ collisions, and they were the first to point out the importance of adiabatic vibronic potential energy curves in determining the dynamics of CT processes at low collision energies.[51] Finally, in 1985 the present authors, from the University of Paris at Orsay and the University of Illinois at Chicago, began a collaboration, which continues to the present time. During this period we have studied collisions of $Ar^+ + N_2$ and its reverse,[52–54,56,57] $Ar^+ + CO,$[55] and $H^+ + O_2$[58] using the classical path procedure. All of this work is based on potential energy surfaces computed by our colleagues at Orsay.[19,59,60] We have also carried out calculations[61] on the reactive system $Ar^+ + H_2$, where the charge-transfer process

$$Ar^+(^2P_J) + H_2 \rightarrow Ar + H_2^+ \tag{3a}$$

competes at low collision energy with the reaction

$$Ar^+(^2P_J) + H_2 \rightarrow ArH^+ + H. \tag{3b}$$

Henri et al.[62] have shown that a model which assumes that small impact-parameter collisions lead to reaction and that CT comes only from large impact-parameter collisions is consistent with the available experimental data. Based on this model we have calculated CT cross sections[61] for reaction (3a) as well as reactive cross sections[63] for reaction (3b).

In recent years another area of ion–molecule CT reactions has become very active. This is the field of dissociative CT. A typical process is

$$H_2^+ + Mg \rightarrow H_2^*(b\,^3\Sigma_u^+) + Mg^+(^2S) \rightarrow 2H(1s) + Mg^+. \qquad (4)$$

Sidis[12,64] has reviewed the experimental and theoretical work on these systems. A theoretical treatment of dissociative CT is difficult, because the product states lie in the vibrational continuum of the molecule. Nevertheless, the classical path procedure has proved to be very useful in studying these reactions. At high collision energies, one can use the sudden approximation, which assumes that the molecule does not rotate or vibrate during the collision.[65] At lower energies, however, the molecule is in the process of dissociating while the collision is occurring. In this energy range Sidis and co-workers have developed a local complex potential approximation to treat the dissociation[66] as well as the vibrational excitation[67] that occurs. Very recently, Gauyacq and Sidis[68] have reported a coupled wave packet procedure which, in principle, can be used to treat reaction (4) exactly.

In this chapter we begin by reviewing the computational procedures for the classical path method. This includes calculations of both total and differential state-to-state cross sections. Next we give a thorough discussion of results that have been obtained using this procedure by the various groups named. We conclude by describing three general properties of CT processes that have emerged from the calculations carried out in our laboratory. First, we discuss the breakdown of the FC principle at low collision energies. Then we describe the importance of adiabatic vibronic potential energy curves for determining nonadiabatic behavior at low energies. Finally, we present a general theory of CT reactions that describes the change in mechanism as the collision energy is varied. This theory summarizes all of the work we have done to date.

II. CLASSICAL PATH FORMULATION

In this section for simplicity we will discuss only charge-transfer systems of the form $(A + BC)^+$, although the presentation can be extended with no difficulty to diatom–diatom encounters.

As mentioned in the Introduction, state-specific experiments on CT are now run in several laboratories.[1–8] These results require interpretation at

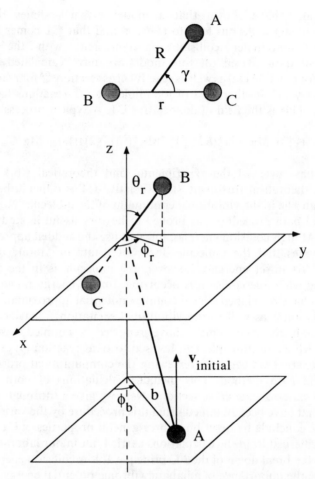

Figure 1. Coordinates for collision of $A + BC$ (Upper panel) Internal coordinates $\mathbf{R}$, $\mathbf{r}$, and γ. (Lower panel) Coordinates in the laboratory reference frame θ_r, ϕ_r, b, and ϕ_b.

the level of individual (electronic and vibrational) quantum states of the collision partners. Despite its success in interpreting many molecular dynamics experiments[69], the classical trajectory method is not well suited to provide this kind of information. One reason is that, if only a small number of product quantum states are populated, the (discrete) product distribution is not well represented by the histogram obtained by binning the corresponding classical (continuous) quantities.[70] Another reason is directly related to the fact that CT is a nonadiabatic process. In terms of classical trajectories, such a process can be treated by the trajectory-surface-hopping

(TSH) method,[9,10,71,72] which allows for hops between potential energy surfaces (PES) along the trajectory. For computational convenience, the transition probabilities are generally obtained by means of analytic models, such as the Landau–Zener[73] and the Demkov[74] models. Another important limitation of the TSH method is that the transitions can take place only on a well-specified hypersurface, generally the avoided intersection (*seam*) of two PES. No provision is made for nonlocalized transitions, although some authors[75–79] have recently developed modifications of the TSH method aimed at relaxing this constraint.

Considering these limitations of the TSH technique, it appears desirable in the calculations to quantize degrees of freedom corresponding to the state-selected information obtained from experiments. This means that we will treat—in addition to the electronic motion of the whole system—the vibration of the diatomic target quantum mechanically, retaining the simplicity of classical mechanics for the relative translation and for the rotation of the diatom. In the following, the quantized coordinates will be called *internal*, and the classical coordinates, *external*.[9] This selection of internal/external coordinates is not the only choice available.[10,80] In particular, the dependencies of the cross sections on the rotational degrees of freedom of the target *BC* can be handled more rigorously if the rotational motion of *BC* is also quantized.[12,17]

As we shall see, the semiclassical procedure developed in this section allows one to calculate vibronic transition probabilities systematically, without referring to any dynamical analytical models and without the need to localize transitions. The vibronic states of the diatomic molecule before and after the collision will be, of course, quantized. The method works extremely well for collision energies above 10 eV, and with care good results can be obtained down to 1 eV.

A. Classical Path Equations

Let us consider the collision of an atomic projectile A with a diatomic target BC (Fig. 1). We denote by $\mathbf{r}$ the B–C vector, by $\mathbf{R}$ the vector joining A to the center of mass of BC, and by γ the angle between $\mathbf{r}$ and $\mathbf{R}$. The classical path (CP) method consists in solving the quantum motion of the internal coordinates $\mathbf{q}$ in the field of the external coordinates $\mathbf{Q}$, which are constrained to follow a classical trajectory $\mathbf{Q}(t)$. With the present choice of coordinates, $\mathbf{Q}$ stands for $\mathbf{R}$ and the orientation $\hat{\mathbf{r}}$ of BC in the space-fixed frame. This results in the following time dependent Schrödinger equation:[9]

$$H_{\mathbf{q}}(\mathbf{q}, \mathbf{Q})\psi(\mathbf{q}, \mathbf{Q}(t)) = i\hbar \frac{\partial}{\partial t} \psi(\mathbf{q}, \mathbf{Q}(t)). \qquad (5)$$

It can be shown[81-87] that, under the JWKB conditions, this equation is equivalent to a fully quantum-mechanical stationary Schrödinger equation. In Eq. (5), H_q is the hamiltonian corresponding to the internal (that is, electronic and vibrational) coordinates, defined by

$$H_q = H - T_Q, \tag{6}$$

where H is the total hamiltonian and T_Q is the kinetic energy operator for the external coordinates. This operator does not figure in Eq. (5), since it refers to *classical* coordinates. With the present choice of internal/external coordinates, the wavefunction ψ, which describes the internal state of the system during the collision, can be expanded in a set of orthonormal basis functions as

$$\psi(\mathbf{q}, \mathbf{Q}(t)) = \sum_{\alpha',v'} c_{\alpha'v'}(\mathbf{Q}(t))\varphi_{\alpha'}(\{\boldsymbol{\rho}_i\}; r, R, \gamma)\chi_{\alpha'v'}(r; R, \gamma). \tag{7}$$

In this equation, α' identifies the electronic state and v' the vibrational level (in state α'). In addition, $\{\boldsymbol{\rho}_i\}$ are the electronic coordinates in the ABC body-fixed frame, and the $\varphi_{\alpha'}$'s are the electronic basis functions, which may depend parametrically on the internuclear distance of the diatom, r, and on the relative coordinates R and γ. The vibrational basis functions $\chi_{\alpha'v'}$ may also depend parametrically on R and γ. Finally, the (complex) probability amplitudes $c_{\alpha'v'}$, which vary continuously along the trajectory, are to be calculated. We shall sometimes simplify $c_{\alpha v}(\mathbf{Q}(t))$ to $c_{\alpha v}(t)$ when the meaning is clear.

At a particular collision energy a given trajectory is characterized by the impact parameter $\mathbf{b} = (b, \phi_b)$, the initial orientation $\hat{\mathbf{r}} = (\theta_r, \phi_r)$ of BC, and the initial rotational angular momentum vector $\mathbf{J}$ of BC. Equation (5) is solved with the condition that only the entrance channel (α, v) has a non zero amplitude at the beginning of the collision:

$$c_{\alpha'v'}(\mathbf{Q}(\mathbf{b}, \hat{\mathbf{r}}, \mathbf{J}, t \to -\infty)) = \delta_{\alpha\alpha'}\delta_{vv'}. \tag{8}$$

The S-matrix element corresponding to the formation of the vibronic state (α', v') is simply equal[12,24,88] to the appropriate probability amplitude at the end of the collision:

$$S_{\alpha'v',\alpha v}(\mathbf{b}, \hat{\mathbf{r}}, \mathbf{J}) = c_{\alpha'v'}(\mathbf{Q}(\mathbf{b}, \hat{\mathbf{r}}, \mathbf{J}, t \to +\infty)). \tag{9}$$

Finally, the various cross sections can be obtained from the S matrix. It is necessary to properly average over the initial parameters $\mathbf{b}, \hat{\mathbf{r}}, \mathbf{J}$. This calculation is discussed in detail in Section IIF.

B. Basis Set

The electronic and vibrational basis functions can be chosen in various ways. The electronic states can be obtained[89-91] by means of quantum-chemistry techniques, by solving the Schrödinger equation for the electronic motion for a number of different fixed nuclear geometries (r, R, γ). These solutions form the *adiabatic* representation.[92] The wave functions φ_α^{ad} satisfy

$$H_{el}\varphi_\alpha^{ad}(\{\boldsymbol{\rho}_i\}; r, R, \gamma) = E_\alpha^{ad}(r, R, \gamma)\varphi_\alpha^{ad}(\{\boldsymbol{\rho}_i\}; R, r, \gamma), \tag{10}$$

where H_{el} is the electronic hamiltonian of the whole system. In this representation the basis states are coupled by derivative coupling terms, which can be calculated as well.[91] Although it is well known, the adiabatic representation is not necessarily the best one to perform dynamics calculations with. This is because the sharp variation of derivative couplings in the vicinity of avoided crossings leads to computational difficulties.[12] The treatment of collisional systems, as opposed to bound systems, is handled better by means of another electronic basis known as a *diabatic* representation. While the definition of adiabatic states is unique [Eq. (10)], several diabatic representations can be defined.[12,93] This subject has been developed at length in the literature and will be discussed only briefly here. It suffices to say that diabatic states are constructed in order to minimize radial derivative couplings as much as possible. As a consequence, they are not eigenfunctions of H_{el}. Transitions between diabatic states are induced by the off-diagonal electronic hamiltonian matrix elements. This definition is closely related to another definition, which requires that diabatic states should conserve as much as possible their main characteristics along the nuclear coordinates. An example of such a representation would be the two states associated with $(A^+ + BC)$ and $(A + BC^+)$ with the charge constrained to stay on A or BC, respectively, as the collision partners get closer.[93] This is the natural choice for diabatic states in ion–molecule systems. Until recently, however, there were no *ab initio* calculations of the diabatic wavefunctions. Now Levy and co-workers have developed an approximate method to directly calculate diabatic PES and couplings in an *ab initio* procedure.[19]

Following these considerations, we will make the assumption that the electronic basis functions do not contain any parametric dependency on the nuclear coordinates. As for the vibrational basis, adiabatic and diabatic representations can also be defined. Similar to our choice for the electronic states, we will choose the *diabatic* basis set to be the vibrational eigenfunctions of the isolated BC (or BC^+) molecule.

C. Coupled Equations

The solution of the time-dependent Schrödinger equation obtained by substituting Eq. (7) into Eq. (5) requires the numerical integration of a set of coupled first-order differential equations. Before discussing that, let us first specify our notation. The internal hamiltonian H_q [Eq. (6)] is written as

$$H_q = T_r + H_{el}, \tag{11a}$$

where H_{el} is the electronic hamiltonian and

$$T_r = -\frac{\hbar^2}{2\mu}\frac{d^2}{dr^2} \tag{11b}$$

represents the vibrational kinetic energy of the diatomic target. The reduced mass of BC is denoted μ. The operator T_r is purely radial, since the rotational coordinates of BC are *external*. The matrix elements of H_{el} are defined as

$$H_{\alpha\alpha'}(r, R, \gamma) = \langle \varphi_\alpha | H_{el} | \varphi_{\alpha'} \rangle, \tag{12}$$

where the integration is done over the electronic coordinates only. The matrix element $H_{\alpha\alpha'}$, for the case $\alpha \neq \alpha'$, is the coupling that generates nonadiabatic transitions (such as CT) between electronic states α and α'. Each diagonal element $H_{\alpha\alpha}$ represents a (diabatic) PES of the system. $H_{\alpha\alpha}$ can be split into two parts, the potential energy function of the isolated molecule BC and the potential interaction between A and BC, denoted by U_α^{diat} and V_α^{int}, respectively:

$$H_{\alpha\alpha}(r, R, \gamma) = V_\alpha^{int}(r, R, \gamma) + U_\alpha^{diat}(r). \tag{13}$$

The function $V_\alpha^{int}(r, R, \gamma)$ goes to zero when $R \to \infty$. The (diabatic) vibrational basis functions $\chi_{\alpha v}(r)$ are solutions of the Schrödinger equation

$$[T_r + U_\alpha^{diat}(r) - \varepsilon_{\alpha v}]\chi_{\alpha v}(r) = 0, \tag{14}$$

with the associated vibrational energies $\varepsilon_{\alpha v}$. Finally, we define the vibronic quantities

$$V_\alpha^{vv'}(R, \gamma) = \langle \chi_{\alpha v} | V_\alpha^{int}(r, R, \gamma) | \chi_{\alpha v'} \rangle, \tag{15a}$$

$$H_{\alpha\alpha'}^{vv'}(R, \gamma) = \langle \chi_{\alpha v} | H_{\alpha\alpha'}(r, R, \gamma) | \chi_{\alpha'v'} \rangle, \tag{15b}$$

and

$$F_\alpha^v(R, \gamma) = V_\alpha^{vv}(R, \gamma) + \varepsilon_{\alpha v}, \tag{16}$$

where the bracket in Eq. (15) means an integration over the variable r. Definition (16) is used to extract the main part of the phase of the probability amplitude $c_{\alpha v}$ as follows:

$$c_{\alpha v}(\mathbf{Q}(t)) = a_{\alpha v}(\mathbf{Q}(t)) \exp(-i\Phi_{\alpha v}(t)), \tag{17a}$$

with

$$\Phi_{\alpha V}(t) = \frac{1}{\hbar} \int^t d\tau F_\alpha^v(R(\tau), \gamma(\tau)). \tag{17b}$$

The exponential term takes into account the rapid oscillations of $c_{\alpha v}$. As a consequence, except in strong coupling regions, the (complex) coefficients $a_{\alpha v}$ should vary slowly with time, thus facilitating the numerical integration of the coupled equations. The total phase of $c_{\alpha v}$ will be needed to calculate the differential cross sections (see Section II F).

The system of coupled differential equations is obtained by insertion of Eq. (7) into Eq. (5), which is then multiplied by a particular vibronic wavefunction $\varphi_\alpha^*(\{\boldsymbol{\rho}_i\})\chi_{\alpha v}(r)$ and integrated over electronic and vibrational coordinates. This results in

$$i\hbar \frac{\partial}{\partial t} a_{av}(t) = \sum_{\alpha' \neq \alpha, v'} a_{\alpha' v'}(t) H_{\alpha\alpha'}^{vv'}(R(t), \gamma(t)) \exp\left[-\frac{i}{\hbar} \int^t d\tau [F_{\alpha'}^{v'}(\tau) - F_\alpha^v(\tau)]\right]$$

$$+ \sum_{v' \neq v} a_{\alpha v'}(t) V_\alpha^{vv'}(R(t), \gamma(t)) \exp\left[-\frac{i}{\hbar} \int^t d\tau [F_\alpha^{v'}(\tau) - F_\alpha^v(\tau)]\right]. \tag{18}$$

In deriving this equation, we used the fact that the vibronic basis states are diabatic, as stated in Section II B. Consequently, the time-derivative operator has no effect on φ_α and $\chi_{\alpha v}$ since they do not depend on R or γ. In addition, φ_α, which is independent of r, is not modified by T_r. With a different choice of basis states the coupled equations would look very similar with, however, additional (derivative) coupling terms.[9] The right-hand side of Eq. (18) exhibits two terms. The first one represents the interaction between the vibrational states of *different* electronic states ($\alpha v \to \alpha' v'$). In particular, in a two-state CT system it would correspond to charge exchange. The matrix element $H_{\alpha\alpha'}(r, R, \gamma)$ is often assumed to be independent of r. Then it can be factored out of the bracket in Eq. (15b), which simply becomes a Franck–Condon overlap. This is the Franck–Condon approximation for vibronic matrix elements. The second term corresponds to vibrational excitation on a *single* PES ($\alpha v \to \alpha v'$). Note that it will be zero if $V_\alpha^{int}(r, R, \gamma)$ happens to be independent of r. In general, the second term is important only at small values of R. Consequently, it is often neglected in systems that involve charge transfer at large distances.

D. Vibrationally Sudden Approximation

It is interesting to examine how the coupled equations simplify when the projectile velocity is so high that the target can be considered as frozen during the collision.[12,17,65] Similarly to what we did in the preceding section, let us insert the expansion (7) into the Schrödinger equation (5) and multiply by a particular electronic wavefunction φ_α^*. After integrating over the electronic coordinates we obtain:

$$i\hbar \sum_{v'} \frac{\partial c_{\alpha v'}(t)}{\partial t} \chi_{\alpha v'}(r) = \sum_{v'} \varepsilon_{\alpha v'} c_{\alpha v'}(t) \chi_{\alpha v'}(r)$$
$$+ V_\alpha^{int}(r, R, \gamma) \sum_{v'} c_{\alpha v'}(t) \chi_{\alpha v'}(r)$$
$$+ \sum_{\alpha' \neq \alpha} H_{\alpha \alpha'}(r, R, \gamma) \sum_{v'} c_{\alpha' v'}(t) \chi_{\alpha' v'}(r). \quad (19)$$

To simplify this further, we introduce an average vibrational level $\bar{\varepsilon}_\alpha$ for each electronic state α. The vibrationally sudden approximation should be valid when

$$\tau_{coll} \ll \frac{\hbar}{|\varepsilon_{\alpha v} - \bar{\varepsilon}_\alpha|} \quad (20)$$

for all v in set α. When this condition is satisfied, the vibrational wave packet associated to electronic state α will not spread during τ_{coll}. Replacing $\varepsilon_{\alpha v}$ by $\bar{\varepsilon}_\alpha$ in Eq. (19) and defining

$$C_\alpha(\mathbf{Q}(t)) \chi_I(r) = \sum_v c_{\alpha v}(\mathbf{Q}(t)) \chi_{\alpha v}(r), \quad (21)$$

where the subscript I designates the initial vibrational state of BC, we obtain:

$$i\hbar \frac{\partial}{\partial t} C_\alpha(t) = [\bar{\varepsilon}_\alpha + V_\alpha^{int}(r, R, \gamma)] C_\alpha(t) + \sum_{\alpha' \neq \alpha} H_{\alpha \alpha'}(r, R, \gamma) C_{\alpha'}(t). \quad (22)$$

The first term on the right-hand side simply corresponds to the evolution of state α, decoupled from the other states. [No similar term appears in Eq. (18) owing to the redefinition of the probability amplitudes in Eq. (17)]. This system of coupled equations is similar to those encountered in atom–atom collisions. However, it depends on r, the internuclear distance of BC, and must be solved for each fixed value of this parameter. In the vibrationally sudden approximation, the collision time is *a fortiori* much shorter than the rotational period of the target. The orientation of BC will thus be fixed in space. Each trajectory is then characterized by the impact parameter $\mathbf{b}$, the initial orientation $\hat{\mathbf{r}}$, and a fixed internuclear distance r. After solving Eq. (22) for a set of r values, the probability amplitude for a product vibronic state αv can be obtained by a quadrature over r:

$$c_{\alpha v}(\mathbf{Q}(\mathbf{b}, \hat{\mathbf{r}}, t \to +\infty)) = \langle \chi_{\alpha v} | C_{\alpha}(\mathbf{Q}(\mathbf{b}, \hat{\mathbf{r}}, r, t \to +\infty)) | \chi_I \rangle. \tag{23}$$

It should be noted that this result is more general than the Franck–Condon approximation, which makes the additional assumption that the matrix elements $H_{\alpha\alpha'}$ do not depend on r. In this latter case Eq. (23) would simplify to

$$c_{\alpha v}(\mathbf{Q}(\mathbf{b}, \hat{\mathbf{r}}, t \to +\infty)) = C_{\alpha}(\mathbf{Q}(\mathbf{b}, \hat{\mathbf{r}}, \bar{r}, t \to +\infty)) \langle \chi_{\alpha v} | \chi_I \rangle, \tag{24}$$

where $\bar{r}$ is some mean value of r.

E.   Classical Trajectory

In the classical path method the external coordinates execute a classical trajectory denoted $\mathbf{Q}(t)$. This determines the time behavior of the internal coordinates. As noted by Sidis, "the common trajectory itself is merely a device to extract the probability amplitudes $c_{\alpha v}$"[17] (from the coupled equations). We shall see (Section II F) that the semiclassical derivation of the differential cross section $\sigma_{\alpha' v', \alpha v}(\theta, \phi)$ using the sudden approximation for rotation does not even make use of the classical deflection angle obtained in the trajectory. Much of the scattering information is contained in the phase of $c_{\alpha v}$. Sidis's implication that the particular method to determine the trajectory is unimportant appears to be correct at collision energies above $20 \, \text{eV}$,[46] but at lower collision energies care must be exercised in this part of the calculation.

Three procedures have been widely used to determine the classical trajectory. The first is to use straight-line trajectories. This method, which has been widely used in the past for high-energy collisions of all types, is commonly referred to as the impact-parameter method[9]. Test calculations by Lee et al.[46] indicate that total state-selected CT cross sections for $H_2^+(v) + H_2(v = 0)$ are computed correctly by the impact-parameter method at relative energies above $10 \, \text{eV}$. Thus, we expect even state-to-state cross sections to be accurate above 20 eV. There may be a numerical problem with this method at small impact parameters, because the interaction potentials and couplings that appear in the coupled equations (see Section II C) can become very large when the nuclei get close.

A second procedure is to compute the trajectory along the diabatic entrance channel potential energy surface $V_\alpha^{vv}(R, \gamma)$ defined in Eq. (15). This method was first used for CT processes by Parlant and Gislason.[52] The surface $V_\alpha^{vv}(R, \gamma)$ is typically weakly attractive at large R, owing to ion-induced dipole and ion-quadrupole forces, and it is repulsive in the region where the electron clouds overlap.[19] This procedure leads to a more realistic trajectory. For example, the nuclei never get too close, so there are fewer numerical problems with the solution of the coupled equations. It has never been tested against exact calculations, but it is likely that it leads to accurate state-to-state cross sections at collision energies above a few eV.

A problem with the two trajectory methods described previously is that no effort is made to conserve the total energy of the system. For example, if the total vibronic energy of the system, as determined by the coupled equations, increases at some point along the trajectory, we would expect that the kinetic energy along the trajectory should decrease to conserve energy. Since a single trajectory couples many states with a wide range of energies, it is only possible to conserve energy on the average.

Moran et al.[38] and DePristo[42] have developed a third method to compute the classical trajectory. DePristo calls this the "semiclassical energy-conserving trajectory technique" (SCECT). The total average energy of all the internal coordinates at any instant t is equal to

$$E(R, \gamma) = \sum_{\alpha', v', \alpha, v} c_{\alpha'v'}^*(t) c_{\alpha v}(t) \langle \phi_{\alpha'} \chi_{\alpha'v'} | H_q | \phi_\alpha \chi_{\alpha v} \rangle. \qquad (25)$$

Here the vibronic coefficients $c_{\alpha v}$ are defined in Eq. (7) and H_q is given in Eq. (11). Then $E(R, \gamma)$ in the SCECT procedure is the potential that determines the classical trajectory. DePristo has tested the classical path procedure using the SCECT method for the CT system $O_2^+-O_2$ against the exact quantum-mechanical calculations of Becker.[21] He obtained perfect agreement at $E_{rel} = 8$ eV, and his calculations at 1 eV suggest that the total CT cross section is accurate there. We expect that state-to-state CT cross sections obtained with this method should be correct at energies of 2 eV and above.

As we have emphasized several times, the classical path method should work well at high energies and can be extended down to the region of a few eV. It would be extremely useful to have an indicator during the calculations of how accurate the state-to-state cross sections are. One such indicator is the principle of microscopic reversibility.[25] This relates the cross sections for the state-to-state processes $\alpha v \to \alpha' v'$ and $\alpha' v' \to \alpha v$ at the same *total* energy E. The relationship (for nonreactive scattering) is

$$g_{\alpha v}(E - E_{\alpha v})^{1/2}\sigma_{\alpha' v', \alpha v} = g_{\alpha' v'}(E - E_{\alpha' v'})^{1/2}\sigma_{\alpha v, \alpha' v'}. \tag{26}$$

Here $E_{\alpha v}$ and $E_{\alpha' v'}$ are the internal energies and $g_{\alpha v}$ and $g_{\alpha' v'}$ are the degeneracies of the two states. The left-hand side refers to the process $\alpha v \to \alpha' v'$ and the right-hand side to the reverse process. It is straightforward to show that microscopic reversibility must be satisfied when straight-line trajectories are used; in this case it serves as a useful check on the computations[14]. For the other two methods described here the results obtained from the classical path calculations will not strictly verify the principle of microscopic reversibility, since the classical trajectory is not the same for the process $\alpha v \to \alpha' v'$ and $\alpha' v' \to \alpha v$, even if all initial conditions (impact parameter, molecular orientations) are chosen to be the same. Consequently, a comparison of cross sections that should be related by Eq. (26) provides a powerful test of the classical path method. The available data for ion–molecule CT systems gives us considerable confidence in the classical path procedure using either of these trajectory methods. To cite one example, Parlant and Gislason determined that microscopic reversibility for $Ar^+ + N_2$ collisions is satisfied to better than 6% at $E_{rel} = 1\,eV$ and to better than 1% at $E_{rel} = 3\,eV$ and higher.[52]

It would be useful to extend classical path calculations below $E_{rel} = 1\,eV$. At this point, however, any energy difference between the reactant state and product state of interest becomes a serious problem. A number of techniques have been suggested to correct approximatively for this problem. The procedures are based on the need to satisfy the principle of microscopic reversibility, and they require separate calculations for each pair of reactant–product channels, which greatly increases the computational effort. The various methods have been reviewed by Billing.[25] To date none of these methods have been applied to CT processes.

Finally, it is interesting to compare the trajectory calculations in the classical path method, discussed above, with those used in the TSH procedure.[9,10,71,72] In the classical path method one common trajectory gives rise to a large number of product states. The advantage is that all internal coordinates are treated quantum mechanically and, in particular, all of the phase information is retained. A disadvantage is that the total energy is not conserved except on the average. By comparison, a typical TSH trajectory branches into many subtrajectories, and total energy is conserved along each subtrajectory. A disadvantage, however, is that all of the phase information is lost. In addition, most TSH calculations have not quantized the vibrational motion, so many interesting experimental results cannot be reproduced. One exception to this for CT processes is the calculation by Nikitin et al.[94] who treated the $(ArN_2)^+$ system. In the future, we expect more TSH calculations on *vibronic* potential energy surfaces to be carried out for CT.

F. Differential Cross Sections

The quantum scattering amplitude matrix[95,96] $f(\theta, \phi)$ for a transition from reactants to products is:

$$f(\theta, \phi) = - i(\pi/kk')^{1/2} \sum_{ll'm'} i^{-\Delta l}(2l + 1)^{1/2}(S - I)Y_{l'm'}(\theta, \phi). \tag{27}$$

Here θ and ϕ are the product polar deflection and azimuthal scattering angles. In addition, S is the scattering matrix, I is the unit matrix, and $Y_{l'm'}(\theta, \phi)$ is a spherical harmonic function. The initial and final relative translational states are specified by the linear momenta $\hbar k$ and $\hbar k'$; the collisional angular momentum quantum numbers l and $l'(\Delta l = l' - l)$; and the corresponding azimuthal quantum numbers $m = 0$ and m', respectively, which refer to projections along the initial relative velocity vector (chosen as the z axis). It is understood that S depends on l, l', and m' as well as the initial and final rovibronic states. The expression in Eq. (27) is a generalization of the result for elastic atom–atom scattering:

$$f(\theta) = \frac{1}{2ik} \sum_{l=0}^{\infty} (2l + 1)[\exp(2i\eta_l) - 1]P_l(\cos \theta); \tag{28}$$

where $P_l(\cos \theta)$ is a Legendre polynomial and η_l is the phase shift.

We now restrict the discussion to atom–diatom scattering and consider a transition from an initial rovibrational state $|nJM\rangle$ to a final state $|n'J'M'\rangle$. Here n identifies the initial electronic state α and the initial vibrational level v; similarly, n' identifies the product values α' and v'. A particular element of the f matrix is

$$f_{n'J'M',nJM}(\theta, \phi) = \langle n'J'M'|f|nJM\rangle. \tag{29}$$

The state-to-state differential cross section is

$$\sigma_{n'J'M',nJM}(\theta, \phi) = \frac{k'}{k}|f_{n'J'M',nJM}(\theta, \phi)|^2. \tag{30}$$

In Eq. (27) the sum over m' is only formal. The total angular momentum is conserved, so $M = M' + m'$. Since M and M' are fixed, m' is fixed, and the sum over m' can be omitted. In addition, since the entire ϕ dependence of $Y_{l'm'}(\theta, \phi)$ is $e^{im'\phi}$, Eq. (27) can be written $\mathbf{f}(\theta, \phi) = e^{im'\phi}\mathbf{F}(\theta)$. When this is substituted into Eq. (30), it is apparent that the state-to-state differential cross section is independent of ϕ.[97] In fact this result holds true for any differential cross section provided that there is no polarizing field in the experiment, which establishes some other axis of quantization.

If a classical path CT calculation is ever carried out that solves the time-dependent Schrödinger equation for all rovibronic states of the system (a monumental undertaking), Eqs. (27), (29), and (30) can be used to evaluate the state-to-state differential cross sections. In many cases one is not interested in the rotational states of the reactants and products. In that case, one can sum Eq. (30) over the product states (J', M') and average over the initial (thermal) distribution of (J, M) to obtain $\sigma_{n',n}(\theta, \phi)$. An important example of this is the state-selected differential cross section

$$\sigma_{n',nJ}(\theta, \phi) = (2J + 1)^{-1} \sum_{J'M'M} \sigma_{n'J'M',nJM}(\theta, \phi) \qquad (31)$$

from a particular (n, J) reactant state.

The scattering formalism in Eqs. (27)–(31) is exact. When one uses it in conjunction with the classical path approximation,[12,13,16-18,27] it is customary to make a number of semiclassical approximations. The orbital angular momentum quantum numbers l and m are replaced by the classical two-dimensional impact parameter $\mathbf{b} = (b, \phi_b)$. Here b has the usual meaning, and ϕ_b is the azimuthal angle that initially locates the incoming projectile in the space-fixed coordinate system. In particular the relationship

$$b = (l + \tfrac{1}{2})/k \qquad (32)$$

is used. In addition, it is assumed that there is a relatively small transfer of translational momentum, $|k' - k| \ll k$; and a relatively small transfer of collisional angular momentum, $|l' - l| \ll l$. The latter approximation guarantees that the molecular and orbital angular momenta are decoupled.

Finally, it is assumed that the scattering angle is relatively small. In this case, an asymptotic Bessel function approximation [97,98] can be used for the spherical harmonics,

$$Y_{lm}(\theta, \phi) \cong (-1)^m [(2l + 1)/4\pi]^{1/2} \cos^m(\theta/2) e^{im\phi} J_m[(2l + 1)\sin(\theta/2)]. \qquad (33)$$

This approximation requires $l \gg m$. In most applications, the small-angle assumption is used to drop the $\cos^m(\theta/2)$ factor. In its simplest form, the approximation can be written

$$\mathscr{Y}_{lm}(\theta, \phi) = (-1)^m (l/2\pi)^{1/2} e^{im\phi} J_m(l\theta). \tag{34}$$

It is remarkable that Eq. (34) preserves the orthonormality of the spherical harmonics in the semiclassical sense, since[97]

$$\int_{\theta=0}^{\infty} \int_{\phi=0}^{2\pi} \mathscr{Y}_{l'm'}^*(\theta, \phi)\mathscr{Y}_{lm}(\theta, \phi)\theta d\theta d\phi = \delta_{m'm}\delta(l' - l), \tag{35}$$

where $\delta(l' - l)$ is the delta function. (Note that the θ range has been extended from zero to infinity.) For ease of computation, the Bessel function is usually written in its integral representation.[96]

When all of these approximations are substituted into Eq. (27), the classical path expression for the scattering amplitude becomes

$$f_{n'J'M',nJM}(\theta, \phi) = \frac{-ik}{2\pi} \int_{b=0}^{\infty} \int_{\phi_b=0}^{2\pi} \exp[-2ikb\sin(\theta/2)\cos(\phi - \phi_b)]$$

$$\cdot(S_{n'J'M',nJM}(b, \phi_b) - \delta_{n'n}\delta_{J'J}\delta_{M'M})bdbd\phi_b. \tag{36}$$

The differential cross section can be obtained using Eq. (30) with $k = k'$. The S-matrix elements are obtained, as described in the previous section, by integrating the time-dependent Schrödinger equation along the classical path.

All of the CT calculations done with the classical path procedure have made the sudden approximation for molecular rotation.[99-102] In this high-energy approximation it is assumed that the molecule does not rotate during the collision. From a quantum-mechanical point of view the rotationally sudden approximation is equivalent to assuming that $\sigma_{n',nJ}(\theta, \phi)$, defined in Eq. (31), is independent of J. In that case, it is simplest to carry out any calculations of $\sigma_{n',n}(\theta, \phi)$ assuming that $J = 0$. The semiclassical equivalent of the rotationally sudden approximation is to assume that the diatomic molecule's orientation in space is fixed throughout the collision. The angles (θ_r, ϕ_r), defined in Fig. 1, describe this orientation. The differential cross section is computed for fixed values of (θ_r, ϕ_r), and then the results are averaged over all orientations. It should be emphasized that the sudden approximation for rotation does not assume that the complete state-to-state cross sections $\sigma_{n'J'M',nJM}(\theta, \phi)$ are independent of the various rotational quantum numbers. In fact, they are not. We shall discuss their calculation later.

The final scattering variables depend upon the two-dimensional impact parameter $\mathbf{b} = (b, \phi_b)$ and on the two orientation angles (θ_r, ϕ_r). Note that the z axis is defined by the initial relative velocity vector and that the same space-fixed axes are used to define (b, ϕ_b) and (θ_r, ϕ_r) (see Fig. 1). In this case we can write[101-103]

$$\theta = \theta(b, \theta_r, |\phi_b - \phi_r|), \tag{37a}$$

$$\phi = \phi(b, \phi_b, \theta_r, \phi_r), \tag{37b}$$

$$S_{n',n} = S_{n',n}(b, \theta_r, |\phi_b - \phi_r|). \tag{37c}$$

These functional relationships are implicit in the following equations. When all of the approximations and substitutions are made, the expression for the scattering amplitude for one orientation becomes[99,101,102]

$$f_{n',n}(\theta, \phi, \theta_r, \phi_r) = \frac{-ik}{2\pi} \int_{b=0}^{\infty} \int_{\phi_b=0}^{2\pi} \exp[-2ikb\sin(\theta/2)\cos(\phi - \phi_b)]$$

$$\cdot (S_{n',n} - \delta_{n'n})b\,db\,d\phi_b. \tag{38}$$

Note that the time-dependent Schrödinger equation is only solved for vibronic transitions $(n \to n')$ along the trajectory. The rotational state-to-state differential cross section is obtained by projecting $f_{n',n}$ onto the rotational wave functions[99,101,102]

$$\sigma_{n'J'M',nJM}(\theta, \phi) = |\langle J'M'|f_{n',n}(\theta, \phi, \theta_r, \phi_r)|JM\rangle|^2. \tag{39}$$

This involves a quadrature over θ_r and ϕ_r. Equation (31) can then be used to compute $\sigma_{n',nJ}(\theta, \phi)$. The calculation is simplified considerably by the closure properties of the rotational wavefunctions.[99] The result is

$$\sigma_{n'nJ}(\theta, \phi) = \frac{1}{4\pi} \int_{\theta_r=0}^{\pi} \int_{\phi_r=0}^{2\pi} |f_{n',n}(\theta, \phi, \theta_r, \phi_r)|^2 \sin\theta_r d\theta_r d\phi_r$$

$$= \sigma_{n',n}(\theta, \phi). \tag{40}$$

As expected, the result is independent of J. To date only one complete calculation of $\sigma_{n',n}(\theta, \phi)$ using Eqs. (38) and (40) for charge transfer has been carried out,[58a] but we expect it will be widely used in the future.

The total state-to-state cross section $\sigma_{n',n}$ can be obtained by integrating $\sigma_{n',n}(\theta, \phi)$ over all θ and ϕ or, alternatively, by using

$$\sigma_{n',n} = \frac{1}{2} \int_{\theta_r=0}^{\pi} \int_{\phi_r=0}^{2\pi} \int_{b=0}^{\infty} bP_{n',n}(b, \theta_r, \phi_r)\sin\theta_r d\theta_r d\phi_r db, \tag{41a}$$

$$P_{n',n}(b, \theta_r, \phi_r) = |S_{n',n} - \delta_{n'n}|^2. \tag{41b}$$

The function $P_{n',n}(b, \theta_r, \phi_r)$ is the probability that scattering from reactant state n with initial conditions (b, θ_r, ϕ_r) will give the product state n'. The total cross section $\sigma_{n',n}$ is much easier to compute than the differential cross section, because it is only necessary to integrate over the three variables in Eq. (41). By contrast, the evaluation of $\sigma_{n',n}(\theta, \phi)$ using Eqs. (38) and (40) requires a four-dimensional quadrature. The integral over ϕ_b in Eq. (38) is the most difficult, because there is a rapidly oscillating phase factor.[101] Numerical techniques to evaluate this integral exactly are discussed later in this section. This oscillating phase problem does not arise in Eq. (41).

Here we summarize various approximation techniques that can be used to evaluate Eq. (38). Cross[104] has used the stationary-phase approximation to simplify this integral. The most accurate approximation available to date is that of Grimbert et al.[105]. They approximated Eq. (38) for the case $\phi = 0$ [recall that $\sigma_{n',n}(\theta, \phi)$ is independent of ϕ] by making the change of variables $u = 2kb \sin(\theta/2)\cos\phi_b$. It is then clear that the major contribution to $f_{n',n}(\theta, \phi)$ comes from the regions $\cos\phi_b \cong \pm 1$. The integral is expanded in the limits l large, θ small and $l\theta \gg 1$ to give

$$\sigma_{n',n}(\theta, \phi, \theta_r, \phi_r) = \left(\frac{1}{4k}\right)^2 |J|^2, \tag{42a}$$

$$J = \sum_l (2l+1)\left\{\left[P_l(\cos\theta) - \frac{i}{l}P_l^1(\cos\theta) \right][S_{n',n}(b, \theta_r, |\pi - \phi_r|) - \delta_{n'n}] \right.$$

$$\left. + \left[P_l(\cos\theta) + \frac{i}{l}P_l^1(\cos\theta) \right][S_{n',n}(b, \theta_r, |\phi_r|) - \delta_{n'n}] \right\}. \tag{42b}$$

Here $P_l^1(\cos\theta)$ is an associated Legendre polynomial. Thus, $\sigma_{n',n}(\theta, \phi)$ is obtained by carrying out this sum and averaging over θ_r and ϕ_r [see Eq. (40)].

Even simpler results can be obtained if one assumes straight-line trajectories. In this case, it is possible to show using a limiting process on Eq. (27) with $\theta \to 0$ and $\Delta l \to 0$ that Eq. (27) becomes:

$$f_{n',n}(\theta, \phi, \theta_r, \phi_r) = \frac{1}{2ik} \sum_{l=0}^{\infty} (2l+1)P_l(\cos\theta)(S_{n',n} - \delta_{n'n}). \tag{43}$$

This is very similar to Eq. (28). Here $S_{n',n}$, defined in Eq. (37), is the S-matrix element evaluated along the straight-line trajectory. This result is easier to evaluate than Eq. (38), but it will only be valid at high collision energies where the scattering angles are very small. Spalburg et al.[106] used this procedure to compute differential cross sections for chemiionization processes. A similar expression has been given by Flannery and co-workers,[13,39] but where the Legendre polynomial has been approximated by a zeroth-order Bessel function. Flannery's result has the advantage that it accounts to first order for the momentum transferred to the atom perpendicular to the trajectory.

The simplest procedure to extract the differential cross section from a classical path method would be to use the deflection angle coming from the classical trajectory. It should be kept in mind that in the classical path procedure, a single trajectory generates a large number of product states. The most consistent way to proceed is to assume that all of these states appear at the same angle. It is not clear that this is physically reasonable, however. For example, suppose the potential energy surface that determines the trajectory has a rainbow at some angle θ_R. Then the state-to-state cross sections will all show rainbows there. Studies of rotational rainbows, however, show that different product states have different rainbow angles. In spite of these concerns, it is likely that this procedure in conjunction with Monte Carlo sampling of reactant variables and the semiclassical energy-conserving trajectory technique[42] will be used extensively in the future. A big advantage of this procedure is that there is no need to make the sudden approximation for rotational motion, so it can be used at lower energies. In addition, it is computationally simpler than the procedure in Eqs. (38) and (40).

Cole and DePristo[47] have suggested a different method to avoid the problem of all product states from a particular trajectory appearing at the same angle. They assume that the scattering angle associated with each product channel is determined by the kinetic energy of the product in that channel; in general, exothermic channels are scattered to smaller c.m. angles than endothermic channels. The implementation of this procedure requires solving transcendental equations for each energy and set of initial conditions. In their paper, Cole and DePristo restricted the calculation to a single orientation in the space-fixed coordinate system for the two molecules. Consequently, the scattering was restricted to a plane, and the only important initial variable was the impact parameter b. In this case, the well-known formula

$$\sin \theta \sigma_{n',n}(\theta) = \sum_i b_i P_{n',n}(b_i) \left| \frac{d\theta}{db} \right|_i^{-1} \tag{44}$$

could be used. Here $P_{n',n}(b_i)$ is the probability that a collision at a particular value of b_i will give the product state n', and the sum is over all impact parameters that give scattering at angle θ (with the understanding that for a given value of b, θ also depends on n and n'). It is not clear how to implement this method for more general systems, such as atom–diatom collisions, where there are four independent variables to contend with. Thus, it is unlikely that this procedure will be used for the more general problem in the near future.

G. Numerical Procedures

As discussed in Section II A, the time-dependent Schrödinger equation must be solved for a particular trajectory, characterized by a set of initial conditions. This calculation is repeated for a large number of trajectories with different initial conditions and the results are finally averaged in order to obtain the cross sections, as explained in section II G 2 below. The time-dependent problem is actually transformed into a set of coupled first-order differential vibronic equations (see Section II C), which is solved step-by-step on a computer. Since the trajectory provides the time dependency for the vibronic equations, and is propagated simultaneously with them, in the following the computation of the total system of equations (vibronic plus trajectory) will often be simply referred to as the "trajectory calculation." The main ingredients of the problem are the matrix elements of the couplings, interaction potentials and their derivatives, which must be available at each step of the integration. Fortunately, by making some approximations, the quadratures involving vibrational wavefunctions can be done before running the trajectories. This point is developed in Section II G 1.

Although the semiclassical calculations discussed in this chapter can be done faster than fully quantum calculations, the effort in terms of computer time must not be underestimated. For example, for the $Ar^+ + CO$ system,[55] at a relative collision energy of 2 eV, including 75 vibronic states, the calculation of total state-to-state cross sections takes several hours on an IBM 3090, depending on the required precision. Although the computer time can vary widely from one system to another, one sees that it is worth searching for a fast (and accurate) integrator. The choice of a numerical integrator is discussed in Section II G 3.

1. Vibrational Matrix Elements

Central to the set of coupled equations (18) are the matrix elements $H_{\alpha\alpha'}^{vv'}$ and $V_{\alpha}^{vv'}$ of the couplings and interaction potentials, given in Eq. (15). The vibrational wavefunctions of BC used in the evaluation of these vibrational matrix elements (VME) are computed first. This is usually done[107] by using a Numerov algorithm. For each vibrational level, the Numerov method is used to propagate a vibrational wavefunction for a trial eigenvalue, and the calculation is repeated iteratively to converge on the closest true eigenvalue. The vibrational eigenvalues are also utilized in the calculations of the phases [Eq. (17)]. Then, the quadratures in Eq. (15) can be evaluated by means of an extended Simpson rule.[108] Other methods have been used to obtain vibrational wavefunctions.[107] One interesting technique is the so-called distributed gaussian basis (DGB) method.[109] It consists in expanding the wavefunctions over a basis of gaussian functions. For better results, the gaussians are distributed unevenly along r, following a semiclassical criterion that ensures that the density of gaussians is everywhere sufficient, even if the vibrational wavefunctions exhibit rapid oscillations.[109] The eigenvalues are obtained by diagonalization of the vibrational hamiltonian. A great advantage of this method is that no iteration is necessary. The integration over r is also made easier because the functions to be integrated contain products of gaussians. In this case, the quadratures can be carried out with an excellent precision by a very low-order Gauss–Hermite procedure.[108,109]

Now consider the calculation along a single trajectory. As apparent from Eq. (18) the VME $H_{\alpha\alpha'}^{vv'}(R(t), \gamma(t))$ and $V_{\alpha}^{vv'}(R(t), \gamma(t))$ must be available at any point $(R(t), \gamma(t))$. After the vibrational wavefunctions have been computed and stored, it would be, in principle, possible to calculate the couplings and potential interaction functions (CPIF), $H_{\alpha\alpha'}(r, R, \gamma)$ and $V_{\alpha}^{int}(r, R, \gamma)$, and perform the quadratures to obtain the VME *at each step* along the trajectory. Obviously, the quadratures would considerably slow down the integration of the coupled equations. This is why, generally, interpolations and/or simplifications of the CPIF have to be implemented.

A major problem in using couplings and potential interactions is to build accurate representations of these interactions which can be used relatively easily in dynamics calculations.[110] In some situations, the number of points available from electronic structure calculations is large enough to allow for interpolation of the CPIF at any point (r, R, γ). This was done recently by Sizun et al.[58b] for the charge-transfer collision $H^+ + O_2(X\ ^3\Sigma_g^-) \rightarrow H + O_2^+(X\ ^2\Pi_g)$. For each point i of a grid (R_i, γ_i), the quadratures over r were carried out for 15 vibrational states of O_2 and O_2^+, using an interpolation between 15 points over the relevant domain of r. The VME were evaluated on a grid containing 30 points along R and 10 points along, γ, taking into account the symmetry of the oxygen molecule. The VME were stored on a computer file. Then it was possible to obtain the VME at any time along a trajectory from an interpolation over the (R_i, γ_i) grid.

The calculation of *ab initio* data at every point needed in the dynamics calculation is rarely feasible. To stress this point, it should be noted here that not only the values of the CPIF are needed but also their derivatives, which govern the classical motion. Generally, the couplings and potential interactions are supplied as analytic formulas, resulting from a fit to a restricted number of *ab initio* points and/or from semiempirical calculations.[110] The γ dependency is often expressed as an expansion over Legendre polynomials.[59] Thus, the CPIF can be easily computed from analytic formulas at any point (r, R, γ) but, if no other simplification is made, the VME still have to be interpolated over a grid (R_i, γ_i). In order to avoid this interpolation, the r dependency of the CPIF can be approximated by an expansion about some mean value of r, denoted $\bar{r}$. The vibrational matrix elements of $(r - \bar{r})$, $(r - \bar{r})^2, \ldots$, are evaluated prior to the calculation, and the VME can be obtained along the trajectory by simple multiplications. In many previous calculations the r dependence was not known, and therefore it was simply ignored[52,55] (see end of Section II C).

2. Average Over Initial Conditions

For a specified collision energy, the determination of the vibrational state-to-state cross section consists in evaluating an integral over $b, \phi_b, \theta_r, \phi_r$ [Eqs. (38) and (40)], if the sudden approximation for the rotation of BC is assumed. (If the diatomic target is rotating classically, two more initial parameters are needed: the rotational angular momentum J and its orientation angle[70]. The calculation of this four-dimensional integral is greatly simplified by the fact that the S-matrix element $S_{n',n}$ depends on the difference $\delta\phi = |\phi_r - \phi_b|$ and not on ϕ_r and ϕ_b individually. This fact is exploited by Sizun et al.[58b] in their CT calculation on the $H^+ + O_2$ system. First, in order to evaluate the integral (38), for a fixed pair of variables (θ_r, ϕ_r), they carry out the dynamics calculations on a regular grid of points $(b, \delta\phi)$. For b, 200 points are used between 0 and 7 a.u. and $\delta\phi$ is scanned between 0 and $180°$ for symmetry reasons, with a stepsize of $15°$. The values of $S_{n',n}$ are then interpolated on the $(b, \delta\phi)$ grid and the integral (38) is evaluated by a modified Simpson rule, using step sizes of 0.02 a.u. and $2.5°$ for b and $\delta\phi$, respectively. The small step size for $\delta\phi$ is made necessary by the rapidly oscillating phase factor contained in Eq. (38). Krüger and Schinke[101] use an expansion of the integrand into a Fourier series to carry out this quadrature. Finally, taking into account the symmetry of O_2, the integral over θ_r and ϕ_r [Eq. (40)] is also performed by means of an extended Simpson rule, using the following ranges and stepsizes: $0° \leqslant \theta_r \leqslant 90°$, $0° \leqslant \phi_r \leqslant 180°$, $\Delta\theta_r = 30°$, $\Delta\phi_r = 2.5°$.

It must be emphasized here that $S_{n',n}$ depends on three parameters, not four. Nevertheless, the differential cross section is effectively a four-dimensional integral because ϕ_b appears in the exponential term of Eq. (38). Moreover, this term is rapidly oscillating and, consequently, more points are needed along ϕ_b than along the other three variables.

As for the state-to-state total cross sections [Eq. (41)], the calculation is much easier because one has to evaluate an integral that is only three-dimensional, and that does not contain the oscillating term of Eq. (38). This three-dimensional integral could also be evaluated by means of a Monte Carlo algorithm,[70] better adapted to the calculation of high-dimensional integrals. As opposed to trapezoidal-type methods, it uses random initial conditions for the trajectories. This gives a reasonably good estimation of the integral with a limited number of trajectories. The result can be further improved by calculating more trajectories. This technique is widely used in standard classical trajectory calculations,[70] and it can be implemented here exactly in the same way (for an example, see Gislason et al.[55]). In the calculation of the differential cross section, however, the Monte Carlo method has not been used and would have to be tested.

3. Numerical Solution of Coupled Equations

The set of vibronic coupled differential equations (18) is solved simultaneously with the hamilton equations for the classical trajectory. These latter equations will not be given here, since they can be found in standard papers on classical trajectory calculations.[70] (For the particular case of the SCECT method, see DePristo and Sears.[41]) The phase $\Phi_{\alpha v}(t)$ [Eq. (17b)] is differentiated to give $\partial\Phi_{\alpha v}(t)/\partial t = F_\alpha^v(t)/\hbar$, and this is added to the set of vibronic and trajectory differential equations. In order to organize conveniently the program, the total set of (complex) equations is rewritten in real form and the quantum numbers (α, v) are remapped onto a single index.[42]

In principle, these differential equations present no serious stability problem. The standard integrators[108] used to calculate classical trajectories have been discussed in detail by Truhlar and Muckerman[70] and the same discussion should apply here. Since the speed of integration is an important parameter, an adaptive step-size integrator is highly recommended. In this kind of integrator, the stepping algorithm returns the truncation error and tries to keep it within some desired bounds by varying the integration step. In practice, this means that we demand that the coordinates, momenta, phases, and probability amplitudes stay within a specified tolerance. An excellent discussion of this adaptive mechanism can be found in Numerical Recipes.[108] In our experience, it is essential that this algorithm is implemented properly to avoid accuracy problems. A reason why accuracy problems could occur is that the integrator is used under highly variable conditions. For example, the integrator must work correctly for collisions energies varying from 1 eV to several keV. As an additional precaution, it is useful to check the conservation of total energy, total angular momentum, and total probability at the end of the integration.

As discussed by Truhlar and Muckerman,[70] the choice of the optimum integrator depends on each specific problem, and in particular on the complexity of the couplings and potential interactions. The most popular integrator is certainly the Runge–Kutta integrator.[108] Although it is slower than others, it is simple and robust and should work in every case. More sophisticated integrators, like Adams–Moulton and Bulirsch–Stoer, are supplied by most mainframe-computer libraries. Recently, we implemented (unpublished work) the Bulirsch-Stoer integrator described by Press et al.[108]. It was found to be much faster than the Runge–Kutta. It was also observed that the Bulirsch–Stoer takes most of its advantage in regions of slowly varying potentials. It was concluded that a mixed algorithm using Runge–Kutta at short distances and Bulirsch–Stoer at large distances would be an efficient integrator.

Finally, it is very important to test the convergence of the calculation. The results should stay unchanged after redoing the calculation with more vibronic states.

4. Examples

As discussed earlier the phases of the coefficients $c_{\alpha v}(t)$ vary rapidly with time and, as in any quantum-mechanical calculation, the phase behavior can give rise to a number of interesting interference effects. These are best seen in intermediate calculations before the averaging over molecular orientation is carried out. One example of this is the state-to-state probability $P_{n',n}(b, \theta_r, \phi_r)$, defined in Eq. (41b), plotted against b for a fixed orientation (θ_r, ϕ_r). Plots of this type are discussed in detail by Spalburg et al.[111] for a chemiionization process, and Lee and DePristo[44] have used them to illustrate their work on H_2^+/H_2.

In Fig. 2 we show such plots computed by Sizun et al.[58b] for collisions of $H^+ + O_2$. It should be emphasized that their work includes the full r dependence for the electronic couplings $H_{\alpha \alpha'}(r, R, \gamma)$ and for the diabatic potentials $V_\alpha^{int}(r, R, \gamma)$ defined in Eqs. (12) and (13), respectively. Consequently, vibrational excitation of O_2 (or O_2^+) on a single (diabatic) electronic PES can occur. The calculations in Fig. 2 were carried out for the orientation $(\theta_r = 0, \phi_r = 0)$. Thus, the three atoms are collinear when $b = 0$, but not for other impact parameters. The first two panels show the probabilities for the transitions $H^+ + O_2(v = 0) \rightarrow H^+ + O_2(v = 1)$ and $H^+ + O_2(v = 0) \rightarrow H + O_2^+(v = 2)$ obtained from a full solution of the coupled equations. We denote these two processes $(\alpha, 0) \rightarrow (\alpha, 1)$ and $(\alpha, 0) \rightarrow (\alpha', 2)$. The probabilities die off for $b > 6$ a.u., because the couplings are too weak there. Two distinct types of oscillations are observed, fast and slow. To investigate where these oscillations come from, two additional calculations were carried out for the same molecular orientation. Both were two (vibronic) state calculations, using only the four appropriate vibronic matrix elements from the full hamiltonian matrix. The first involved the direct vibrational excitation process $(\alpha, 0) \rightarrow (\alpha, 1)$. Here the two vibronic PES are parallel, and vibrational excitation is a Demkov-type process induced by the r dependence of the potential $V_\alpha^{int}(r, R, \gamma)$. The resulting probability distribution is shown in Fig. 2(c), and should be compared to Fig. 2(a). An inset shows a schematic of the two potential surfaces. One might expect that the slow oscillations observed in $P_{n',n}$ come from the phase difference shown in the second term in Eq. (18), which is approximately equal to $\exp[-i(\varepsilon_{\alpha 0} - \varepsilon_{\alpha 1})t/\hbar]$. However,

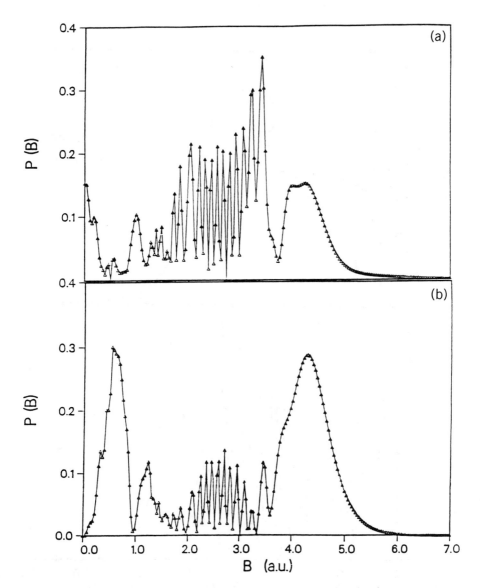

Figure 2. State-to-state probability $P_{n'n}(B, \theta_r = 0, \phi_r = 0)$ plotted against impact parameter B for collisions of $H^+ + O_2(v = 0)$ at $E_{rel} = 23$ eV. (a) Results of full solution of coupled equations to give $H^+ + O_2(v = 1)$. (b) Results of full solution to give $H + O_2^+(v = 2)$. (c) Result of two-state calculation to give $H^+ + O_2(v = 1)$. The inset shows a schematic of the two relevant diabatic potential energy curves. (d) Result of two-state calculation to give $H + O_2^+(v = 2)$. The insert shows the two relevant diabatic potential energy curves. See text for further details. (Taken from Ref. 58b, with permission.)

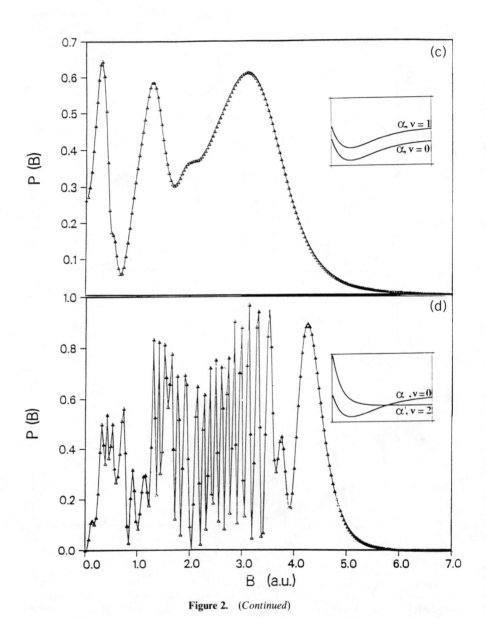

Figure 2. (*Continued*)

Sizun et al.[58b] have shown that this is not the case at this collision energy. In fact, if the interference were due to an exponential term of this type, the oscillations would go to zero at the minima. This is not observed in Fig. 2c. Rather, Sizun et al.[58b] have shown that vibrational excitation for vibronic curves such as those shown in Fig. 2c is related to the linearly forced harmonic oscillator system. In addition, the slow oscillations observed in Fig. 2c are related to the semiclassical phase shift for scattering by the vibrational coupling potential $V_\alpha^{vv'}$. They also showed for this system that the couplings $V_\alpha^{vv'}$ can be estimated quite accurately from V_α^{int} using the harmonic oscillator model.

The second two-state calculation is for the CT process $(\alpha, 0) \rightarrow (\alpha', 2)$. The results are shown in Fig. 2(d), along with a schematic of the two diabatic surfaces, which cross in this case. This corresponds to a Landau–Zener transition. The $P_{n',n}$ curve should be compared to the one in Fig. 2(b). For this process the energy difference, which appears in the phase factor in the first term of Eq. (18), can be very large, so the phase changes rapidly, and this generates the fast (Stueckelberg) oscillations in Fig. 2(d). The same type of oscillations are seen in the full calculation in Fig. 2(b), but other slow oscillations damp them out somewhat. It is apparent that the full calculations involve many Demkov transitions, induced by parallel diabatic PES, and, simultaneously, many Landau–Zener transitions, induced by diabatic PES which cross. The final result is sensitive to the relative importance of each crossing, and it is not apparent before doing the calculation what the final result will be. It should also be emphasized that many of the oscillations observed in Fig. 2 will be lost when the average over initial molecular orientation is carried out.

A second example of quantum oscillations can be seen if one plots a specific probability for populating a state as a function of the time, along the trajectory. Such a figure is useful to see the regions of importance for the mixing of the probability amplitude and to see if there are one or more localized transitions. Examples of such plots have been shown for CT processes at high energy by Kimura and co-workers[49,50] and by Klomp et al.[112] for a chemiionization process.

III.   CLASSICAL PATH CALCULATIONS—TOTAL CROSS SECTIONS

A.   Paper of Bates and Reid

As discussed earlier, the first application of the classical path method to molecular CT reactions was the paper of Bates and Reid[26]. They calculated cross sections for the processes

$$H_2^+(v_0') + H_2(v_0'' = 0) \rightarrow H_2^+(v') + H_2(v'') \quad \text{(VE)}, \qquad (45a)$$

$$\rightarrow H_2(v'') + H_2^+(v') \quad \text{(CT)}. \qquad (45b)$$

The first process involves inelastic vibrational excitation, whereas the second is a true CT process. The two reactions can be distinguished experimentally, because (45a) produces fast H_2^+ ions, whereas (45b) generates ions with thermal velocities. For this system it should be kept in mind that at collision energies below $E_{rel} = 2$ eV chemical reaction can occur to give $H_3^+ + H$. Bates and Reid used simple, isotropic potential interactions, and they made some approximations when calculating the cross sections. Nevertheless, they demonstrated that the cross sections for both processes are quite large. Their results reproduced the experimental data fairly well in the energy range of 5 to 500 eV (Lab).

B.   Work of Moran, Flannery, and Co-workers

The group of Moran, Flannery, and co-workers has carried out a wide range of theoretical and experimental studies of symmetric molecular CT processes such as reactions (2) and (45).[28-38] In addition, they studied several atom–molecule CT reactions involving Ar^+ ions.[39] This work has been described in detail in a comprehensive review by Moran,[27] and Flannery[13] has given an extensive discussion of the theoretical procedures used in their calculations. Because of this, we shall only highlight their work on reaction (45). The reader is referred to the original references for more details.

The calculated CT cross sections for reaction (45b), with $H_2^+(v_0')$ initially in vibrational level $v_0' = 0$ or $v_0' = 1$, are shown as a function of the H_2^+ ion energy in Fig. 3.[27,32] The scattering at low energy is dominated by the presence of the symmetric CT product state, which has exactly the same energy as the reactant state. As an example, the initial state $(v_0' = 1, v_0'' = 0)$ is resonant with the final state $(v' = 1, v'' = 0)$. The symmetric state is the major CT product channel at low collision energies. The H_2^+–H_2 system also has a number of accidental near resonances, owing to the fact that the vibrational frequency of H_2^+ is approximately one half that of H_2 (see Table I). Thus, the reactant state $(v_0' = 2, v_0'' = 0)$, which is resonant with $(v' = 2, v'' = 0)$, is also only 0.01 eV off-resonance with the product state $(v' = 0, v'' = 1)$. This also means that there is an *inelastic* product channel [Reaction (45a)], namely, $(v' = 0, v'' = 1)$, which is also only 0.01 eV away from the reaction state. This has important implications for the inelastic scattering. In addition to the resonances and near resonances, the low-energy CT cross section depends on the product of the Franck–Condon factors for the transition indicated

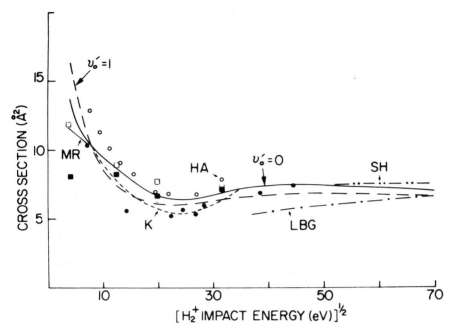

Figure 3. Absolute cross sections for charge transfer in $H_2^+(v_0') + H_2(v_0'' = 0)$ collisions as a function of the square root of the H_2^+ beam energy. The experimental data are dashed-dot-dot curve, Ref. 113; dashed curve, Ref. 114; solid curve at low energies, Ref. 115; open circles, Ref. 116; dashed-dot curve, Ref. 117; solid circles, Ref. 118. In addition, the closed and open squares are data for $v_0' = 0$ and 1, respectively, from Ref. 119. The solid and dashed curves spanning a wide energy range are the computations from Ref. 32. (Taken from Ref. 27, with permission.)

in Eq. (45b):

$$P_{fi} = |\langle v_0'|v''\rangle|^2|\langle v_0''|v'\rangle|^2. \tag{46}$$

Unfortunately, there are no simple rules to predict the CT cross sections without doing the calculations. Figure 3 also shows a number of experimental measurements[113-119] of the CT cross sections. The agreement between theory and experiments is quite good.

At high collision energies the presence of the symmetric CT product state becomes much less important. In the limit where perturbation theory should apply[41], the state-to-state cross sections become proportional to the Franck–Condon product P_{fi}. In addition, the total CT cross section for a particular $H_2^+(v_0')$ reactant state becomes independent of v_0'.[41,53] The transition from low-energy to high-energy behavior is shown in Fig. 4,[27,32]

TABLE I

Energy Defects and Franck–Condon Factors for the Charge-Transfer Reaction
$H_2^+ (v_0' = 0, 1) + H_2(v_0'' = 0) \rightarrow H_2(v'') + H_2^+ (v')^a$

Product levels (v'', v')	$v_0' = 0$		$v_0' = 1$	
	Energy Defect (eV)	Franck–Condon Factor	Energy Defect (eV)	Franck–Condon Factor
0,0	0.000	0.0082	−0.272	0.0145
0,1	0.272	0.0145	0.000	0.0256
1,0	0.516	0.0252	0.244	0.0155
0,2	0.528·	0.0158	0.256	0.0278
0,3	0.768	0.0139	0.497	0.0244
1,1	0.788	0.0443	0.516	0.0272
0,4	0.994	0.0108	0.722	0.0191
2,0	1.003	0.0306	0.731	0.0003
1,2	1.044	0.0482	0.772	0.0296
0,5	1.206	0.0080	0.934	0.0140
2,1	1.275	0.0538	1.003	0.0005
1,3	1.284	0.0422	1.013	0.0260

[a] Based on Table I of Ref. 27. Columns three and five give the product of the two relevant Franck–Condon factors for the transition; this product is defined in Eq. (46).

which presents the state-to-state CT cross sections for $H_2^+ (v_0' = 0) + H_2(v_0'' = 0)$ as a function of the ion energy. The relevant energy defects and Franck–Condon overlaps are summarized in Table I. It is particularly remarkable that the symmetric product state ($v' = 0$, $v'' = 0$) becomes a minor product channel above 1000 eV, due to its small value of P_{fi}.

Flannery et al.[33] have also calculated cross sections for vibrational excitation (VE) of H_2^+ ions by H_2 [Reaction (45a)]. Their results for $H_2^+ (v_0' = 0) + H_2(v_0'' = 0)$ are shown in Fig. 5. The total cross section for VE rises to $6 \, \text{Å}^2$ at an ion energy of 400 eV; this is nearly as large as the cross section for CT at that energy. Flannery et al. emphasize that the large VE cross sections are due to the presence of charge exchange in this system. Thus, if the electron transfers an odd number of times between the two molecules, CT results; an even number of hops leads to VE. The energy dependence of the total VE cross section in Fig. 5 is different from that of the CT process, shown in Fig. 3, at both low and high energies. As we have seen, CT at low energies is dominated by the presence of the symmetric product state. The equivalent state for VE is the reactant state itself, and this state contributes only to the *elastic* scattering. Therefore, the VE cross section for $H_2^+ (v_0' = 0)$ is quite low at ion energies below 20 eV. As would be expected, reactant states such as $H_2^+ (v_0' = 2) + H_2(v_0'' = 0)$, which have near-resonant VE product states, have much larger VE cross sections at low

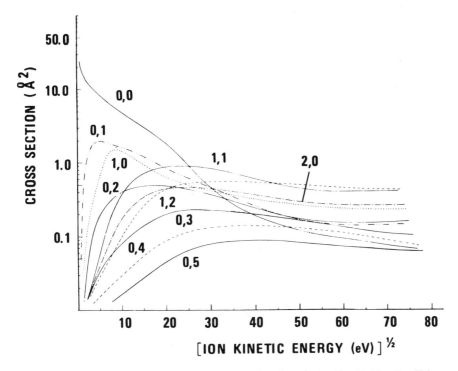

Figure 4. State-to-state charge-transfer cross sections for $H_2^+ (v_0' = 0) + H_2(v_0'' = 0)$ collisions as a function of the square root of the H_2^+ beam energy. The vibrational quantum numbers of the respective neutral and ion products are shown (see also Table I). (Taken from Ref. 27, with permission.)

energy. At high energies the short collision time reduces the effective charge exchange coupling between the two molecules. It becomes difficult for the active electron to make even one hop to the other molecule. Consequently, the CT cross sections begin to decline at ion energies above 2500 eV; this decline can be seen in Fig. 3. By comparison, VE requires at least two electron hops, and this is very improbable at 2500 eV. This explains the rapid decline of the VE cross section observed in Fig. 5.

C.   DePristo's Work

The initial work by DePristo on molecular CT processes involved the development of scaling relationships to predict state-to-state cross sections from a limited set of cross section data.[40,41] He then adapted the "semiclassical energy conserving trajectory technique" (SCECT) to CT processes.[42] This is a classical path method that conserves energy, on the

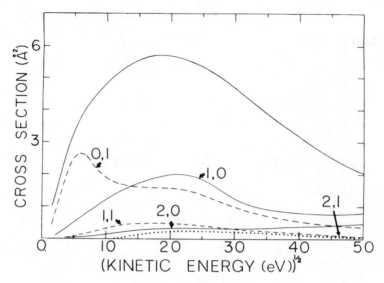

Figure 5. State-to-state cross sections for vibrational excitation in collisions of $H_2^+ (v_0' = 0) + H_2(v_0'' = 0)$ as a function of the square root of the H_2^+ beam energy. The vibrational quantum numbers of the respective neutral and ion products are shown. The solid curve at the top of the figure is the total cross section for vibrational excitation (summed over all product states, but not including the elastic channel). (Taken from Ref. 33, with permission.)

average, along the trajectory. A similar procedure was developed by Moran et al.[38] DePristo pointed out that in the paper by Bates and Reid[26] and in the early work by Moran et al.[31] a number of simplifying approximations had been made, and it was not clear what effect they had on the accuracy of the calculations. Now that absolute state-selected CT cross sections were available, these approximations would be severely tested. Therefore, it was important to carefully check them.

The approximations made by Bates and Reed[26] were (1) straight line trajectories were used. (2) No effort was made to properly treat energy flow between classical (translation) and quantal (vibronic) degrees of freedom. (3) The electronic coupling between the two molecules was assumed to be independent of the molecular orientations. (4) This coupling was also assumed to be independent of the H_2 and H_2^+ bond lengths. This is the Franck–Condon approximation for coupling elements. In later work Moran and Flannery added the approximation that (5) the interaction potentials were made up of simple Morse and anti-Morse potentials.[28] And, finally, there was the general question of over what energy range the classical path technique should give accurate results.

DePristo[42] began by comparing calculations done with the SCECT against the exact quantal calculations of Becker.[21] The work was done for the processes

$$O_2^+(v_1 = 1) + O_2(v_2 = 0) \rightarrow O_2^+(v_1') + O_2(v_2') \quad \text{inelastic,} \quad (47a)$$

$$\rightarrow O_2(v_2') + O_2^+(v_1') \quad \text{CT,} \quad (47b)$$

using the potentials constructed by Becker. A detailed comparison of the state-to-state cross sections at $E_{rel} = 8\,eV$ is given in Table II.[42] The agreement between the two methods is excellent. Similar agreement was obtained at $E_{rel} = 36\,eV$, which was expected, because the SCECT is expected to work best at high energies. DePristo also made a comparison at $E_{rel} = 1\,eV$, where he found that the agreement was poor for off-resonance transitions. However, these transitions do not contribute significantly to the total CT cross section, so we expect that the state-selected CT cross sections would be fairly accurate. We conclude that the SCECT works very well at collision energies of 2 eV and higher.

The state-to-state cross sections shown in Table II are quite interesting. As expected, the symmetric CT product channel dominates the cross section

TABLE II
State-to-State Cross Sections at $E_{rel} = 8\,eV^a$

(v_1', v_2')	ΔE (eV)	$O_2^+(v_1 = 1) + O_2(v_2 = 0) \rightarrow O_2^+(v_1') + O_2(v_2')$ $\rightarrow O_2(v_2') + O_2^+(v_1')$ Direct		Direct Exchange Exchange	
		S^b	Q^c	S^b	Q^c
0,0	−0.232	0.11	0.11	0.14	0.14
0,1	−0.039	2.33	2.23	1.81	1.80
1,0	0.000	—^{d}		26.34	26.40
0,2	0.151	0.89	0.90	1.33	1.16
1,1	0.193	1.08	1.05	1.27	1.23
2,0	0.228	0.03	0.03	0.05	0.05
0,3	0.338	0.07	0.07	0.05	0.04
1,2	0.383	0.03	0.02	0.10	0.10
2,1	0.421	0.02	0.02	0.02	0.02
3,0	0.452	0.01	0.01	0.01	0.01

^{a}This table is based on Table VI of Ref. 42. ΔE is the energy defect between the reactant and product states. The cross sections are in $\mathring{A}^2$.
^{b}Semiclassical results from Ref. 42.
^{c}Quantal results from Ref. 21.
^{d}The elastic cross section is not computed.

at this collision energy. The cross sections to other channels are approximately the same for both inelastic scattering and CT, but overall the CT cross section (31.1 Å^2) is seven times larger than that for inelastic scattering (4.6 Å^2). State-to-state cross sections for both product channels correlate well with the energy defects, but the pattern is by no means perfect.

In the same paper DePristo[42] compared his SCECT calculations to the work of Moran et al.[38] on this system. During the course of this work he discovered that Moran et al. had neglected the diagonal elements of the diabatic potential between O_2^+ and O_2. This potential is important in determining the energy-conserved trajectory. In particular, its neglect leads to differential cross sections that are too strongly peaked at small angles. However, DePristo's work shows that neglecting these potential terms has only a moderate effect on the state-to-state and state-selected CT cross sections at $E_{rel} = 8$ eV. DePristo also determined that there were some numerical inaccuracies in the calculations by Moran et al.[38] Nevertheless, the overall conclusion from this work is that the SCECT works remarkably well down to collsion energies of a few eV.

In the next paper on O_2^+/O_2 DePristo[43] compared his calculated cross sections for reaction (47b) with the experimental results of Baer et al.[120] for O_2^+ vibrational levels $v = 0$–8. The comparison of the state-selected cross sections for $E_{rel} = 15$ eV is shown in Fig. 6. The cross sections computed from the original potential of Flannery et al.[28] are too large and do not have the proper dependence on the vibrational level v_1 of O_2^+. A second set of calculations was carried out with the electronic coupling between the two molecules reduced by a factor of 2.2. Figure 6 shows that this improves the overall agreement, but the dependence on v_1 is still incorrect. DePristo[43] concluded that the potential energy surfaces for this system are not correct and, in particular, the orientation dependence of the interactions cannot be neglected. Thus, approximations (3) and (5) made in the early work on this system (see preceding discussion) must be rethought.

DePristo has also carried out several studies of reaction (45). Lee and DePristo[48] developed a simple model for the H_2^+/H_2 interaction potentials, which agrees well with more accurate calculations.[121] As part of this work they showed that the Franck–Condon approximation for coupling elements, assumption (4) of Bates and Reid,[26] works very well for this system. The H_2^+/H_2 system presents a severe test of this approximation, because the bond lengths of H_2^+ and H_2 are quite different. It should work even better for other molecular systems. These potential surfaces were then used to carry out a series of calculations on the systems H_2^+/H_2, H_2^+/D_2, D_2^+/H_2, and D_2^+/D_2.[44–46] State-to-state CT cross sections were computed at several collision energies for five vibrational levels of the reactant molecular ion. The calculations were done for three space-fixed orientations of the two

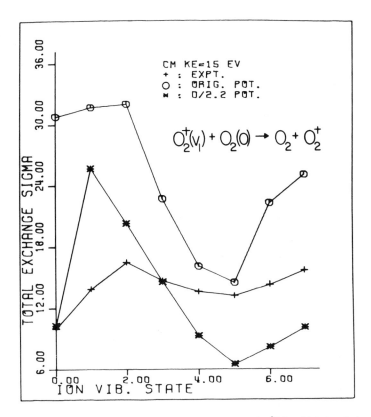

Figure 6. Total state-selected charge transfer cross sections in Å^2 for $O_2^+(v_1) + O_2(v_2 = 0) \rightarrow$ $O_2 + O_2^+$ plotted against v_1 at $E_{rel} = 15\,\text{eV}$. The calculated cross sections for the original potential and for the potential divided by 2.2 are shown as open circles and stars, respectively. The experimental cross sections measured in Ref. 120 are denoted by the "plus" symbols. (Taken from Ref. 43, with permission.)

molecules. They determined that the effect of different orientations was to broaden the product vibrational distributions. They also observed that the relative contribution of the symmetric CT product state varied with the initial vibrational level of the H_2^+ or D_2^+ ion. The relative state-selected CT cross sections as a function of the reactant state for the four systems are shown in Fig. 7.[45] It is seen that the dependence on vibrational level v_1 is fairly weak at all three collision energies. Nevertheless, this dependence on v_1 persists even at $E_{rel} = 400\,\text{eV}$. At sufficiently high energies the cross sections should become independent of v_1,[41,53] when Franck–Condon behavior is obtained. The calculations at $E_{rel} = 16\,\text{eV}$ are compared with the experiments of Cole et al.[122] for H_2^+/H_2 in Fig. 8. The agreement is very good.

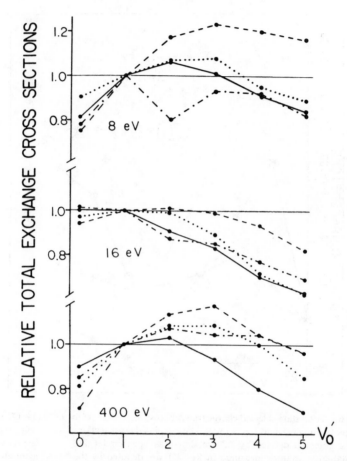

Figure 7. Relative total charge transfer cross sections versus the initial ion vibrational state for three relative collision energies. The systems are solid line, H_2^+/H_2 from Ref. 44; dotted line, H_2^+/D_2; dashed-dot line, D_2^+/H_2; dashed line, D_2^+/D_2. (Taken from Ref. 45, with permission.)

Figure 9 shows a comparison of *absolute* theoretical and experimental cross sections for the H_2^+/H_2 system.[46] The experimental results for $H_2^+(v_0' = 0, 1)$ are shown as well as for H_2^+ ions with a Franck–Condon distribution. In addition, SCECT calculations for $v_0' = 0$ are given. The agreement is not as good as for the relative cross sections shown in Fig. 8, but the theoretical and experimental cross sections do lie within the mutual uncertainties. In addition, the energy dependence of the two curves is very similar. Also shown are the results of a calculation using straight-line trajectories. It is seen that these results are similar to those for the SCECT

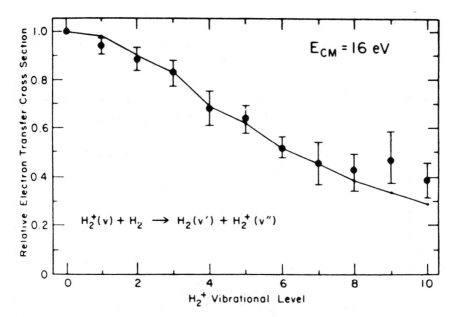

Figure 8. Relative charge transfer cross section for $H_2^+(v) + H_2$ for $v = 0$–10 at a relative translational energy of $16 \pm 1.4\,eV$. The points are the experimental data of Cole et al.[122] The solid line segments are the calculated cross sections of Lee and De Pristo.[44] (Taken from Ref. 122, with permission.)

calculations at energies above $E_{rel} = 10\,eV$. We also note that Cole and DePristo computed angular distributions for CT in the H_2^+/H_2 system and confirmed that the scattering distributions extend to fairly large c.m. angles due to the strong interactions in this system.[47] This work is discussed further in Section IV.

State-to-state cross sections for a CT process like reaction (45) are expected to depend primarily on the energy defect ΔE for the transition and on the product of Franck–Condon factors P_{fi} defined in Eq. (46). Figure 10 shows the various cross sections computed for the process

$$D_2^+(v_0' = 2) + H_2(v_0'' = 0) \rightarrow D_2(v'') + H_2^+(v')$$

plotted against the energy defect at two collision energies.[45] It is seen that the data fall approximately on a straight line for both energies. Also shown are the cross sections divided by P_{fi}. At sufficiently high energy the CT cross sections σ_{fi} should be proportional to the Franck–Condon factors. In that case, the ratios σ_{fi}/P_{fi} are expected to fall on a single curve. As noted above,

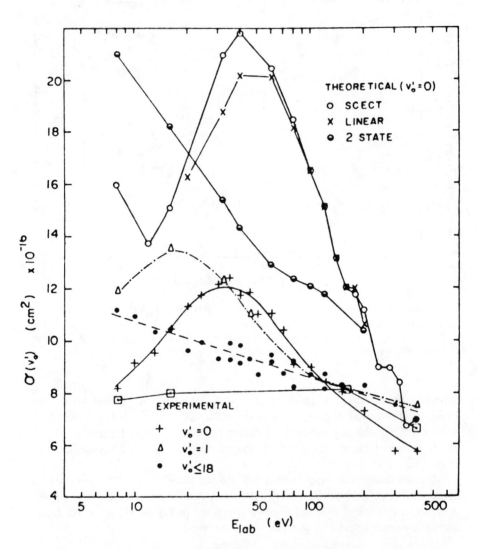

Figure 9. Experimental and theoretical state-selected charge-transfer cross sections for $H_2^+(v_0') + H_2(v_0'' = 0)$.[46] Experimental values: dashed curve, electron impact values recommended by Barnett et al.[123]; "plus" symbols, $v_0' = 0$; triangles, $v_0' = 1$; solid circles, $v_0' \leqslant 18$; squares, $v_0' = 0$ results scaled from Fig. 2 of Ref. 124. Theoretical values: open circles, $v_0' = 0$ with SCECT; X's, $v_0' = 0$ with linear trajectories; half-darkened circles, $v_0' = 0$ with a two-state calculation. (Taken from Ref. 46, with permission.)

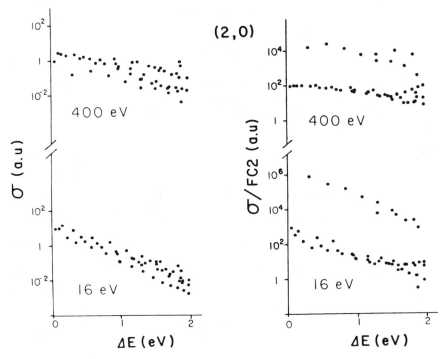

Figure 10. State-to-state cross sections in a.u. for the charge-transfer reaction $D_2^+(v_0' = 2) + H_2(v_0'' = 0) \rightarrow D_2(v'') + H_2^+(v')$ at two c.m. kinetic energies as a function of the energy gap between reactant and product states.[45] The right-hand panel shows the cross section divided by the product P_{fi} of the Franck–Condon factors for the transition. P_{fi} is defined in Eq. (46). (Taken from Ref. 45, with permission).

however, even 400 eV is not high enough to be in the Franck–Condon limit. The wide scatter in the data points confirm this. We conclude that at low collision energy the state-to-state CT cross sections are primarily determined by the energy defect ΔE.

In conclusion, the SCECT has been shown to work extremely well at energies of 2 eV and higher. Using this procedure, DePristo was able to test the approximations that had been made in previous molecular CT calculations. He confirmed that earlier potential energy surfaces were not accurate, and he was able to demonstrate that the Franck–Condon approximation for coupling elements works very well. Finally, he was able to obtain good agreement between experiment and *ab initio* theory for the H_2^+/H_2 system.

D. Work of Kimura and Co-workers

Kimura was one of the first to point out the importance of electron-translation factors in theoretical treatments of CT processes.[20,24] These factors properly treat the momentum of the electron, which is transferred, and they are important for collision energies in the keV range. Without them, the scattering wavefunction does not display the proper Galilean invariance. Kimura has used his methodology to study CT in two ion–molecule systems, $H^+ + H_2 \rightarrow H + H_2^+$ [49] and $Ar^+ + H_2 \rightarrow Ar + H_2^+$ [50]. In both cases, diatomics-in-molecules (DIM) potential energy surfaces[49,125] were used in the calculations. At low energies these systems undergo chemical reaction, but the calculations were carried out in the keV range, and Kimura made the sudden approximation for both rotational and vibrational motion. The calculations paid particular attention to the variation in the CT cross section with molecular orientation.

The calculated total CT cross sections for $Ar^+ + H_2$ are shown in Fig. 11. The results agree very well with the experimental cross sections[39,126–128] for this system. The cross sections are large, because the ionization potentials of the two species are very similar. The decline from $18 \, \text{Å}^2$ at 400 eV (lab) to $12 \, \text{Å}^2$ at 10 keV reflects the fact that it is more difficult for the electron transfer to occur as the collision time decreases. Note that the theoretical cross sections of Hedrick et al.[39] are considerably higher than the

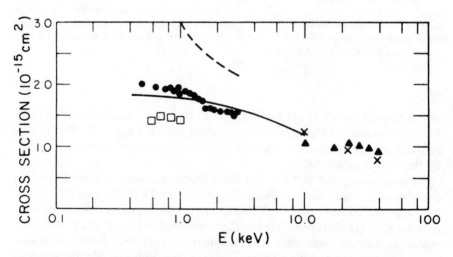

Figure 11. Charge-transfer cross sections for $Ar^+ + H_2(v=0) \rightarrow Ar + H_2^+(v'=0\text{–}20)$. Theory: solid curve, Ref. 50; dashed curve, Ref. 39. Experiments: solid circles, Ref. 39; X's, Ref. 126; squares, Ref. 127; solid triangles, Ref. 128 (Taken from Ref. 50, with permission.)

experimental values for this system. This is probably because the potential energy surfaces used in their work were not as accurate as the DIM surfaces.[125]

E. Work of Spalburg and Co-workers

The second classical path study of an atom–molecule CT process, following the pioneering work of Hedrick et al.,[39] was that of Spalburg and Gislason.[14] They studied the $(Ar + N_2)^+$ system. At moderate collision energies a large number of state-to-state processes can occur; these are summarized in Table III. In addition to CT, vibrational excitation or deexcitation can also occur in every collision. The $N_2^+(A)$ state lies about 1 eV above the $N_2^+(X)$ state, and transitions between them, known as intersystem crossing, can be observed. Finally, the two spin–orbit states of Ar^+ are only separated by 0.18 eV, so the cross sections for fine-structure transitions are generally large. For reference the relative energies of the five lowest vibrational levels for each electronic state are shown in Table IV. Spalburg and Gislason[14] omitted the $N_2^+(A)$ state in their work; otherwise, all of the processes indicated in Table III were studied.

At the time they did this work, a number of intriguing experimental results had appeared. Several laboratories[129–131] had determined that the CT cross

TABLE III
Possible Processes in $(Ar + N_2)^+$ System[a]

State #1		State #2	Process[b]
$N_2^+(X;v) + Ar$	$\leftrightarrow$	$N_2(v') + Ar^+(^2P_{3/2})$	CT
$N_2^+(X;v) + Ar$	$\leftrightarrow$	$N_2(v') + Ar^+(^2P_{1/2})$	CT
$N_2^+(A;v) + Ar$	$\leftrightarrow$	$N_2(v') + Ar^+(^2P_{3/2})$	CT
$N_2^+(A;v) + Ar$	$\leftrightarrow$	$N_2(v') + Ar^+(^2P_{1/2})$	CT
$N_2^+(X;v) + Ar$	$\leftrightarrow$	$N_2(X;v') + Ar$	VE
$N_2^+(A;v) + Ar$	$\leftrightarrow$	$N_2(A;v') + Ar$	VE
$N_2(v) + Ar^+(^2P_{3/2})$	$\leftrightarrow$	$N_2(v') + Ar^+(^2P_{3/2})$	VE
$N_2(v) + Ar^+(^2P_{1/2})$	$\leftrightarrow$	$N_2(v') + Ar^+(^2P_{1/2})$	VE
$N_2^+(X;v) + Ar$	$\leftrightarrow$	$N_2^+(A;v') + Ar$	ISC
$N_2(v) + Ar^+(^2P_{3/2})$	$\leftrightarrow$	$N_2(v') + Ar^+(^2P_{1/2})$	FST

[a]The two states of N_2^+ are the $X\ ^2\Sigma_g^+$ ground state and the $A\ ^2\Pi_u$ excited state, and v and v' indicate vibrational levels of N_2 or N_2^+. Only the $^1\Sigma_g^+$ ground state of N_2 is considered here. The ground state of Ar^+ is the $^2P_{3/2}$ state.

[b]CT = charge transfer, VE = vibrational excitation (or deexcitation), ISC = intersystem crossing, and FST = fine-structure transition.

TABLE IV

Energies of Vibronic States of the $Ar^+ + N_2 \leftrightarrow Ar + N_2^+$ System[a]

v	$Ar^+(3/2) + N_2(v)$	$Ar^+(1/2) + N_2(v)$	$Ar + N_2^+(X,v)$	$Ar + N_2^+(A,v)$
0	0.000	0.178	−0.178	0.940
1	0.289	0.467	0.092	1.172
2	0.574	0.752	0.358	1.401
3	0.856	1.034	0.619	1.625
4	1.134	1.312	0.876	1.847

[a]The notation $Ar^+(3/2)$ and $Ar^+(1/2)$ means $Ar^+(^2P_{3/2})$ and $Ar^+(^2P_{1/2})$. The energy of the vibronic state $Ar^+(^2P_{3/2}) + N_2(v = 0)$ is taken to be the zero of energy. The data were taken from Ref. 53. The energies are in electron volts.

section for $N_2^+(X; v = 0) + Ar$ at collision energies below 25 eV was nearly zero, whereas other vibrational levels of N_2^+ had CT cross sections in excess of 20 Å^2. This surprising result is obtained, even though the $Ar^+(^2P_{3/2}) + N_2(v = 0)$ product state is nearby ($\Delta E \approx 0.18$ eV), and the Franck–Condon factor for that transition, $|\langle 0|0 \rangle|^2$, is 0.9. In addition, the measured CT cross sections for the two spin–orbit states of $Ar^+(^2P_J)$ are quite different at low collision energies.[18,132-134] Spalburg and Gislason[14] gave an extensive review of earlier theoretical procedures to study CT processes and concluded that only the classical path method would be able to properly explain the new state-selected data.

Their calculations used very simple potential energy surfaces.[14] The diabatic surfaces $V_\alpha^{int}(r, R, \gamma)$ [see Eq. (13)] were assumed to be constant, regardless of the atom–molecule configuration. This meant that long-range terms in the potential, such as the ion-induced dipole potential, were ignored, and the short-range repulsions were also omitted. In addition, they assumed that the electronic coupling $H_{\alpha\alpha'}$ between the two $^2A'$ states corresponding to $N_2^+(X) + Ar$ and $N_2 + Ar^+(^2P)$, defined in Eq. (12), was given by the simple formula[135]

$$H_{\alpha\alpha'}(R) = 27.6R \exp(-1.746R), \tag{48}$$

where $H_{\alpha\alpha'}$ is in eV and R is in Å. Thus, $H_{\alpha\alpha'}$ was assumed to be independent of molecular orientation and the vibrational coordinate. Straight-line trajectories were used to compute the dynamics. An important part of the calculations was that the spin–orbit splitting of $Ar^+(^2P_J)$ was treated correctly.[136] Thus, Clebsch–Gordan coefficients were used to transform the hamiltonian from the (L, S) coupling case to the (j, m_j) coupling case. A total angular momentum of $j = 1/2$ gives rise to $m_j = \pm 1/2$, whereas the value

$j = 3/2$ gives $m_j = \pm 1/2, \pm 3/2$. For the hamiltonian used in this work

$$\Omega = |m_j| \tag{49}$$

is a good quantum number. The electronic state $N_2^+(X) + Ar$ corresponds to $\Omega = 1/2$. Consequently, all collisions of $Ar^+(^2P_{1/2}) + N_2$ can undergo CT, but only one-half of the $Ar^+(^2P_{3/2}) + N_2$ collisions can. In fact, the $\Omega = 3/2$ states of $Ar^+(^2P_{3/2}) + N_2$ are completely inert in their calculations. Thus, Spalburg and Gislason[14] carried out calculations only for $\Omega = 1/2$, and the results for $Ar^+(^2P_{3/2}) + N_2$ had to be divided by 2 before comparing with experiments.

Some of the state-to-state CT cross sections obtained in their calculations are summarized in Table V. The principle of microscopic reversibility, summarized in Eq. (26), has been used to compress the results. Charge-transfer cross sections for $Ar + N_2^+(v)$ are read in the various rows, and cross sections for $Ar^+(^2P_{3/2}, {}^2P_{1/2}) + N_2(v')$ are read in the various columns. However, as discussed above, the CT cross sections for $Ar^+(^2P_{3/2}) + N_2$ must be divided by 2. As an example, the cross section for the process $Ar + N_2^+(v=2) \rightarrow Ar^+(^2P_{3/2}) + N_2(v'=1)$ at 10.3 eV is 24.4 Å^2, but the value for $Ar^+(^2P_{3/2}) + N_2(v'=1) \rightarrow Ar + N_2^+(v=2)$ is 12.2 Å^2. It is seen that the total CT cross section for $N_2^+(v=0) + Ar$ is quite small at $E_{rel} = 10.3$ eV and is only 2.9 Å^2 at 41.2 eV. By comparison, CT from the other $N_2^+(v)$ states gives much larger cross sections. This is consistent with the experimental results.[129-131] There was, however, a problem with the cross section computed for $N_2^+(v=4) + Ar$; the calculated value of 46 Å^2 at 10.3 eV greatly exceeds the experimental value of 27 Å^2.[130] There are two likely explanations for the discrepancy. The first is that the electronic coupling term in Eq. (48) is too large. It was based on the formula of Olson et al.,[135] who emphasized that their results are typically uncertain by a factor of 2. Reducing H_{12} by two gives better agreement with the experiments.[14] The other likely explanation is that the calculations omit the $N_2^+(A)$ state. The $N_2^+(X, v=4)$ state is very close in energy to $N_2^+(A, v=0)$ (see Table IV). Thus omitting the $N_2^+(A)$ state is likely to lead to inaccurate CT cross sections.

Another remarkable result of the Spalburg calculations is that the cross sections for fine-structure transitions are quite large.[14] The calculated results at four collision energies are summarized in Table VI. It should be emphasized that there is no direct coupling between $Ar^+(^2P_{3/2})$ and $Ar^+(^2P_{1/2})$ in the interaction Hamiltonian, so the transition can only occur because of the presence of the charge-exchange state $Ar + N_2^+$. The total fine-structure transition cross section for $Ar^+(^2P_{1/2}) + N_2(v=0)$ exceeds 4 Å^2 at all collision energies; at energies of 4.1 eV and higher this product channel is even more important that the CT channel.

TABLE V

State-to-State Cross Sections (Å^2) at $E_{rel} = 10.3$ and $41.2\,\text{eV}$ for
$\text{Ar} + \text{N}_2^+(v) \rightleftarrows \text{Ar}^+(^2P_{3/2}, {}^2P_{1/2}) + \text{N}_2(v')^a$

v	v'				
	0	1	2	3	4
$E_{rel} = 10.3\,\text{eV}$					
0	0.15	—	—	—	—
	0.01	—	—	—	—
1	24.36	0.50	—	—	—
	1.87	0.02	—	—	—
2	0.16	24.43	0.60	0.01	—
	1.24	2.64	0.03	0.00	—
3	0.00	0.39	26.96	1.36	0.01
	0.02	3.09	2.98	0.05	0.00
4	—	0.01	0.91	34.41	1.64
	—	0.10	5.78	3.22	0.06
$E_{rel} = 41.2\,\text{eV}$					
0	2.37	0.03	—	—	—
	0.56	0.01	—	—	—
1	15.19	1.29	0.04	—	—
	4.29	0.33	0.01	—	—
2	0.21	23.11	1.06	0.05	—
	0.51	5.65	0.23	0.01	—
3	0.00	0.64	27.22	1.13	0.05
	0.01	1.50	5.34	0.20	0.01
4	—	0.01	1.35	28.48	1.24
	—	0.06	3.13	4.45	0.19

[a]Two cross sections are given for each pair of vibrational quantum numbers. The upper value refers to $\text{Ar}^+(^2P_{3/2})$; the lower value refers to $\text{Ar}^+(^2P_{1/2})$. If no numbers are shown, both cross sections are less than $0.01\,\text{Å}^2$. Because of microscopic reversibility the cross sections are the same for the forward and reverse reactions. However, as discussed in the text, cross sections shown here for the process $\text{Ar}^+(^2P_{3/2}) + \text{N}_2(v') \rightarrow \text{Ar} + \text{N}_2^+(v)$ must be divided by two. The tables can be read from left to right to obtain cross sections for $\text{Ar} + \text{N}_2^+$ or from top to bottom to obtain cross sections for $\text{Ar}^+ + \text{N}_2$. The data are taken from Ref. 14.

A number of other interesting conclusions arise from the work of Spalburg and Gislason.[14] They showed that the cross sections for vibrational excitation and deexcitation were small at all collision energies. It should be emphasized that the potential energy surfaces V_α^{int} used by Spalburg were independent of the vibrational coordinate r, so that vibrational excitation, like

TABLE VI
State-to-State Cross Sections (Å^2) for Fine-Structure Transitions[a]

E_{cm}(eV)	v'		
	0	1	2
$Ar^+(^2P_{1/2}) + N_2(v = 0) \rightarrow Ar^+(^2P_{32}) + N_2(v')$			
1.2	0.05	3.83	0.31
4.1	0.89	3.38	0.08
10.3	3.95	2.62	0.02
41.2	13.17	1.31	0.02
$Ar^+(^2P_{3/2}) + N_2(v = 0) \rightarrow Ar^+(^2P_{1/2}) + N_2(v')$			
1.2	0.02	0.00	0.00
4.1	0.44	0.00	0.00
10.3	1.97	0.01	0.00
41.2	6.59	0.04	0.00

[a] Data taken from Ref. 14.

fine-structure transitions, can only occur via the charge exchange inter-mediate. They also confirmed the conclusion of DePristo[43] that the long-range ion-induced dipole and ion-quadrupole forces do not play an important role in determining CT cross sections. Finally, they determined that the CT cross section ratio $\sigma(1/2)/\sigma(3/2)$ for $Ar^+(^2P_{1/2}, {}^2P_{3/2}) + N_2(v = 0)$ collisions was much less than one in the energy region considered by them. The fascinating story of this ratio and of the many theoretical and experimental attempts to pin it down is discussed in the following section.

Spalburg, Los, and Gislason also published a paper[51] that discussed for the first time the calculation of adiabatic *vibronic* potential energy surfaces and their importance for low energy CT processes. This work is discussed in Section V B.

F. Work of Parlant and Gislason on $(Ar + N_2)^+$

1. *Introduction*

The work of Spalburg and co-workers[14,51] gave considerable insight into the system $(Ar + N_2)^+$. However, they made a number of questionable approximations in their calculations. These include the neglect of any repulsive forces and the use of straight-line trajectories. In addition, the electronic coupling in Eq. (48) was only approximate, and it was assumed to be independent of the orientation angle γ. Perhaps the biggest problem, however, was their neglect of the $N_2^+(A)$ state in their work. In 1986 Archirel and Levy[19] computed diabatic potential energy surfaces and couplings for the $(Ar + N_2)^+$ system including the A state. Thus, for the first time it was possible to carry out fully *ab initio* calculations of the cross sections for this reaction. This was done by Parlant and Gislason.[52-54,56,57]

The computations were carried out at relative collision energies between 1 and 4000 eV. At the higher energies the *Franck–Condon principle* is expected to be valid. This principle, which is discussed further in Section V A, assumes that for a process such as

$$Ar^+(^2P_{3/2}) + N_2(v) \rightarrow Ar + N_2^+(X, v') \tag{50}$$

the state-to-state cross section is given by

$$\sigma(v, v') = \sigma_0 |\langle v | v' \rangle|^2. \tag{51}$$

Here $|\langle v | v' \rangle|^2$ is the Franck–Condon factor for the transition $N_2(v) \rightarrow N_2^+(v')$, and σ_0, the cross section for the electron transfer, is independent of v and v'. If we define $\sigma(v)$ to be the state-selected total cross section for producing $Ar + N_2^+(X)$, regardless of the product vibrational level v', then Eq. (51) gives

$$\sigma(v) = \sum_{v'} \sigma(v, v') = \sigma_0. \tag{52}$$

Here the closure property of the Franck–Condon factors has been used. Thus, whenever the Franck–Condon principle is valid, the cross section $\sigma(v)$ must be independent of v. Both Eqs. (51) and (52) have been tested in the work of Parlant and Gislason.

Another important principle used to interpret their calculations is the concept of a dynamic energy range δE around the reactant state energy. To a first approximation all states within $\pm \delta E$ of the reactant energy are populated by the scattering event, but outside that range the product population is very small. Spalburg et al.[51] have used a weak-coupling model to show that if the charge-exchange coupling is given by the formula

$$H_{\alpha\alpha'}(R) = A \exp(-aR), \tag{53}$$

then the energy width is given by

$$\delta E = a\hbar g. \tag{54}$$

Here g is the relative velocity. For the $(Ar + N_2)^+$ system a reasonable choice for a gives

$$\delta E = 0.038(E_{rel})^{1/2}, \tag{55}$$

where δE and E_{rel} are in eV. Thus, $\delta E = 0.04$ eV at $E_{rel} = 1$ eV, but it grows to 1.2 eV at 1000 eV. The Franck–Condon principle is only expected to be valid when δE is large compared to the range of states populated in a Franck–Condon transition.

2. Theoretical Procedures

The calculations were carried out as described in Section II. However, certain approximations were made. First, the computations were carried out for fixed values of the orientation angle γ, and then the results were averaged over γ. Second, the classical trajectory was propagated on the reactant (diabatic) potential energy surface. Third, the Franck–Condon approximation was made to evaluate the vibronic matrix elements (see Section II C). Other approximations as well as the full expressions for the hamiltonian matrix are detailed in the first paper by Parlant and Gislason.[52] The two spin–orbit states were treated correctly, as described in Section III E.

3. Potential Energy Surfaces and Couplings

Parlant and Gislason used simple analytic fits to the potential energy surfaces and couplings calculated by Archirel and Levy.[19] The fits are summarized in Ref. 52. In principle, there are six potential energy surfaces for $(Ar + N_2)^+$, but Archirel and Levy[19] showed that one surface does not couple to the other five. They carried out calculations for $\gamma = 0$ and $\gamma = 90°$ at $r = r_e(N_2)$. Parlant and co-workers fit these calculations to an expansion in Legendre polynomials and then diagonalized the five-dimensional electronic hamiltonian matrix at several values of R and γ.[59] The resulting adiabatic potential energy surfaces are shown in Fig. 12 for $\Omega = \frac{1}{2}$. [The $\Omega = \frac{3}{2}$ curves are less interesting, because only the $Ar^+(^2P_{3/2}) + N_2$ and $Ar + N_2^+(A)$ states can give $\Omega = \frac{3}{2}$. In addition, the electronic couplings are weaker.] It is seen that the ground state, which correlates to $Ar + N_2^+(X)$, has an appreciable well in all orientations. The maximum bond energy is 0.785 eV at 0°. The best experimental estimate of this bond energy is 0.91 ± 0.06 eV,[59] in fairly good agreement with the theoretical value. By comparison, the first excited state, which correlates to $Ar^+(^2P_{3/2}) + N_2$, has a negligible well in every orientation. It is also apparent that the three lowest potential energy surfaces are quite anisotropic.

4. Calculations for $N_2^+(X, A) + Ar$

In their first paper Parlant and Gislason[52] presented extensive calculations at $E_{rel} = 20$ eV as well as a smaller number at 8 eV. The various processes that can occur in this system are summarized in Table III. A large number of product states are populated in a typical collision at $E_{rel} = 20$ eV. A typical example is shown in Table VII, which gives all state-to-state cross sections

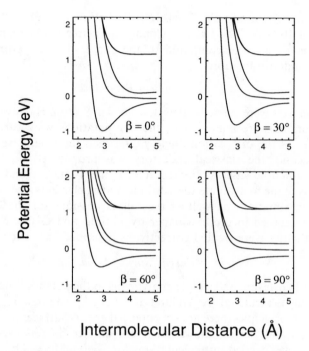

Figure 12. Potential energy curves for the five lowest electronic states of $(ArN_2)^+$ for $\Omega = \frac{1}{2}$, plotted against the $Ar-N_2$ distance R for fixed values of the orientation angle β. The N_2 bond length is 1.098 Å. At infinite separation the states correspond, in order of increasing energy, to $Ar + N_2^+(X)$, $Ar^+(^2P_{3/2}) + N_2$, $Ar^+(^2P_{1/2}) + N_2$, and $Ar + N_2^+(A)$, which is doubly degenerate. The zero of energy corresponds to $Ar^+(^2P_{3/2}) + N_2$ at $R = \infty$. (Taken from Ref. 59, with permission.)

which exceed 0.1 Å^2 for collisions of $N_2^+(A, v = 2) + Ar$ at 20 eV. [Note that $N_2^+(A) + Ar$ can give both $\Omega = \frac{3}{2}$ and $\frac{1}{2}$, and the results in Table VII have been properly averaged.] Also shown is the average energy transfer $\langle \Delta E \rangle$ and the standard deviation $\delta \Delta E$ for the energy transfer. It is seen that the state-to-state cross sections are largest when the energy gap ΔE is small, but even in the limit of a very close energy resonance the cross sections are not extremely large. In addition, the cross sections tend to be larger if the product state is directly coupled to the reactant state in the hamiltonian matrix. Thus, cross sections to produce $Ar^+(^2P_{3/2}) + N_2(v')$, which is directly coupled, are larger than those to produce $Ar + N_2^+(X, v')$, which is not, for comparable values of ΔE.

Table VII also shows that the average value of ΔE is very small, and the standard deviation $\delta \Delta E$ is 0.24 eV. This latter value, which corresponds to

TABLE VII

State-to-State Cross Sections for $N_2^+(A, v = 2) + Ar$ at $E_{rel} = 20\,eV^a$

Product Ion State	v'	ΔE (eV)	Cross Section (Å^2)
$Ar^+(^2P_{3/2})$	7	0.548	0.9
$N_2^+(X)$	8	0.463	0.3
$Ar^+(^2P_{1/2})$	6	0.457	0.8
$N_2^+(A)$	4	0.446	0.5
$Ar^+(^2P_{3/2})$	6	0.280	3.0
$N_2^+(A)$	3	0.225	3.4
$N_2^+(X)$	7	0.223	0.9
$Ar^+(^2P_{1/2})$	5	0.186	1.4
$Ar^+(^2P_{3/2})$	5	0.009	5.5
$N_2^+(X)$	6	−0.023	2.5
$Ar^+(^2P_{1/2})$	4	−0.089	4.4
$N_2^+(A)$	1	−0.229	3.5
$Ar^+(^2P_{3/2})$	4	−0.266	1.7
$N_2^+(X)$	5	−0.271	0.4
$Ar^+(^2P_{1/2})$	3	−0.367	0.4
$N_2^+(A)$	0	−0.461	0.2
Total cross section^b			30.3
$\langle \Delta E \rangle$ (eV)^c			0.05
$\delta \Delta E$ (eV)^c			0.24

^{a}The various product states can be $Ar^+(^2P_{3/2}) + N_2(v')$, $Ar^+(^2P_{1/2}) + N_2(v')$, $N_2^+(X, v') + Ar$, or $N_2^+(A, v') + Ar$; these are identified in the first two columns. ΔE is the energy difference between reactant and product states. The data are taken from Ref. 52.

^{b}Total cross section for producing all states other than the reactant state.

$^c\langle \Delta E \rangle$ is the average energy transferred during the collision and $\delta \Delta E = \langle \Delta E^2 \rangle - \langle \Delta E \rangle^2$.

about one vibrational quantum, is comparable to the dynamic energy width $\delta E = 0.17\,V$ computed from Eq. (55). The small value of $\langle \Delta E \rangle$ justifies the use of the classical path method for this system down to relative energies of $1\,eV$ or so. The values of $\langle \Delta E \rangle$ and $\delta \Delta E$ are definitely not consistent with the Franck–Condon principle. Parlant and Gislason[52] showed that a Franck–Condon distribution of products would correspond to $\langle \Delta E \rangle = -0.41\,eV$ and $\delta \Delta E = 0.66\,eV$. We shall see that the Franck–Condon principle is not expected to be valid at such a low collision energy.

Partial state-to-state cross sections for collisions of $N_2^+(X, A; v) + Ar$ are given in Table VIII for several reactant states at a relative energy of $20\,eV$. A more complete set of results is given in Table VIII of Ref. 52. It is seen that the CT cross section for $N_2^+(X, v = 0) + Ar$ is much smaller than for the other reactant states; this is consistent with the experimental data[129-131]. Parlant and Gislason[52] argue that this occurs because $\delta E = 0.17\,eV$ at

TABLE VIII

Partial State-to-State Cross Sections for $N_2^+(X, A; v) + Ar^a$

Initial State	Product State				
	$Ar^+(^2P_{3/2})$	$Ar^+(^2P_{1/2})$	VE	VD	ISC
X 0	2.6	0.6	0.7	0.0	0.0
X 2	21.5	6.5	1.3	1.1	0.1
X 4	21.2	8.5	1.8	1.6	2.8
X 6	15.9	8.5	2.5	2.0	9.0
A 0	7.4	4.3	2.0	0.0	1.4
A 2	11.2	7.2	3.9	3.8	4.4
A 4	8.4	4.6	3.3	3.7	3.5

^{a}All cross sections are in Å. The initial state of N_2^+ is specified by the electronic state (X or A) and the vibrational quantum number v. The calculations were done for $E_{rel} = 20$ eV. The various product states are $Ar^+(^2P_{3/2}) + N_2$, $Ar^+(^2P_{1/2}) + N_2$, vibrational excitation (VE) or deexcitation (VD), and intersystem crossing (ISC) between the $N_2^+(X)$ and $N_2^+(A)$ electronic states. In each case, the cross sections have been summed over all vibrational levels of the products. The data are taken from Ref. 52.

$E_{rel} = 20$ eV, whereas the nearest product state for $N_2^+(X, v = 0) + Ar$ is 0.18 eV away (Table IV). All other $N_2^+(X, v) + Ar$ reactant states have CT product states within 0.09 eV. Table VIII also shows that the CT cross sections for producing $Ar^+(^2P_{3/2})$ are always larger than for making $Ar^+(^2P_{1/2})$. For the $N_2^+(X, v)$ reactants this occurs because the nearest product state is always $Ar^+(^2P_{3/2}) + N_2(v')$ (see Table IV). For collisions of $N_2^+(A, v) + Ar$ the $Ar^+(^2P_{3/2})$ products are favored because CT collisions along the $\Omega = \frac{3}{2}$ surfaces can only give $Ar^+(^2P_{3/2})$. Intersystem crossing (ISC) is an important process for $N_2^+(X, v \geqslant 4) + Ar$ and for all $N_2^+(A, v) + Ar$ collisions, since the energy gap (Table IV) is quite small. Cross sections for vibrational excitation and deexcitation are also fairly large. The values for the $N_2^+(A)$ state are larger, reflecting the smaller vibrational quantum in the A state. Parlant and Gislason[52] also carried out calculations at 8 eV, and the results were similar.

The total CT cross sections can be obtained by adding the values in the second and third column of Table VIII together. The results at $E_{rel} = 20$ eV are shown in Fig. 13 for the eight lowest levels of $N_2^+(X)$ and the seven lowest levels of $N_2^+(A)$. Also shown are the experimental data of Govers et al.[130] It must be emphasized that the theoretical calculations, which are based on *ab initio* potential energy surfaces,[19] involve no adjustable parameters. Overall, the agreement between theory and experiment is quite good.

After the first paper on $N_2^+ + Ar$ had been published by Parlant and Gislason,[52] Ng and co-workers carried out some very exciting experiments[131]

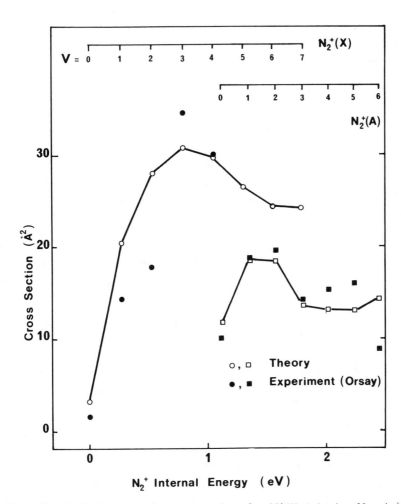

Figure 13. Total charge-transfer cross sections for $N_2^+(X, A; v) + Ar \rightarrow N_2 + Ar^+$ at $E_{rel} = 20\,eV$ are plotted against the internal energy of the N_2^+ ion. Theoretical results from Ref. 52 for $v = 0$–7 of the X state and $v = 0$–6 of the A state are shown as open circles and squares, respectively. Experimental results from Ref. 130 are shown as solid circles and squares. The lines drawn through the theoretical values are only a guide for the eye. (Taken from Ref. 52, with permission.)

on the reactions

$$N_2^+(X; v = 0, 1, 2) + Ar \rightarrow N_2(v') + Ar^+(^2P_{3/2}, {}^2P_{1/2}). \qquad (56)$$

In particular, they were able to determine the CT branching ratio for producing the two spin–orbit states of Ar^+ at a range of collision energies. This allowed them to present absolute state-to-state cross sections. Because of this new work Parlant and Gislason[57] decided to calculate the cross sections at the energies studied by Ng. The procedure was exactly the same as that used in the earlier paper.[52]

State-to-state CT cross sections were computed for reaction (56) at 12 relative collision energies between 1.2 and 320 eV.[57] The state-selected cross sections, summed over product vibrational levels, are shown in Fig. 14, where they are compared with the measurements of Govers et al.[130] and Shao et al.[131b] Overall, the agreement between theory and experiment is good, although the computed cross sections are systematically somewhat higher than the experimental values for $v = 1$ and 2. The cross section for $v = 0$ is quite small at low energy but rises monotonically up to 24 Å^2 at 320 eV. As discussed earlier, this behavior is seen because the nearest CT product state is 0.18 eV away, and the dynamic energy width δE, defined in Eq. (55), only equals 0.18 eV at $E_{rel} = 22$ eV. Thus, it is not surprising that the CT cross section is so small at lower collision energies.

By comparison, both $N_2^+(v = 1)$ and $N_2^+(v = 2)$ colliding with Ar have product states $Ar^+(^2P_{3/2}) + N_2(v')$, which are much closer in energy (Table IV). The vibronic transition $N_2^+(v = 1) + Ar \rightarrow N_2(v' = 0) + Ar^+(^2P_{3/2})$ has been shown to behave like a two-state Landau–Zener system.[54] The state-to-state cross section rises to a maximum at 4 eV and then declines at higher energies. As a consequence, the calculated cross section in Fig. 14 for $N_2^+(v = 1)$ goes through a maximum of 26 Å^2 at 6 eV, declines to 14 Å^2 at 140 eV, and then rises again due to the growing importance of the two product channels $Ar^+(^2P_{3/2}, {}^2P_{1/2}) + N_2(v' = 1)$. (These states are favored by the Franck–Condon factors.) The behavior of $N_2^+(v = 2) + Ar$ is similar. At low energy the primary product channel is $Ar^+(^2P_{3/2}) + N_2(v' = 1)$, but at high energy the two product states $Ar^+(^2P_{3/2}, {}^2P_{12}) + N_2(v' = 2)$ become favored due to Franck–Condon considerations.

The relative CT cross sections for the three reactant states are shown in Fig. 15 for the 12 collision energies considered by Parlant and Gislason.[57] The agreement between the theory and the experimental data of Liao et al.[131a] is quite good. At low energies the cross sections usually satisfy the inequalities $\sigma(v = 0) < \sigma(v = 1) < \sigma(v = 2)$. This is consistent with the ordering of the energy gaps between the reactant states and the nearest CT product states (Table IV). Above $E_{rel} = 100$ eV the three cross sections are about the same (although

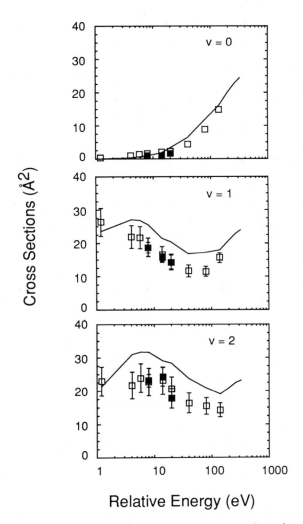

Figure 14. Absolute state-selected charge-transfer cross sections for the process $N_2^+(v) + Ar \rightarrow N_2 + Ar^+$ plotted against the relative collision energy for $v = 0, 1$, and 2. The theoretical results from Ref. 57 are shown as points connected by straight-line segments. The open squares are the experimental data of Shao et al.[131b]; the solid squares are the cross sections measured by Govers et al.[130]. (Taken from Ref. 57, with permission.)

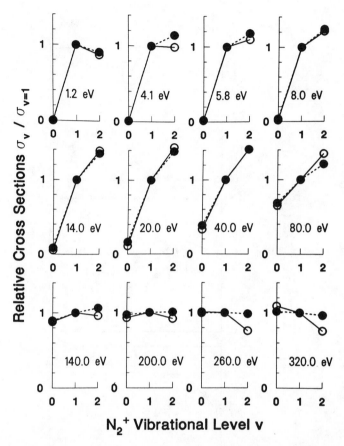

Figure 15. Relative charge-transfer cross sections for $N_2^+(v) + Ar$ plotted as a function of vibrational quantum number v for various relative collision energies. The theoretical results from Ref. 57 are shown as solid circles; the experimental results of Liao et al.[131a] are shown as open circles. (Taken from Ref. 57, with permission).

there is some variation in the experimental results), in agreement with the prediction of the Franck–Condon principle in Eq. (52). Clearly, the Franck–Condon principle is violated at lower energies.

A number of experimental measurements have been made of the CT cross section for reaction (56), but using non state-selected beams of $N_2^+(X)$.[126,137-144] The results are shown in Fig. 16. In most cases, the ions were made using electron bombardment ionization, so the vibrational distribution of the N_2^+ ions is uncertain. Nevertheless, Henri et al.[62] have made a reasonable estimate of $v = 0$, 77%; $v = 1$, 14%; $v = 2$, 5%; and higher

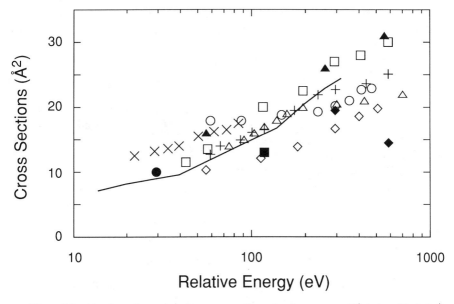

Figure 16. Absolute charge-transfer cross sections for the process $N_2^+ + Ar \rightarrow N_2 + Ar^+$ plotted against the relative collision energy. The theoretical results from Ref. 57 are shown as points connected by heavy straight-line segments. It was assumed that the $N_2^+(v)$ vibrational distribution was $v = 0$, 77%; $v = 1$, 14%; and $v = 2$, 9%. The experimental data are denoted by solid triangles, Ref. 126; open circles, Ref. 137; solid diamonds, Ref. 138; open squares, Ref. 139; crosses, Ref. 140a; open diamonds, Ref. 140b; solid diamonds, Ref. 141; solid circle, Ref. 142; X's, Ref. 143; and solid squares, Ref. 144. (Taken from Ref. 57, with permission.)

levels 4%. For the purposes of comparison Parlant and Gislason[57] assumed a distribution of 77%, 14%, and 9%, and their calculated cross sections based on this distribution are also shown in Fig. 16. Given the spread in the experimental data and the rough estimate of the vibrational distribution used in the theory, the agreement is very good.

One problem did arise in the calculations of Parlant and Gislason.[57] Ng and co-workers[131] measured the branching ratio for producing the two spin–orbit states of Ar^+ in reaction (56). For all three reactant states the fraction of $Ar^+(^2P_{1/2})$ products is small, primarily because the $Ar^+(^2P_{3/2})$ + $N_2(v')$ product states are closer in energy to the reactant states (Table IV). However, the fraction of $Ar^+(^2P_{1/2})$ products increases with energy, and it is expected to ultimately reach one. Both theory and experiments agree on this general behavior, but the experimental fraction of $Ar^+(^2P_{1/2})$ products[131] is only two-thirds of the theoretical value[57] at most collision energies. It is disappointing that the agreement is so poor. It is quite possible that assumptions made in the theoretical calculations are causing errors in this calculation. It is also true that the experimental results are very sensitive to the CT cross section ratio for the reactions

$$Ar^+(^2P_{3/2}, {}^2P_{1/2}) + H_2(v=0) \rightarrow Ar + H_2^+$$

at $E_{lab} = 20\,eV$. The difference in CT cross sections was used by Ng to differentiate the two spin–orbit states of Ar^+. Three measurements of the ratio $\sigma(1/2)/\sigma(3/2)$ at this energy have given 3.3,[134] 4.1,[61] and 6.1.[145] It is possible that part of the disagreement in the branching ratio for reaction (56) is due to the uncertainty in this ratio.

Overall, the agreement between experiments and the theoretical calculations by Parlant and Gislason[52,57] on the $N_2^+ + Ar$ system is quite good. The disagreement over the branching ratio for reaction (56) is disturbing, however, and more work is called for on this question.

5. Charge-Transfer Collisions Between $Ar^+(^2P_J) + N_2$

In addition to their work on $N_2^+ + Ar$, Parlant and Gislason[54] also carried out a series of calculations on the reverse reaction

$$Ar^+(^2P_{3/2}, {}^2P_{1/2}) + N_2(v=0) \rightarrow Ar + N_2^+(X, A; v'). \tag{57}$$

The calculations were carried out in the energy range between 1 and 4000 eV. This system has been studied experimentally by many groups, and this work has been reviewed several times.[1,2,6,8,18] One of the most interesting questions that has arisen is the CT cross section ratio $\sigma(1/2)/\sigma(3/2)$ for reaction (57). Different laboratories[18,132–134] have obtained quite different results for this ratio. The calculations of Parlant and Gislason[54] used the same potential energy surfaces and couplings and the same theoretical procedure as for the $N_2^+ + Ar$ computations.[52,57] In particular, the $N_2^+(A)$ state was included in all the calculations discussed here.

The theoretical state-to-state and total CT cross sections for $Ar^+(^2P_{3/2})$ and $Ar^+(^2P_{1/2})$ colliding with $N_2(v=0)$ are shown in Fig. 17. The results for $Ar^+(^2P_{3/2})$ show a maximum near 3 eV, a minimum at 30 eV, and then a plateau above 300 eV. The major product state at low collision energies is $N_2^+(X, v'=1) + Ar$. Parlant and Gislason[54] have argued that these two vibronic states behave very much like a two-state Landau–Zener system with a peak in the cross section at $E_{rel} = 5\,eV$. At energies above 5 eV the production of $N_2^+(v'=1) + Ar$ declines, so the total CT cross section also decreases. The subsequent increase in the total cross section is due to the growth of other product channels. At $E_{rel} = 100\,eV$, $N_2^+(X, v'=0) + Ar$, which is favored by Franck–Condon considerations, becomes the major product,

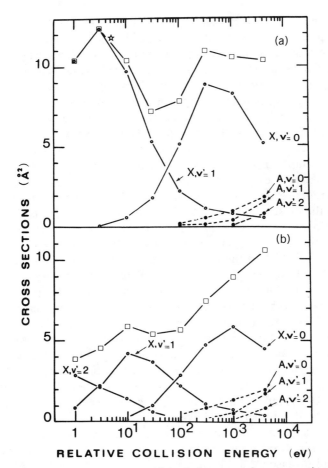

Figure 17. Theoretical state-to-state and total charge-transfer cross sections for (a) the $Ar^+(^2P_{3/2})$ reactant state and (b) the $Ar^+(^2P_{1/2})$ reactant state are plotted as a function of relative collision energy.[54] Cross sections for the processes $Ar^+(^2P_{3/2}, ^2P_{1/2}) + N_2(v = 0) \rightarrow Ar + N_2^+(X, A; v')$ are shown as open circles and solid circles for the X and A product states, respectively. The total cross sections are shown as open squares. The lines through the points are guides for the eyes. The maximum of the cross section calculated by means of a two-state Landau–Zener model is shown as a star in part (a). (Taken from Ref. 54, with permission.)

and above 300 eV the $N_2^+(A)$ states also are important products. The opening up of new product channels as the collision energy rises can be attributed to the increase of the dynamical energy width δE, defined earlier in Eq. (55). The value of δE is 0.04 eV at $E_{rel} = 1$ eV, 0.38 eV at $E_{rel} = 100$ eV, and 2.40 eV at $E_{rel} = 4000$ eV. It is instructive to compare the values of δE with the energy gaps for this system, summarized in Table IV.

The results for $Ar^+(^2P_{1/2}) + N_2(v = 0)$ shown in Fig. 17 are somewhat more complex. At $E_{rel} = 1$ eV the major product state is $N_2^+(X, v' = 2) + Ar$. This interesting result occurs even though the $N_2^+(X, v' = 1) + Ar$ state is both closer in energy to the reactants (see Table IV) and is more favored by Franck–Condon considerations. This result can be understood by examining the *adiabatic vibronic* potential energy surfaces for this system (see Section V B). By $E_{rel} = 10$ eV, $N_2^+(X, v' = 1) + Ar$ has become the primary product state, and then at $E_{rel} = 100$ eV the state favored by Franck–Condon considerations, $N_2^+(X, v' = 0) + Ar$, is the major product channel. This remains the main product up to $E_{rel} = 4000$ eV, although various $N_2^+(A, v')$ states become significant above 40 eV. As in the case of $Ar^+(^2P_{3/2}) + N_2(v = 0)$, this behavior reflects the increasing dynamic width δE as the relative energy rises.

One of the most interesting questions about reaction (57) is the CT cross section ratio for the two spin–orbit states of Ar^+. A number of measurements of the ratio $\sigma(1/2)/\sigma(3/2)$ at various collision energies have been made,[18,132–134,146] and the results exhibit a large dispersion. The data are shown in Fig. 18. The explanation of the dispersion of the data is not clear, but it may involve fine-structure transitions involving the $Ar^+(^2P_J)$ ion, which are not recognized by the experimentalists.[18] The two measurements by Ng's group[133] differ considerably from each other, but both show a minimum near 10 eV. The data from Guyon's laboratory[18] lie between the Ng data. Finally, Kato et al.[132] and Lindsay and Latimer[134] measured ratios which lie above the results of Ng and Guyon. By comparison, the theoretical calculations[54] give a ratio of 0.4 at energies below 10 eV. The ratio then rises to 0.75 at 100 eV and to 1.0 at 4000 eV. It is encouraging that measurements of Lindsay and Latimer,[134] which are the most recent results and where great care was taken to minimize the target thickness to avoid any problems of fine-structure transitions, are in good agreement with the calculations of Parlant and Gislason[54] at all energies.

In an experimental tour de force, Liao et al.[133b] were able to determine absolute state-to-state cross sections for reaction (57) by measuring the vibrational distribution of the $N_2^+(X)$ products at several collision energies. The results are shown in Fig. 19. Also shown are the calculations of Spalburg and Gislason[14] and of Parlant and Gislason.[54] Overall, the agreement between experiments and the two theoretical calculations is excellent. As discussed earlier the major product channel for $Ar^+(^2P_{3/2}) + N_2$ collisions is $Ar + N_2^+(v = 1)$ at all four collision energies, whereas for $Ar^+(^2P_{1/2}) + N_2$ the favored product channel is $Ar + N_2^+(v = 2)$ at the two low energies considered here. There is a greater scatter in the results for $Ar^+(^2P_{1/2}) + N_2$, but Liao et al.[133b] have reported larger uncertainties in these data. In our opinion, Fig. 19 provides striking evidence of how the field of ion–molecule reactions has matured.

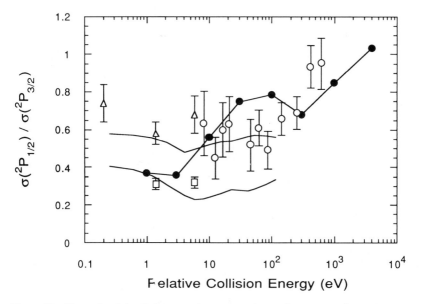

Figure 18. The ratio of the charge-transfer cross sections $\sigma(^2P_{1/2})$ and $\sigma(^2P_{3/2})$ for the two spin–orbit states of Ar^+ colliding with N_2 is plotted as a function of the relative collision energy. The theoretical results of Parlant and Gislason (Ref. 54) are shown as solid circles; the lines between points are only guides for the eye. The upper and lower full curves are the experimental results of Liao et al.[133b] and of Liao et al.[133a], respectively. The experimental uncertainties are estimated to be $\approx 20\%$. The other experimental results are denoted by triangles, Ref. 132; squares, Ref. 18; and open circles, Ref. 134.

6. Inelastic Collisions of $Ar^+(^2P_J)$ and N_2

Parlant and Gislason[56] also carried out theoretical calculations of state-to-state cross sections for inelastic collisions of $Ar^+(^2P_{3/2}, {}^2P_{1/2}) + N_2(v = 0)$. The computations were carried out at the same time as the CT work described in the previous section using exactly the same procedure.

Two important processes studied were the fine-structure transitions

$$Ar^+(^2P_{1/2}) + N_2(v = 0) \rightarrow Ar^+(^2P_{3/2}) + N_2(v'), \qquad (58a)$$

$$Ar^+(^2P_{3/2}) + N_2(v = 0) \rightarrow Ar^+(^2P_{1/2}) + N_2(v'). \qquad (58b)$$

The first process is denoted deexcitation or quenching; the second is excitation. Fine-structure transitions in *neutral* atoms induced by collisions with molecules such as N_2 have been well studied.[147] By comparison, only a few measurements have been made for ion-molecule systems. However, three recent studies[148–150] have been made of reaction (58).

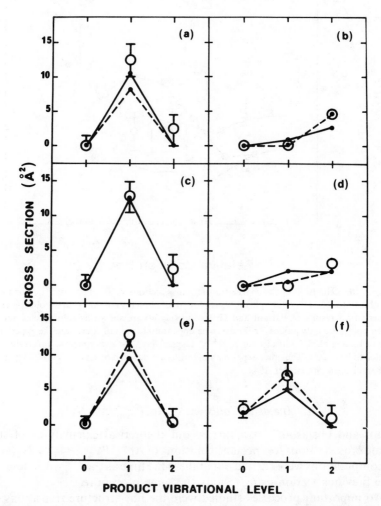

Figure 19. Absolute state-to-state cross sections for the charge-transfer processes $Ar^+(^2P_{3/2}) + N_2(v=0) \rightarrow Ar + N_2^+(X, v')$ [panels (a), (c), (e), and (f)] and $Ar^+(^2P_{1/2}) + N_2(v=0) \rightarrow Ar + N_2^+(X, v')$ [panels (b) and (d)] are plotted for various collision energies. The experimental results from Liao et al.[133b] are shown as open circles. The theoretical calculations by Parlant and Gislason[54] are shown as solid circles connected by solid lines; the calculations by Spalburg and Gislason[14] are shown as circles connected by dashed lines. In (c) the two theoretical results are indistinguishable. The experimental relative collision energies are 1.2 eV [(a) and (b)], 4.1 eV [(c) and (d)], 10.3 eV [(e)], and 41.2 eV [(f)]. The theoretical energies are either the same[14] or close in energy.[54] (Taken from Ref. 54, with permission.)

The theoretical state-selected cross sections for reactions (58a) and (58b) calculated by Parlant and Gislason[56] are shown in Fig. 20. The cross sections are surprisingly large. For example, at $E_{rel} = 1$ eV the cross section for the quenching of $Ar^+(^2P_{1/2})$ by $N_2(v = 0)$ is 1.6 Å^2; this is about one-half of the calculated CT cross section for the same system.[54] Between 10 and 4000 eV

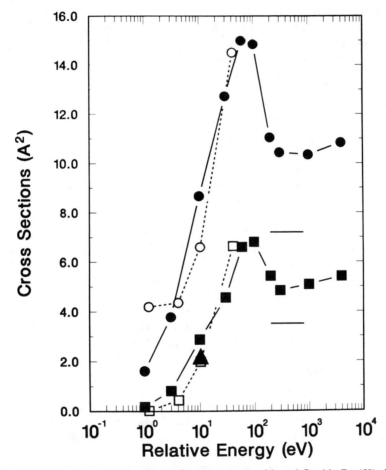

Figure 20.   Total cross sections for the fine-structure transitions defined in Eq. (58) plotted against the relative collision energy. The solid circles are the cross sections for deexcitation of $Ar^+(^2P_{1/2})$ by $N_2(v = 0)$ computed by Parlant and Gislason;[56] the solid squares are the calculated cross sections for excitation of $Ar^+(^2P_{3/2})$ by $N_2(v = 0)$. The lines between points are guides for the eye. The open circles and squares are the analogous cross sections by Spalburg and Gislason.[14] The triangle is the experimental excitation cross section measured by Shao et al.[148] The short solid lines are the deexcitation (upper) and excitation (lower) cross sections measured by Nakamura et al.[149]. (Taken from Ref. 56, with permission.)

the quenching cross sections, of the order of $12 \, \text{Å}^2$, actually exceed the values for charge transfer. The cross sections for excitation of $Ar^+(^2P_{3/2})$ by $N_2(v=0)$ are considerably smaller than the deexcitation cross sections, particularly at low energies. As a consequence, reaction (58b) is not the major process for $Ar^+(^2P_{3/2})$ at any collision energy. Nevertheless, it is important at all energies above $E_{rel} = 10 \, \text{eV}$. The calculations by Spalburg and Gislason[14] for reactions (58a) and (58b) give very similar results. The one exception is the quenching reaction at 1.2 eV, where Spalburg's result is twice that obtained by Parlant. Also shown in Fig. 20 is the total cross section for reaction (58b) measured by Shao et al.[148] Their measured values of $2.18 \pm 0.25 \, \text{Å}^2$ agrees well with the calculated values.[14,56]

Parlant and Gislason[56] also computed the vibrational distributions for the products in reactions (58a) and (58b). Since there is no change in the electronic state of the N_2 product, the Franck–Condon principle predicts that the major product state should be $v' = 0$. This is, in fact, found for reaction (58b) at all collision energies. For the quenching of $Ar^+(^2P_{1/2})$ by $N_2(v=0)$ the product state $Ar^+(^2P_{3/2}) + N_2(v'=1)$, which is the state closest in energy to the reactants (Table IV), is favored below $E_{rel} = 10 \, \text{eV}$. At $E_{rel} = 10 \, \text{eV}$ and higher, the major product is $Ar^+(^2P_{3/2}) + N_2(v'=0)$. By $E_{rel} = 200 \, \text{eV}$, $N_2(v'=0)$ is the dominant product channel for both fine-structure transitions. Microscopic reversibility [Eq. (26)] then guarantees that the two cross sections shown in Fig. 20 should differ by a factor of 2, since the degeneracy of $Ar^+(^2P_{3/2})$ is twice that of $Ar^+(^2P_{1/2})$. The vibrational distributions computed by Spalburg and Gislason[14] are very similar to those obtained by Parlant and Gislason.[56]

Futrell and co-workers[150] have measured the velocity vector distribution of the products of the two fine-structure transitions. They are able to resolve individual vibrational levels of the N_2 molecular product. The differential cross sections cannot be directly compared to the total cross sections computed by Parlant and Gislason.[56] Nevertheless, it is interesting to see which vibrational levels they observe. For the quenching process Futrell observes only $N_2(v'=0)$ products at $E_{rel} = 0.78 \, \text{eV}$ in the backward direction, but at $E_{rel} = 1.54 \, \text{eV}$ he sees $v' = 0, 1, 2$, and 3. By comparison, the theoretical calculations at $E_{rel} = 1 \, \text{eV}$ give mainly $v' = 1$ with appreciable populations in $v' = 0$ and 2.

Parlant and Gislason[56] also computed state-to-state cross sections for vibrational excitation for collisions of $Ar^+(^2P_{3/2}, {}^2P_{1/2}) + N_2(v=0)$. The cross sections initially increase with collision energy, reaching $1 \, \text{Å}^2$ at $E_{rel} = 30 \, \text{eV}$, peaking at $\sim 2 \, \text{Å}^2$ at 300 eV, and then declining at higher energies. The calculations by Spalburg and Gislason[14] give considerably smaller cross sections. As discussed earlier, for the potential energy surfaces calculated by Archirel and Levy,[19] both fine-structure transitions and vibrational excitation are second-order processes that go through the intermediate charge exchange states $N_2^+(X, A) + Ar$. The fact that both processes have substantial cross sections shows how strongly coupled the $(Ar + N_2)^+$ system is. At sufficiently high collision energies only first-order processes can occur, so the inelastic cross sections must fall off. The vibrational excitation cross sections are declining by $E_{rel} = 1000\,eV$, but the fine structure transitions are not, even at $E_{rel} = 4000\,eV$ (Fig. 20).

Cross sections for the inelastic processes considered by Parlant and Gislason[56] were measured by Nakamura et al.[149] in the energy range of 200–800 eV. They determined the total scattered intensity of a particular product state entering their detector set nominally at $0°$ in the laboratory. The acceptance angle of the detector was $0.45°$, which corresponds to $1.09°$ in the c.m. system. Nakamura stated that this acceptance angle allowed them to detect all products, but the work of Birkinshaw et al.[150] as well as the calculations indicate that a certain fraction of the inelastically scattered Ar^+ products miss the detector. Consequently, Parlant and Gislason[56] carried out two calculations on this system at $E_{rel} = 300\,eV$. The first included all c.m. scattering angles $0 \leqslant \theta \leqslant 180°$, and the second only included products produced along trajectories where $\theta \leqslant 1.09°$. The results are compared in Table IX with Nakamura's experimental values. It is seen that the calculated cross sections for $\theta \leqslant 1.09°$ are often much smaller than the full cross sections, especially for collisions involving transfers of more that one vibrational quantum. The agreement between Nakamura's data and the theoretical values corrected for the angular resolution are remarkably good. The experimental cross sections for fine-structure transitions[149] are also shown in Fig. 20. Since they are being compared with the uncorrected cross sections, the agreement is poor.

G.   Work of Archirel, Gislason, Parlant, and Sizun on $(Ar + CO)^+$

1.   *Introduction*

The molecules CO and N_2 are isoelectronic. As a consequence, many properties of the molecules are quite similar. Thus, the Franck–Condon factors for the transition $CO(X, v) \to CO^+(X, v')$ are nearly diagonal, as is the case for N_2, since the bond lengths of $CO(X)$ and $CO^+(X)$ are very similar. However, the ionization potential of CO is 1.6 eV lower than for N_2. This causes the CT process to be completely different for the two systems. The relative energies of selected states of the $(Ar + CO)^+$ system are summarized in Table X. It is seen that $Ar^+(^2P_{3/2}) + CO(v = 0)$ is nearly isoenergetic with the state $Ar + CO^+(X, v' = 7)$. However, the Franck–Condon factor for the transition $CO(v = 0) \to CO^+(v' = 7)$ is less than 10^{-10}. As a consequence the CT cross section at low collision energy is extremely

TABLE IX

Inelastic State-to-State Cross Sections at $E_{rel} = 300\,eV^a$

| | Cross Sections, This Work | | Cross Sections |
Transition[b]	Total	$\theta < 1.09°$	Nakamura[c]
	Fine-Structure Transitions		
$\frac{1}{2} \to \frac{3}{2}, 0 \to 0$	8.52	4.90	6.0
$\frac{1}{2} \to \frac{3}{2}, 0 \to 1$	1.42	0.70	0.67
$\frac{1}{2} \to \frac{3}{2}, 0 \to 2$	0.36	0.12	0.23
$\frac{1}{2} \to \frac{3}{2}, 0 \to 3$	0.09	0.01	—
$\frac{3}{2} \to \frac{1}{2}, 0 \to 0$	4.26	2.45	2.8
$\frac{3}{2} \to \frac{1}{2}, 0 \to 1$	0.43	0.11	0.20
$\frac{3}{2} \to \frac{1}{2}, 0 \to 2$	0.12	0.01	—
	Vibrational Excitation[d]		
$v' = 1$	2.26	0.46	0.61
$v' = 2$	0.65	0.04	0.076
$v' = 3$	0.18	0.005	—

[a] All cross sections are in $Å^2$. The results in column 3 were obtained for trajectories where the product scattering angle was less than 1.09° c.m. See the text for further details.

[b] The notation $1/2 \to 3/2$, $0 \to 1$ identifies the process $Ar^+(^2P_{1/2}) + N_2(v = 0) \to Ar^+(^2P_{3/2}) + N_2(v' = 1)$.

[c] Reference 149.

[d] Cross sections for vibrational excitation. The calculated crosss sections represent a weighted average for the two reactant states $Ar^+(^2P_{3/2}) + N_2(v = 0)$ and $Ar^+(^2P_{1/2}) + N_2(v = 0)$. Experiments cannot distinguish the two processes.

TABLE X

Selected Energy Levels for the $(Ar + CO)^+$ System^a

State	Energy (eV)
$Ar + CO^+(A, v = 7)$	2.040
$Ar + CO^+(A, v = 0)$	0.778
$Ar^+(^2P_{1/2}) + CO(v = 0)$	0.178
$Ar^+(^2P_{3/2}) + CO(v = 0)$	0.000
$Ar + CO^+(X, v = 7)$	0.046
$Ar + CO^+(X, v = 0)$	−1.745

[a] Energies defined relative to $Ar^+(^2P_{3/2}) + CO$ $(v = 0)$. The data are taken from Ref. 55.

small.[151-153] The cross section grows as the collision energy increases, but it has only reached 3 Å^2 at 200 eV.[39,128]

Gislason et al.[55] have carried out calculations for the CT processes

$$Ar^+(^2P_{3/2}, ^2P_{1/2}) + CO(v = 0) \rightarrow Ar + CO^+(X, A; v'). \qquad (59)$$

A considerable amount of experimental data are available for comparison. In addition to the cross section measurements,[39,128,151-153] information about the vibrational distribution of $CO^+(X, v')$ at low collision energies has been obtained by Marx and co-workers[154-156] and by Leone and co-workers.[157,158] In addition, Kato et al.[132] and Lindsay and Latimer[134] have measured the ratio $\sigma(1/2)/\sigma(3/2)$ of the CT cross sections for reaction (59).

2. Theoretical Procedures

The theoretical procedures used by Gislason et al.[55] were very similar to those of Parlant and Gislason.[52] The one difference was that the molecular orientation γ (see Fig. 1) was no longer assumed to be fixed. Rather, the sudden approximation for molecular rotation, described in Section II F, was made. Thus, the molecular orientation was fixed in space for each trajectory. The trajectory itself was propagated along the reactant (diabatic) potential energy surface. The results were averaged over all molecular orientations, using Monte Carlo sampling, to obtain the state-to-state cross sections $\sigma_{n'n}$, following the procedure in Eq. (41). To be certain that the calculations were vibrationally converged it was necessary to use 15 levels for each electronic state.

3. Potential Energy Surfaces and Couplings

The potential energy surfaces and couplings used in the calculations[55] were semiempirical results computed from diabatic electronic wavefunctions using the procedure of Archirel and Levy.[19] The calculations of the surfaces are described in the paper by Parlant et al.[59] The spin–orbit splitting of Ar^+ was properly taken into account. As in the case of $(Ar + N_2)^+$, the quantum number Ω, defined in Eq. (49), is a good quantum number for $(Ar + CO)^+$. Consequently, transitions between $\Omega = \frac{1}{2}$ and $\Omega = \frac{3}{2}$ states are not allowed. The Franck–Condon approximation was used to evaluate the vibronic matrix elements.

The five-dimensional electronic hamiltonian matrices can be diagonalized to obtain the adiabatic potential energy surfaces.[59] The results for $\Omega = \frac{1}{2}$ for fixed values of the molecular orientation angle β (this angle is denoted γ in the rest of this review) are shown plotted against R in Fig. 21. (The results for $\Omega = \frac{3}{2}$[59] are less interesting, since the couplings between diabatic states are much weaker.) The vibrational coordinate was fixed at $r = r_e$

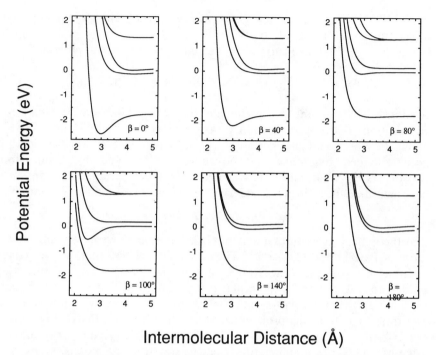

Intermolecular Distance (Å)

Figure 21. Potential energy curves for the five lowest electronic states of $(Ar + CO)^+$ for $\Omega = \frac{1}{2}$, plotted against the distance R between Ar and the center of mass of CO for fixed values of the orientation angle β. Linear Ar–C–O corresponds to $\beta = 0°$. The CO bond length is 1.128 Å. At $R = \infty$ the states correspond, in order of increasing energy, to $Ar + CO^+(X)$, $Ar^+(^2P_{3/2}) + CO$, $Ar^+(^2P_{1/2}) + CO$, and $Ar + CO^+(A)$, which is doubly degenerate. The zero of energy corresponds to $Ar^+(^2P_{3/2}) + CO$ at infinite separation. (Taken from Ref. 59, with permission.)

$(CO) = 1.128$ Å. For the case $\Omega = \frac{1}{2}$ there is strong electronic coupling between the Ar^+–CO states and the two Ar–CO^+ states. As discussed by Gislason and Ferguson,[159] this coupling can create a considerable bond energy in the ground state, which correlates at infinite separation to $Ar + CO^+(X)$. In addition, Baker and Buckingham[160] have shown that the center of charge in $CO^+(X)$ resides at the C end of the molecule. These considerations explain why the complex has a deep well (0.82 eV) as Ar approaches $CO^+(X)$ at the carbon end. By comparison, the electronic coupling is much smaller when Ar moves near the oxygen end of $CO^+(X)$, and the computed well depth is only 0.03 eV. The experimental measurement of the Ar–CO^+ well depth gave 0.70 ± 0.06 eV,[161] which is in reasonable agreement with the theoretical value.

The first excited electronic state dissociates to $Ar^+(^2P_{3/2}) + CO$. The calculations of Parlant et al.[59] show that this state is weakly coupled to $Ar + CO^+(X)$ for $\beta \geqslant 80°$, whereas the coupling to $Ar + CO^+(A)$ peaks near $\beta = 90°$. As a consequence, the potential energy surface for this excited state has a substantial well at $\beta = 100°$. The minimum occurs at $R = 2.51$ Å, where the well depth is 0.53 eV. It is remarkable that the two lowest surfaces of the same symmetry actually have deep wells. For comparison the isoelectronic system $(Ar + N_2)^+$ does not have a well on the first excited potential energy surface (see Fig. 12). There is no direct experimental measurement of the well depth of the excited state of $(Ar + CO)^+$, since any of these ions would be formed in the continuum of the electronic ground state, and they would predissociate to give $Ar + CO^+(X, v')$. However, Norwood et al.[161] were able to obtain a rough estimate of 0.66–0.97 eV for the bond energy. This is in fair agreement with the calculated value.

As discussed by Gislason and Ferguson,[159] ion–molecule potential energy surfaces are expected to be quite anisotropic due to the orientation dependence of the electronic coupling terms $H_{\alpha\alpha'}$. The variation of the potential energy curves with angle shown in Fig. 21 confirms that this is the case for $(Ar + CO)^+$. This anisotropy can be seen even more clearly in the contour map[59] for the ground adiabatic potential energy surface shown in Fig. 22. The surface is dominated by the deep well at the carbon end of CO

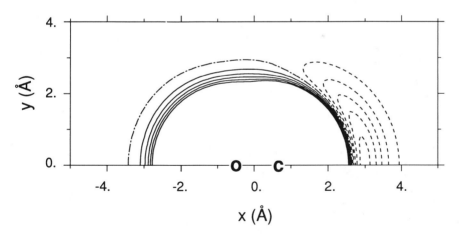

Figure 22. Potential energy contour map for the ground electronic state of $(Ar + CO)^+$ for $\Omega = \frac{1}{2}$. The CO molecule is fixed as shown ($r = 1.128$ Å); the contours indicate the potential energy when Ar is located at a particular point. The zero of energy corresponds to $Ar + CO^+(X)$ at infinite separation; the dashed-dot curve also shows $E = 0$. The dashed curves indicate negative energies, running from $E = -0.15$ eV to -0.75 eV in steps of 0.15 eV. The solid curves indicate positive energies, running from 0.15 eV to $E = 0.75$ eV in steps of 0.15 eV. (Taken from Ref. 59, with permission.)

which extends out to $\beta = 60°$. At all other orientations the interaction is essentially repulsive. This strong angular dependence of the electronic couplings plays an important role in the CT process, discussed below.

4. Calculations for $Ar^+(^2P_J) + CO$ Collisions

State-to-state cross sections have been calculated for collisions of $Ar^+(^2P_J)$ with CO ($v = 0$) at a relative energy of 2 eV. The results are summarized in Table XI.[55] Consider first the fine-structure transitions. The excitation cross

TABLE XI
Calculated Total Cross Sections for $Ar^+(^2P_J) + CO$ ($v = 0$) Collisions[a]

Reactant State	Product State	Total Cross Section (Å^2)
Charge transfer		
$Ar^+(^2P_{3/2})$	$CO^+(X)$	0.64 ± 0.09
	$CO^+(A)$	0.04 ± 0.005
	total CO^+	0.68 ± 0.09
		$(< 0.01)^b$
		0.62^c
$Ar^+(^2P_{1/2})$	$CO^+(X)$	0.20 ± 0.02
	$CO^+(A)$	1.63 ± 0.25
	total CO^+	1.83 ± 0.25
		$(< 0.01)^b$
Fine-structure transitions		
$Ar^+(^2P_{3/2})$	$Ar^+(^2P_{1/2})$	0.33 ± 0.03
		$(0.41 \pm 0.03)^b$
$Ar^+(^2P_{1/2})$	$Ar^+(^2P_{3/2})$	6.29 ± 0.30
		$(1.03 \pm 0.17)^b$
Vibrational excitation		
$Ar^+(^2P_{3/2})$	$CO(v' > 0)$	0.13 ± 0.01
		$(< 0.01)^b$
$Ar^+(^2P_{1/2})$	$CO(v' > 0)$	0.64 ± 0.07
		$(< 0.01)^b$

[a]Cross sections are calculated for $Ar^+(^2P_{3/2}, {}^2P_{1/2}) + CO$ ($v = 0$) at 2 eV relative energy. The cross sections are summed over all product vibrational levels. For the calculated values of Gislason et al.[55] the 68% confidence limits are shown.

[b]Theoretical cross sections where the $CO^+(A)$ state has been omitted from the calculation (Ref. 55).

[c]Experimental value (Ref. 153).

section for $Ar^+(^2P_{3/2}) + CO$ is only $0.33\,\text{Å}^2$, and the major product vibrational level is CO $(v = 0)$. The small cross section can be explained by the relatively large energy defect $(0.18\,\text{eV})$ for this process. By comparison, the dynamic energy width δE is only $0.054\,\text{eV}$. [This number was obtained from Eq. (55), which should also be appropriate for $(Ar + CO)^+$.] The cross section for the quenching process

$$Ar^+(^2P_{1/2}) + CO(v = 0) \rightarrow Ar^+(^2P_{3/2}) + CO(v') \tag{60}$$

is much larger $(6.3\,\text{Å}^2)$. In fact, this is the major product channel for $Ar^+(^2P_{1/2}) + CO(v = 0)$ collisions. The CO product in Eq. (60) is formed mainly in the level $v' = 1$. This product state is only $0.09\,\text{eV}$ away from the reactant state.

The calculated CT cross section for $Ar^+(^2P_{3/2}) + CO$ is $0.68\,\text{Å}^2$,[55] which compares well with the experimental value of $0.62\,\text{Å}^2$.[153] The computed CT cross section for $Ar^+(^2P_{1/2})$ is considerably larger $(1.63\,\text{Å}^2)$, but there is no direct measurement available for comparison. The ratio $\sigma(1/2)/\sigma(3/2) = 2.4$ is consistent with the experimental data of Lindsay and Latimer,[134] but it is somewhat larger than the measurement by Kato et al.[132] The experience with the system $(Ar + N_2)^+$ suggests that this ratio is very hard to measure (see Section III F 6). This is particularly the case for $(Ar + CO)^+$, because the cross section for reaction (60) is more than three times larger than the CT cross section. The quenching process, which is not observed in the CT experiments, converts $Ar^+(^2P_{1/2})$ ions to $Ar^+(^2P_{3/2})$ in the beam. This has the effect of lowering the CT ratio $\sigma(1/2)/\sigma(3/2)$. It would be useful to have additional experiments done on this reaction.

One of the most interesting predictions from the calculations[55] is that for CT collisions of $Ar^+(^2P_{3/2}) + CO$, more than 90% of the products appear as $CO^+(X)$, whereas for collisions of $Ar^+(^2P_{1/2}) + CO$, 85% of the CT products appear as $CO^+(A)$. This occurs even though the spin–orbit states of Ar^+ are separated by only $0.18\,\text{eV}$ (see Table X), whereas the two electronic states of CO^+ differ by $2.5\,\text{eV}$. This remarkable result is made more plausible by an inspection of the potential energy curves for $(Ar + CO)^+$ shown in Fig. 21. The curve that goes asymptotically to $Ar^+(^2P_{1/2}) + CO$ comes very close to the two higher curves at small R when $\beta \approx 0°$. By comparison the $Ar^+(^2P_{3/2}) + CO$ curve approaches the ground state curve in the angular region around $\beta = 100°$. It would be very interesting to test this prediction experimentally.

The authors[55] also carried out a second set of calculations but with the $CO^+(A)$ state omitted. The results are shown in parentheses in Table XI. Under these conditions CT and vibrational excitation do not occur. This is particularly interesting for collisions of $Ar^+(^2P_{3/2}) + CO$, where CT with the

A state in the calculations gives mainly $CO^+(X)$ products. When the A state is removed, no CT occurs. It is clear that any theoretical treatment of this system which omits the $CO^+(A)$ state will be seriously incomplete. Gislason et al.[55] believe that the presence of the $CO^+(A)$ state, whose r_e value is shifted relative to $r_e(CO)$, allows the molecule to vibrate during the collision, thereby allowing the system to overcome the very poor Franck–Condon overlaps with the product states which are close in energy to the reactant state. A similar conclusion was drawn by Hamilton et al.[158]

The various theoretical results for $(Ar + CO)^+$ show how quantum mechanical these collisions are at low energy. The cross section for reaction (60) is large, proving that the $Ar^+(^2P_{1/2}) + CO$ and $Ar^+(^2P_{3/2}) + CO$ states are strongly coupled during the collision. Nevertheless, CT from these two reactant states give completely different product states. At first glance, the two results contradict each other. They can occur only because the system's wavefunction retains a strong coherence throughout the collision. Thus, even though the two spin–orbit states of Ar^+ are strongly mixed, the wavefunction remembers which state was the reactant state.

The calculated vibrational distribution[55] for the $CO^+(X)$ products produced in CT are shown in Fig. 23. By comparison, the Franck–Condon factors for the transition $CO(X, v = 0) \rightarrow CO^+(X, v')$ are 0.964 and 0.036 for $v' = 0$ and 1, respectively. The factor for $v' = 2$ is 1×10^{-4}, and all of the others are much smaller. The distribution for $CO^+(X)$ produced from $Ar^+(^2P_{3/2})$ peaks at $v' = 1$ and 2, which represents a compromise between energy resonance and Franck–Condon considerations. Two types of experiments have been carried out to determine this distribution for collisions at thermal or near-thermal energies. The ICR experiments of Marx and co-workers[154,155] indicate that almost all of the ions are produced in $v' = 4$. The laser-induced fluorescence experiments of Leone,[157,158] however, show a broader distribution which peaks at $v' = 5$. This distribution is also shown in Fig. 23.

Both experimental distributions peak at higher vibrational levels than does the theoretical calculation.[55] The most likely explanation of this difference is that the mechanism for CT is different at thermal energies where the experiments were carried out. This is certainly suggested by the behavior of the rate constant for CT, which rises rapidly as the collision energy decreases.[153] This suggests that CT at low energy involves the formation of a long-lived $(ArCO)^+$ complex. The long life of the complex would allow highly forbidden but nearly energy-resonant processes such as $Ar^+(^2P_{3/2}) + CO(v = 0) \rightarrow Ar + CO^+(v' = 5)$ to occur. A similar mechanism has been invoked by Ferguson[162] to explain vibrational quenching of small molecular ions. The failure of Norwood et al.[161] to observe $(ArCO)^+$ ions in the first excited electronic state also supports this complex mechanism. By contrast,

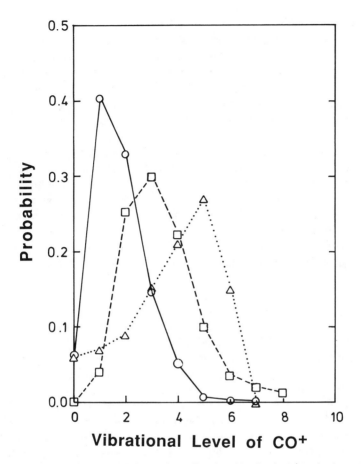

Figure 23.   Probability of populating various vibrational levels of $CO^+(X)$ in charge transfer collisions of $Ar^+(^2P_J) + CO(v = 0)$. The circles and squares are the calculated values of Gislason et al.[55] at $E_{rel} = 2\,eV$ for $Ar^+(^2P_{3/2})$ and $Ar^+(^2P_{1/2})$, respectively. The triangles are the experimental results at thermal energy of Hamilton et al.[158]. (Taken from Ref. 55, with permission.)

no complex could be formed in collisions at $E_{rel} = 2\,eV$ where the calculations[55] were carried out. The collision complex mechanism is discussed further in Section V B.

H.   Work of Sidis and Co-workers on Dissociative Charge Transfer

In this chapter we have seen that the theory of atom–diatom CT processes using the classical path method is well developed. The calculations typically involve a moderate number of vibronic states (25–75), and it is not difficult to obtain exact, vibrationally converged solutions to the scattering problem. The classical path method becomes less satisfactory for systems where chemical reaction or molecular dissociation is occurring, since it is not clear how to treat the vibrational continuum. Nevertheless, some progress has been made on reactive systems by making certain approximations.[61]

In principle, the question of dissociative CT is even more difficult, since there is no way to avoid including the vibrational continuum in the classical path procedure. In spite of this difficulty, very good progress has been made on this problem by Sidis and co-workers. This work has been reviewed in detail recently,[12,17,64,67] and we shall only briefly discuss their work here.

The first theoretical approach used by Sidis and DeBruijn[65] was the vibrationally sudden approximation. This approximation, which assumes that all nuclear motion is frozen during the collision, should work well for collisions in the kilovolt energy range. The sudden approximation for molecular vibration is summarized in Section II D. The fundamental result for the transition probability from state I to state F is given by [see Eq. (23)]

$$P_{IF} = |\langle \chi_F | A(\hat{r}, r) | \chi_I \rangle|^2. \tag{61}$$

Here $|\chi_I\rangle$ and $|\chi_F\rangle$ are the initial and final nuclear wavefunctions, and A is the transition amplitude obtained by solving the electronic equations of motion for fixed nuclear orientation $\hat{r}$ and fixed bond length r. The vibrational part of the final wavefunction can be approximated by an Airy function or, more simply, by using the reflection approximation (or δ-function approximation).[65,163] In addition, if the CT process is close to energy resonant, the Demkov model can be used to solve for the amplitude A.[164] Sidis and De Bruijn[65] have applied the vibrationally sudden approximation to dissociative CT collisions of $H_2^+ + Mg$. The results for the differential cross section as a function of the kinetic energy of the H atom fragments are shown in Fig. 24. The experimental results[165] are shown as a dashed curve. The theoretical curve labeled "b" gives the cross section for the direct dissociative CT process shown in reaction (4). The other important process that contributes to the experimental cross section is CT to produce the bound $a\,^3\Sigma_g^+$ state, which then radiates to the $b\,^3\Sigma_u^+$ state and dissociates. The calculated cross section for that process is shown as curve "a" in Fig. 24. The agreement between theory and experiment is quite good at all energies and is excellent above 4 eV.

In the collision energy range of 1–100 eV, it is no longer appropriate to make the vibrationally sudden approximation. At these energies the collision time is sufficiently long that the molecule can vibrate and, in fact, at least partially dissociate during the collision. The process of dissociative CT at these energies has been treated by Sidis and Courbin–Gaussorgues[66] using

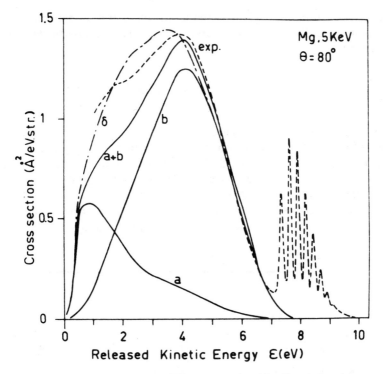

Figure 24. Kinetic energy spectrum of H atoms produced by dissociative charge transfer in collisions of H_2^+ + Mg at a lab energy of 5 keV. The two H atoms are scattered at an angle of 80° with respect to the incident direction. The dashed curve is the experimental curve.[165] The theoretical results of Sidis and DeBruijn[65] are shown as solid and dashed-dot curves. The former used the Airy function approximation for the vibrational continuum wavefunction, and the latter was done with a δ-function approximation for that wavefunction. Curves a and b are described in the text. The vibrational energy distribution of the H_2^+ beam is that given in Ref. 166. (Taken from Ref. 65, with permission.)

a local complex potential approximation. The physical picture treated in their paper is that of a bound reactant vibrational state embedded in the dissociative continuum of the products. As the H_2^+ approaches Mg, the electronic coupling induces an interaction with the continuum, which can be characterized by an energy width given by

$$\Gamma_v(R, \gamma) = 2\pi |\langle F | H_{12}(R, \gamma, r) | G_v \rangle|^2. \tag{62}$$

Here G_v and F are the initial bound vibrational wavefunction of $H_2^+(v)$ and the continuum wavefunction representing the molecular dissociation of H_2, respectively. In addition, H_{12} is the electronic coupling between the initial and final states. The survival probability at time t along a particular trajectory is then given by

$$P_v(t) = \exp\left(- \int_{-\infty}^{t} \Gamma_v(R, \gamma) dt' \right). \tag{63}$$

Sidis and Courbin–Gaussorgues[66] carried out a number of model calculations for $H_2^+ (v) + Mg$ collisions using this formalism. A typical example of their results is shown in Fig. 25. This shows the kinetic energy spectrum of the H atoms produced in dissociative CT for a collision with a large impact parameter (15 a.u.). The calculations were carried out at $E_{lab} = 1$, 10, and 100 eV. The exchange interaction was assumed to be isotropic, and the energy difference between bound and continuum states was assumed to be inde-

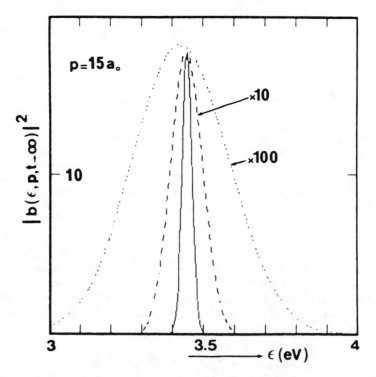

Figure 25. Kinetic energy spectrum of H atoms produced by dissociative charge transfer for a ($H_2^+ + Mg$)-like prototype collision with impact parameter 15 a.u. for $E_{lab} = 1$ eV (solid curve), 10 eV (dashed curve), and 100 eV (dotted curve). (Taken from Ref. 66, with permission.)

pendent of R. At low energies the spectrum is very sharp, but it broadens with increasing collision energy. At this large impact parameter the system should be in the weak-coupling limit; in this case the electronic equations of motion can be solved analytically. The model predicts that the peak height should scale as E_{lab}^{-1}, and the width of the spectrum should increase as $E_{lab}^{1/2}$. Both predictions are shown to be satisfied in Fig. 25. The authors also observed a number of other interesting effects when the assumption of an isotropic exchange interaction was dropped, and when the energy difference was assumed to vary with R.

In dissociative CT there are many vibrational states of the reactants embedded in the continuum. Sidis et al.[67] have treated the case of a few vibrational states of H_2^+ interacting with the same dissociative continuum but not otherwise coupled to each other in a collision of $H_2^+ + Mg$. Vibrational excitation in this case results via the coupling of each state with the continuum. A sample calculation at $E_{lab} = 100\,eV$ for a $(H_2^+ + Mg)$-like prototype collision is shown in Fig. 26. The model parameters are the same as in Fig. 25, except that the exchange interaction has been multiplied by 0.75. At large impact parameters neither dissociative CT nor vibrational

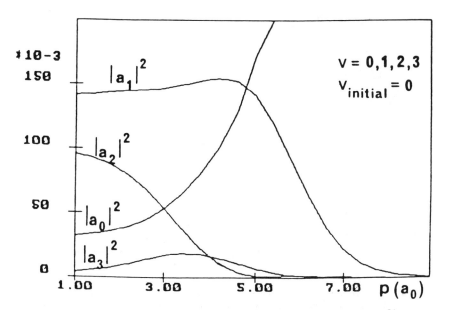

Figure 26. Survival $|a_0|^2$ and vibrational excitation $|a_v|^2$ ($v = 1, 2, 3$) probabilities for a $(H_2^+ + Mg)$-like prototype collision plotted against impact parameter at a collision energy $E_{lab} = 100\,eV$. The H_2^+ was initially in $v = 0$. (Taken from Ref. 67, with permission.)

excitation occurs. At smaller values of b, CT becomes the major product channel, but at the same time vibrational excitation becomes important. For the system considered in Fig. 26 both $v' = 1$ and 2 are more populated than $v' = 0$ for impact parameters less than 3 a.u.

Although the local complex interaction method has proven to work very well, it is not clear that this procedure can be used to treat a case where the interaction couplings are very large. Because of this Gauyacq and Sidis[68] have developed a more general procedure to treat dissociative CT. They refer to this procedure as the "coupled wave packets" description. The calculations are carried out by expanding the vibrational wavefunction as a sum of wave packets. The wave packets are then propagated in time using a finite difference method. In principle, their procedure is exact. Calculations for an $(H_2^+ + Mg)$-like prototype dissociative CT system are shown in Fig. 27 for a collision at $E_{lab} = 20\,eV$.[68] This shows the survival probability of the initial $v = 0$ level, $|a_0|^2$, as well as the total vibrational excitation probability plotted against impact parameter. Also shown is the calculation of $|a_0|^2$ for the same system using the local complex potential method.[66] The coupling for this system is large at small values of R, and it is seen that the local complex

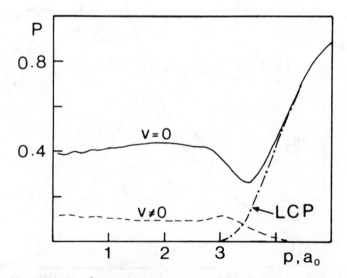

Figure 27. **Survival** $|a_0|^2$ and total vibrational excitation probabilities at $E_{lab} = 20\,eV$ for an $(H_2^+ + Mg)$-like prototype collision plotted against impact parameter. The calculations using the coupled wave packets procedure are shown as solid and dashed curves.[68] The survival probability using the local complex potential approximation[66] is shown as a dashed-dot curve. (Taken from Ref. 68, with permission.)

potential procedure does not work well under these circumstances. To date, Gauyacq and Sidis have only published a few model calculations using the method of coupled wave packets, but it is clear that this procedure will be widely used in the future. There are obvious applications to the process of collision-induced dissociation, and Sidis[64] also expects the method will work well on the more conventional CT problem dissussed in detail in this review. In addition, it should be possible to extend the coupled wave packets formalism to larger systems.

IV. CLASSICAL PATH CALCULATIONS—DIFFERENTIAL CROSS SECTIONS

The calculation of differential cross sections for both CT and nonreactive scattering presents a more severe test for the theory than total cross sections. A considerable number of differential cross sections have been measured for ion–molecule systems.[2] This work is reviewed in the chapters by Futrell[167] and by Niedner–Schatteburg and Toennies[168] in the companion volume. Only a small number of theoretical calculations using the classical path method have been carried out.

A. Work of Grimbert, Sidis and Sizun

To date there is only one theoretical calculation done in the semiclassical framework which can be compared directly to experiment. This is the work of Sizun et al.[58]. The procedure they followed is described in Section II. They used the classical path method to determine vibrational excitation and charge-transfer differential cross sections for the reactions

$$H^+ + O_2(X\ ^3\Sigma_g^-, v = 0) \rightarrow H^+ + O_2(X\ ^3\Sigma_g^-, v'), \qquad \text{(VE)}$$

$$\rightarrow H + O_2^+(X\ ^2\Pi_g, v'). \qquad \text{(CT)}$$

Experimental vibrationally state-resolved measurements of the differential cross sections $\sigma_{n',n}(\theta)$ are available at $E_{rel} = 23\ \text{eV}$.[169] The experimental results show for the two channels a decrease with scattering angle which is rather steep in the region out to $\theta = 7°$. The cross sections become somewhat flatter at higher scattering angles. A weak structure at about $10–13°$ corresponds to a rainbow maximum. The total CT curve reaches a plateau near $\theta = 0°$. At $\theta = 0°$ the major product state is $v' = 3$. This result should be compared to the Franck–Condon distribution, which peaks at $v' = 1$, and resonant CT, which will give $v' = 6$.

An unique aspect of the HO_2^+ collisional system lies in the availability of potential energy surfaces and exchange interactions which are functions of the three internal coordinates (R, r, γ). Grimbert et al.[60] provided a grid of points from a model potential which can be fit by splines and introduced into the dynamical calculation. The two lowest adiabatic $^3A''$ potentials are involved in the collision; they have an avoided crossing at about $R \approx 4.7$ a.u. when $r = r_e(O_2)$ and $\gamma = 45°$. This crossing facilitates CT. The ground state has a potential well of about 2.31 eV at $\gamma = 45°$. The depth varies somewhat with the geometry of the molecule. This well gives rise to the rainbow scattering observed experimentally in the two channels. Grimbert et al.[60] also show the dependence on r of the potential energy surfaces and the couplings. The diabatic state $H^+ + O_2$ undergoes bond elongation of O_2 when R decreases to about 3 a.u. In addition, the coupling between the two electronic states increases when the O_2 bond length decreases between 2.77 and 1.87 a.u. for $\gamma = 45°$ by almost a factor of 2. Contrary to other CT studies these dependences were taken into account in the dynamical calculation. Thus their effects on the dynamics can be examined.

The evolution of the system's vibronic wavefunction along the classical trajectory, chosen to move on the diabatic entrance channel potential, is described using an expansion over two asymptotic vibrational basis sets (15 states each) associated with $O_2(^3\Sigma_g^-)$ and $O_2^+(^2\Pi_g)$. The calculations were done as discussed in Section II F, using Eq. (18) for the coupled equations, Eq. (41) to evaluate the total state-to-state cross sections, and Eqs. (38) and (40) to compute differential vibronic cross sections. The authors[22c] also did quantum IOS calculations (QIOS) to study the same reaction, and this study allows a comparison between the two methods.

There have been no measurements to date of the total state-to-state cross sections to compare to the theoretical results. Figure 2 shows two state-to-state probability functions $P_{n',n}(b, \theta_r, \phi_r)$ computed for a fixed molecular orientation. These results were discussed earlier. Figure 28 presents the differential cross sections summed over product vibration levels for CT and vibrational excitation (VE). They are compared with earlier calculations and experimental results. A gratifying feature of Fig. 28 is the agreement in the overall shapes and magnitudes among the two theoretical and the experimental cross sections. It is not possible to determine which theoretical method gives the better agreement with experiment. Both work quite well. Nevertheless, several differences with the experiments are noticeable, namely, (1) the rainbow structure in the VE curve extends over a broader angular range than in the experimental data; (2) the classical path VE and CT rainbows occur at smaller angles than do the experimental and QIOS values; the differences are nearly 1° and 2°, respectively.

The relative state-to-state probabilities for CT processes are displayed in Fig. 29 as a function of scattering angle. These are calculated from the

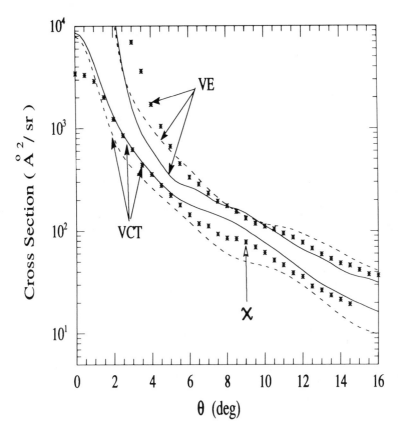

Figure 28. Differential cross sections for vibrational excitation (VE) and charge transfer (VCT) plotted against the c.m. scattering angle for collisions of $H^+ + O_2(v = 0)$ at $E_{rel} = 23\,eV$. The cross sections in each case have been summed over all final vibrational levels. The solid curves show the classical path calculations of Sizun et al.,[58] and the dashed curves give the quantal results of Sidis et al.[22c] The experimental data of Noll and Toennies[169] are shown as asterisks. The experimental VE curve has been calibrated to the theoretical result at $\theta = 8°$. In principle, the experimental VCT curve is also set by this calibration, but the mark of caution (χ) indicates that the estimated 1% neutral detection efficiency[169] may be in error by a factor of 2, which would then shift the VCT curve up from what is shown. (Taken from Ref. 58a, with permission.)

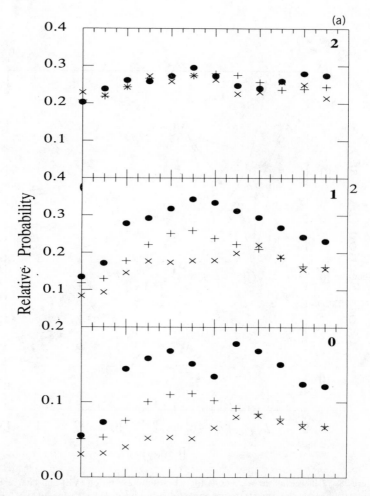

Figure 29. Relative state-to-state probabilities for charge transfer, defined in Eq. (64), plotted against the the c.m. scattering angle for collisions of $H^+ + O_2(v = 0)$ at $E_{rel} = 23$ eV. The "plus" signs show the classical path calculations from Ref. 58a, the x's are the quantal calculations from Ref. 22c, and the circles are the experimental data from Ref. 169. The final O_2^+ vibrational level is indicated in each panel. (Taken from Ref. 58a, with permission.)

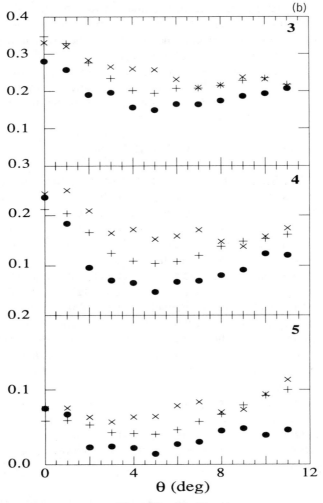

Figure 29.   *(Continued)*

expression

$$P_{v'}(\theta) = \sigma_{v'}(\theta) \bigg/ \sum_{v'} \sigma_{v'}(\theta), \tag{64}$$

where only CT states are included in the sum. The classical path results are generally closer to the experimental data than the QIOS values are. Nevertheless, the two theories underestimate charge transfer into low

$(v' = 0, 1)$ vibrational states and overestimate it for high $(v' = 4, 5)$ states. It is possible that this disagreement could arise from the experimental detection efficiency of low-energy neutrals, since the lower the O_2^+ vibrational level, the more energetic the scattered H atom will be.

Sizun et al.[58] have only reported preliminary results so far, and full papers are being prepared. These will provide more detailed analyses of the role played by the angle γ in the QIOS approach and of the influence of the r dependence on the CT and VE processes.

B. Work of DePristo

As explained in Section II F, Cole and DePristo[47] carried out calculations of the state-to-state differential cross sections for CT processes for H_2^+ on H_2 using the SCECT method. From the average scattering angle obtained from each energy-conserving trajectory, the average kinetic energy, and the state-to-state probabilities, they used a procedure to extract a different scattering angle for each vibronic channel. They assume that if a channel gives more kinetic energy than the average, its scattering angle will be smaller than the average angle. It is then possible to compute the classical differential cross section for each vibronic channel using Eq. (44).

The main result for the H_2^+/H_2 system is strong forward scattering at $E_{rel} = 8$ and $16 \, eV$ with the range of scattering angles decreasing for the higher reactant vibrational states. A more general conclusion is that exoergic and slightly endoergic product channels will scatter at smaller angles than the average angle. This result is due to the fact that the calculation gives an average internal energy for the products, which is bigger than the reactant energy. Strongly endoergic products appear at larger angles.

Unfortunately, no experimental results exist for the H_2^+/H_2 state-to-state angular distributions. Thus, no test of the assumptions made by Cole and DePristo[47] is possible.

V. GENERAL FEATURES OF CHARGE-TRANSFER COLLISIONS

The considerable theoretical work on CT processes reviewed in this chapter have led to certain general conclusions about the CT process which are the subject of this section. These include the range of validity of the Franck–Condon principle, the importance of adiabatic vibronic potential energy surfaces for describing low-energy collisions, and a general theory of CT collisions as a function of collision energy. These three topics are discussed in this section.

A.   The Franck–Condon Principle

The Franck–Condon principle (FCP) for nonadiabatic transitions is based upon the (questionable) assumption that electronic transitions occur so rapidly that the molecular vibrational coordinate does not have time to adjust. In that case, the state-to-state cross section for a process such as reaction (50) is given by

$$\sigma(v, v') = \sigma_0 |\langle v | v' \rangle|^2. \tag{51}$$

This equation was given earlier, where it was noted that σ_0 is independent of v and v' and that the last factor is the Franck–Condon factor for the transition. If $\sigma(v, v')$ is summed over all product levels to obtain the state-selected cross section $\sigma(v)$, Eq. (52) shows that $\sigma(v) = \sigma_0$, independent of the reactant vibrational quantum number v.

The experimental and theoretical work for the $(Ar + N_2)^+$ system shows clearly that the FCP is not valid for collisions at low energy, but it does appear to be valid at higher energies. Sidis and co-workers[17,170] have discussed the necessary conditions for the FCP to be valid. The first condition is that the electronic transition take place in such a short time that the vibrationally sudden approximation is valid. This normally means that the collision time must be much shorter than a vibrational period. In this case, the probability amplitude for a particular vibronic transition can be written as shown in Eq. (23). The second condition is that the electronic coupling(s) which induces the transition must be a weak function of the vibrational coordinate r. In this case, Eq. (24) becomes valid. Finally, there should be no vibrational excitation or deexcitation prior to or after the electronic transition occurs. This excludes both impulsive excitation[171] as well as any excitation due to long-range interactions. (However, the effect of any long-range interaction will diminish at high collision velocities[172].) We shall assume that the second and third conditions are always satisfied in what follows. Thus Eq. (51) is valid at sufficiently high collision energy.

To examine the approach of a system to Franck–Condon behavior it is useful to consider a model calculation carried out by Spalburg et al.[51] They treated a system with two electronic states whose diabatic potential energy surfaces are parallel. If the electronic coupling is given by the expression in Eq. (53), then the weak-coupling approximation gives the following result for the state-to-state cross sections:

$$\sigma_w(v, v') = \sigma^* |\langle v | v' \rangle|^2 [1 + (\Delta E / a\hbar g)^2]^{-3}, \tag{65a}$$

$$\sigma^* = (16\pi/3) A^2 / a^2 (a\hbar g)^2. \tag{65b}$$

Here g is the initial relative velocity of the collision, and ΔE is the energy difference between the reactant state (identified by vibrational level v) and the product state (identified by v'). This result was used to estimate the energy width δE given in Eq. (54). For this system the FCP is not, in general, valid because ΔE depends on the vibrational levels v and v'. However, it is apparent that at higher energy, where

$$g \gg \Delta E / a\hbar$$

is satisfied, the state-to-state cross section reduces to

$$\sigma_w = \sigma^* |\langle v | v' \rangle|^2,$$

which is equivalent to Eq. (51). The weak coupling approximation is normally not satisfied in ion–molecule CT systems. Nevertheless, Eq. (65) does give a good qualitative description of the cross section behavior. From these considerations Spalburg et al.[51] concluded that the FCP will be satisfied at high collision energies, but at low energies states with small ΔE values will have the largest cross sections.

To test these predictions Gislason and Parlant[53,54] carried out a series of calculations on the CT process

$$\text{Ar}^+(^2P_{3/2}) + \text{N}_2(v=0) \rightarrow \text{Ar} + \text{N}_2^+(X, A; v') \qquad (66)$$

over a wide range of collision energies. The computational procedures were the same as those described in Section III F. The reader is referred to Table IV for a listing of the system's vibronic energies. Table XII gives the product vibrational distributions for the X and A states at the various collision energies; these can be compared with the Franck–Condon factors in the last column. At the lowest energy studied ($E_{\text{rel}} = 1 \, \text{eV}$) only the near-resonant $\text{N}_2^+(X, v' = 1)$ state is formed. By $E_{\text{rel}} = 10 \, \text{eV}$ small amounts of $v' = 0$ and 2 are produced as well. The state favored by Franck–Condon considerations, $\text{N}_2^+(X, v' = 0)$, becomes the major product at $E_{\text{rel}} = 100 \, \text{eV}$, even though the distribution is far from Franck–Condon. At this energy a small amount of $\text{N}_2^+(A)$ products is formed as well. The vibrational distribution favors states such as $v' = 0$ and 1 which are closest in energy to the reactant state. It is interesting that $\text{N}_2^+(X, v' = 3 \text{ and } 4)$ are not formed, even though they are closer in energy to the reactants than any $\text{N}_2^+(A)$ state. By $E_{\text{rel}} = 1000 \, \text{eV}$ the distribution of vibrational levels for $\text{N}_2^+(X)$ is in agreement with the FCP, but the $\text{N}_2^+(A)$ state is still skewed toward the lower vibrational levels. This effect persists at $E_{\text{rel}} = 4000 \, \text{eV}$, although the distribution for $\text{N}_2^+(A)$ is closer to Franck–Condon.

Table XII

Relative State-to-State Cross Sections for $Ar^+(2P_{3/2}) + N_2(v=0) \rightarrow Ar + N_2^+(X, A; v')^a$

N_2^+ Product Ion State		Relative Collision energy (eV)					Franck–Condon Factor
		1	10	100	1000	4000	
X	0		0.060	0.690	0.910	0.908	0.916
X	1	1.000	0.935	0.296	0.086	0.092	0.079
X	2		0.005	0.013	0.004		0.005
X	3						0.0003
A	0			0.706	0.671	0.391	0.264
A	1			0.294	0.264	0.347	0.319
A	2				0.064	0.174	0.220
A	3					0.066	0.115
A	4					0.022	0.051
A	5						0.020
$\sigma(A)/\sigma(X)$		0.000	0.000	0.046	0.154	0.798	

^{a}The first two columns identify the electronic state (X or A) and the vibrational level of the N_2^+ product. The next five columns give, at each collision energy, the fraction of the products in each vibrational level in the X state (first four rows) and in the A state (next six rows). These fractions can be compared directly with the Franck–Condon factors for the transition $N_2(v=0) \rightarrow N_2^+(X, A; v')$ in the last column. If no entry is given, the calculated cross section was less than 0.01 Å^2. The last row gives the ratio of the total cross sections for producing the A state compared to producing the X state.

A comparison of the $N_2^+(X)$ and $N_2^+(A)$ vibrational distributions at $E_{rel} = 1000 \text{ eV}$ leads to a surprising conclusion. For the model used to obtain Eq. (65) the collision time can be estimated as

$$\tau_{coll} = 2\pi/ag. \qquad (67)$$

Naively, one would expect that the FCP will be satisfied when τ_{coll} is much shorter than the vibrational period

$$\tau_{vib} = 1/\nu_{vib}, \qquad (68)$$

where ν_{vib} is the vibrational frequency. Since the vibrational frequencies of $N_2^+(X)$ and $N_2^+(A)$ are nearly the same, one would expect both states to exhibit Franck–Condon behavior at the same collision energy ($E_{rel} = 1000 \text{ eV}$ for $Ar^+ + N_2$). In fact, this is not observed. An examination of Eq. (65) suggests that it is also necessary to consider the electronic time

$$\tau_{el} = h/\Delta E_{el}, \qquad (69)$$

where ΔE_{el} is a typical energy difference between reactant and product electronic states.

A comparison of the three times τ_{el}, τ_{vib}, and τ_{coll} determines the type of collision being considered. In the absence of an avoided crossing between two electronic states of the same symmetry (i.e., ΔE_{el} large) it is normally the case that

$$\tau_{el} \ll \tau_{vib}.$$

In this case, the Born–Oppenheimer approximation is valid. At low collision energies, when $\tau_{vib} \ll \tau_{coll}$, the vibrational motion will be adiabatic. At higher energies, when $\tau_{vib} \approx \tau_{coll}$, vibrational excitation will occur on the single electronic potential energy surface. At very high energy τ_{coll} becomes comparable to τ_{el}, and electronic transitions occur. However, the cross sections will be small, because σ^* in Eq. (65) goes as E_{rel}^{-1}.

Cross sections for nonadiabatic transitions only become large when ΔE_{el} is comparable to a few vibrational quanta. This is the case for reaction (66) giving $N_2^+(X)$. In that case,

$$\tau_{el} \approx \tau_{vib}.$$

Now there are three possible ranges of collision energy. When $\tau_{coll} \approx \tau_{el}$ [$0.1\,eV \leqslant E_{rel} \leqslant 100\,eV$ for reaction (66)], nonadiabatic transitions occur with large cross sections, and the vibrational distribution is not Franck–Condon-like. When $\tau_{coll} \ll \tau_{el}(E_{rel} > 300\,eV)$, the FCP is obeyed. Finally, at very low energies [thermal for reaction (66)], $\tau_{coll} \gg \tau_{el}$, and the cross sections for nonadiabatic transitions should become very small. We shall see, however, that a new mechanism for ion–molecule CT opens up at these low energies. To summarize, the FCP is only valid when both

$$\tau_{coll} \ll \tau_{el}$$

and

$$\tau_{coll} \ll \tau_{vib}$$

are satisfied. Since ΔE_{el} for reaction (66) giving $N_2^+(A)$ is about $1\,eV$, the FCP will not be satisfied until $E_{rel} \gtrsim 40\,keV$.

B. Adiabatic Vibronic Potential Energy Surfaces

At sufficiently low collision energies the dynamic energy width δE, defined in Eqs. (54) and (55), becomes comparable to a vibrational quantum. In that case, calculations show that CT leads to at most a few vibronic product states that are close in energy to the reactant state. Spalburg et al.[51] have argued that in this energy range the best zero-order picture for CT is given in terms of adiabatic vibronic (AV) states. The AV potential energy surfaces can be obtained by diagonalizing the vibronic hamiltonian matrix defined in Eq. (15). There is one surface for each vibrational level of each electronic state. These surfaces have been computed by Spalburg et al.,[51] Sonnenfroh and co-workers,[23,173] and by Parlant et al.[59]. The implications of these surfaces for CT processes have been discussed in detail by these authors. In particular, Sonnenfroh and Leone[173] have explained the rotational distribution of the vibronic products measured for the CT process $Ar^+({}^2P_{3/2}) + N_2(v = 0) \rightarrow Ar + N_2^+(X, v')$ using the AV surfaces. Their qualitative discussion is supported by the quantum-mechanical calculations of Clary and Sonnenfroh.[23] In this section we shall review two examples of AV surfaces and show how they relate to classical path calculations discussed earlier.

The AV potential energy curves for $(Ar + N_2)^+$ with a molecular orientation $\gamma = 90°$ are shown in Fig. 30 for the case $\Omega = \frac{1}{2}$. These were calculated by Parlant et al.[59] using the surfaces of Archirel and Levy.[19] Results at other angles are similar, but the ground-state well is deepest at $\gamma = 0°$. It is instructive to compare the curves in this figure with the electronic potential energy curves shown in Fig. 12. The ground vibronic level in Fig. 30, which correlates at infinite separation to $N_2^+(X, v = 0) + Ar$, does not approach other vibronic levels at any value of R. This explains the very small CT cross section for this state (Fig. 14), even at relative collision energies as high as 20 eV. The behavior of this state is unique, however. All other vibronic curves show one or more avoided crossings where the system can easily hop from one AV curve to another. The end result of these hops as the particles separate can be CT or fine-structure transitions or vibrational excitation or deexcitation. It is apparent that the nature of the asymptotic state is lost at small R where the electronic coupling is strong.

The most dramatic avoided crossing in Fig. 30 is that between the first and second excited vibronic curves, which correlate to $Ar^+({}^2P_{3/2}) + N_2(v = 0)$ and $Ar + N_2^+(X, v = 1)$, respectively. As discussed earlier, these two curves behave very much like a two-state Landau–Zener system,[54] and the state-to-state CT cross section shown in Fig. 17 is well fit by this model at energies below 20 eV. The third excited state, which correlates to $Ar^+({}^2P_{1/2}) + N_2(v = 0)$, is seen to rise as R becomes smaller. This explains why CT from this state at $E_{rel} = 1$ eV gives $Ar + N_2^+(X, v' = 2)$ as the major product (see Fig. 17). This occurs even though $Ar + N_2^+(X, v' = 1)$ is closer in energy (Table IV) and has a larger Franck–Condon overlap with the reactants; this product state is not favored because it lies below the reactants. Another important product state which lies above $Ar^+({}^2P_{1/2}) + N_2(v = 0)$ is

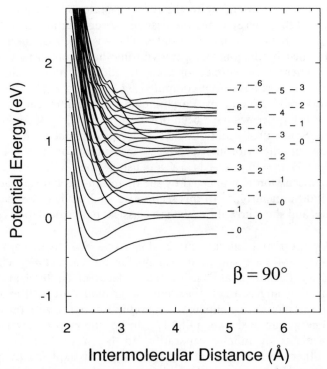

Figure 30. Adiabatic vibronic potential energy curves as a function of R for the $(Ar + N_2)^+$ system with $\Omega = \frac{1}{2}$ for the orientation angle $\beta = 90°$. The zero of energy corresponds to $Ar^+(^2P_{3/2}) + N_2(v = 0)$. The asymptotic states are indicated by the energy limits shown to the right. From left to right the four columns correspond to $Ar + N_2^+(X, v)$, $Ar^+(^2P_{3/2}) + N_2(v)$, $Ar^+(^2P_{1/2}) + N_2(v)$, and $Ar + N_2^+(A, v)$. In each case the vibrational level v is indicated. (Taken from Ref. 59, with permission.)

$Ar^+(^2P_{3/2}) + N_2(v' = 1)$. The avoided crossing between the two curves gives rise to a large fine-structure transition cross section (see the discussion of Fig. 20).

Parlant et al.[59] have also computed AV potential energy surfaces for the $\Omega = \frac{1}{2}$ states of $(Ar + CO)^+$. The results for the orientation angle $\gamma = 100°$ are shown in Fig. 31. Collisions with this orientation play an important role in CT at low collision energies. To interpret the results it is useful to review the electronic curves for $\gamma = 100°$ shown in Fig. 21. The lowest electronic curve is purely repulsive, while the first excited state, which correlates to $Ar^+(^2P_{3/2}) + CO(X)$, gives rise to a curve with a deep well which approaches the ground curve at R values less than 2.8 Å. Returning to Fig. 31 it is apparent that there is a set of *diabatic* vibronic curves based upon the repulsive ground

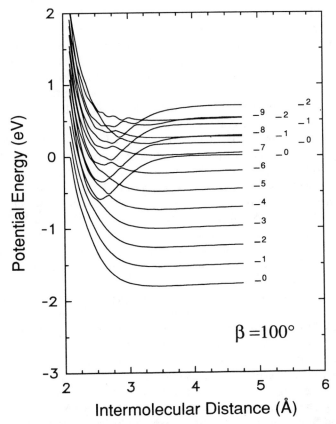

Figure 31. Adiabatic vibronic potential energy curves as a function of R for the $(Ar + CO)^+$ system with the molecular orientation angle $\beta = 110°$ and $\Omega = \frac{1}{2}$. The zero of energy corresponds to $Ar^+(^2P_{3/2}) + CO(v = 0)$. Linear Ar–C–O corresponds to $\beta = 0°$. The asymptotic states are indicated by the energy limits shown to the right. From left to right the three columns correspond to $Ar + CO^+(X,v)$, $Ar^+(^2P_{3/2}) + CO(v)$, and $Ar^+(^2P_{1/2}) + CO(v)$. In each case the vibrational level v is indicated. Curves which correlate to $CO^+(A,v)$ states are higher in energy and are not shown. (Taken from Ref. 59, with permission.)

electronic curve; there is one curve for each asymptotic vibrational level of $CO^+(X)$. These curves are cut by a second set of attractive diabatic curves that are based on the excited electronic curve. The crossing between the two sets of curves are avoided, but the energy gaps are quite small. It is instructive to follow the lowest attractive vibronic curve, which correlates to $Ar^+(^2P_{3/2}) + CO(v = 0)$. At $R = 5$ Å this curve is parallel to, but just below, the repulsive vibronic curve which dissociates to $Ar + CO^+(X,v = 7)$. For R

values less than 4 Å, however, it becomes attractive and cuts repulsive curves which correlate to $Ar + CO^+(X; v = 6, 5, 4,$ and 3).

The behavior of this attractive diabatic vibronic curve suggests a likely mechanism[59] for CT in thermal energy collisions of $Ar^+(^2P_{3/2}) + CO(X, v = 0)$. The mechanism is similar to Ferguson's model[162] for vibrational relaxation of molecular ions. The distribution of orientation angles γ is initially random, but the anisotropic nature of the vibronic potential energy surface puts a considerable torque on the CO molecule, favoring angles near $\gamma = 100°$ where the potential well is deepest (see Fig. 21). In many collisions this torque will induce rotational excitation of the CO molecule, which will trap the ion–molecule complex for a period of time. The system will pass through the weakly avoided crossings shown in Fig. 31 many times, until a transition is made to a repulsive curve which correlates to $Ar + CO^+(X; v' = 6, 5, 4,$ or 3). The complex will then quickly dissociate along this curve giving CT. The complex mechanism proposed by Parlant et al.[59] is supported by the fact that the measured rate constant for CT increases with decreasing temperature.[153] In addition, it provides an explanation of Leone's experiments,[157,158] which show that CT of $Ar^+(^2P_{3/2}) + CO(v = 0)$ produces $CO^+(X)$ primarily in vibrational levels $v' = 3, 4, 5,$ and 6. By comparison, the classical path calculations at $E_{rel} = 2 \, eV$,[55] discussed earlier, give mainly $CO^+(X; v' = 1, 2$ and 3). This is an energy where a collision complex could not be formed.

The calculations discussed here confirm Spalburg's suggestion[51] that AV states are the best zero-order states to describe CT at collision energies below 20 eV. Theoretical work based on this framework should be very useful. One example is the quantum-mechanical treatment of the $(Ar + N_2)^+$ system by Clary and Sonnenfroh.[23] An alternative would be to carry out TSH calculations on the AV surfaces. Some work along this line has been done by Nikitin et al.[94].

C. Energy Dependence of Charge-Transfer Cross Sections

Classical path calculations have given considerable insight into the CT process. In this section we summarize what we have learned about the energy dependence of state-to-state CT cross sections. We restrict the discussion to strongly coupled systems such as $(Ar + N_2)^+$, which typically have charge-transfer cross sections in excess of $15 \, Å^2$. By strongly coupled we mean: (1) The electronic coupling between the two charge exchange states is large—this is usually the case if the interaction corresponds to a one-electron exchange; and (2) the reactant state is nearly energy resonant with one or two product vibronic states, and the Franck–Condon factors for the transitions are at least 0.01. We also assume that the electronic couplings are weak functions of the vibrational coordinate r.

The behavior of the state-to-state cross sections at high energies are best understood by reference to the weak-coupling result in Eq. (65). At *high energies* this equation reduces to the Franck–Condon limit shown in Eq. (51). Two other comments can be made. First, in this energy region the CT cross sections are falling off as E_{rel}^{-1}, as shown in Eq. (65). Second, the dynamic energy width δE, defined in Eqs. (54) and (55), is growing with collision energy, which means that other electronic CT states will be populated. For example, when $E_{rel} = 10,000$ eV, Eq. (55) shows that $\delta E = 3.8$ eV. Thus, we expect that all electronic states of $Ar + N_2^+$ within ± 3.8 eV of the reactants will be produced with significant cross sections. This energy range includes the X, A, and B states of N_2^+.

When the collision energy is lowered to between 50 and 1000 eV (*medium energies*), Eq. (65) shows that the state-to-state CT cross sections depend on the energy difference ΔE between reactant and product states. In particular, product states with small energy gaps are favored. The classical path calculations in general do not support the actual dependence on ΔE shown in Eq. (65). Rather, an exponential dependence of the type $\exp(-p\Delta E)$ is normally seen (see Fig. 10). State-to-state cross sections are also expected to be proportional to the Franck–Condon factors for the transition. Equation (65) predicts that if the state-to-state cross sections are divided by the Franck–Condon factors, the ratio should be a simple function of ΔE. This point was explicitly tested by Lee and Depristo[45] at two collision energies, and the result are shown in Fig. 10. It is seen that most of the data lie on a single curve, but some points lie higher. Lee and DePristo pointed out that all of these points correspond to transitions with very small Franck–Condon factors. In our opinion, this result is obtained because even at $E_{rel} = 400$ eV the H_2 and H_2^+ molecules have enough time to undergo some vibrational motion during the collision, and the change in the r values allows transitions with small Franck–Condon factors to be more likely. We conclude that at medium energies the state-to-state cross sections are primarily determined by the energy gap ΔE, and they are approximately proportional to the Franck–Condon factors.

At collision energies between 0.5 and 10 eV (*low energies*) the molecule has time to vibrate several times during the collision. In this case Eq. (65) is no longer useful for understanding the scattering. As discussed in the previous section the best zero-order picture for CT in this energy range is the adiabatic vibronic (AV) formalism. The particles approach on the reactant AV potential energy surface, which has avoided crossings with other AV surfaces. The system can hop from one surface to another, and the probability for hopping can often be estimated from the Landau–Zener formula. Typically, only a few product states are important, and these are close in energy to the reactants. The AV surfaces often give a good clue as to which products are formed.

The final energy range to be considered is thermal, where the collision energy is less than the well depth of the ion–molecule intermediate. In this *thermal energy* range only exothermic processes can occur. The AV formalism remains the best way to consider these collisions. However, at these energies the collision velocities are often so low that the probability of hopping from one AV surface to another becomes very small. Thus CT cross sections are expected to decrease rapidly at low energies, and this is observed for many ion–molecule systems.[153] A good example of this is the CT reaction of $Ar^+(^2P_{3/2}) + N_2(v = 0)$. The data of Lindinger et al.[174] shows that the rate constant for this process decreases by a factor of 100 as the collision energy falls from 3 to 0.03 eV. We have seen that the major product is $Ar + N_2^+(X, v = 1)$, produced in a single curve crossing (Fig. 30), but which is endothermic by 0.092 eV. This product state cannot be made at low energy, so the rate constant might be expected to fall to zero. Instead, recent experiments by Rebrion et al.[175] have shown that the rate constant begins increasing as the collision energy is lowered below 0.02 eV. We believe the explanation for this behavior is that at very low energies collisions of $Ar^+ + N_2$ form a long-lived complex. This complex lives long enough for the system to hop to the lower AV potential energy surface and dissociate as $Ar + N_2^+(X, v = 0)$. A similar argument to explain the thermal CT results for $Ar^+ + CO$ was described in the previous section.

The relative importance of energy defects and Franck–Condon factors in determining state-to-state CT cross sections has been debated for many years. We have seen that at high energy the Franck–Condon factors dominate the scattering. At lower collision energies small energy defects are always favored. The role of Franck–Condon factors is less clear. They do enter into the calculation of the AV potential energy surfaces, because they are part of the vibronic hamiltonian matrix, but there is no direct relationship between cross section and Franck–Condon factor.

VI.   FUTURE DEVELOPMENTS INVOLVING THE CLASSICAL PATH PROCEDURE

As is apparent from the companion volume to this book, it is now possible to carry out a remarkable variety of state-selected and state-to-state measurements of CT in ion–molecule collisions. In addition, we have emphasized in this chapter that classical path calculations of total state-to-state cross sections are fairly routine for nonreactive systems. Thus, we expect that future experimental work will inspire a large number of classical path calculations. There is, however, a need for more *ab initio* calculations of potential energy surfaces and couplings.[91]

The formalism for computing differential and doubly differential cross sections has now been developed by Sizun et al.[58] for scattering at small angles. We expect that a large number of calculations will be carried out using this procedure in the near future. The same can be said for the problem of dissociative CT. The techniques developed by Sidis and co-workers,[64–68] and especially the wave packet procedure,[68] will be applied to many systems in the next few years.

In the future we can expect that the classical path procedure will be extended into new areas where present methods are inadequate. The first is the energy range below 1 or 2 eV. It is quite possible that the TSH method on adiabatic vibronic potential energy surfaces, discussed in Section V B, will prove to be useful in this energy range. The second area is CT collisions involving negative ions. We are not aware of any classical path calculations on this type of system, but in principle the procedures described in this chapter will work there. Any realistic calculation, however, must allow for the possibility of electron detachment. This complicates the calculations, but approximate techniques for treating the detachment have been developed.[176] The third area is to extend the classical path method to treat polyatomic molecules and/or to treat molecular rotations quantum mechanically. Both cases can be treated within the existing formalism with no difficulty, but the computer requirements for these calculations will be extensive. In the case of quantum-mechanical rotations, it is likely that these calculations will wait until (if?) the rotational state-dependence of ion–molecule CT processes is determined experimentally. The fourth and most difficult area will be the calculations of CT cross sections in the presence of chemical reaction. Accurate calculations for this process using the classical path procedure are probably many years away. We note, however, that a classical path approach to reactive scattering on a single potential energy surface using hyperspherical coordinates has recently been developed by Muckerman, Gilbert, and Billing.[177,178] This technique will hopefully be extended in the future to include additional electronic states.

In summary, the classical path method has proven to be a powerful method for studying CT processes. In the next 10 years we can look forward to many more systems being studied, and, in addition, a rapid development of new techniques to increase the range of processes that can be studied.

Acknowledgments

The authors are pleased to acknowledge the many years of helpful discussions with our numerous colleagues at the University of Paris at Orsay. In particular, we would like to thank our many coauthors on various aspects of this work: P. Archirel, O. Dutuit, D. Grimbert, P. M. Guyon, G. Henri, M. Lavollée, B. Levy, J. B. Ozenne, and V. Sidis. We also acknowledge discussions with M. Baer and C. Y. Ng, who, not coincidentally, are the editors of these two volumes. G.P. would like to thank D. Yarkony for his hospitality when this chapter was written and M. Alexander for his earlier hospitality. E.A.G. and G.P. thank A. Kleyn, J. Los, and M. R. Spalburg for getting them started in this field.

Support from the CNRS/NSF Exchange Programs is gratefully acknowledged. Acknowledgment is made to the Donors of the Petroleum Research Fund, administered by the American Chemical Society, for partial support of this work, in the form of Grant 20515-AC6 to D. Yarkony.

References

1. (a) S. R. Leone, *Ann. Rev. Phys. Chem.* **35**, 109 (1984); (b) V. M. Bierbaum and S. R. Leone, in *Structure/Reactivity and Thermochemistry of Ions*, edited by P. Ausloos and S. G. Lias, Reidel, Boston, 1987, p. 23; (c) S. R. Leone and V. M. Bierbaum, *Faraday Discuss. Chem. Soc.* **84**, 253 (1987).

2. (a) J. H. Futrell, in *Gaseous Ion Chemistry and Mass Spectrometry*, edited by J. H. Futrell, Wiley-Interscience, New York, 1986, p. 201; (b) *Int. J. Quant. Chem.* **31**, 133 (1987); (c) in *Structure/Reactivity and Thermochemistry of Ions*, edited by P. Ausloos and S. G. Lias, Reidel, Boston, 1987, p. 57.

3. Z. Herman and I. Koyano, *J. Chem. Soc. Faraday Trans. II* **83**, 127 (1987).

4. F. Linder, in *Invited Papers of the XV ICPEAC*, edited by H. B. Gilbody, W. R. Newell, F. H. Read, and A. C. H. Smith, North-Holland, Amsterdam, 1988, p. 287.

5. N. Kobayashi, in Ref. 4, p. 333.

6. C. Y. Ng, in *Techniques for the Study of Ion-Molecule Reactions*, edited by J. M. Farrar and W. H. Saunders, Wiley, New York, 1988, p. 417.

7. T. Baer, in *Gas Phase Ion Chemistry*, edited by M. T. Bowers, Academic, New York, 1979, Vol. 1, p. 153.

8. (a) P. M. Guyon and E. A. Gislason, in *Topics in Current Chemistry: Synchrotron Radiation in Chemistry and Biology III*, edited by E. Mandelkow, Springer, Berlin, 1989, Vol. 151, p. 161; (b) P. M. Guyon, G. Bellec, O. Dutuit, D. Gerlich, E. A. Gislason, and J. B. Ozenne, *Bull. Soc. Roy. (Liege)* **58**, 187 (1989).

9. J. C. Tully, in *Dynamics of Molecular Collisions*, Part B, edited by W. H. Miller, Plenum, New York, 1976, p. 217.

10. A. W. Kleyn, J. Los, and E. A. Gislason, *Phys. Rep.* **90**, 1 (1982).

11. M. Baer, in *Molecular Collision Dynamics*, edited by J. M. Bowman, Springer-Verlag, Berlin, 1983, p. 117.

12. V. Sidis, in *Collision Theory for Atoms and Molecules*, edited by F. A. Gianturco, Plenum, New York, 1989, p. 343.

13. M. R. Flannery, in *Swarms of Ions and Electrons in Gases*, edited by W. Lindinger, T. D. Mark, and F. Howorka, Springer-Verlag, New York, 1984, p. 103.

14. M. R. Spalburg and E. A. Gislason, *Chem. Phys.* **94**, 339 (1985).

15. E. Pollack and Y. Hahn, *Adv. At. Mol. Phys.* **22**, 243 (1986).

16. G. Parlant and E. A. Gislason, in *Invited Papers of the XV ICPEAC*, edited by H. B. Gilbody, W. R. Newell, F. H. Read, and A. C. H. Smith, North-Holland, Amsterdam, 1988, p. 357.

17. V. Sidis, *Adv. At. Mol. Opt. Phys.* **26**, 161 (1990).

18. P. M. Guyon, T. R. Govers, and T. Baer, *Z. Phys. D* **4**, 89 (1986).

19. P. Archirel and B. Levy, *Chem. Phys.* **106**, 51 (1986).

20. M. Kimura and N. F. Lane, *Adv. At. Mol. Opt. Phys.* **26**, 79 (1990).

21. C. H. Becker, *J. Chem. Phys.* **76**, 5928 (1982).

22. (a) M. Baer and H. Nakamura, *J. Chem. Phys.* **87**, 4651 (1987); (b) M. Baer, G. Niedner-Schatteburg, and J. P. Toennies, *J. Chem. Phys.* **91**, 4169 (1989); (c) V. Sidis, D. Grimbert, M. Sizun, and M. Baer, *Chem. Phys. Lett.* **163**, 19 (1989); (d) M. Baer, C. L. Liao, R. Xu, S. Nourbahksh, G. D. Flesch, C. Y. Ng, and D. Neuhauser, *J. Chem. Phys.* **93**, 4845 (1990); (e) M. Baer and C. Y. Ng, *J. Chem. Phys.* **93**, 7787 (1990).

23. D. C. Clary and D. M. Sonnenfroh, *J. Chem. Phys.* **90**, 1686 (1989).

24. (a) D. S. F. Crothers, *Adv. At. Mol. Phys.* **17**, 55 (1981); (b) J. B. Delos, *Rev. Mod. Phys.* **53**, 287 (1981).

25. G. D. Billing, *Comput. Phys. Rept.* **1**, 237 (1984).

26. D. R. Bates and R. H. G. Reid, *Proc. Roy. Soc. (London) A* **310**, 1 (1969).

27. T. F. Moran, in *Electron-Molecule Interactions and Their Applications*, edited by L. G. Christophorou, Academic, New York, 1984, Vol. 2, p. 1.

28. M. R. Flannery, P. C. Cosby, and T. F. Moran, *J. Chem. Phys.* **59**, 5494 (1973).

29. M. R. Flannery, P. C. Cosby, and T. F. Moran, *Chem. Phys. Lett.* **27**, 221 (1974).

30. P. C. Cosby, T. F. Moran, and M. R. Flannery, *J. Chem. Phys.* **61**, 1259 (1974).

31. T. F. Moran, M. R. Flannery, and P. C. Cosby, *J. Chem. Phys.* **61**, 1261 (1974).

32. T. F. Moran, M. R. Flannery, and D. L. Albritton, *J. Chem. Phys.* **62**, 2869 (1975).

33. M. R. Flannery, K. J. McCann, and T. F. Moran, *J. Chem. Phys.* **63**, 1462 (1975).

34. T. F. Moran, K. J. McCann, and M. R. Flannery, *J. Chem. Phys.* **63**, 3857 (1975).

35. K. J. McCann, M. R. Flannery, J. V. Hornstein, and T. F. Moran, *J. Chem. Phys.* **63**, 4998 (1975).

36. T. F. Moran, K. J. McCann, M. R. Flannery, and D. L. Albritton, *J. Chem. Phys.* **65**, 3172 (1976).

37. T. F. Moran and M. R. Flannery, *J. Chem. Phys.* **66**, 370 (1977).

38. T. F. Moran, K. J. McCann, M. Cobb, R. F. Borkman, and M. R. Flannery, *J. Chem. Phys.* **74**, 2325 (1981).

39. A. F. Hedrick, T. F. Moran, K. J. McCann, and M. R. Flannery, *J. Chem. Phys.* **66**, 24 (1977).

40. S. B. Sears and A. E. DePristo, *J. Chem. Phys.* **77**, 290 (1982).

41. A. E. DePristo and S. B. Sears, *J. Chem. Phys.* **77**, 298 (1982).

42. A. E. DePristo, *J. Chem. Phys.* **78**, 1237 (1983).

43. A. E. DePristo, *J. Chem. Phys.* **79**, 1741 (1983).

44. C. Y. Lee and A. E. DePristo, *J. Chem. Phys.* **80**, 1116 (1984).

45. C. Y. Lee and A. E. DePristo, *J. Chem. Phys.* **81**, 3512 (1984).

46. C. Y. Lee, A. E. DePristo, C. L. Liao, C. X. Liao, and C. Y. Ng, *Chem. Phys. Lett.* **116**, 534 (1985).

47. S. K. Cole and A. E. DePristo, *J. Chem. Phys.* **85**, 1389 (1986).

48. C. Y. Lee and A. E. DePristo, *J. Am. Chem. Soc.* **105**, 6775 (1983).

49. M. Kimura, *Phys. Rev. A* **32**, 802 (1985).

50. M. Kimura, S. Chapman, and N. F. Lane, *Phys. Rev. A* **33**, 1619 (1986).

51. M. R. Spalburg, J. Los, and E. A. Gislason, *Chem. Phys.* **94**, 327 (1985).

52. G. Parlant and E. A. Gislason, *Chem. Phys.* **101**, 227 (1986).

53. E. A. Gislason and G. Parlant, *Comm. At. Mol. Phys.* **19**, 157 (1987).

54. G. Parlant and E. A. Gislason, *J. Chem. Phys.* **86**, 6183 (1987).

55. E. A. Gislason, G. Parlant, P. Archirel, and M. Sizun, *Faraday Discuss. Chem. Soc.* **84**, 325 (1987).

56. G. Parlant and E. A. Gislason, *J. Chem. Phys.* **88**, 1633 (1988).

57. G. Parlant and E. A. Gislason, *J. Chem. Phys.* **91**, 5359 (1989).

58. (a) M. Sizun, D. Grimbert, and V. Sidis, *J. Phys. Chem.* **94**, 5674 (1990); (b) M. Sizun, D. Grimbert, and V. Sidis (to be published).

59. G. Parlant, P. Archirel, and E. A. Gislason, *J. Chem. Phys.* **92**, 1211 (1990).

60. D. Grimbert, B. Lassier-Govers, and V. Sidis, *Chem. Phys.* **124**, 187 (1988).

61. G. Henri, M. Lavollée, G. Parlant, and P. Archirel (unpublished).

62. G. Henri, M. Lavollée, O. Dutuit, J. B. Ozenne, P. M. Guyon, and E. A. Gislason, *J. Chem. Phys.* **88**, 6381 (1988).

63. E. A. Gislason and G. Parlant, *J. Chem. Phys.* **94**, 6598 (1991).

64. V. Sidis, *J. Phys. Chem.* **93**, 8128 (1989).

65. V. Sidis and D. P. DeBruijn, *Chem. Phys.* **85**, 201 (1984).

66. V. Sidis and C. Courbin-Gaussorgues, *Chem. Phys.* **111**, 285 (1987).

67. V. Sidis, D. Grimbert, and C. Courbin–Gaussorgues, in *Invited Papers of the XV ICPEAC*, edited by H. B. Gilbody, W. R. Newell, F. H. Read, and A. C. H. Smith, North-Holland, Amsterdam, 1988, p. 485.

68. J. P. Gauyacq and V. Sidis, *Europhys. Lett.* **10**, 225 (1989).

69. D. G. Truhlar and D. A. Dixon in *Atom-Molecule Collision Theory: A Guide for the Experimentalist*, edited by R. B. Bernstein, Plenum, New York, 1979, p. 595.

70. D. G. Truhlar and J. T. Muckerman, in *Atom-Molecule Collision Theory: A Guide for the Experimentalist*, edited by R. B. Bernstein, Plenum, New York, 1979, p. 505.

71. J. R. Stine and J. T. Muckerman, *J. Phys. Chem.* **91**, 459 (1987).

72. S. Chapman, in this volume.

73. J. R. Stine and J. T. Muckerman, *J. Chem. Phys.* **65**, 3975 (1976).

74. M. Sizun and E. A. Gislason, *J. Chem. Phys.* **91**, 4603 (1989).

75. N. C. Blais and D. G. Truhlar, *J. Chem. Phys.* **79**, 1334 (1983).

76. C. W. Eaker, *J. Chem. Phys.* **87**, 4532 (1987).

77. G. Parlant and E. A. Gislason, *J. Chem. Phys.* **91**, 4416 (1989).

78. G. Parlant and M. H. Alexander, *J. Chem. Phys.* **92**, 2287 (1990).

79. J. C. Tully, *J. Chem. Phys.* **93**, 1061 (1990).

80. W. R. Gentry, in *Atom-Molecules Collision Theory: A Guide for the Experimentalist*, edited by R. B. Bernstein, Plenum, New York, 1979, p. 391.

81. D. R. Bates and A. R. Holt, *Proc. R. Soc. London, Ser. A* **292**, 168 (1966).

82. I. J. Berson, *Latv. PSR Zinat. Akad. Vestis. Fiz. Teh. Zinat. Ser. N.* **4**, 47 (1968).

83. R. J. Cross, *J. Chem. Phys.* **51**, 5163 (1969).

84. D. R. Bates and D. S. F. Crothers, *Proc. R. Soc. London, Ser. A* **315**, 465 (1970).

85. J. B. Delos, W. R. Thorson, and S. K. Knudson, *Phys. Rev. A* **6**, 709 (1972).

86. C. Gaussorgues, C. Le Sech, F. Masnou-Seeuws, R. McCarroll, and A. Riera, *J. Phys. B* **8**, 239 (1975).

87. K. J. McCann and M. R. Flannery, *J. Chem. Phys.* **69**, 5275 (1978).

88. M. S. Child, *Molecular Collision Theory*, Academic Press, London, 1974.

89. T. H. Dunning Jr. and L. B. Harding, in *Theory of Chemical Reaction Dynamics*, edited by M. Baer, CRC Press, Boca Raton, FL, 1985, Vol. 1, p. 1.

90. P. J. Kuntz, in *Theory of Chemical Reaction Dynamics*, edited by M. Baer, CRC Press, Boca Raton, FL, 1985, Vol. 1, p. 71.

91. B. H. Lengsfield and D. R. Yarkony, in this volume.

92. M. Born and J. R. Oppenheimer, *Ann. Phys.* **84**, 457 (1927).

93. V. Sidis, in this volume.

94. E. E. Nikitin, M. Y. Ovchinnikova, and D. V. Shalashilin, *Chem. Phys.* **111**, 313 (1987).

95. N. F. Mott and H. S. W. Massey, *The Theory of Atomic Collisions*, Clarendon Press, Oxford, 1965, Chap. XIV.

96. R. J. Cross, *J. Chem. Phys.* **50**, 1036 (1969).

97. E. A. Gislason and D. R. Herschbach, *J. Chem. Phys.* **64**, 2133 (1976).

98. G. N. Watson, *A Treatise on the Theory of Bessel Functions*, Cambridge University, London, 1966, p. 158.

99. M. A. Wartell and R. J. Cross, *J. Chem. Phys.* **55**, 4983 (1971).

100. J. R. Stallcop, *J. Chem. Phys.* **62**, 690 (1975).

101. H. Krüger and R. Schinke, *J. Chem. Phys.* **66**, 5087 (1977).

102. R. Schinke, *Chem. Phys.* **24**, 379 (1977).

103. E. A. Gislason, *Chem. Phys. Lett.* **42**, 315 (1976).

104. R. J. Cross, *J. Chem. Phys.* **52**, 5703 (1970).

105. D. Grimbert, M. Sizun, and V. Sidis, *J. Chem. Phys.* **93**, 7530 (1990).

106. M. R. Spalburg, M. G. M. Vervaat, A. W. Kleyn, and J. Los, *Chem. Phys.* **99**, 1 (1985).

107. See J. Tellinghuisen, *Adv. Chem. Phys.* **60**, 299 (1985), and references therein.

108. W. H. Press, B. P. Flannery, S. A. Teukolsky, and W. T. Vetterling, *Numerical Recipes: The Art of Scientific Computing*, Cambridge University Press, New York, 1986.

109. I. P. Hamilton and J. C. Light, *J. Chem. Phys.* **84**, 306 (1986), and references therein.

110. G. C. Schatz, *Rev. Mod. Phys.* **61**, 669 (1989).

111. M. R. Spalburg, J. Los, and V. Sidis, *Chem. Phys. Lett.* **96**, 14 (1983).

112. U. C. Klomp, M. R. Spalburg, and J. Los, *Chem. Phys.* **83**, 33 (1984).

113. J. B. H. Stedeford and J. B. Hasted, *Proc. Roy. Soc. (London)* A **227**, 466 (1955).

114. D. W. Koopman, *Phys. Rev.* **154**, 79 (1967).

115. T. F. Moran and J. R. Roberts, *J. Chem. Phys.* **49**, 3411 (1968).

116. H. C. Hayden and R. C. Amme, *Phys. Rev.* **172**, 104 (1968).

117. C. J. Latimer, R. Browning, and H. B. Gilbody, *J. Phys. B* **2**, 1055 (1969).

118. H. L. Rothwell, B, VanZyl, and R. C. Amme, *J. Chem. Phys.* **61**, 3851 (1974).

119. F. M. Campbell, R. Browning, and C. J. Latimer, *J. Phys. B* **14**, 3493 (1981).

120. T. Baer, P. T. Murray, and L. Squires, *J. Chem. Phys.* **68**, 4901 (1978).

121. R. F. Borkman and M. J. Cobb, *J. Chem. Phys.* **74**, 2920 (1981).

122. S. K. Cole, T. Baer, P. M. Guyon, and T. R. Govers, *Chem. Phys. Lett.* **109**, 285 (1984).

123. C. F. Barnett, J. A. Ray, E. Ricci, M. I. Wilker, E. W. McDaniel, E. W. Thomas, and H. B. Gilbody, *Atomic Data for Controlled Fusion Research*, Oak Ridge National Laboratory Report 5206 (1977).

124. R. W. Nicholls, *J. Phys. B* **1**, 1192 (1968).

125. P. J. Kuntz and A. C. Roach, *J. Chem. Soc. Faraday Trans. II* **68**, 259 (1972).

126. H. B. Gilbody and J. B. Hasted, *Proc. Roy. Soc. (London) A* **238**, 334 (1957).

127. R. C. Amme and J. F. McIlwain, *J. Chem. Phys.* **45**, 1224 (1966).

128. C. J. Latimer, *J. Phys. B* **10**, 515 (1977).

129. T. Kato, K. Tanaka, and I. Koyano, *J. Chem. Phys.* **77**, 834 (1982).

130. T. R. Govers, P. M. Guyon, T. Baer, K. Cole, H. Fröhlich, and M. Lavollée, *Chem. Phys.* **87**, 373 (1984).

131. (a) C. L. Liao, R. Xu, and C. Y. Ng, *J. Chem. Phys.* **85**, 7136 (1986); (b) J. D. Shao, Y. G. Li, G. D. Flesch, and C. Y. Ng, *J. Chem. Phys.* **86**, 170 (1987).

132. T. Kato, K. Tanaka, and I. Koyano, *J. Chem. Phys.* **77**, 337 (1982).

133. (a) C. L. Liao, R. Xu, and C. Y. Ng, *J. Chem. Phys.* **84**, 1948 (1986); (b) C. L. Liao, J. D. Shao, R. Xu, G. D. Flesch, Y. G. Li, and C. Y. Ng, *J. Chem. Phys.* **85**, 3874 (1986).

134. B. G. Lindsay and C. J. Latimer, *J. Phys. B* **21**, 1617 (1988).

135. R. E. Olson, F. T. Smith, and E. Bauer, *Appl. Opt.* **10**, 1848 (1971).

136. F. Rebentrost, in *Theoretical Chemistry: Advances and Perspectives*, edited by D. Henderson, Academic, New York, 1981, Vol. 6B, p. 1.

137. S. N. Ghosh and W. F. Sheridan, *J. Chem. Phys.* **26**, 480 (1957).

138. J. C. Abbe and J. P. Adloff, *Bull. Soc. Chim. Fr.* **1964**, 1212.

139. R. C. Amme and H. C. Hayden, *J. Chem. Phys.* **42**, 2011 (1965).

140. (a) J. B. Homer, R. S. Lehrle, J. C. Robb, and D. W. Thomas, *Trans. Faraday Soc.* **62**, 619 (1966); (b) *Adv. Mass Spectrosc.* **3**, 415 (1966).

141. R. S. Lehrle, J. E. Parker, J. C. Robb, and J. Scarborough, *Int. J. Mass Spectrosc. Ion Phys.* **1**, 455 (1968).

142. D. L. Smith and L. Kevan, *J. Am. Chem. Soc.* **93**, 2113 (1971).

143. A. Rosenberg, H. Bregman-Reisler, and S. Amiel, *Int. J. Mass Spectrosc. Ion Phys.* **11**, 433 (1973).

144. E. W. Kaiser, A. Crowe, and W. E. Falconer, *J. Chem. Phys.* **61**, 2720 (1974).

145. C. L. Liao, C. X. Liao, and C. Y. Ng, *J. Chem. Phys.* **82**, 5489 (1985).

146. P. M. Guyon, cited in Ref. 54.

147. D. L. King and D. W. Setser, *Ann. Rev. Phys. Chem.* **27**, 407 (1976).

148. J. D. Shao, Y. G. Li, G. D. Flesch, and C. Y. Ng, *Chem. Phys. Lett.* **132**, 58 (1986).

149. T. Nakamura, N. Kobayashi, and Y. Kaneko, *J. Phys. Soc. Jpn.* **55**, 3831 (1986).

150. K. Birkinshaw, A. Shukla, S. Howard, and J. H. Futrell, *Chem. Phys.* **113**, 149 (1987).

151. V. G. Anicich and W. T. Huntress, *Ap. J. Suppl.* **62**, 553 (1986).

152. N. Kobayashi, *J. Phys. Soc. Jpn.* **36**, 259 (1974).

153. I. Dotan and W. Lindinger, *J. Chem. Phys.* **76**, 4972 (1982).

154. R. Marx, G. Mauclaire, and R. Derai, *Int. J. Mass Spec. Ion Phys.* **47**, 155 (1983).

155. R. Marx, in *Ionic Processes in the Gas Phase*, edited by M. A. Almoster Ferreira, Reidel, Boston, 1984, p. 67.

156. J. Danon and R. Marx, *Chem. Phys.* **68**, 255 (1982).

157. G. H. Lin, J. Maier, and S. R. Leone, *J. Chem. Phys.* **82**, 5527 (1985).

158. C. E. Hamilton, V. M. Bierbaum, and S. R. Leone, *J. Chem. Phys.* **83**, 2284 (1985).

159. E. A. Gislason and E. E. Ferguson, *J. Chem. Phys.* **87**, 6474 (1987).

160. J. Baker and A. D. Buckingham, *J. Chem. Soc. Faraday Trans. II* **83**, 1609 (1987).

161. K. Norwood, J. H. Guo, G. Luo, and C. Y. Ng, *Chem. Phys.* **129**, 109 (1989).

162. E. E. Ferguson, *J. Phys. Chem.* **90**, 731 (1986).

163. E. A. Gislason, *J. Chem. Phys.* **58**, 3702 (1973).

164. R. E. Olson, *Phys. Rev. A* **6**, 1822 (1972).

165. D. P. DeBruijn, J. Neuteboom, and J. Los, *Chem. Phys.* **85**, 233 (1984).

166. F. Von Busch and G. H. Dunn, *Phys. Rev. A* **5**, 1726 (1972).

167. J. H. Futrell (companion volume).

168. G. Niedner-Schatteburg and J. P. Toennies (companion volume).

169. M. Noll and J. P. Toennies, *J. Chem. Phys.* **85**, 3313 (1986).

170. D. Dhuicq, J. C. Brenot, and V. Sidis, *J. Phys. B* **18**, 1395 (1985).

171. G. Parlant, M. Schröder, and S. Goursaud, *Chem. Phys.* **75**, 175 (1983).

172. E. A. Gislason and E. M. Goldfield, *Phys. Rev. A* **25**, 2002 (1982).

173. D. M. Sonnenfroh and S. R. Leone, *J. Chem. Phys.* **90**, 1677 (1989).

174. W. Lindinger, F. Howorka, P. Lukac, S. Kuhn, H. Villinger, E. Alge, and H. Ramler, *Phys. Rev. A* **23**, 2319 (1981).

175. C. Rebrion, B. R. Rowe, and J. B. Marquette, *J. Chem. Phys.* **91**, 6142 (1989).

176. M. Sizun, E. A. Gislason, and G. Parlant, *Chem. Phys.* **107**, 311 (1986).

177. J. T. Muckerman, R. D. Gilbert, and G. D. Billing, *J. Chem. Phys.* **88**, 4779 (1988).

178. G. D. Billing and J. T. Muckerman, *J. Chem. Phys.* **91**, 6830 (1989).

THE CLASSICAL TRAJECTORY-SURFACE-HOPPING APPROACH TO CHARGE-TRANSFER PROCESSES

SALLY CHAPMAN

Department of Chemistry
Barnard College, Columbia University
New York, NY

CONTENTS

State-Selected and State-to-State Ion–Molecule Reaction Dynamics, Part 2: Theory, Edited by Michael Baer and Cheuk-Yiu Ng. Advances in Chemical Physics Series, Vol. LXXXII. ISBN 0-471-53263-0 © 1992 John Wiley & Sons, Inc.

I. INTRODUCTION

Classical trajectory methods have a number of attractive features for the study of the dynamics of molecular collisions. Classical mechanics is intuitive, the equations are relatively straightforward, and the size and complexity of the computer programs and demands on machine time increase much less dramatically with the size of the problem than with quantum methods. The quasiclassical trajectory method for simple bimolecular reactions is well established, and has been thoroughly reviewed.[1-3] There are obvious shortcomings: classical trajectories do not tunnel, so calculations near threshold must be interpreted with care, and classical trajectories carry no phase information, so phase sensitive phenomena (interferences and some types of resonances) are absent. Semiclassical methods[4-5] overcome these problems and, for simple systems, have provided and excellent picture of why purely classical methods work as well as they do. If quantum mechanics is "classical dynamics plus quantum superposition,"[4] then bimolecular collisions, where phase information is usually largely lost by extensive averaging over initial conditions, are well described classically.

The potential energy surface (more precisely, for systems with three or more atoms, hypersurface of dimension $m = 3N - 6$) lies at the core of a trajectory calculation. For many systems, the Born–Oppenheimer approximation is valid over all relevant regions of configuration space: a single potential surface governs the dynamics. However, there are many interesting systems whose overall dynamics involves more than one surface. Such nonadiabatic behavior is particularly common in ion–molecule reactions: charge transfer between two fragments at a particular geometry is intrinsically nonadiabatic, the transfer of charge corresponding to a change in electronic configuration. It is often found that the breakdown of the Born-Oppenheimer approximation is *local*: that over most of configuration space the potential surfaces are well separated in energy. It was for such cases that Preston and Tully developed the trajectory-surface-hopping (TSH) method,[6] and applied it to the simplest of ion molecule reactions, $H^+ + H_2$. In earlier work, Bjerre and Nikitin[7] had introduced a similar approach, but with reduced dimensionality, in a study of electronic to vibrational energy transfer in the collision of excited sodium with nitrogen molecules.

The TSH method is simple in concept: in those regions of space where the Born–Oppenheimer approximation is valid, standard quasiclassical trajectory methods are applied, while in the localized regions of strong nonadiabatic coupling, the trajectories are permitted to hop between the coupled surfaces. With this method one can describe in a consistent framework competing processes that are frequently treated separately: for example, charge transfer and chemical reaction (that is, processes in which chemical bonds are broken or made). For this reason, while the title of this review emphasizes charge transfer, the multiplicity of processes which may occur in molecular collisions involving charged particles will be included.

Some of the basic concepts and equations for nonadiabatic processes are reviewed in Section II. The TSH method is described in Section III, including a number of alternate procedures which have been developed for handling parts of the calculation. Applications are reviewed in Section IV, with emphasis on ion–molecule and ion–pair formation reactions. Some comments and speculations about future developments are made in Section V.

II.   NONADIABATIC MOLECULAR COLLISIONS: BACKGROUND

A.   The Semiclassical Description of Nonadiabatic Transitions

The theory of nonadiabatic transitions has been reviewed many times. Garrett and Truhlar[8] give an excellent survey of the subject with a particularly useful discussion of various diabatic representations. Delos[9] focuses on atom–atom and ion–atom collisions. Baer[10] emphasizes quantum-mechanical treatments of nonadiabatic processes in atom- and ion–molecule collisions. The work of Nikitin[11-13] provides a very clear exposition of the semiclassical theory of nonadiabatic dynamics. Following his notation, we give below a highly simplified version, in order to establish notation and terminology.

We start with the electronic hamiltonian for fixed nuclear geometry $H_e(\mathbf{r}, \mathbf{R})$ which depends on both the electronic coordinates $\mathbf{r}$ and the nuclear coordinates $\mathbf{R}$. H_e includes potential energy terms for both electrons and nuclei, but kinetic energy terms for electrons only. The solutions to the time-independent Schrödinger equation

$$H_e(\mathbf{r}, \mathbf{R})\Phi_m(\mathbf{r}; \mathbf{R}) = U_m(\mathbf{R})\Phi_m(\mathbf{r}; \mathbf{R}) \tag{2.1}$$

are a set of adiabatic wavefunctions $\Phi_m(\mathbf{r}; \mathbf{R})$ that depend parametrically on the nuclear coordinates $\mathbf{R}$, and functions $U_m(\mathbf{R})$ that are the adiabatic potential energy surfaces.[14] Conventional electronic structure calculations may be used to solve Eq. (2.1).

The semiclassical theory begins with the assumption that the nuclei follow some path in time, $\mathbf{R}(t)$. We postpone the definition of this trajectory. The time-dependent electronic Schrödinger equation

$$i\hbar \frac{\partial \Psi}{\partial t}(\mathbf{r}, t) = H_e(\mathbf{r}, \mathbf{R})\Psi(\mathbf{r}, t) \tag{2.2}$$

gives a wavefunction for the entire system Ψ that depends on the nuclear coordinates through $\mathbf{R}(t)$. Because the solutions to Eq. (2.1) form a complete set, the total wavefunction may be written

$$\Psi_m(\mathbf{r}, t) = \sum_i a_i(t)\Phi_i(\mathbf{r})\exp\left[-\frac{i}{\hbar}\int^t U_i(\mathbf{R})\,dt\right]. \tag{2.3}$$

Substituting in (2.2) and using the orthonormality of the Φ_i gives a set of coupled equations for the time-dependent expansion coefficients $a_i(t)$:

$$i\hbar\dot{a}_f = \sum_i a_i\left\langle \Phi_i\left| -i\hbar\frac{\partial}{\partial t}\right|\Phi_f\right\rangle\exp\left[-\frac{i}{\hbar}\int^t (U_i - U_f)\,dt\right], \tag{2.4}$$

where the overdot indicates a time derivative. Since the functions Φ_i depend on time only through $\mathbf{R}$, the matrix element in (2.4) can be rewritten

$$\left\langle \Phi_i\left| -i\hbar\frac{\partial}{\partial t}\right|\Phi_f\right\rangle = i\hbar\langle \Phi_i|\nabla_\mathbf{R}|\Phi_f\rangle \cdot \dot{\mathbf{R}}. \tag{2.5}$$

The final form for the coupled semiclassical equations is

$$\dot{a}_f = -\sum_i a_i\langle \Phi_i|\nabla_\mathbf{R}|\Phi_f\rangle \cdot \dot{\mathbf{R}}\exp\left[-\frac{i}{\hbar}\int^t (U_i - U_f)\,dt\right]; \tag{2.6}$$

Eq. (2.6) is very revealing. Aside from the phase factor, the nonadiabatic behavior depends on the term $\Omega_{if} = \langle \Phi_i|\nabla_\mathbf{R}|\Phi_f\rangle \cdot \dot{\mathbf{R}}$. $\langle \Phi_i|\nabla_\mathbf{R}|\Phi_f\rangle$ is the nonadiabatic coupling matrix element, which can be obtained in an electronic structure theory calculation, and $\dot{\mathbf{R}}$ is the nuclear velocity. If the nonadiabatic coupling matrix element or the nuclear velocity is sufficiently small, the system will behave adiabatically: beginning in a particular electronic state Φ_i, and therefore on the energy surface $U_i(\mathbf{R})$, it will remain on that surface for all time.

Solving Eqs. (2.6) to first order in perturbation theory gives a multidimensional generalization of the well-known Massey[15] criterion for adiabatic behavior:

$$\xi_{if} = \Delta U_{if}/\hbar \langle \Phi_i | \nabla_R | \Phi_f \rangle \cdot \dot{R} = \Delta U_{if}/\hbar \Omega_{if} \gg 1; \qquad (2.7)$$

ΔU_{if} is the splitting between the levels i and f; the dimensionless quantity ξ_{if} is known as the Massey parameter. If $\xi \gg 1$ throughout, the system is adiabatic: a single potential surface governs the dynamics. However, there are many circumstances where ξ is small only in well-defined regions; it is for such situations that the TSH method was designed.

Observe in Eq. (2.6) that $\langle \Phi_i | \nabla_R | \Phi_f \rangle$ is a vector whose dot product with the nuclear velocity $\dot{R}$ governs the coupling. To understand nonadiabatic behavior for systems with more than two atoms, it is critical to consider both the magnitude and direction of this coupling.

Equation (2.6) is the basis for many semiclassical calculations of nonadiabatic processes. To use it, one must define a nuclear trajectory $R(t)$. At high energies when the relative motion is much faster than any internal nuclear motion, a simple straight-line collision may be appropriate; this is the basis for the well-studied impact parameter method.[9] At lower energies, a more detailed trajectory is preferred. When the system is adiabatic, $R(t)$ is logically a trajectory that obeys classical mechanics: a classical trajectory on the potential $U_i(R)$. For weak nonadiabatic coupling, a reasonable choice is again a trajectory that moves on surface i; the semiclassical coefficients are initiated accordingly ($a_k = \delta_{ik}$) and after the trajectory has passed through a region of strong coupling, Eq. (2.6) yields transition probabilities to other surfaces. However, when the coupling is strong, so that a_i falls to zero as some other a_f rises while the trajectory passes through the interaction region, $R(t)$ might more properly begin on surface i and end up on surface f: a nonadiabatic path. In the region of strong nonadiabatic coupling, the very concept of motion in a single surface is breaking down; there is no unique $R(t)$. This is a fundamental source of uncertainty in any trajectory-based semiclassical procedure.

B. Two-State Systems

1. Adiabatic and Diabatic States

In the preceding discussion, no assumption was made about the nature of the potentials or the number of interacting states. Much can be learned about nonadiabatic dynamics by examining the case of two interacting states. Even in real systems with several interacting states, the regions of nonadiabatic coupling between pairs of states may be separate; when this is so, the dynamics can be viewed as evolving through a sequence of two-state interactions. The study of two-state problems has a long history; we again borrow heavily from the work of Nikitin.[11-13] For simplicity, we consider a single nuclear coordinate R.

Consider adiabatic energy levels $U_-(R)$ and $U_+(R)$ with corresponding wavefunctions $\Phi_-(\mathbf{r}, R)$ and $\Phi_+(\mathbf{r}, R)$. It is convenient to view these levels as the eigenvalues of a real symmetric 2×2 diabatic matrix $H_{ij}(R)$. The relationship between these is formally simple:

$$U_\pm = \tfrac{1}{2}[(H_{11} + H_{22}) \pm ((H_{22} - H_{11})^2 + 4H_{12})^{1/2}]. \qquad (2.8)$$

The matrices U and H are related by a mixing angle $\theta(R)$:

$$\begin{pmatrix} U_- & 0 \\ 0 & U_+ \end{pmatrix} = \begin{pmatrix} \cos\theta & \sin\theta \\ -\sin\theta & \cos\theta \end{pmatrix}\begin{pmatrix} H_{11} & H_{12} \\ H_{21} & H_{22} \end{pmatrix}\begin{pmatrix} \cos\theta & -\sin\theta \\ \sin\theta & \cos\theta \end{pmatrix}, \qquad (2.9)$$

where

$$\tan 2\theta(R) = 2H_{12}/(H_{22} - H_{11}). \qquad (2.10)$$

The nuclear derivative coupling between the adiabatic levels U_+ and U_- is conveniently written in terms of the mixing angle $\theta(R)$:

$$\left| \left\langle \Phi_+ \left| \frac{d}{dR} \right| \Phi_- \right\rangle \right| = \left| \frac{d\theta}{dR} \right|. \qquad (2.11)$$

While there exists this formal relationship between 2×2 adiabatic and diabatic matrices, it is important to observe that there is no unique way to specify three elements of the diabatic matrix $H_{ij}(R)$ given two adiabatic curves $U_\pm(R)$. Hence, the definition of a diabatic representation is arbitrary. A useful criterion for a diabatic representation is that, as the system passes through a region of strong coupling, the diabatic states should change very little while the electronic character of the adiabatic states should change quite abruptly. One commonly used diabatic representation is that proposed by Smith:[16] the diabatic states are taken to be those for which the nuclear derivative coupling $\langle \Phi_1 | d/dR | \Phi_2 \rangle$ vanishes. The concept of diabatic states is convenient when discussing nonadiabatic collisions; for a thorough discussion of problems with various diabatic representations, the reader should consult Ref. 8 or 9.

2. The Landau–Zener Model

It is instructive to consider nonadiabatic behavior for certain assumed 2×2 diabatic matrices. The first is the well-known Landau[17]–Zener[18] model, which describes the typical situation at an avoided crossing. One starts with a diabatic matrix in which H_{11} and H_{22} are linear functions of R and H_{12} is constant. If R_c is the position of the crossing of H_{11} and H_{22} and ΔF is the difference of their slopes, then $H_{22} - H_{11} = (R - R_c)\Delta F$. The curves U_+ and U_- exhibit a classic avoided crossing: the energy gap $W(R) = U_+(R) - U_-(R)$ passes through a minimum at R_c; $W(R_c) = 2H_{12}$. If the total energy is large compared to the energy at the crossing, Eq. (2.6) leads to the Landau–Zener transition probability[11]:

$$P_{12} = |a_2(\infty)|^2 = \exp\left(-\frac{2\pi H_{12}^2}{\hbar v \Delta F} \right), \qquad (2.12)$$

where v is the nuclear velocity at R_c.

Consider the evolution of the mixing angle $\theta(R)$ for this model, as a trajectory passes through the region of strong nonadiabatic coupling. θ starts at zero away from the crossing, passes through $\pi/4$ at R_c, and reaches $\pi/2$ on the other side. Far from the crossing on one side, the diabatic and adiabatic states coincide; at the crossing the adiabatic states are an even admixture of the diabatic states; and, on the other side of the crossing, the states again coincide, but with opposite labeling. The nuclear derivative nonadiabatic coupling for this model is a Lorentzian with area $\pi/2$ and width $\Gamma = 4H_{12}/\Delta F$:

$$\langle \Phi_+ | \frac{d}{dR} | \Phi_- \rangle = \frac{\Gamma/4}{(R - R_c)^2 + \Gamma^2/4}. \qquad (2.13)$$

If H_{12} is small, we observe two signatures of a localized nonadiabatic transition: the energy gap W at R is small and the nuclear derivative coupling is very sharply peaked. The Lorentzian shape predicted by this simple model has been observed in *ab initio* calculations of avoided crossings in molecular systems: for example, the calculations of Deharing et al.[19] on H_2O^+ and Desouter–Lecomte et al.[20] on CH_2^+.

3. The Demkov Model

Localized nonadiabatic coupling can occur in a quite different situation, where there is no avoided crossing between the adiabatic levels. Demkov[21] gave the first semiclassical description of this model, devised to describe near-resonant charge-transfer processes. The model again begins with an assumed 2×2 diabatic matrix. H_{11} and H_{22} are parallel curves; the spacing Δ between the diabatic levels is small. The coupling $H_{12}(R)$ decreases exponentially: $H_{12} = A \exp(-\lambda R)$. The diabatic levels do not cross, and there is no minimum in the gap $W(R)$ between the adiabatic curves. At large R,

$\Delta \gg H_{12}$ and the adiabatic curves are parallel; for smaller R, H_{12} increases and the gap grows.

Although there is no avoided crossing, there is a well-defined region of strong nonadiabatic coupling where the term $\langle \Phi_+ | d/dR | \Phi_- \rangle$ has a maximum. Consider again the mixing function $\theta(R)$. At large R, θ and $d\theta/dR$ are zero: the adiabatic and diabatic states coincide. As R decreases, θ increases. For sufficiently small R, $H_{12} \gg \Delta$, $\theta \approx \pi/4$, and $d\theta/dR$ is nearly zero: the adiabatic levels are an even admixture of the diabatic states. The coupling term $d\theta/dR$ passes through a maximum when $H_{12} = \Delta/2$. If one linearizes H_{12} at this point R_c, the shape is again Lorentzian, its width sensitive to the parameter λ, the steepness of the interaction.

In the limit where the velocity v at R_c is large, the transition probability takes a simple form:

$$P_{12} = \exp(-\pi\Delta/\hbar\lambda v) \tag{2.14}$$

with a somewhat more complicated semiclassical result at lower energies. These equations have been tested against exact quantum-mechanical results;[22] the simple equations work quite well at higher energies.

Mixing of spin–orbit states during a collision is an example of a Demkov-type process; a sharp coupling term with a Lorentzianlike shape was found for $F(^2P_J) + H^+$ (or Xe) by Preston et al.[23] and for $F(^2P_J) + H_2$ by Tully.[24]

III.   TRAJECTORY-SURFACE-HOPPING METHODS

A.   Foundations

1.   The General Concept

The TSH method was reviewed previously by Tully in 1973[25] and 1979.[26] The central idea common to all TSH calculations was described by Tully and Preston[6] nearly 20 years ago. Classical trajectories are propagated on a single potential energy surface in the regions where the system is adiabatic. When a trajectory reaches a region of strong nonadiabatic coupling, the trajectory may hop to another surface; the hopping is vertical, and the momenta are adjusted to compensate for the discontinuous change in potential. Energy conservation governs the magnitude of the correction; conservation of total angular momentum is guaranteed if the correction is applied in the internal coordinate frame. The trajectory continues on the new surface; if it again encounters a region of strong coupling, it may hop again, and so on until the collision is over. If the number of possible hops along one trajectory is not too large, each branch of a trajectory may be followed (one branch continues adiabatically, the other hops), the outcome of all branches weighted according to their accumulated probability. Preston and Tully called this the "ants" mode: many branches swarm across the manifold of surfaces. If the trajectories pass through regions of strong coupling too many times, the "anteater" mode may be preferred: a single trajectory results from each set of starting conditions. The trajectory either hops or not at each crossing, the decision made randomly weighted by the transition probability.

For an N atom system with $N \geqslant 3$, trajectories move in a $m = 3N - 6$ dimensional space; a three-dimensional space for $N = 3$. The function $|\langle \Phi_i | \nabla_\mathbf{R} | \Phi_f \rangle|^2$ is likewise m dimensional. The maximum of this function defines an $(m-1)$-dimensional object; a surface for $N = 3$. There may be several surfaces linking states i and f. Tully and Preston call such a surface a "seam," and permit surface hopping only when a trajectory passes through the seam. Variations on the TSH method described below include slightly different definitions of the seam, and in some cases permit transitions in regions near the seam. There is, of course, no guarantee that the seam as defined above implies well-localized transitions: one can imagine situations in which regions of strong coupling extend in all directions through space; in such cases the TSH approach is probably not advisable. Finally, symmetry may place constraints on the function $|\langle \Phi_i | \nabla_\mathbf{R} | \Phi_f \rangle|^2$ at points, lines, or even surfaces in the m-space; these must be considered with care.[27]

The major differences between versions of the TSH method discussed below are the answers to two questions: (1) when and how the seam is defined, and (2) when and how the hopping probability is determined. To a large extent the choice is dictated by the problem: careful preliminary examination of the surfaces and couplings is essential; nevertheless, a number of useful strategies have been proposed. These are discussed in the remainder of this section.

There are other variations in TSH procedures. Different choices have been made for the direction for the momentum correction when a hop occurs. Some workers have chosen to propagate the trajectories on diabatic surfaces. Various hybrid ants-and-anteater schemes have been used. These differences will be noted as individual applications are described in Section IV.

2. Tully and Preston

Tully and Preston[6] (TP) studied the $H^+ + H_2$ reaction. The potential surfaces were obtained by the valence-bond diatomics-in-molecules (DIM) method; the DIM method is particularly convenient for TSH calculations because it gives a consistent set of surfaces and couplings.[28] Two potential surfaces are involved in the H_3^+ problem, one correlating asymptotically to $H^+ + H_2$, the other to $H + H_2^+$. Preston and Tully observed that, for energies relevant to their study, the nuclear coupling function $|\langle\Phi_i|\nabla_\mathbf{R}|\Phi_f\rangle|$ is large only in narrow regions near the crossing of the diabatic surfaces. (The choice of diabatic basis is implicit in the DIM method.) When R, the separation between the atomic and diatomic fragments, is infinite, the diatomic $H_2(^1\Sigma_g)$ and $H_2^+(^2\Sigma_g)$ potential curves intersect at a point r_c; see Fig. 1. As R decreases, interactions begin to couple these states, and the adiabatic curves in r exhibit an avoided crossing. For large R, the energy gap between the adiabatic levels is very small, and the nonadiabatic coupling very sharp. If R is sufficiently large, the dynamics is diabatic: when r passes through r_c, the trajectory changes surfaces. As R decreases further, the interaction increases, the gap increases, and the coupling is broader. Trajectories may now pass r_c

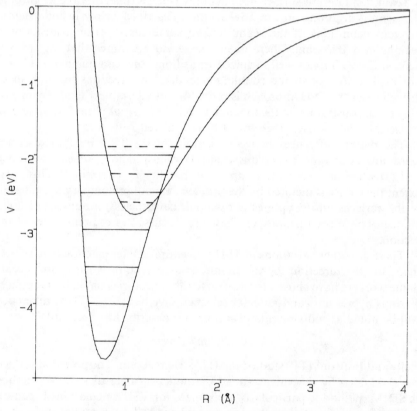

Figure 1. Slices at infinite R of H_3^+ potential energy surfaces, showing asymptotic H_2 and H_2^+. The zero of energy is at $H^+ + H + H$. Note that the crossing is well above the zero point energy for H_2, only slightly above H_2^+.

adiabatically: charge-transfer occurs. To reiterate: a surface hop (that is, transition from U_1 to U_2) means retention of the electronic configuration, adiabatic passage through the seam implies charge transfer. Preston and Tully[6a] found that, in the region where the Massey criterion suggests possible nonadiabatic behavior, the position of maximum coupling does not vary with R: the seam is well defined by the equation $(r - r_c) = 0$. They demonstrated that at the seam the vector $\langle \Phi_i | \mathbf{V_R} | \Phi_f \rangle$ points along r; the direction of maximum coupling is normal to the seam. Therefore, vibrational energy is primarily responsible for inducing nonadiabatic transitions in this system, and the momentum correction is applied along the vibrational coordinate.

Tully and Preston[6b] used semiclassical methods to confirm the dynamical picture inferred from examination of the surfaces and to investigate the hopping probability. Using Eq. (2.6), they ran semiclassical trajectory fragments through the seam for various initial conditions. They found that the Landau–Zener formula, Eq. (2.11), gives excellent agreement with the semiclassical results. In the Landau–Zener equation, H_{12} is taken as one-half the energy gap at the seam, v is the velocity component normal to the seam, and ΔF is a constant, chosen to fit the semiclassical results.

Based on this preliminary work, TP implemented the TSH method as follows: (1) the seam was defined by a simple equation before the trajectories were begun and (2) the hopping probability was calculated using the Landau–Zener formula. The results will be described in Section IV A.

3. *Miller and George*

Shortly after the Tully and Preston[6] work, Miller and George[29] (MG) developed a general semiclassical theory of multidimensional nonadiabatic transitions. As their theory offers a more formal underpinning for some aspects of the TSH approach, and with some simplification suggests an alternate way of implementing the TSH model, we discuss it here briefly.

Using the approach first introduced by Stuckelberg[30] in his investigation of the Landau–Zener model, Miller and George exploit the analytic properties of the potential energy surfaces in the complex plane to induce nonadiabatic transitions. A trajectory does not hop to make a transition between surfaces; rather it follows a continuous path from the initial surface on the real axis, into the multidimensional complex plane, passing smoothly through the branch point where the pair of interacting surfaces is degenerate, and arrives back on the real axis, now on the other surface. The trajectories are still governed by Hamilton's equations, but propagate in a carefully controlled way into the complex plane by choosing a complex time path. The transition probability comes naturally from the semiclassical equations. In a very rough sense, the farther a branch point is from the real axis, the larger the accumulated action in reaching the branch cut, and therefore the smaller the probability.

This method takes skill to implement: simple vibrations grow exponentially depending on the complex time path, so the path must be chosen with great care. Moreover, there are constraints on the route a trajectory may take through a sequence of branch cuts.[31] The search for root trajectories satisfying asymptotic boundary conditions, a standard part of the classical S-matrix[4] method, is complicated by the possibility of many alternate routes through the sequence of branch cuts. Uniform techniques may be required[32] if multiple root trajectories undergo several transitions close to one another in time. Implementation of the Miller–George approach for real systems without some simplification is probably impractical in most cases; root trajectories at sample energies were found for $H^+ + H_2$ by Lin et al.[33,34]

MG theory has the interesting feature that it does not use the coupling terms $\langle \Phi_i | \nabla_R | \Phi_f \rangle$; all the necessary information is implicit in the analytic structure of the adiabatic surfaces. This surprising fact is discussed in Refs. 31 and 35; that it is rigorously so for a two-state system was proven by Dhykne.[36] This has some advantages; couplings are not always available. Yet it is not without cost: one needs reliable complex potentials, generally obtained by analytic continuation of the real functions.

Comparisons have been made of analytically continued potentials with those obtained by direct solution of the electronic structure problem in the complex plane;[37] unless the functions on the real axis are very precise, direct analytic continuation may be perilous. The results are more satisfactory for Landau–Zener-type interactions than for Demkov-type interactions.

One promising simplification of the Miller–George approach uses an approximate decoupling scheme.[38] The coordinate most responsible for the transition (in Tully–Preston language: the direction normal to the seam) is identified first, and the complex part of the trajectory involves this coordinate alone. If quasiclassical quantization is imposed only at the beginning of the trajectory, this is effectively a variant of the TSH method. When the trajectory reaches a seam, the transition probability is determined by a one-dimensional complex integral. This technique was applied to the three-dimensional $F(^2P_J) + H_2$ problem.[39]

MG give a compact equation for the momentum correction to be applied when a trajectory hops in conventional TSH theory. If $\mathbf{p}$ is the momentum vector before the hop and $\mathbf{p}'$ after the hop

$$\mathbf{p}' = \mathbf{p} - \hat{\mathbf{n}}\left(\frac{\hat{\mathbf{n}} \cdot \mathbf{M}^{-1} \cdot \hat{\mathbf{p}}}{\hat{\mathbf{n}} \cdot \mathbf{M}^{-1} \cdot \mathbf{n}}\right)\left(1 - \left[1 - 2\Delta U\left(\frac{\hat{\mathbf{n}} \cdot \mathbf{M}^{-1} \cdot \hat{\mathbf{n}}}{(\hat{\mathbf{n}} \cdot \mathbf{M}^{-1} \cdot \hat{\mathbf{p}})^2}\right)\right]^{1/2}\right), \quad (3.1)$$

where ΔU is the energy gap between the final and initial surfaces, M^{-1} is the matrix of inverse masses, and $\hat{n}$ is a unit vector in the direction in which the correction is to be applied. A sign error has been corrected.[40]

B. Alternate Methods for Locating the Seam and Predicting Hopping

1. *Stine and Muckerman*

The Tully–Preston implementation of TSH is convenient where functional representations of the seam and hopping probabilities are easily found. In general, however, the seam may be have a complicated shape, and there may be several energetically accessible seams. Stine and Muckerman[40] (SM) devised a TSH procedure in which the location of the seam and the hopping probability are determined along with the dynamics. Their method requires no coupling functions; only the adiabatic surfaces are needed. While their approach is both convenient and powerful, it is applicable only to systems where the nonadiabatic coupling may be viewed as arising from an avoided crossing: the energy gap W between the interacting adiabatic surfaces must have a local minimum in one coordinate. The function $W(R)$ is said to have a "troughlike" minimum.[41]

The analysis of Stine and Muckerman begins with the multidimensional two-state problem. One assumption is made: that the off-diagonal diabatic coupling H_{ij} is (locally) constant. This defines a diabatic representation. It follows that the multidimensional surface intersection problem is reduced to an equivalent one-dimensional curve-crossing problem. From the adiabatic surfaces alone, it is possible to locate the seam and calculate the Landau–Zener transition probability, using information calculated along the trajectory. There are important practical advantages with this method: there is no need to define a functional form for the seam, and nonadiabatic coupling terms are not needed. The method has limitations; this has been discussed in the literature.[27,42]

The central assumption in the SM analysis is a constant diabatic coupling. The seam is defined as the $(m-1)$-dimensional surface on which the (implicit) diabatic surfaces cross: $\Delta(R) = H_{22} - H_{11} = 0$. On this seam the adiabatic energy gap function $W(R) = U_2 - U_1$ has zero gradient, and β, the hessian matrix of W, $(\beta_{ij} = \partial^2 W / \partial R_i \partial R_j)$, has a single nonzero eigenvalue, which is positive. Thus the interaction is one-dimensional: there is a unique direction along which W has nonzero (positive) cuvature. SM further demonstrate that this direction, specified by any nontrivial row of β evaluated at the seam, coincides with the vector normal to the seam. As a trajectory evolves, the seam is identified by the condition $\nabla_R W = 0$ and its direction lies along the eigenvector corresponding to the unique nonzero eigenvalue of β. It is not even necessary to diagonalize β; the unique eigenvalue $\bar{\beta}$ is simply the trace of the hessian matrix. In the one-dimensional case, $\bar{\beta}$ corresponds to ΔF, the difference in the diabatic force constants. The direction normal to the seam is

$$\mathbf{n} = \nabla_{\mathbf{R}}\Delta(R) = (2H_{12}/\beta_{11})(\beta_{11}, \beta_{12}, \ldots, \beta_{1n}), \tag{3.2}$$

where $(\beta_{11}, \beta_{12}, \ldots, \beta_{1n})$ is any nontrivial row of β, and all terms are evaluated in the seam. H_{12}, the coupling, is one-half of the adiabatic energy gap W at the seam.

Since the multidimensional interaction has been reduced to an effective one-dimensional process, the transition probability is calculated using the Landau–Zener formula, Eq. (2.11). The denominator for the 1D case, $v\Delta F$, is replaced with its vector equivalent, $\dot{\mathbf{R}} \cdot \nabla_{\mathbf{R}}\Delta$. Using notation corresponding to Eq. (3.1) and $\mathbf{n}$ defined by Eq. (3.2), SM show that

$$\dot{\mathbf{R}} \cdot \nabla_{\mathbf{R}}\Delta(\mathbf{R}) = \mathbf{n} \cdot \mathbf{M}^{-1} \cdot \mathbf{p}\left(2 - \left[1 \pm 2H_{12}\left(\frac{\hat{\mathbf{n}} \cdot \mathbf{M}^{-1} \cdot \hat{\mathbf{n}}}{(\hat{\mathbf{n}} \cdot \mathbf{M}^{-1} \cdot \hat{\mathbf{p}})^2} \right) \right]^{1/2} \right). \tag{3.3}$$

The SM formalism is based on the assumption of constant H_{12}. How reasonable is this assumption? Globally, it is clearly wrong. Consider the troughlike intersection in H_3^+ studied by Preston and Tully[6a]: the interaction between the two asymptotic electronic systems $(H^+ + H_2)$ and $(H + H_2^+)$ obviously increases as the fragments approach one another. However, such variaton of H_{12} along a seam is often gradual. What about H_{12} normal to the seam? Desouter-Lecomte et al.[20] carried out *ab initio* electronic structure calculations for CH_2^+, in the region of an avoided intersection. Defining a diabatic basis using the nuclear derivative coupling matrix elements, they observed that H_{12} is very nearly constant through the coupling region.

If H_{12} is slowly varying, modified versions of the Stine–Muckerman equations can be applied; this is the method as actually implemented. The bottom of the troughlike function $W(\mathbf{R})$ is not flat, but may slope and curve somewhat. The condition $\nabla_{\mathbf{R}}W = 0$ is therefore relaxed; instead one looks for a minimum in W along the trajectory, that is, $\dot{W} = 0$. Since $\dot{W} = \nabla_{\mathbf{R}}W \cdot \dot{\mathbf{R}}$, W may pass through zero for three reasons: $\nabla_{\mathbf{R}}W$ is zero, $\dot{\mathbf{R}}$ is zero, or the vectors are orthogonal.[40] The first two conditions are very improbable; the third is satisfied when the trajectory passes through the seam, but it is also satisfied elsewhere. A second condition is added: the magnitude of $\nabla_{\mathbf{R}}W$ must be below some tolerance. Finally, to improve efficiency, these criteria are applied only in those regions where the Massey parameter indicates the likelihood of a nonadiabatic transition; the nonadiabatic couplings are required for this test.

If H_{12} is not constant, the hessian matrix β has more than one nonzero eigenvalue. The vector corresponding to the largest eigenvalue defines the direction normal to the seam. That one eigenvalue is significantly larger than all others should be a useful check on the validity of the method. Using DIM surfaces for H_4^+, Stine and Muckerman compared this definition to the Tully–Preston definition of the direction of the seam, $\langle \Phi_i | \nabla_R | \Phi_f \rangle$; the results are similar.

The Stine–Muckerman procedure is designed for systems with avoided crossings. Can it be applied when there are conical intersections? Mead and Truhlar[27] raise this question. In practice, it depends of the system. Consider first a symmetric triatomic, A_3. In equilateral configurations, it has intrinsically degenerate (E) electronic states. Motion along either of the two symmetry breaking nuclear coordinates lifts the degeneracy: the familiar conical intersection. Contours of the energy difference W around the equilateral point are roughly circular. This gap is not troughlike, so the Stine–Muckerman method would not apply. The conical intersection in H_3 was studied by Blais, Truhlar, and Mead[43] using a variation of the TSH method described in Section II C.

AB_2 molecules also exhibit conical intersections. The electronic states are in general nondegenerate, but states of different symmetry may have an accidental degeneracy. Σ and Π states may cross in collinear $(C_{\infty v})$ and A_1 and B_1 states in perpendicular (C_{2v}) configurations. When the symmetry is broken, these state can mix. As in A_3, there is a curve in configuration space where such surfaces are degenerate; they are split in the two transverse directions. However, what is different in this case is that the two degeneracy-breaking coordinates are not equivalent. Contours of ΔW in the symmetry-breaking coordinates around the point of degeneracy in this system may be quite elliptical, so the troughlike function required for the Stine–Muckerman procedure is a reasonable approximation.

The $Ar + H_2^+$ system[44] for which there are DIM surfaces[45] has such a conical intersection. Define R as the Ar^+ to H_2 distance, r as the H_2 bond length, and γ as the angle between them. At large R, the curve along which there is a conical intersection is parallel to R. The gap between the two A' states which are Σ and Π in collinear $(\gamma = 0°)$ geometries remains very small for any value of the symmetry breaking coordinate γ. On the other hand, the gap increases rapidly moving away from the crossing along r. Except for the numerical challenge of locating very sharp seams, the SM procedure should apply.

2. Kuntz, Kendrick, and Whitton

Kuntz, Kendrick, and Whitton[46] (KKW) devised another variant to the TSH method, which, like SM, allows the location of the seam and the hopping probability function to be determined along the trajectory. The KKW method is not restricted to systems with troughlike energy gaps, but it does require nuclear coupling terms $\langle \Phi_i | \mathbf{V_R} | \Phi_f \rangle$.

The KKW method allows a trajectory to branch between surfaces i and f at the time t_c when the quantity $\Omega_{if}(t) = \langle \Phi_i | \mathbf{V_R} | \Phi_f \rangle \cdot \dot{\mathbf{R}}(t)$ passes through a maximum. Recall that in the TP prescription, the seam is at the maximum of $|\langle \Phi_i | \mathbf{V_R} | \Phi_f \rangle|$. Since the latter is sharply peaked, the KKW criterion is most easily met at the TP seam. However, it may also be met in different circumstances: (1) the trajectory passes close to the TP seam, but is deflected away, that is, a classical turning point near the TP seam; (2) away from the TP seam, in a region where the magnitude of $\mathbf{R}$ and $\langle \Phi_i | \mathbf{V_R} | \Phi_f \rangle$ are slowly varying, the angle between them passes through a minimum; (3) away from the seam, the magnitude of $\langle \Phi_i | \mathbf{V_R} | \Phi_f \rangle$ and the angle between the vectors are slowly varying, but $\dot{\mathbf{R}}$ passes through a maximum; or (4) some combination of (2) and (3). To eliminate (2) through (4), a second condition is imposed at t_c:

$$\hbar |\langle \Phi_i | \mathbf{V_R} | \Phi_f \rangle| / \Delta U_{if} < 0.2.$$

Situation (1) is more interesting: the trajectory has a turning point near the seam. In many cases, energy conservation will prevent a hop: if a trajectory on the lower surface has insufficient energy to reach the seam, is usually will have insufficient energy to reach the upper surface. However, in the case of an avoided crossing of two surfaces that slope in the same direction, or in the case of a Demkov seam, the KKW method may permit hopping near the TP seam. This may well be an advantage: the same criterion is used in the TSH procedure of Parlant and Gislason,[47] discussed in section II. C. 2, and the results for model systems were excellent.

The hopping probability at t_c is calculated using the equation[48]

$$P_{if} = \exp(-\tfrac{1}{2} \pi \Delta U_{if} \Delta t / \hbar), \tag{3.4}$$

where Δt is the half-width at half-height of the function $\Omega_{if}(t)$. This is determined assuming that $\Omega_{if}(t)$ is Lorentzian, with parameters found along the trajectory. Note that the width Δt is sensitive to the velocity normal to the seam. Equation (3.4) reduces to the Landau–Zener and Demkov formulas when applied to those model systems.

In applying this procedure to the $Ne + He_2^+$ system, with has a seam running along the entrance valley, Kuntz et al.[46] experienced numerical difficulties at large R when the coupling function $\langle \Phi_i | \mathbf{V_R} | \Phi_f \rangle$ is very narrow. A simple but effective solution was offered: when the neon was more than $4.2 \, \text{Å}$ from either helium, the trajectory was forced to follow the diabatic surfaces.

C. Combining TSH with the Time-Dependent Semiclassical Equations

1. Blais and Truhlar

Blais and Truhlar[49] (BT) introduced another TSH procedure in their study the electronic quenching reaction $Na\,(3p^2P) + H_2 \rightarrow Na(3s^2S) + H_2$. Whereas Tully and Preston[6b] used the semiclassical equations (2.6) in preliminary work only, Blais and Truhlar incorporate these equations in the trajectory calculation. If the system starts in state i, the semiclassical coefficients are initiated as $a_k = \delta_{ik}$. The trajectory code integrates both Hamilton's equations on surface U_i and the semiclassical equations. At the first step after the coefficient $|a_i(t)|^2$ falls below 0.5, a random decision is made whether to hop, weighted by the probabilities $|a_j|^2$. If a hop is made to surface f, a momentum correction is applied, along the vector $\mathbf{V}_R W_{if}(\mathbf{R})$. Whether a hop is made or not, the coefficients are reset. The process continues until the trajectory is done.

In a study of the effects of the conical intersection on collision-induced dissociation in H_3, Blais, Truhlar, and Mead[43] introduced some refinements to this method: locating accurately the exact position where $|a_i|^2 = 0.5$ and applying the hop there, and making the correction for a hop between i and f along the vector $\langle \Phi_i|\mathbf{V}_R|\Phi_f \rangle$ rather than along $\mathbf{V}_R W_{if}$. The former ensures that the hopping is not sensitive to the integration step size; the latter, now the same as TP, was motivated by analytical work by Herman.[50] For a small sample of trajectories, they compared the two choices of coupling direction: interestingly, even though the vectors were often nearly perpendicular, the results were not very different.

The Blais–Truhlar procedure has a number of attractive features. Unlike SM, it can be applied for any kind of coupling; there is no restriction to troughlike energy gaps. Unlike TP, there is no need to define the seam before running the trajectories. While the semiclassical coefficients change significantly only in regions of strong coupling, the condition $|a_i|^2 = 0.5$, and therefore the hop, will not necessarily occur at the maximum in $\langle \Phi_i|\mathbf{V}_R|\Phi_f \rangle$. BT argue that this is more realistic: electronic transitions in quantum mechanics are not strictly localized.

Nevertheless, the BT method is not completely general; Stine and Muckerman[51] discuss this when they compare several TSH procedures. The decision to hop when $|a_i|^2 = 0.5$ is quite arbitrary, and reinitializing the coefficients in the middle of a region of strong coupling is counterintuitive: contribution to a hopping probability from part of a strong-coupling region may end up forcing a hop at a different seam. This is not necessarily a problem: averaging over many crossings and trajectories may produce the more quantumlike nonlocalized hopping, which BT argue is realistic. There are two cases when the BT procedure would seem to give incorrect results. The first is at a seam where passage between the relevant surfaces is essentially diabatic; for example, at large R for $H^+ + H_2$ or $Ar^+ + H_2$. In this region, the semiclassical equations, integrated all the way through the seam coupling states i and f, would show essentially complete exchange between a_i and a_f. The TP or SM procedures would both force a hop, while the BT procedure would make only half of the trajectories hop, charge transfer occurring over physically unreasonable distances. This is, of course, an extreme example with a simple remedy; even without the unphysical adiabatic passage, considerations of numerical accuracy in solving the semiclassical equations would mandate that strictly diabatic behavior must be imposed in this region.

The second situation where the BT method would also underestimate surface hopping is when many trajectories pass only once through a seam with moderately weak coupling. Imagine, for example, a two-state system for which the population in the initial state, $|a_i|^2$, falls to about 0.75 through the seam. The BT method would result in no hopping, instead of the 25% hopping which the semiclassical equations imply.

Eaker[52] compared the SM and BT variants of the TSH methods for quenching of the electronic excitation $Na(^2P) + H_2 \rightarrow Na(^2S) + H_2$. The coupling is entirely in the interaction region; there is no essentially diabatic region. Based on contour plots, the energy gaps appear to be troughlike, at least for the coordinates shown. Eaker found that the choice for the direction of the momentum correction was not important. The two methods use comparable computer time. However, hopping probabilities (and therefore quenching cross sections) did differ, by as much as 30%. SM trajectories hopped to one of the three surfaces much more readily. Which is better? Eaker argues in favor of BT, believing that the picture of nonlocal hopping is more realistic for this problem. However, uncertainties in the potentials make it impossible to use comparisons with experiment to make a definitive choice.

2.   *Parlant and Gislason*

Quite recently Parlant and Gislason[47] (PG) have developed what they term an "exact" TSH method. Similar in spirit to BT, it avoids some of the problems of that method. It is only slightly more complicated to implement, and looks promising. The recipe for locating the seam is identical to that of KKW: the trajectory branches when the quantity $\Omega_{if} = \dot{\mathbf{R}} \cdot \langle \Phi_i | \nabla_{\mathbf{R}} | \Phi_f \rangle$ passes through a maximum. For reasons discussed previously, it is necessary to supplement this condition with the Massey condition. If the trajectory hops, the momentum is corrected along $\langle \Phi_i | \nabla_{\mathbf{R}} | \Phi_f \rangle$.

The PG procedure differs from KKW in the method for calculating the hopping probability, and therefore the weights associated with the two branches. As in the Blais–Truhlar procedure, the semiclassical equations are propagated along with the trajectory. The novel feature is that the coeffficients a_i are initialized at the point on a trajectory when Ω is at a minimum, and their evolution is followed until, having passed through a maximum, Ω again reaches a minimum. The transition probabilities are therefore determined by semiclassical equations on a trajectory segment that represents one complete passage through a region of strong coupling. PG applied this method to one-dimensional Landau–Zener and Demkov models, with impressive results.[47]

Parlant and Alexander[53] applied this method to the system He + $CN(A^2\Pi_J)$. They eliminated some branches, using a modified "anteater" algorithm. They relaxed slightly the requirement that Ω pass through a minimum, replacing it with the requirement that it falls to 10^{-4} times the energy gap; that is, that the Massey parameter exceeds 10^4. The results were compared to quantum-scattering calculations, the agreement generally quite good. One problem they cite is the failure of the TSH results to satisfy microscopic reversibility; this is a common problem with classical trajectory methods.

The PG method appears to eliminate some of the inconsistencies of the Blais–Truhlar procedure. There may be some ambiguities when several states interact (several Ω_{ij} exist, with separate maxima and minima). Nevertheless, it is an interesting development.

D.   Electron-Detachment Processes

Problems of electron detachment (Penning ionization, associative ionization, and so on) may be studied with a TSH-like approach if some assumption is made about the mechanism of the electron loss. MacGregor and Berry[54] studied the following reaction, important in hydrocarbon flames:

$$O + CH \rightarrow HCO^+ + e^-$$

The assumption that the detached electron carries away no kinetic energy establishes the relative positions of the CHO and CHO^+ surfaces. The electron is assumed to detach only at the intersection of these surfaces. The Landau–Zener equation was used for the hopping probabilities. In this problem, the system can pass through a seam adiabatically only one time: once the electron detaches, it is gone. In this study the dynamics was simplified by neglecting CH rotation and vibration. A related strategy, but with full 3D dynamics, was used for electron detachment in $Cl^- + H_2$ collisions by Sizun, Gislason, and Parlant[55]; this study will be discussed in Section IV.

Preston and Cohen[56] devised a model, called "trajectory surface leaking," which allows electron detachment to occur over a broad region. They were interested in ionization processes in $He^*(2\,^3S) + H_2$ collisions. At any point on a trajectory where the neutral surface (HeH_2^*) is above the ionic surface (HeH_2^+), there is a probability for the electron to detach, that is, for the system to leak to the lower surface. Assuming a function $\Gamma(R)$, the probability for the system to leak in a time interval Δt is $P_{leak} = \Gamma(R)\Delta t/\hbar$. At each integration step of length Δt, if a random number is less that P_{leak}, the electron detaches, and carries off the energy difference as kinetic energy. With this model, they were able to simulate rate constants and branching ratios for the various ionized products quite well.

IV. APPLICATIONS

A. Ion–Molecule Reactions: Charge-Transfer and Competing Processes

1. H_3^+

The first detailed application of the TSH method was the study of the simplest ion–molecule reaction $H^+ + H_2$ (and isotopic variants). Tully and Preston[6] reported results for $H^+ + D_2$ at 4 eV; Preston and Cross[57] for $H + D_2^+$ at 2 eV; and Krenos et al.[58] compared theory and crossed-beam experiment for $H^+ + D_2$ at energies 3, 4.5, and 6 eV, and $D^+ + HD$ at enrgies 3, 4, 5.5, and 7 eV. Comparisons included total cross sections, product energy distributions, and velocity angle contour plots. The agreement between theory and experiment was very good; particularly in view of the fact that the comparisons were based on absolute cross sections: there are no adjustable parameters in this theory.

The complexity of this "simple" system is illustrated by the range of processes that can occur at energies above about 2 eV:

$$H^+ + D_2 \rightarrow D^+ + HD \qquad \Delta H = 0.04\,eV$$
$$\rightarrow H + D_2^+ \qquad \Delta H = 1.87\,eV$$
$$\rightarrow D + HD^+ \qquad \Delta H = 1.85\,eV$$

with dissociative channels opening at higher energy. Moreover, some experimentally observable processes may include several microscopic mechanisms. The TSH method, distinguishing such contributions, can contribute significantly to the interpretation of the experimental results.

The seam, located in the entrance (and by symmetry in the exit) channel normal to the vibrational coordinate, is inaccessible for the early part of an $H^+ + H_2(v_i)$ trajectory unless the molecule is in a high vibrational state (Fig. 1). Motion in the strongly interacting region (where only one surface is energetically accessible) determines which valley the trajectory exits, that is, the final arrangement of the nuclei. The nascent molecule may have considerable vibrational excitation, causing the trajectory to cross the seam as the products separate. The final position of the charge depends on the number of diabatic or adiabatic crossings of the seam. Therefore, products in the same valley (e.g., $D^+ + HD$ and $D + HD^+$) have similar angular and energy distributions.

$H + H_2^+(v_i)$ collisions are quite different: the seam (Fig. 1) is only slightly above $v_i = 0$ reagents, so trajectories will cross the seam as the molecules approach. The deep H_3^+ potential well correlates with reagents $H^+ + H_2$, not with $H + H_2^+$. Charge transfer must therefore precede rearrangement, so they will have a similar energy dependence. Both are strongly sensitive to the vibrational state. Energy must be channeled into vibration before the $v_i = 0$ trajectories can undergo charge transfer, while for higher vibrations, charge transfer occurs as soon as the coupling is strong enough.

New experimental methods capable of determining cross sections for ionic reactions with unprecedented precision inspired a recent resurgence of interest in the $H^+ + H_2$ system. Schlier, Nowotny, and Teloy[59] report an extensive comparison of TSH calculations with guided-beam experimental integral cross sections. The TSH calculation is essentially as in Ref. 6: it includes a DIM potential (with slightly altered diatomic curves), Landau–Zener hopping probability, and a hybrid branching algorithm (both branches generated only when $0.05 < P_{LZ} < 0.95$, with anteaters evoked after four branches were started). Three isotopic combinations ($H^+ + D_2, D^+ + D_2$, and $D^+ + H_2$) were studied for a range of translational energies, with the molecule initially in the $v = 0, J = 1$ state; 4000 trajectories were generated for each point. Agreement with the earlier TSH study[58] was excellent; the newer results significantly better resolved because of a much larger number of trajectories.

Figure 2 shows the cross sections for the observable channels in $H^+ + D_2$. The power of the TSH method (and the beautiful precision of the guided-beam results[60]) are evident. Recall that these are comparisons on an absolute scale, with no adjustable parameters. The overall shapes (and sizes) of each function $\sigma(E)$ are all well represented. Moreover, more-detailed structure is represented (albeit at slightly shifted energies): the doubly peaked structure in the HD^+ channel between 2.5 and 6 eV, and a shoulder on the rising D_2^+ channel at about 3 eV. The agreement is not perfect: the threshold behavior for the D^+ and HD^+ channels is not correct. Classical trajectories are known to be unreliable near threshold because of tunneling; this is probably the explanation here. However, in this case the tunneling is not through an overall barrier to reaction on the potential energy surface (there is no such barrier), but rather through the barrier to the vibrational coordinate that the lower adiabatic surface has at the seam. A second serious discrepancy is the falloff

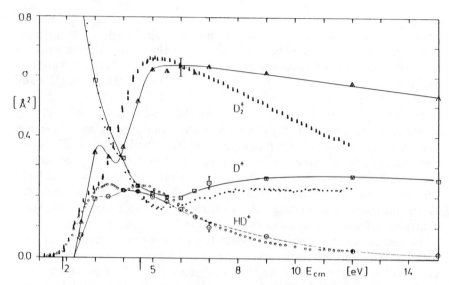

Figure 2. Integral cross sections for the $H^+ + D_2$ reaction, labeled according to the ionic product. The TSH results of Schlier et al.[59] are connected with a line. Error bars indicate typical TSH statistical uncertainty. The experimental results of Ochs and Teloy[60] are shown as points. The comparison between theory and experiment is on an absolute scale. (Reproduced with permission from Ref. 59.)

of the D_2^+ (charge-transfer) cross section at higher energy; the experimental falloff is markedly steeper than the TSH result. At this energy, the use of classical mechanics should not be a problem, so the fault is probably with the DIM potentials. As was argued previously, the shape of the repulsive wall on the ground-state surface in the interaction region has a profound effect on the charge-transfer process, since it determines the vibrational energy the receding molecule and therefore the curve crossing. Better H_3^+ potentials exist,[61] particularly for the lowest surface, so future calculations should resolve this discrepancy.

Lower energy $H^+ + H_2$ collisions are less likely to include nonadiabatic effects, and more likely to involve long-lived complexes. Very long-lived trajectories may present problems for the integration. Gerlich et al.[62] studied lower-energy collisions on a single surface for various isotopic combinations, and devised a useful criterion to identify a true complex, that is, one whose dissociation properties may quite reasonably be treated statistically. This criterion was used in the TSH calculations. Complex formation is a significant process (cross sections on the order of 0.5 Å^2 or more) for energies below 3.0 eV; by 4.5 eV is negligible.

Niedner, Noll, Toennies, and Schlier[63] studied charge transfer and inelastic collision in $H^+ + H_2(v_i = 0)$ at 20 eV, both with experiments and by the TSH method. Doubly differential cross sections for scattering angles out to $18°$ in the center-of-mass frame were obtained in a crossed-beam apparatus for both channels $[H^+ + H_2(v_i)$ and $H + H_2^+(v_i)]$; separate peaks for the product vibrational states (v_f) are clearly resolved. These results give direct experimental evidence for the two-step mechanism for charge transfer discussed previously: vibrational excitation to a level $v_f \geqslant 4$ on the ground-state surface followed by charge transfer between the receding products. This is seen is two ways; (1) the similarity of the angular distributions for the $H + H_2^+$ and $H^+ + H_2$ (higher v_i) channels, and (2) nearly identical angular-dependent average vibrational energy transfer to the molecular product for H_2^+ and $H_2(v_f \geqslant 4)$; the equivalent function for H_2 in lower vibrational states is markedly different.

The TSH procedure was implemented as in Ref. 59. The calculation was done twice: one set with 50,000 trajectories with impact parameters selected randomly to a maximum of 4 Å, and a second set with 2000 trajectories each at fixed b values $0, 0.1, \ldots, 1.0, 1.25, \ldots, 3$ Å. The latter set gives a higher density at small b, and therefore a better resolved $P(b)$. It was found that all trajectories for $b \geqslant 0.8$ Å were confined to the lowest surface. The results were compared as a check in the statistical error in the calculation.

TSH and experimental results were compared for state-to-state differential cross sections; the results are shown in Fig. 3. For the inelastic channel, the results are satisfactory for the total cross section (all v_f) and for lower vibrational states $(v_f \leqslant 3)$; somewhat less so for $v_f = 4$. Both the overall shape and the rainbow features (slightly displaced) are reasonably well reproduced. The comparison is poorer for the charge-transfer channel: while the overall and relative magnitudes for the different v_f's are reasonably good, there are significant disparities; in particular, the pronounced rainbow structure in the experiment is not seen in the TSH results. The angular dependence of average vibrational energy transfer is likewise in good agreement for the inelastic channel, less so for charge transfer. The discrepancies are larger in the comparison of integral vibrational cross sections. For the inelastic channel, TSH and experiment agree for $v_f = 1$, but deviations increase quite significantly for larger v_f's, where the TSH values fall of faster than experiment. For the CT channel, the experimental cross sections fall rather steeply with v_f, while the TSH results peak at $v_f = 1$ and fall off more gradually above. While the discrepancy for $v_f = 0$ may be attributed to tunneling, it is more difficult to understand the higher v_f results. It should be noted that a recent IOSA calculation by Baer et al.[64] using the same potential gets significantly better agreement with experiment for some of these more detailed results, clearly indicating that the failure is in the TSH dynamics, not the surface.

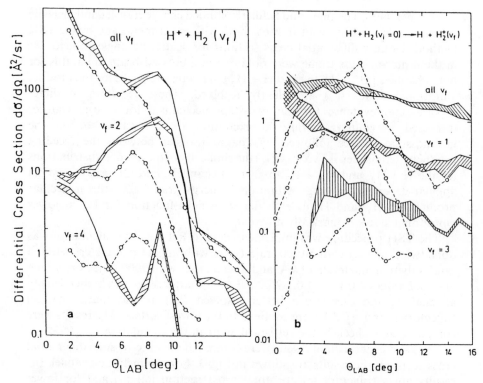

Figure 3. TSH and experimental state-to-state differential cross sections for $H^+ + H_2$ at $E_{CM} = 20$ eV. Panel (*a*) shows representative inelastic channels, panel (*b*) the charge-transfer channels. The two versions of the TSH calculation bound the hatched areas, and the experimental results are shown by open circles. (Reproduced with permission from Ref. 62.)

The H_3^+ system provides a stringent and largely successful test for the TSH method. As might be expected, as one examines more resolved data (vibrational, angular, or both), discrepancies with experiment increase, nevertheless, the classical model succeeds to a remarkable degree. The atoms are light, the experimental data are very precise, and highly accurate potential surfaces are possible (although not yet used). The successes are very promising. Yet in one important respect, this sytem does not challenge the TSH method. One might expect that results of a surface-hopping study will be sensitively dependent on precise definition of the surface hopping probability. However, as stressed in Refs. 6 and 59, this is not so for H_3^+. The seams are in the entrance valleys, parallel to the radial coordinate R. There is a rather narrow range of R, 4.5 ± 1.0 Å, where trajectories may branch: at larger R the crossing is diabatic, and at smaller R adiabatic (the upper surface rises steeply out of range). Depending on the translational energy, a typical trajectory passes through this branching window once or twice on its way to products. A change in the Landau–Zener coefficient ΔF shifts this window a bit (a factor of 2.5 shifts the window about 0.7 Å[59]) but this has little effect on the results; this was tested in both studies. The effect might be somewhat larger for charge transfer cross sections in $H + H_2^+$ collisions where the seam is encountered on the approach; but even here the effect of a 2.5 change in ΔF on charge transfer cross sections would probably be no larger that 33%.

2. ArH_2^+

Study of the ion–molecule reaction $Ar^+ + H_2$ has a long history, both experimentally and theoretically. At energies of a few eV, cross sections are quite large for both charge transfer (CT) and chemical reaction (R):

$$Ar^+ + H_2 \rightarrow Ar + H_2^+ \qquad \Delta H = -0.31\,eV \qquad (CT)$$

$$\rightarrow ArH^+ + H \qquad \Delta H = -1.53\,eV \qquad (R)$$

Kuntz and Roach[45] generated DIM potential energy surfaces for ArH_2^+. The interacting surfaces are in some ways similar to H_3^+: a seam in the entrance channel arises from the crossing of the $Ar^+ + H_2$ and $Ar + H_2^+$ states; but because Ar and H_2 have very similar ionization potentials, this crossing in the ArH_2^+ system is near the bottom of the H_2 well; see Fig. 4a. Thus, in contrast to $H^+ + H_2$, $Ar^+ + H_2$ trajectories encounter the seam as the reactants approach even for $v_i = 0$. Like $H^+ + H_2$, the seam is normal to the vibrational coordinate r, and is well represented by a simple function: $r - r_c = 0$.

Chapman and Preston[44] used the Kuntz and Roach surfaces as the basis of a TSH study of the reaction $Ar^+ + H_2(v_i = 0)$. While spin–orbit coupling is absent from these surfaces (and therefore from the dynamics), the directional properties of the Ar^+ ion give rise to new complexities in the surfaces. In the presence of (but still distant from) the H_2 molecule, the degeneracy of the ^2P states of the Ar^+ is lifted. Of the three states formed in arbitrary (C_s) symmetry, two are $^2A'$, symmetry with respect to the molecular plane, and one $^2A''$. Ignoring spin–orbit and Coriolis coupling, the $^2A''$ state cannot lead to either product above, and is excluded from the dynamical study. The remaining $^2A'$ states can interact with the $Ar + H_2^+$ state, which is also $^2A'$; the last correlates adiabatically with products. Thus there are *three* interacting states in the entrance channel, with two seams between them (Fig. 3b).

The semiclassical equations (2.6) were used to study the behavior of the coupled three-state system as trajectories moved through the coupling region. It was observed that one of the couplings (labeled 2–3 in Fig. 4b) is essentially diabatic, so the three-state problem is effectively a two state problem. The

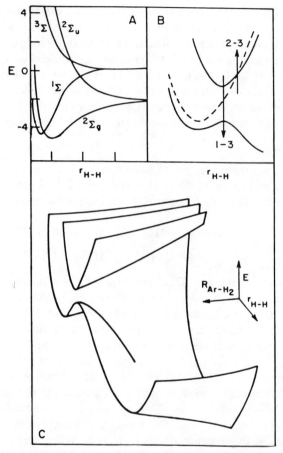

Figure 4. Views of the ArH$_2^+$ potential surfaces in the entrance (Ar–H$_2$)$^+$ valley. Panel a shows the asymptotic H$_2$ and H$_2^+$ curves. The zero of energy is Ar$^+$ + H + H; the H$_2^+$ curves dissociate to Ar + H + H$^+$. Note that the crossing is near the bottom of both wells. Panel b shows the three-lying curves in the region where hopping begins, $R \approx 10$ bohr. States 1 and 3 (solid curves) interact strongly, while state 2 (dashed) interacts weakly. Trajectories all hop at the 2–3 crossing. Panel c is a schematic picture of the two strongly interacting surfaces in the region where nonadiabatic behavior begins. (Reproduced from Ref. 68.)

two interacting surfaces are shown schematically in Fig. 4c. Symmetry, as discussed in Section III B, plays an important role in this decoupling: the state (shown with a dashed line in Fig. 4b), which is uncoupled by symmetry in both C_{2v} and $C_{\infty v}$, interacts so weakly in other geometries that the coupling can be ignored. As in the H$_3^+$ system, the Landau–Zener formula with fixed coefficient ΔF provided an excellent fit to the semiclassical transition probabilities for the remaining two interacting states.

The initial TSH calculation was carried out at a single collision energy, $E_{CM} = 3.36$ eV. Absolute cross sections for charge exchange and chemical reaction agreed quite well with experiment. Charge exchange dominated for larger impact-parameter collisions and rearrangement for smaller, as often assumed in simple models, but there was a significant region of overlap, underscoring the need for a self-consistent treatment of these processes.

Because of several beautiful new experimental studies[63,65–67] on the ArH$_2^+$ system, the earlier TSH calculation was extended[68] more that a decade later. The surface and the TSH procedure were as before. Ar$^+$ + H$_2$ was studied from low (0.13 eV) to higher (3.44 eV) translational energies; cross sections were compared with a number of experiments. At energies above 1 eV, the agreement is satisfactory: CT cross sections (≈ 20 Å^2) roughly constant, and reaction cross sections falling. Angular distributions of the ArH$^+$ product were strongly peaked at higher energy, reflecting a stripping mechanism, while forward–backward symmetry (and formation of short-lived complexes) was seen at lower energy; again in qualitative accord with experiment.[65] Agreement with experiment is less satisfactory at the lower energies, particularly as compared with precise guided-beam results[69] published after the TSH study: at the lowest energy studied the experimental cross section is more than twice the calculated one.

Ar + H$_2^+$(D$_2^+$) was studied for a range of translational (1–9 eV) and vibrational energies ($v_i = 0$–4). Cross sections are compared with the results of Houle et al.[66] in Table I. These are absolute comparisons, with no scaling or flexible parameters. Agreement is better at higher energies than at lower; the calculated values again too small at lower energies. Experimental determination of absolute cross sections is challenging; the calculated cross sections are closer to those of Ref. 67, which are generally smaller than those Ref. 66. One possible source of error in the calculations at lower energies is the shape of the attractive part of the potential: at long range the ion–molecule interaction should go as R^{-4}; the DIM surface lack this.

The trends in the cross sections are interesting. As in Ar$^+$ + H$_2$, both experiment and theory show that the CT channel has a weak translational energy dependence, while the reactive channel decreases. Vibrational energy effects are more pronounced. For the reactive channel, the dependence of the total cross section on v_i is rather weak: $\sigma(v_i)$ generally increases from $v_i = 0$ to 2, and levels off thereafter; the TSH results generally capture this trend well. The charge-transfer channel exhibits a more striking vibrational dependence: increasing by an order of magnitude from $v_i = 0$ to $v_i = 1$. This was observed both in the experiment and the calculation. In the experiments the cross sections were higher still for $v_i = 2$, falling off somewhat for higher

v_i; the TSH results decreased after $v_i = 1$. For Ar + $D_2^+(v_i)$, both experimental and TSH cross sections were largest for $v_i = 2$.

The strong vibrational effect can be understood in terms of the interacting potential surfaces. The seam is normal to the vibrational coordinate, so vibrational energy governs hopping. The crossing of the asymptotic curves is between $v_i = 0$ and $v_i = 1$ for H_2, and between $v_i = 1$ and $v_i = 2$ for D_2.

TABLE I
Ar$^+$ + $H_2(D_2)$ Cross Sections (Å^2)

v	Ar + $H_2^+(v) \rightarrow$ ArH$^+$ + H		Ar + $H_2^+(v) \rightarrow$ Ar$^+$ + H_2	
	TSH[a]	Expt[b]	TSH[a]	Expt[b]
		1 eV		
0	15.0 ± 1.3	51.8	1.9 ± 0.5	3.06
1	28.6 ± 3.1	66.1	28.6 ± 3.1	17.7
2	26.2 ± 2.9	69.1	22.4 ± 3.6	47.2
4	22.5 ± 2.6	73.4	21.1 ± 3.6	24.4
		3 eV		
0	5.1 ± 0.8	12.3	1.4 ± 0.6	1.64
1	8.9 ± 1.3	30.4	25.3 ± 2.1	20.6
2	11.1 ± 1.7	44.6	24.1 ± 3.7	44.9
		6 eV		
0	4.1 ± 0.8	4.95	2.0 ± 0.8	1.51
1	4.4 ± 0.8	12.5	30.3 ± 4.2	24.8
8	4.1 ± 0.8	18.2	17.5 ± 2.9	38.9
		9 eV		
0	3.4 ± 0.7	3.17	$2.1 + 0.8$	2.06
v	Ar + $D_2^+(v) \rightarrow$ ArD$^+$ + D		Ar + $D_2^+(v) \rightarrow$ Ar$^+$ + D_2	
	TSH[a]	Expt[b]	TSH[a]	Expt[b]
		1 eV		
0	14.4 ± 2.1	50.7	1.5 ± 0.6	2.32
1	20.7 ± 2.5	57.2	6.8 ± 1.7	13.3
2	29.3 ± 3.2	60.8	30.1 ± 4.3	32.8
3	27.7 ± 3.1	57.1	25.4 ± 3.9	30.3
4	22.2 ± 2.6	58.1	24.9 ± 3.7	25.5
		3 eV		
0	4.4 ± 0.8	9.74	1.0 ± 0.4	3.0
1	5.1 ± 1.1	17.5	8.3 ± 2.5	13.3
2	10.7 ± 1.6	27.1	24.6 ± 3.8	38.0
3	10.8 ± 1.8	26.5	28.8 ± 4.0	35.9
		6 eV		
0	4.4 ± 0.8	3.03	1.1 ± 0.4	1.89
2	3.9 ± 0.7	9.0	22.7 ± 3.3	15.1

[a] Reference 68. The uncertainities represent ± 1 standard deviation.
[b] Reference 66.

When the vibrational energy exceeds the crossing energy, facile charge transfer occurs as soon as the coupling permits adiabatic passage through the seam: the cross sections are large (and largely independent of translational energy). When the vibrational energy is below the crossing energy, such long-range charge transfer is prohibited.

This important energy relationship offers a simple explanation of the main discrepancy between experiment and theory. Because the surfaces omit spin–orbit coupling, the energy of the state correlating to $Ar^+ + H_2$ represents an average $Ar^+(^2P)$ energy. If spin–orbit were included, this surface would split into two: the lower for $^2P_{3/2}$ and the upper for $^2P_{1/2}$: see Fig. 5. While Ar^+ spin–orbit coupling is small (0.1775 eV), its effect on the relative energies of the crossings and the vibrational levels is significant. A TSH calculation, using surfaces which incorporate these spin–orbit splittings, is clearly in order.

Both Tanaka et al.[67] and Houle et al.[66] emphasize the importance of a resonance: the particularly large charge-transfer cross section for $Ar + H_2^+(v_i = 2)$ results from a near energy resonance with $Ar^+(^2P_{1/2}) + H_2(v_f = 0)$. A TSH calculation, even with spin–orbit coupling, would fail to give enhancement for such a resonance. Baer et al.[70] recently reported a very promising new method to treat such problems quantum mechanically, an

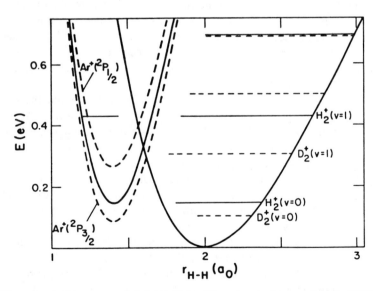

Figure 5. ArH_2^+ curves at infinite R. The solid curves are those used in the DIM surfaces which neglect spin–orbit coupling. The dashed curves represent the locations of the corrected surfaces if spin–orbit were included. Note the shift in the crossings with respect to the energy levels of H_2^+ or D_2^+. (Reproduced from Ref. 68.)

applied it to the 3D $Ar^+ + H_2$ reaction. Agreement with the TSH results using the Kuntz and Roach surfaces is quite satisfactory, and a three-state calculation, including the spin–orbit coupling has been completed. Detailed comparison with a TSH study on these improved surfaces would be interesting. The ArH_2^+ system will challenge us for some time to come.

3. HeH_2^+

Nonadiabatic effects in the HeH_2^+ system are very different from ArH_2^+, since the ionization potential of He greatly exceeds that of Ar. Low-energy $He + H_2^+$ collisions are adiabatic: the $He^+ + H_2$ surfaces are higher in energy. The $He^+ + H_2$ reaction has nonadiabatic effects, but in this case, the asymptotic H_2 curve intersects the dissociative $(^2\Sigma_u)$ state of H_2^+, rather than the bound $(^2\Sigma_g)$ state. Preston, Thompson, and McLaughlan[71] devised a simple model, using surface-hopping ideas but circumventing a detailed TSH calculation, to predict the effects of molecular vibration on the dissociative charge-transfer process:

$$He^+ + H_2(v) \rightarrow He + H + H^+.$$

They used *ab initio* electronic structure techniques to calculate properties of the surfaces near the seam. They found that the coupling is weak: for collinear geometries even at $R_{He-H} = 3$ bohr, the gap at the avoided crossing is small, leaving a residual barrier to dissociation in the H_2 coordinate. The interaction is weak in HeH_2^+, as compared to that in H_3^+, because these two states have a symmetry allowed crossing in C_{2v} geometry: a conical intersection. Recall the similar situation in ArH_2^+.

The dynamical model developed for this problem was based on several assumptions: (1) energy transfer between translation and vibration or translation and rotation is negligible, (2) the curve-crossing dynamics involves the vibrational coordinate only, and (3) the collision induces transitions through the R dependence of the coupling. The one-dimensional predissociation coordinate was treated using a semiclassical solution to the nearly diabatic coupling problem of Nikitin,[11] which, in effect, includes a tunneling correction to the Landau–Zener transition probability. Once a trajectory has crossed the seam adiabatically, the H_2^+ flies apart, so there is no need to continue the trajectories. The cross sections were obtained by a two-dimensional quadrature over impact parameter b and separation R.

The vibrational effect, which can be obtained in closed form, is pronounced: a 94-fold enhancement on the cross section from $v_i = 0$ to $v_i = 1$. This is another example of the strong effect of vibrational energy when a seam is normal to the vibrational coordinate. Thermal rate constants and isotope effects were in qualitative agreement with experiment, although it was suggested that a more accurate treatment of the dynamics would be needed to obtain quantitative agreement.

At low energies the ion–molecule reaction $He + H_2^+ \rightarrow HeH^+ + H$ is adiabatic, amenable to conventional quasiclassical trajectory methods. However, the process of collision-induced dissociation

$$He + H_2^+ \rightarrow He + H^+ + H \qquad \Delta H = +2.66\,eV \qquad \text{(CID)}$$

can exhibit some very interesting nonadiabatic behavior, as studied in work by Sizun and Gislason,[72] and Sizun, Parlant, and Gislason.[73] When the three fragments are infinitely separate, the ground-state surface is degenerate, corresponding to the arrangements $He + H^+ + H$ and $He + H + H^+$. As the two hydrogens approach, the degeneracy is lifted, corresponding to the $^2\Sigma_g$ (bound) and $^2\Sigma_u$ (dissociative) H_2^+ states. The CID process starts on the lower of these surfaces, but as the fragments recede, there is the possibility of a nonadiabatic transition to the upper surface. There is no avoided crossing; this is a Demkov-type nonadiabatic event (see Section II B).

This is an interesting phenomenon theoretically, but one might guess that it is not observable experimentally: the products and asymptotic energy are the same. Such a guess would be incorrect: Gislason and Guyon[74] argue that the He atom breaks the degeneracy of the two potential energy surfaces, even if the hydrogen atoms are very far apart. At long range the He–H$^+$ attraction (which has an r^{-4} dependence) is stronger than that of He–H (which falls off as r^{-6}). The relative positions of the fragments thus label the final surface: the system is in the ground state if the charge is on the hydrogen atom nearer to the helium. This in turn can be observed: the fast protons in the center of mass frame imply upper state CID.

Sizun and Gislason[72] used the TSH method for this problem. The DIM potential energy surfaces for the two relevant states were developed by Whitton and Kuntz,[75] but modified to give the proper long-range attractive form for the diatomic functions, essential for CID. The multidimensional generalization of the Demkov relationship

$$2H_{12}(R) = H_{22}(R) - H_{11}(R) \qquad (4.1)$$

defines the surface of maximum nonadiabatic interaction (the term "seam" is not entirely appropriate here, since the diabatic surfaces are not stitched together, but will be used in for convenience). The seam cuts diagonally through the ridge at large R_{He-H} and R_{H-H} where the upper surface is energetically accessible, roughly dividing the He–H and H–H regions.

The trajectories begin on the lower surface. At each step, Eq. (4.1) is used to determine whether the seam has been crossed. If so, the trajectory branches, with hopping probability calculated using the equation

$$P_{12} = \tfrac{1}{2} \text{sech}^2(-\pi H_{12}/\hbar v_\perp \alpha |\cos \theta|), \tag{4.2}$$

where $\alpha = |d \ln 'H_{12}/dr_{H-H}|$, $v_\perp$ is the component of velocity normal to the seam, and θ is the angle between r_{H-H} and the normal to the seam. Since H_{12} depends on r_{H-H} only, the $\cos \theta$ term projects the component of the coupling in the direction normal to the seam. A simple procedure is used to predict the contribution to CID products from quasibound trajectories, those trapped behind a centrifugal barrier with sufficient energy to dissociate. These represented as much as 15% of the product.

Cross sections for reaction ($HeH^+ + H$ products) and for CID were calculated for initial vibrational states $v_i = 0, 3, 6$, and 10 at energies 3.1, 5, and 10 eV. The results are shown in Table II. The reactive cross sections are small and decrease with translational energy. The total cross sections for CID increase with trranslational energy for $v_i = 0$, and increase with v_i at all three translational energies. The results are in qualitative accord with previous single-surface trajectory calculations, with experiment, and with simple models for CID. The $v_i = 10$ results show an interesting opposite trend: they are quite large, and decrease with translational energy. This was attributed to a different mechanism: most CID, requiring large energy transfer, involves small impact-parameter collisions. However, for $v_i = 10$, only 0.53 eV below CID products, larger impact-parameter collisions, drawn in by the attractive force, can lead to dissociation of the highly extended H_2^+.

TABLE II

He + $H_2^+(v)$ cross sections (Å^2)[a]

E(eV)	v	σ_R	σ_{D1}	σ_{D2}
3.1	0	0.51	0.031	0.016
	3	0.65	0.27	0.14
	6	0.61	0.99	0.59
	10	0.48	3.00	1.95
5.0	0	0.39	0.27	0.19
	3	0.30	0.55	0.34
	6	0.24	0.97	0.66°
	10	0.14	2.39	1.98
10.0	0	0.08	0.60	0.42
	3	0.05	0.73	0.57
	6	0.04	0.89	0.72
	10	0.02	1.73	1.51

[a] Reference 72. R = reaction; $D1$ = dissociation on the ground surface; $D2$ = dissociation on the excited surface.

What is particularly interesting in this study is the role of the upper potential surface. At $E = 3.1$ eV and $v_i = 0$, roughly one-third of the CID products emerged on the upper surface. This fraction increases with both translational and vibrational energy. Both trends can be explained in terms of the location of the Demkov seam.

Velocity-angle differential scattering contour maps for the ionic product (as would be measured in a crossed-beam apparatus) were constructed from the TSH results at 5 eV for $v_i = 0, 3$, and 10. The reactive scattering was sharply forward peaked for $v_i = 10$, and shifts to wider angles with lower vibrational energy. The CID products show a similar trend: for $v_i = 10$ the angular distribution is quite sharp, with H^+ scattered forward with respect to the initial H_2^+ velocity vector. This is a further consequence of the higher impact-parameter contribution to the CID process at this high vibration. For lower vibrational states, the H^+ scatters more widely, still mainly in the forward hemisphere for $v_i = 3$, but sidewise peaked for $v_i = 0$. However, it is again the contributions from the two surfaces which is of particular interest: the two contributions are barely distinguishable for $v_i = 10$, whereas there are distinguishable peaks for the two components for lower v_i. These results were analyzed further and compared with experimental results in Ref. 73.

4. $NeHe_2^+$

Kuntz, Kendrick, and Whitton[46] studied competing reactive (ion transfer: IT) and dissociative (collision-induced dissociation: CID, and collision-induced predissociation: CIP) processes for the reaction $Ne + He_2^+$:

$$Ne + He_2^+ \rightarrow Ne + He^+ + He \qquad \Delta E = \quad 2.47 \text{ eV} \qquad \text{(CID)}$$

$$\rightarrow NeHe^+ + He \qquad\qquad\quad = -0.88 \text{ eV} \qquad \text{(IT)}$$

$$Ne^+ + He + He \qquad\qquad\quad = -0.55 \text{ eV} \qquad \text{(CIP)}$$

By simple correlation arguments, it is clear that the first two processes are adiabatic: both may occur without surface hopping. CIP is nonadiabatic. This processes is equivalent to dissociative charge transfer described in $He^+ + H_2$.

Diatomics-in-molecules surfaces for thus system were constructed and analyzed by Kuntz and Whitton.[76] The surfaces for this system are similar in some ways to those described previously. As in H_3^+, ArH_2^+, and HeH_2^+, an important avoided-crossing seam lies in the entrance channel, normal to the vibrational coordinate. Like the HeH_2^+ system, this asymptotic crossing is between a bound (He_2^+) and dissociative ($He + He$) state. When the neon approaches, these states are mixed; an electron jump (at the first adiabatic passage through the seam) results in immediate dissociation of the He_2. As

TABLE III
Ne + He$_2^+$ Cross Sections (Å)^a

E_{tot} (eV)	E_{vib} (eV)	CIP	CID	IT
2.7	0.2	2.7	—	0.48
	0.8	31.8	—	0.02
	1.7	24.4	0.68	0.06
6.0	0.2	17.5	0.16	0.18
	0.8	24.8	1.74	—
	1.7	14.6	4.54	0.01
10.7	0.2	16.0	0.50	0.01
	0.8	22.3	2.64	—
	1.7	12.3	5.13	0.01

^{a}Reference 46; symbols defined in text.

in ArH$_2^+$, the picture is in principle complicated by the existence of three components of the Ne$^+$(^2P) state. However, as in that case, consideration of symmetry and approximate symmetry makes it possible to reduce the problem to one which effectively involves only two interacting states. Thus for the TSH calculations, the more elaborate DIM matrix[76] was replaced with a simpler reduced one, effectively treating the Ne$^+$ ion as if it were ^2S. As noted by the authors,[46] it is exactly this simplification which produces the very popular LEPS semiempirical potential energy function[1-3] for adiabatic $A + BC$ reactions.

The specific TSH procedure used by these authors was described in Section III B 2. They studied the Ne + He$_2^+$ process at three total energies (2.7, 6.0 and 10.7 eV) each for three vibrational energies (0.2, 0.8, and 1.7 eV); 0.2 eV in vibration is below the asymptotic crossing; 0.8 and 1.7 are above it. Cross sections for the three processes are given in Table III. The reactive channel is very minor: NeHe$^+$ is weakly bound, so it is difficult to stabilize it. The results are in qualitative accord with experiment (in which the initial vibrational state is not well resolved).

5. HeN$_2^+$

At high collision energies, the vibrational distribution in the products of the collision-induced electronic excitation reaction

$$He + N_2^+ (X\ ^2\Sigma_g^+) \rightarrow He + N_2^+ (B\ ^2\Sigma_u^+)$$

is Franck–Condon. At lower energies, deviations from FC are observed. Kelley and Harris[77] used a TSH-like method to study this process. Diabatic surfaces for the two states govern the dynamics: the system changes from one to the other at the crossing with a calculated probability. Since the diabatic surfaces cross at the seam, there is no discontinuous hop in the potential; hence, no need for a momentum correction. The forces change discontinuously if the trajectory changes diabatic surfaces, so if the integration routine uses past values, it must be restarted; this is a trivial matter. The probability for changing diabatic states at the seam (and therefore moving from surface X to surface B) was calculated using a Landau–Zener-based equation

$$P = 1 - \exp(- H_{XB}^2 \cos^2 \theta /(\dot{\mathbf{R}} \cdot \nabla_{\mathbf{R}} \Delta U),$$

where the denominator may be recognized as the velocity normal to the seam. The numerator is the coupling strength. The $\cos^2 \theta$ term makes the coupling vanish for C_{2v} geometries, where the two states have different symmetries: A_1 and B_2. H_{XB} was taken to be constant, chosen to give a convenient amount of reaction. Two pairs of diabatic surfaces were used; one with the crossing 4 eV above reactants, the other 15 eV.

Trajectories were run on both sets of surfaces for a wide range of translational energies, with a 300 K thermal distribution of internal states in the N_2^+. Cross sections rose steeply from threshold, but fell off at higher energies. As energy increases for both sets of surfaces, the $N_2^+(B)$ vibration changes from broad to very narrow, moving towards a FC distribution. The product rotational distribution exhibits similar sharpening.

6. ClH_2^-

Sizun, Gislason, and Parlant[55] used the TSH method to study the $Cl^- + H_2$ reaction. With the simple assumption that an electron (with a specified energy) will immediately detach when a trajectory reaches the diabatic crossing between the ClH_2^- and ClH_2 systems, they describe a multiplicity of possible processes: elastic collision (EL), reaction (R), dissociative charge transfer (DCT), dissociation (D), reactive detachment (RD), simple detachment (SD), and dissociative detachment (DD):

$$
\begin{array}{lll}
Cl^- + H_2 \rightarrow Cl^- + H_2 & \Delta H = 0.0\,eV & (EL) \\
\rightarrow HCl + H^- & = 2.91 & (R) \\
\rightarrow Cl + H + H & = 7.34 & (DCT) \\
\rightarrow Cl^- + H + H & = 4.49 & (D) \\
\rightarrow HCl + H + e^- & = 3.66 & (RD) \\
\rightarrow Cl + H_2 + e^- & = 3.60 & (SD) \\
\rightarrow Cl + H + H + e^- & = 8.09 & (DD)
\end{array}
$$

Barat et al.[78] have carried out extensive experiments on this system, at collision energies ranging from 5.6 to 12 eV. They found that the detached electron has energy in the range 0.0–0.4 eV. Three sets of surfaces were therefore constructed, with asymptotic energies between the ionic and neutral surfaces corresponding to 0.0, 0.2, and 0.4 eV. If a trajectory moves (with no change in momentum) from the ionic diabatic surface to the neutral surface at a seam where they cross, this implies ejection of a 0.0, 0.2, or 0.4 eV electron.

A LEPS function describes the neutral ClH_2 surface, while the ClH_2^- surfaces were generated using a 5×5 DIM matrix. Three surfaces are relevant to the dynamics. Asymptotically, the neutral surface lies between the ionic surfaces; the crossings are in the interaction region. Since the trajectories run on diabatic surfaces, a procedure was devised to reduce the 5×5 DIM matrix for ClH_2^- to a 2×2 whose eigenvalues (i.e., adiabatic surfaces) were a good approximation to the eigenvalues of the full matrix.

Collinear slices of the ionic surfaces, including the locations of the seams for zero kinetic energy electrons, are shown in Fig. 6. The diabatic surface on which the $Cl^- + H_2$ system begins correlates to dissociative HCl^-: if the trajectory stays on this surface only processes EL or D are possible. The other diabatic ionic surface allows the R and DCT channels. The seam between them runs along the product valley (i.e., normal to the HCl coordinate), and cuts diagonally through the interaction region. The neutral surface includes the detachment channels (RD, SD, and DD). The neutral surface crosses the lower ionic surface on a seam which runs parallel to the entrance valley (normal to the H–H coordinate), but at high enough energy that it is not encountered by incoming trajectories; this seam also cuts diagonally through the interaction region. If the electron carries away energy, the relative positions of the ionic and neutral surfaces are shifted.

The trajectories follow diabatic surfaces. When a trajectory on an ionic surface encounters a seam with the neutral surface, it switches with unit probability (and irreversibly) to the neutral surface. When a trajectory encounters a crossing between ionic surfaces it branches, with probabilities calculated using the Landau–Zener formula, as implemented by Stine and Muckerman.[40] All branches are carried to completion ("ants"). To test the sensitivity to the branching probability, results were compared in which the coupling strength in the potential between the ionic surfaces was doubled.

The reaction was studied at 9.7 eV. All of the processes indicated previously

---

Figure 6. The two ionic diabatic surfaces for collinear $(H–H–Cl)^-$. Coutours are shown on 0.5 eV intervals. Reactants $Cl^- + H_2$ begin at the lower right in panel (*a*); products $HCl^- + H$ exit on the upper left of panel (*b*). The intersection between the ionic surfaces is shown with a dashed curve. Dotted curves show the intersection with the neutral ClH_2 surface, assuming the ejection of a zero kinetic energy electron. (Reproduced with permission from Ref. 55a.)

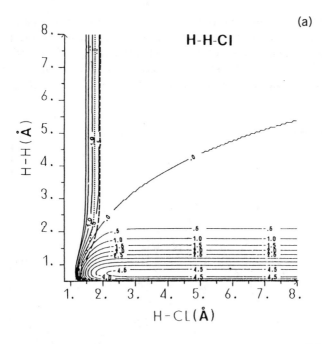

(a)

H-H-Cl

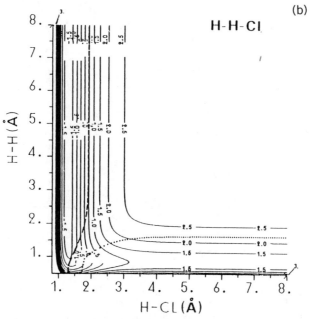

(b)

H-H-Cl

are quite endoergic, and all cross sections were found to be small, in accord with experiment. The surfaces that corresponded to 0.2 eV electrons was deemed most realistic, and for that case agreement with experiment was "fairly good." The effect of doubling the coupling H_{12} is quite dramatic: the R and SD cross sections nearly double, while the RD channel falls. This is

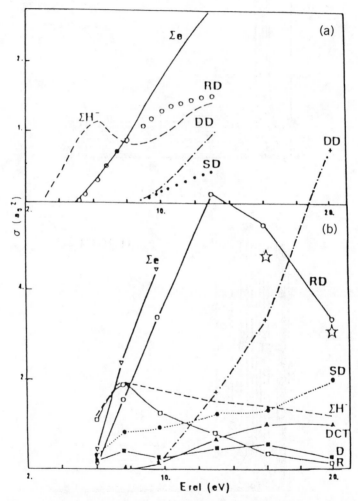

Figure 7. Total cross sections for $Cl^- + H_2$ products. The upper panel shows experimental results (Ref. 78) and the lower panel results of the TSH calculation (Ref. 56). Σe^- is the sum of all electron detachment channels, ΣH^- all channels which produce H^-. (Reproduced with permission from Ref. 56.)

an important point: in the H_3^+ and ArH_2^+ studies, it was observed that the results were not particularly sensitive to the hopping probability; here (and probably in many cases) the results are quite sensitive to the probability.

Angle and internal energy distributions as well as energy-angle contour maps were examined and compared to experiment; agreement was quite satisfactory, including such interesting features as a sharp sidewise peak at about 35° for the SD channel. Details from the trajectories shed light on the mechanisms for the various processes. For example, a very distinctive dependence was found between the angle γ (the angle between the reactant radial and diatomic vectors) at the first encounter of the ionic seam and the eventual outcome: near perpendicular orientations strongly favoring the detachment channels. Since the ionic states are uncoupled in C_{2v} geometries, the explanation is quite simple. Trajectories start on the lower ionic surface (Fig. 6a). They first encounter the (dashed) seam with the other ionic surface. If the geometry is near C_{2v}, they continue diabatically, and almost immediately reach the neutral (dotted) seam, where their fate is sealed: the electron detaches. If the geometry is far from C_{2v}, they are more likely to move to the other ionic surface (Fig. 6b) where they bounce off the repulsive wall in any number of directions.

Sizun, Parlant, and Gislason[55b] studied the same system at a number of energies between 6 and 20 eV, comparing the results to experiment.[78] Their total cross sections are shown in Fig. 7. The agreement is generally good, the TSH calculation again providing excellent interpretive value.

7. H_2O^-

Herbst, Payne, Champion, and Doverspike[79] studied a similar array of ion–molecule and electron-detachment processes in $O^- + D_2$ collisions:

$$\begin{aligned} O^- + D_2 &\rightarrow OD^- + D & \Delta E &= -0.2\,\text{eV} \\ &\rightarrow OD + D^- & &= +0.9\,\text{eV} \\ &\rightarrow O^- + D + D & &= 4.6\,\text{eV} \\ &\rightarrow O + D + D^- & &= 5.3\,\text{eV} \\ &\rightarrow D_2O + e^- & &= -3.5\,\text{eV} \\ &\rightarrow O + D_2 + e^- & &= 1.5\,\text{eV} \\ &\rightarrow O + OD + e^- & &= 1.6\,\text{eV} \end{aligned}$$

They used the trajectory surface leaking model, described in Section III D, to describe the electron detachment. The electron can detach anywhere when the ionic surface is above the neutral; if it does so, the trajectory drops to the lower surface with its momentum unchanged, the electron implicitly carrying off the energy difference.

A LEPS surface was used for D_2O^-, and a Sorbie–Murrell[80] surface for D_2O. As a simple approximation, Herbst et al. assume that the probability of detachment is independent of position: Γ is constant. They further simplify the calculation by not following trajectories on the neutral surface; thus they can obtain only the total cross section for detachment processes. With this simple model, they compare cross sections and simple properties of the velocity-angular distribution with experiment. Results at 6.9 and 8.6 eV agree satisfactorily with experiment; the agreement is rather poor at 1.2 eV, suggesting the need for either an improved treatment of the detachment process or for better surfaces.

8. H_4^+

The reaction $H_2 + H_2^+ \to H_3^+ + H$, exothermic by about 2 eV, is of widespread theoretical, experimental, and applied interest. It is a facile process, with a thermal cross section on the order of 100 Å^2, and is known by isotope studies to proceed either by proton or atom transfer.

DIM surfaces for the H_4^+ system have been constructed by several groups.[81-83] The surfaces are the eigenvalues of an 8×8 DIM matrix; the studies differ only in the representation of the four required diatomic curves, $H_2(^3\Sigma_u^+$ and $^1\Sigma_g^+)$, and $H_2^+ (^2\Sigma_g$ and $^2\Sigma_u)$. The DIM surfaces agree reasonably well with *ab initio* results: the minimum energy path for the reaction above goes smoothly and adiabatically to products; there is no barrier, but there is a shallow H_4^+ minimum. There are several crossings and avoided crossings between the lower surfaces; symmetry alone demands this.

There are two low-energy surfaces in the entrance valley. The seam between them is similar to that in H_3^+ and ArH_2^+; it connects the diabatic valleys representing the two charge arrangements. Using isotopic labeling to distinguish them, the two low-energy states correspond to $H_2^+ + D_2$ and $H_2 + D_2^+$. By symmetry, the equation for this seam is $r_{H-H} = r_{D-D}$. Thus motion normal to the seam is diatomic vibration. As R, the separation between the molecules, decreases, the gap at the avoided crossing grows, and the coupling becomes broader. Therefore, as in H_3^+ and ArH_2^+, the diabatic surfaces will govern the dynamics for sufficiently large R, the adiabatic surfaces for sufficiently small R. As before, the transition from diabatic to adiabatic behavior is found to be complete in a rather small range of R. Let us denote R_t as the midpoint in this range.

Making the simple assumption that trajectories will pass through the seam diabatically if $R > R_t$, and adiabatically otherwise, Eaker and Schatz[84] studied $H_2 + H_2^+$, and Eaker and Muzyka[85] $H_2 + D_2^+$ and $H_2^+ + D_2$. In the few cases where a trajectory was not on the lowest adiabatic surface as it crossed R_t, it continued diabatically until it reached the lowest surface. The dynamics is much simplified; this is now effectively a single surface calculation.

For most cases R_t was taken to be 8 bohr, in Ref. 84 it was found that the H_3^+ products were quite insensitive to this choice.

Eaker and Schatz[84] studied the $H_2^+(v_+, J = 2) + H_2(v = 0, J = 1)$ reaction at three translational energies (0.11, 0.46, and 0.93 eV) for $v_+ = 0$ and $v_+ = 3$. The H_3^+ product can be identified as resulting from either proton (PT) or atom transfer (AT); the contributions were nearly identical at all energies. The cross sections decrease with translational energy, in accord with experiment.[86] Except at the highest energy, the calculated cross sections were larger than experiment. This discrepancy may represent incorrect long-range forces in the DIM potential. The effects of vibration are less pronounced: decreasing the cross sections at lower energy, and increasing at higher. Angular distributions showed that the PT and AT contributions peaked in opposite directions, the AT forward with respect to the H_2^+; the PT backward. Both are evidence of a stripping mechanism: the former direct, the latter after an early charge transfer.

The product energy distributions were examined in several ways. Most of the available energy emerged as vibration. Compared with a microcanonical statistical prediction, the trajectories had less translational and more rotational energy; vibration was roughly statistical for $v_+ = 0$, and higher for $v_+ = 3$.

Detailed internal-state distributions for the H_3^+ product were calculated by applying histograms to actions generated by a Fourier transform method developed by Schatz and co-workers,[87,88] and extended in Ref. 84 to rotating molecules. This method can be applied only to lower energy H_3^+ molecules whose motion is quasiperiodic. The results are decidedly nonstatistical; particular combinations of quanta in the symmetric and degenerate vibrational modes are preferred. The trajectories exhibited a strong (and nonstatistical) propensity for vibrational angular momentum $l = 0$, indicating that the preferred motion is more librational than precessional.

Eaker and Muzyka[85] focused on the translational and vibrational energy dependence of the cross sections. Agreement with experiment[89] was quite good.

The preceding treatment[84,85] simplifies the problem in several critical ways. it treats the seam in the entrance valley with a hopping probability that is a step function in R, and ignores any other surface interactions. In a careful analysis of their DIM surfaces, Stine and Muckerman[40,43] presented a number of slices through these six-dimensional functions. They argue that, while the lowest state in a perfect tetrahedral geometry is triply degenerate, the trajectories are very unlikely to explore this neighborhood, since the steepest descent reaction path (which falls steeply downhill) is planar. Thus, a dynamical treatment based on sequential pairwise interactions is deemed satisfactory. Nevertheless, depending on the available energy, surfaces other than the lowest two may be encountered.

Several full TSH studies of this system have been made. Using their version of TSH (as described in Section III B) Stine and Muckerman[40] showed a representative trajectory, and Muckerman[90] reported cross sections. Eaker and Schatz[84] compared the results of their simplified model to Muckerman's data and to experiment[86]; these results are shown in Table IV. While the agreement for $v_+ = 0$ is excellent, for $v_+ = 3$ the cross sections are significantly larger for the simple trajectories. The results of the full TSH study are in better accord with experiment.

Eaker and Schatz[91] subsequently reported a more detailed TSH study for this system, using the Stine–Muckerman version of TSH. The full TSH studies[90,91] differ only in details of the input to the DIM surfaces, and the results agree. In comparing the simpler trajectory study to the full TSH, Eaker and Schatz show that the main problem in the simpler study was the treatment of the trajectories that are not on the ground-state surface as they pass R_t. The simple trajectory program allowed these to fall diabatically to the lowest surface at the first opportunity, and therefore contribute to H_3^+ products (making its cross section too large). In the full TSH study, these trapped trajectories continued on the upper surface until they eventually bounce back to give inelastic or charge transferred products. When $v_+ = 3$, many more trajectories are trapped in this manner. Eaker and Schatz compare TSH results for $H_2^+(v_+) + D_2$ and $D_2^+(v_+) + H_2$ to experiment.[89] Results are shown in Table V. Even at the lowest energy, 0.23 eV, the agreement is good.

There is currently a great deal of interest in properties of highly excited H_3^+. In an experiment in which excited H_3^+ moleules are fragmented with 800–1100 cm^{-1} photons, Carrington and Kennedy[92] observed some 27,000 lines, which, when coarse-grained, exhibit a simple structure. Schatz, Badenhoop, and Eaker[93,94] investigated the issue of highly excited H_3^+ from the perspective of its formation, from $H_2^+(v_+, J = 2) + H_2(v = 0, J = 1)$ collisions; extending v_+ much higher than in their earlier work. Extra care was taken with metastables: H_3^+ products whose internal energy exceeded the dissociation energy. These were aged for an additional time T (during which some dissociate). The dependence of the cross sections on T (which was varied from 0 to 9.9 ps) gives a measure of the stability of these metastables.

The H_3^+ internal energy distributions from 0.5 eV translational energy collisions (with $T = 0.36$ ps) depend strongly on v_+: the maxima are at energies above dissociation for $v_+ = 13$ and $v_+ = 17$. These results are shown in Fig. 8. The metastables show substantial decay as they are aged up to 9.9 ps; the decay appeared to have both a short- and a long-term component. The distribution of rotational angular momentum J peaks near $J = 25$ for

TABLE IV

Cross Sections for $H_2^+(v_+, J = 2) + H_2(v = 0, J = 1)$

E(eV)	v_+	PT[a]	AT[a]	Total[a]	Total[b]	Total[c]
0.11	0	137	127	261 ± 8		182 ± 7
	3	124	125	249 ± 8		150 ± 11
0.46	0	55	54	109 ± 5	117 ± 6	89 ± 9
	3	58	61	120 ± 5	100 ± 6	71 ± 7
0.93	0	33	32	65 ± 3	61 ± 4	64 ± 7
	3	39	43	82 ± 3	69 ± 6	61 ± 7

[a] TSH calculation (Ref. 84) PT = proton transfer; AT = atom transfer.
[b] TSH calculation (Ref. 90); values as interpolated in Ref. 84.
[c] Experiment (Ref. 86).

TABLE V

Cross Sections (bohr^2) for Isotopic Variants of $H_2^+(v_+) + H_2$

E(eV)	v_+	$H_2^+(v_+) + D_2 \rightarrow D_2H^+ + H(PT), H_2D^+ + D(AT)$				
		PA[a]	PT[b]	PT[c]	AT[a]	AT[b]
0.23	0	31	31	29	17	17
	3	29	23	25	17	16
1.1	0	8.2	8.2	13	6.7	6.7
	3	11	10	13	7.2	5.5
2.1	0	5.3	5.3	5.8	4.6	4.6
	3	7.3	7.0	9.9	5.0	3.3
4.1	0	2.1	2.1	3.0	2.4	2.4
	3	3.9	4.0	5.0	2.4	2.1
6.1	0	1.6	1.6	2.2	1.4	1.4
	3	2.4	2.3	2.6	1.5	1.2

E(eV)	v_+	$D_2^+(v_+) + H_2 \rightarrow H_2D^+ + D (PT), D_2H^+ + H(AT)$				
		PT[a]	PT[b]	PT[c]	AT[a]	AT[b]
0.23	0	31	31	38	17	17
	3	30	26	28	15	14
1.1	0	10	10	12	6	5
	3	14	12	15	5.5	5.0
2.1	0	6.6	6.6	5.0	2.2	2.2
	3	8.8	8.1	7.9	2.4	2.4
4.1	0	4.5	4.5	2.1	0.6	0.6
	3	5.3	4.3	2.1	1.0	1.2
6.1	0	3.3	3.1	1.1	0.1	0.1
	3	3.5	3.1	0.7	1.0	0.8

[a] TSH calculation (Ref. 91).
[b] Single surface trajectory calculation (Ref. 85).
[c] Experiment (Ref. 89.)

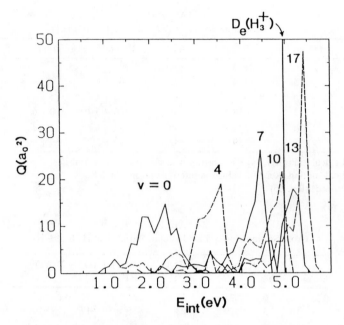

Figure 8. Cross sections (summed over J) for $H_2^+(v) + H_2 \rightarrow H_3^+ + H$ versus the internal energy in the H_3^+. A histogram bin size of 0.124 eV was used to generate the distributions. The vertical line at 4.95 eV indicates the dissociation energy of H_3^+, products above the are metastable. (Reproduced with permission from Ref. 93.)

TABLE VI

Cross Sections for Isotopic Variants of $H_2^+(v) + H_2 \rightarrow H_3^+ + H$

System	v	$D(H^+)^b$	$D(D^+)^b$	Q^c
$H_2^+ + H_2 \rightarrow H_3^+ + H$	13	36 ± 6		18 ± 5
	17	98 ± 9		6 ± 3
$H_2^+ + D_2 \rightarrow H_2D^+ + D$	13	10 ± 3	1 ± 1	10 ± 4
	17	7 ± 2	1 ± 1	3 ± 2
$H_2^+ + D_2 \rightarrow D_2H^+ + H$	13	11 ± 3	20 ± 5	8 ± 3
	17	20 ± 4	56 ± 8	3 ± 2
$D_2^+ + H_2 \rightarrow H_2D^+ + D$	18	14 ± 4	13 ± 3	12 ± 4
	24	47 ± 8	30 ± 7	10 ± 3
$D_2^+ + H_2 \rightarrow D_2H^+ + H$	18	2 ± 1	14 ± 4	6 ± 2
	24	5 ± 2	13 ± 4	5 ± 2
$H_2^+ + H_2 \rightarrow H_3^+ + H$	18		38 ± 6	20 ± 5
	24		95 ± 9	6 ± 2

[a]TSH results of Ref. 94

[b]$D(H^+, D^+)$: H_3^+ dissociates in 0.36–30 ps; (H^*, D^*) given off.

[c]Q = Quasistable; H_3^+ has energy to dissociate, but lasts > 30 ps.

both $v_+ = 13$ and $v_+ = 17$; aging the metastable population does not shift this maximum.

This process was studied in greater detail by Badenhoop, Schatz, and Eaker.[94] They examined other isotopic combinations, and compared 0.05 eV with 0.50 eV collision energies. Metastable products were aged for up to 30 ps. The comparisons are extensive, and rich in interesting detail; some cross sections are shown in Table VI. The fraction of metastables which survived for 30 ps was insensitive to isotopic substitution. The very long-lived metastables have, on average, higher average J than those which dissociate. For highly excited reagents, proton transfer is favored over atom transfer. The excited H_3^+ products were examined in a number of other ways, including power spectra. While the simulation is not able to study lifetimes on the scale relevant to the experiment, the rich detail of this study provides important information about the motions in collisionally prepared highly excited H_3^+.

B. Electron-Transfer Reactions

1. Introduction

Reactions of the type $M + X_2$, where M is an alkali metal and X_2 is a molecule with a positive electron affinity, have a long history in molecular reaction dynamics. The neutral reactants interact weakly (and covalently) at long range. If an electron jumps, forming $M^+ + X_2^-$, there is a strongly attractive (ionic) interaction, drawing the molecules together; the reactive channel, $MX + X$ frequently has a very large thermal cross section. This is the well-known and well-studied "harpooning" mechanism. The electron jump is, of course, the signature of a nonadiabatic process. Some of the earliest quasiclassical trajectory studies focused on such systems,[95–97] using a single potential energy surface which describes the covalent interaction in the asymptotic reactant channel, but switches to the ionic interaction at the crossing.

In 1982, Kleyn, Los, and Gislason[98] reviewed the topic of vibronic coupling at intersections of covalent and ionic states, and discussed very thoroughly the range of theoretical methods which have been brought to bear on this problem. Included in their review were many trajectory-based approaches: classical path calculations where $R(t)$ followed a simplified path through space; calculations in which a radial trajectory follows potentials which combine the vibrational and electronic degrees of freedom (vibronic potentials), as well as TSH calculations. Calculations that were based on a simplified nuclear trajectory have been very important in elucidating the nature of these processes.[99–103] While these may all be characterized as surface-hopping trajectory methods, the focus below will be on TSH calculations in which a trajectory (when it is not hopping) is the exact solution of Hamilton's equations for all the nuclear coordinates, subject to the electronically adiabatic potential.

2.   Alkali Plus Halogen

Among the earliest TSH studies was a 1973 paper by Düren[104] on ion pair formation in $M + X_2$ reactions, with parameters appropriate to $K + Br_2$. The molecule X_2 was taken to be a harmonic oscillator in r. The ionic and covalent potentials, dependent on R and γ only, were a Lennard–Jones (12–6) potential for the covalent surface, and a modified Rittner potential for the ionic. The angular dependence was produced by a Kihara-type core. The diabatic curves cross at a distance R_c. Adiabatic curves were generated by subtracting (or adding) from the lower (or upper) diabatic curve a quantity

$$\delta V = \tfrac{1}{2}\Delta E \exp(-|R - R_c|\Delta F/\Delta E).$$

ΔF is the difference in slopes of the ionic and covalent curves at R_c. The angular dependence of the coupling ΔE was defined to approximate the loss of coupling in C_{2v} geometry: $\Delta E = \Delta E°(1 + b \cos^2 \gamma)$; b was usually taken to be 9; but was sometimes set to zero to suppress the angular dependence. The transition probability at R_c was calculated using the Landau–Zener formula.

Differential cross sections were examined for a number of different sets of parameters: initial energy, coupling strength $\Delta E°$, its angular dependence b, depth of the ionic well, and energy gap between the asymptotic ionic and covalent surfaces. Suggestions were made about which parameters could be inferred from experimental results. In most cases, it was possible to distinguish two channels quite clearly in the ionic products: those who jumped to the ionic surface on the way in from those who jumped on the way out. The former give rise to a broader peak at larger angle, while the latter are sharply forward peaked.

Evers and De Vries[105] used the TSH method to study $K + Br_2$ using a much more detailed and realistic set of potentials. The potentials were the roots of a 2×2 matrix whose matrix elements were the ionic and covalent diabatic functions and the coupling H_{12}. The diabatic functions were written as sums of pairwise interactions: the ionic function including a Morse for the Br_2^-, and distored dipole and ion-induced dipole terms for the K^+–Br^- and K^+–Br, respectively. A three particle interaction term was added. The covalent diabatic potential included a Morse function for the Br_2, and a Lennard–Jones (12,6) function for the K–Br terms. The ionic-covalent coupling term was $H_{12} = A \cos \gamma \, \exp[-c(R_{KBr} - R_c)]$; the crossing radius R_c is strongly dependent on the Br_2 bond length.

The Landau–Zener formula was used to calculate the hopping probability at R_c; the radial velocity was used in this equation. H_{12} and ΔF were evaluated from the potential at R_c. When the Br_2 bond is near its equilibrium length, the crossing radius R_c is small and H_{12} is large; as a consequence, for energies below 10 eV, few trajectories hop to the upper state (this is, remain on the covalent surface) as the particles approach. The Br_2 equilibrium bond length increases on the ionic surface, so forces encountered in the interacting region will stretch this bond. If the trajectory now returns to the entrance valley with a larger r, R_c may be much larger, H_{12} smaller, so hopping can occur; sometimes several times in a trajectory. Trajectories which hop to the upper surface were not followed in this calculation, but the trajectory continuing on the lower surface weighted (at each crossing) by the probability $(1 - P)$.

The results of this somewhat simplified surface-hopping calculation were compared to trajectories which were restricted to the ionic diabatic surface. Cross sections for formation of KBr at energies between 0 and 4 eV were compared for the two calculations, and were compared to (scaled) experimental data. The surface-hopping results did an excellent job of following trends in $\sigma(E)$: fast decrease between 0 and 0.5 eV because of very large impact-parameter events, slower decrease between 0.5 and 1.8 eV due to recrossings, and a fast decrease at 1.8 eV because of the opening up of the ion-pair channel. In contrast, the calculation using only the diabatic ionic curve showed a gradual unstructured decrease over this range. Thus, even though the trajectories in this calculation which reached the upper surface were not followed explicitly, the model did an excellent job of incorporating the relevant nonadiabatic behavior, and demonstrating the importance to the understanding of this reaction of the Br_2 bond length dependence of the crossing redius R_c.

Evers[105] extended the model and applied it to Na, K, and Cs + I_2 collisions in the 10–100 eV range. The forms for the diabatic potentials and coupling were quite similar to the earlier paper; H_{12} had a somewhat more complicated dependence on R_c, and a more precise expression for R_c used to ensure that it correspond to the point where $H_{11} = H_{22}$. The sensitive dependence of R_c on the halogen bond length r remains.

The hopping procedure was also refined. Probabilities are calculated using the Landau–Zener equation, but using for $v\Delta F$ the equivalent vector dot product $\dot{\mathbf{R}}\cdot\Delta\mathbf{F}$ where $F(R) = H_{11}(R) - H_{22}(R)$. A trajectory is allowed to follow both the lower and upper surface, using the "anteater" approach.

In the K + Br_2 study, the stretching of the Br_2^- bond on the ionic surface was shown to have a very important effect on the subsequent surface hopping. This "bondstretching" phenomenon, which had been explored with simplified trajectories[101–103] was further explored in this study. A related effect is "prestretching": the (weaker) stretching of the halogen bond, which occurs on the covalent surface before the first crossing is reached. At sufficiently large energies, if the collision time is too fast for the halogen vibration to

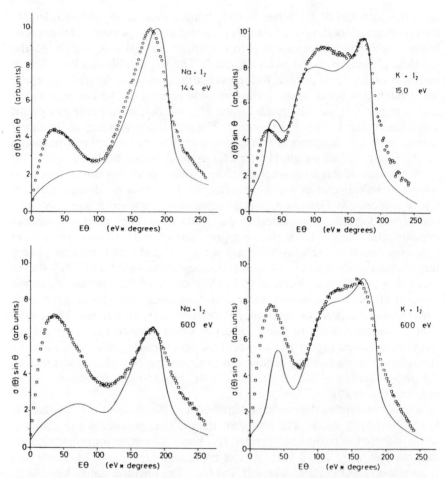

Figure 9. Representative laboratory differential cross sections for ion pair formation from Na(K) + I_2 collisions. E is the laboratory kinetic energy (eV), θ the laboratory scattering angle of M^+. The points are experimental results (Ref. 101), the curves the TSH calculation (Ref. 106). The curves have been scaled to the same maximum value. (Reproduced the permission from Ref. 106.)

respond, these effects will be unimportant. This, however, requires very high energy: more that 100 eV for the systems in this study.

Evers obtained TSH data for the following properties of ion-pair formation which were then compared to experiment:[101] absolute total cross sections, laboratory differential cross sections, and probability of ion dissociation. The agreement with experiment of the differential cross sections is qualitatively good: the positions of the maxima and the energy dependence of their relative heights well reproduced. The relative heights themselves were not always in accord with experiment. Examples are shown in Fig. 9. For $Na + I_2$ there are two peaks. The wider angle (ionic) peak, which results from trajectories which pass the seam adiabatically and follow the ionic surface on the way in, hopping to the ion-pair state on the way out, decreases compared to the smaller angle (covalent) peak as energy increases. In the experiments, this peak was larger at all four energies, while in the calculations the more forward peak was larger at the highest energy. The covalent peaks grows with energy because more trajectories have sufficient velocity to hop at the seam on the way in. Three peaks can be distinguished for $K + I_2$ and $Cs + I_2$: the covalent peak, an ionic rainbow, and the ionic peak. The energy dependence of the total cross sections and the dissociation probabilities agreed well with experiment: deviations were generally no more than 20%. The TSH results are not critically sensitive to the magnitude of the nonadiabatic coupling H_{12}: where a trajectory encounters the seam is the important factor.

Aten, Evers, de Vries, and Los[107] compared their molecular-beam experiments with the same TSH model and potentials for energy transfer and ion-pair formation in the reactions Na, K, and $Cs + I_2$ at around 8 eV. The differential cross section results again show excellent agreement between theory and experiment, particularly in the shapes and positions of maxima (the ionic peaks are sharper than at higher energy). Prestretching plays a significantly larger role in ion-pair formation in $Na + I_2$ and $K + I_2$ than in $Cs + I_2$ because of the smaller coupling H_{12} in the last case.

Evers, de Vries, and Los[108] compared molcular-beam experiments to the same TSH model for $K + Br_2$ at 0–10 eV. Differential cross sections were calculated at two energies $E_{lab} = 1.04$ and 7.04 eV. At each energy, 15,000 trajectories were generated. Total inelastic cross sections, based on 2000 trajectories each, were calculated at intermediate energies. (A collision is deemed inelastic, as opposed to elastic, if at least 0.5 eV of energy has been transferred.)

The TSH calculations give important new insight into the nonreactive scattering process. In both experiment and theory, the function $\log[\sigma(\theta)\theta\sin\theta]$ is linear with $\tau = E\theta$ at low τ (up to a value τ_c); slopes and τ_c values agree to within 20%. There is a peak above τ_c that represents inelastic events. These distributions are interpreted in terms of three kinds of trajectories. The first, "noncrossing" trajectories never reach R_c. They generally contribute to elastic scattering. The crossing trajectories are further divided into two classes: the "ionic" pass R_c adiabatically while the "covalent" hop, remaining on the covalent surface. The covalent trajectories, most of which cross the seam in a near perpendicular configuration where H_{12} is small, contribute to scattering in two regions: small angle scattering for large impact parameters, and larger angle scattering for small impact-parameter collisions. Ionic trajectories are likely to lead to KBr products, particularly at lower energies. However some can contribute to inelastic scattering by returning to the entrance valley, and passing through R_c adiabatically once more. These tend to contribute to scattering at intermediate angles. Covalent trajectories play an increasingly important role as energy increases.

Evers[109] extended the $K + Br_2$ TSH study, covering a range of 22 energies to 8000 eV. Cross sections were calculated for inelastic neutral scattering, reaction, ion-pair formation, dissociation to neutrals, and dissociation to ions. Various combinations of these were compared to experiment. Reactive and ion-pair formation cross sections and the fraction of Br_2^- which dissociate are shown in Fig. 10.

The reactive cross sections are quite similar to those in the earlier study[105] which used a somewhat simpler potential and hopping strategy; the shape of $\sigma_R(E)$ was discussed previously. Inelastic scattering was approximately constant out to about 100 eV, rising at higher energies. Above 100 eV, the collision is sufficiently fast that stretching of the Br_2 no longer enhances hopping, and, owing to the increasing radial velocity, trajectories that hop both on the way in and the way out (that is, stay covalent throughout) increase in number. If such trajectories are deflected toward the product valley in the strongly interacting region, they lead to dissociated neutral atoms; this cross section generally rises with energy above its threshold.

The total cross section for ion formation (ion-pairs and dissociated ions) and the fraction of ions which dissociate (F) both agree very well with experiment.[99,110] The maxima in both functions reflect the energy at which the stretching of the Br_2 is most effective in promoting hopping; the collision time and the molecular vibrational period are roughly the same. This calculation shows that, while a single surface simplification can give satisfactory results for the reactive channel, a full TSH treatment is required to include these competing processes.

3.   Alkali Plus Oxygen

Alkali–oxygen collisions are similar in many ways to alkali–halogen. Ion-pair $(M^+ + O_2^-)$ formation is an important channel. The system is again conveniently described in terms of interacting ionic and covalent diabatic surfaces. Bond stretching is important: the ionic–covalent crossing is strongly dependent on the oxygen bond length whose equilibrium value is larger on the ionic surface. Peaks in the differential cross sections result from different sequences of diabatic and adiabatic passage through the seam. Trajectory models which include the preceding surface features but which uncouple the radial and diatomic motion[111-113] have been very informative. Oscillations

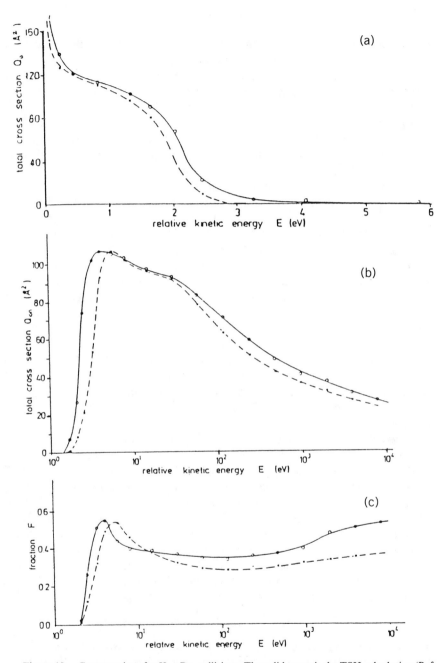

Figure 10. Cross sections for $K + Br_2$ collisions. The solid curve is the TSH calculation (Ref. 109), the dashed curve experimental results. Panel (*a*) shows reactive collisions ($KBr + Br$ products). Panel (*b*) shows total cross section for ion-pair formation. Panel (*c*) shows the fraction of ion product which dissociates to form Br^-. (Reproduced with permission from Ref. 109).

473

in the differential cross sections at energies where the collision time is comparable to the molecular vibrational period can be interpreted as "observation of molecular motion":[114] depending on time, an average trajectory may cross the seam one or more times on the way out.

Parlant, Schroeder, and Goursand[115] carried out full TSH study of the $K + O_2$ reaction at energies up to $100\,eV$. Among the phenomena which they wished to explore was the K^+ recoil energy distribution: two peaks are observed,[116,117] but with uncertainty about their origin; it had been suggested[117] that excited electronic states are involved.

Potential energy surfaces were constructed to correspond to the $^2A''$ states correlating with covalent $K(^2S) + O_2(X\,^3\Sigma)$ and ionic $K^+(^1S) + O_2^-(X\,^2\Pi)$ reagents. The surfaces are the roots of a 2×2 matrix. The covalent surface was a sum of pairwise terms: Morse for the O_2 and Lennard-Jones (12, 6) for the K–O. The ionic surface, based on work by Alexander,[118] includes a Morse for the O_2^- and an exponential-rational form for the K^+ interacting with each partially charged O atom. The polarization of the anion charge is governed by a switching function dependent on the molecular conformation. The coupling term is $H_{12} = c_1 \exp(-c_2 R_c) \sin 2\gamma$. The angular part $(\sin 2\gamma)$ ensures that the coupling vanish both in C_{2v} and in $C_{\infty v}$, as required by symmetry.

The Stine–Muckerman[40] implementation of the Landau–Zener formula. was used to calculate hopping probabilities, and all branches are followed. Most trajectories encounter the seam only twice, once on the way in and once on the way out, giving four branches. However in some cases, the oxygen oscillation is fast enough that more crossings occur.

The differential cross sections show distinct contributions from ionic and covalent trajectories; the ionic peaks are significantly larger at all energies studied. Energy loss spectra, shown in Fig. 11, are double peaked at small scattering angles: the ionic trajectories suffering significantly larger energy losses. At higher angles, the two distributions are similar in shape, but the covalent contribution much less. These results are in excellent agreement with experiment.[116] The TSH calculation demonstrates that bimodal energy loss distributions need not be the result of excited electronic states. Differential cross sections for the neutral (elastic plus inelastic) channel were also presented. The contributions of ionic and covalent trajectories are again distinct, both in angular and energy loss distributions.

C. Neutrals: A Sampling of Systems Studied

The TSH method has been applied to a number of nonadiabatic systems that do not involve transfer of charge; a review of these is beyond the purview of this article, but some examples will be described briefly to indicate the wide applicability of this method.

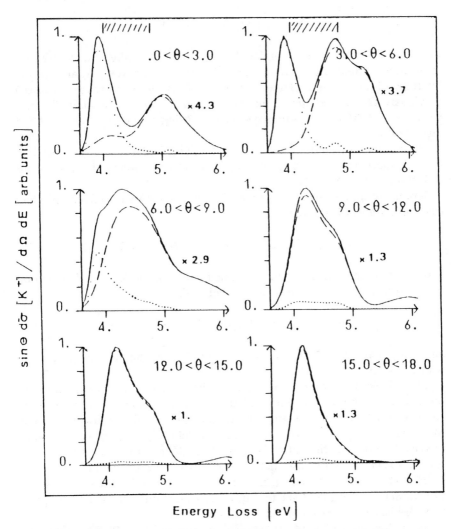

Figure 11. Angular and energy loss differential cross sections for K^+ formation from $K + O_2^+$ collisions at a center of mass collision energy of $14\,eV$. The contributions of the covalent (dotted) and ionic (dashed) trajectories are shown separately. For each angle range the distribution has been scaled to unity; scaling factors are shown. The striped area indicates the O_2^- Franck–Condon region. (reproduced with permission from Ref. 115.)

Quenching of electronically excited sodium in collisions with hydrogen:

$$Na(^2P) + H_2 \rightarrow Na(^2S) + H_2$$

was studied by Blais and Truhlar[49] using an eight configuration DIM set of surfaces calculated by Truhlar et al.[119] with an improved parametrization by Blais, Truhlar, and Garrett.[120] Their TSH procedure was discussed previously. Eaker[52] compared the Stine–Muckerman and the Blain–Truhlar versions of TSH on the same system, but using a slightly different DIM formulation.

Parlant and Alexander[53] used the Parlant and Gislason[47] TSH procedure to study electronic quenching of the fine-structure components in collisions of Helium with $CN(A\,^2\Pi_{3/2})$.

Kinnessly and Murrell[121] studied branching between different product electronic states for the reaction

$$C(^3P) + O_2(^3\Sigma_g^-) \rightarrow O(^3P \text{ or } ^1D) + CO(^1\Sigma).$$

Garetz, Poulsen, and Steinfeld[122] studied collision-induced predissociation of electronically and vibrationally excited iodine:

$$He + I_2(O_u^+, v) \rightarrow He + I_2(O_g^+) \rightarrow He + I + I$$

and compared the three-dimensional TSH results with an optical model and with collinear TSH results.[123]

There have been several TSH studies where the interacting surfaces were the results of interaction with radiation: the laser-dressed (and therefore energy-shifted) surface crossing and interacting with the undressed system. Miret-Artes, Delgado-Barrio, Atabek, and Lefebvre[124] applied such a model to the photodissociation of ICN. Last, Baer, Zimmerman, and George[125] studied the effect on the laser field on spin–orbit level changing collisions in $F(^2P) + H_2$ collisions. Yamashita and Morokuma[126] used the Stine–Muckerman version of TSH to study the transition state spectroscopy in laser-dressed $K + NaCl$ and $Na + KCl$ collisions.

Zahr, Preston, and Miller [127] studied quenching in $O(^1D) + N_2$ collisions, using a model based on a TSH picture; detailed trajectories were used only to determine the cross section for formation of a long lived complex, whose subsequent dissociation was calculated statistically.

V. DISCUSSION AND FUTURE DEVELOPMENTS

The trajectory-surface-hopping model is a very useful generalization of the quasiclassical trajectory method to systems with nonadiabatic behavior.

When potential energy surfaces are available, the TSH method is relatively easy to implement and can provide rich detail for the interpretation of the systems studied. Several different procedures have been proposed; the choice among them depends largely on the information available and the kinds and locations of regions of strong nonadiabatic behavior. If the seams are easily fit to a simple function and preliminary investigation indicates that the Landau–Zener (or some other formula) is adequate, the Tully–Preston approach is very convenient and fast. If the seams are more complicated, but can be characterized as "troughlike" the Stine–Muckerman procedure is very attractive; it may be the method of choice if coupling terms are not available. The Kuntz, Kendrick, and Whitton procedure requires coupling terms, but it is efficient and makes no assumption about the kind of coupling.

Possibly more time-consuming (in that a larger number of coupled equations is required in the trajectory calculation) but conceptually attractive are the combined classical-trajectory–semiclassical approaches of Blais and Truhlar and Parlant and Gislason. The latter seems to improve upon some of the possible inconsistencies of the former.

There are situations in which a TSH approach is inadvisable. When a system must find its way through a complicated series of nearby crossings, the results may become unreliable. Some resonant processes are poorly described in the purely classical TSH model. If the electronic energy gap happens to coincide with some internal energy gap, the TSH method, which is treating the electron and nuclear coordinates in very different ways, will fail to see the strong resonant enhancement. Two important reactions in which this occurs are

$$\text{Ar}^+(^2P_{3/2}) + \text{N}_2(v = 0) \rightarrow \text{Ar}^+(^2P_{1/2}) + \text{N}_2(v = 1)$$

and

$$\text{F}(^2P_{3/2}) + \text{H}(v = 0, J = 2) \rightarrow \text{F}(^2P_{1/2}) + \text{H}_2(v = 0, J = 0).$$

What is required in such a situation is a method which treats the resonant degrees of freedom in the same way. Nikitin[128] treated the ArN_2^+ problem by combining the resonant vibrational and electronic coordinates to generate vibronic surfaces, and (in contrast to many other vibronic-surface based calculations) carried out full TSH dynamics on these surfaces. Miller, McCurdy, and Meyer, in a series of papers[129–133] have shown how to go the other way: how to construct an entirely classical picture for such processes which includes the electronic coordinates as extra classical motions.

The TSH model is based on a number of simplifying assumptions, most importantly the notion of transitions taking place localized on seams between interacting surfaces. Herman,[50,134,135] based on earlier work of Laing and Freed,[136] has studied semiclassical nonadiabatic transitions in a more formally rigorous way; his work has provided insight and guidance for simpler approaches. There has also been extensive relevant work on nonadiabatic effects in unimolecular reactions, some of which was reviewed by Desouter–Lecomte, Dehareng, Leyn–Nihant, Praet, Lorquet, and Lorquet.[137]

The intuitive picture offered by the TSH model is sometimes usefully invoked even without explicit TSH calculations. Staemmler and Gianturco[138] calculated potential energy curves relevant to $H^+ + O_2$ collisions, and interpreted the large vibrational excitation observed in the inelastic channel in terms of trajectory motion on the $H + O_2^+$ surface. Gianturco and Schneider[139] calculated DIM surfaces for $H^+ + BH(X\ ^1\Sigma)$ collisions, indicating the possibility of nonadiabatic pathways. Full dynamical calculations with these surfaces would surely further enrich the interpretation of these systems.

The editors of these volumes encouraged contributors to speculate about fucture developments. Certain statements can be made without risk. With more accurate potentials and faster computers we will be able to use the TSH approach to obtain better agreement with the increasingly detailed experimental data for ion–molecule reactions—and in so doing, to offer useful interpretation of the underlying processes. It is clear from the recent TSH studies on the H_3^+ system that the general approach can be very successful; for that problem, what remains is to use the very best potentials. The potentials need improving for ArH_2^+ also: spin–orbit must be included, and the long-range interaction corrected before one can clearly identify weaknesses in the TSH model. This system (when the three interacting surfaces are all included) has several features which are possible problems for TSH: nearby pairs of surface crossings, turning points near crossing seams, and resonant enhancement of certain state-to-state processes.

The details of the hopping procedure proved to be unimportant in two cases where the seam followed the reactant valley, while for other systems with seams in the strongly three-body region, it was very important. As more systems are studied, and as new potentials and hopping procedures are tested on old and new systems, we should get a clearer understanding of the optimum strategies. Until then, this aspect of a TSH calculation should be examined carefully for each case.

H_4^+ was the only system with $N > 3$ cited in this review. However, it seems that the primary impediment to studying larger systems is the same one which occurs in adiabatic quasiclassical trajectory studies, namely the availability of potential energy surfaces. It is striking that a very large fraction of the TSH studies completed to date are based on DIM surfaces; striking, but not surprising, since the DIM formalism is ideally suited to multiple surface dynamics. While complete DIM surfaces may be impractical for many systems, fitting functions which employ DIM-like matrices may play an important role.

Nevertheless, the richness in detail and the generally satisfactory agreement with experiment which has been obtained even with rather approximate potentials bodes well for future studies.

Acknowledgment

I would like to thank Dr. Mark S. Child and the Theoretical Chemistry Department at the University of Oxford for their generous hospitality for the 1989–90 year during which this review was written.

References

1. D. L. Bunker, *Methods Comput. Phys.* **10**, 287 (1971).

2. D. G. Truhlar and J. T. Muckerman, in *Atom Molecule Collision Theory*, R. B. Bernstein, Ed., Plenum, New York, 1979, p. 505.

3. L. M. Raff and D. L. Thompson, in *Theory of Chemical Reaction Dynamics*, M. Baer, Ed., CRC Press, Boca Raton, 1985, Vol. III, p. 1.

4. W. H. Miller, *Adv. Chem. Phys.* **25**, 69 (1975).

5. M. S. Child, *Theory of Chemical Reaction Dynamics*, M. Baer, Ed., CRC Press, Boca Raton, 1985, Vol. III, p. 279.

6. (a) R. K. Preston and J. C. Tully, *J. Chem. Phys.* **54**, 4297 (1971); (b) J. C. Tully and R. K. Preston, *J. Chem. Phys.* **55**, 562 (1971).

7. A. Bjerre and E. E. Nikitin, *Chem. Phys. Lett.* **1**, 179 (1967).

8. B. C. Garrett and D. G. Truhlar, in *Theoretical Chemistry: Advances and Perspectives*, Academic Press, New York, 1981, Volume 6A, p. 215.

9. J. B. Delos, *Rev. Mod. Phys.* **53**, 287 (1981).

10. M. Baer, in *Theory of Chemical Reaction Dynamics*, M. Baer, Ed., CRC Press, Boca Raton, 1985, Vol II, p. 219.

11. E. E. Nikitin, *Theory of Elementary Atomic and Molecular Processes in Gases*, Clarendon, Oxford, 1974.

12. E. E. Nikitin, in *Chemische Elementarprozesse*, H. Hartmann, Ed., Springer, Berlin, 1968, p. 43.

13. E. E. Nikitin, *Adv. Quantum Chem.* **5**, 135 (1970).

14. The electronic hamiltonian usually excludes the spin–orbit interaction; this can. be added later, as needed. See Ref. 11.

15. N. F. Mott and H. S. W. Massey, *The Theory of Atomic Collisions*, 3rd ed., Oxford, London, 1965, pp. 665, 685.

16. F. R. Smith, *Phys. Rev.* **179**, 111 (1969).

17. L. Landau, *Phys. Z. Sowetunion* **2**, 46 (1932).

18. C. Zener, *Proc. Royal Soc. London A* **137**, 696 (1932).

19. D. Dehareng, X. Chapuisat, J. C. Lorquet, C. Galloy, and G. Raseev, *J. Chem. Phys.* **78**, 1246 (1983).

20. M. Desouter-Lecomte, B. Leyh-Nihant, M. T. Praed, and J. C. Lorquet, *J. Chem. Phys.* **78**, 1246 (1983).

21. Y. N. Demkov, *Sov. Phys. JETP* **18**, 138 (1964).

22. R. E. Olsen, *Phys. Rev. A* **6**, 1822 (1972).

23. R. K. Preston, C. Sloane, and W. H. Miller, *J. Chem. Phys.* **60**, 4961 (1974).

24. J. C. Tully, *J. Chem. Phys.* **60**, 3042 (1974).

25. J. C. Tully, *Ber. Bunsenges Phys. Chem.* **77**, 557 (1973).

26. J. C. Tully, in *Dynamics of Molecular Collisions, Part B*, W. H. Miller, Ed., Plenum, New York, 1976, p. 217.

27. C. A. Mead and D. G. Truhlar, *J. Chem. Phys.* **84**, 1055 (1986).

28. J. C. Tully, *Adv. Chem. Phys.* **42**, 62 (1980).

29. W. H. Miller and T. F. George, *J. Chem. Phys.* **56**, 5637 (1972).

30. E. C. G. Stueckelberg, *Helv. Phys. Acta* **5**, 369 (1932).

31. J.-T. Hwang and P. Pechukas, *J. Chem. Phys.* **67**, 4640 (1977).

32. T. F. George and Y.-W. Lin, *J. Chem. Phys.* **60**, 2340 (1974).

33. Y.-W. Lin, T. F. George, and K. Morokuma, *J. Chem. Phys.* **60**, 4311 (1974).

34. Y.-W. Lin, T. F. George, and K. Morokuma, *Chem. Phys. Lett.* **30**, 49 (1975).

35. P. Pechukas, T. F. George, K. Morokuma, F. J. McLafferty, and J. R. Laing, *J. Chem. Phys.* **64**, 1099 (1976).

36. A. M. Dhynke, *Sov. Phys. JETP* **14**, 941 (1962).

37. R. L. Jaffe, *Chem. Phys.* **23**, 249 (1977).

38. A. Kormornicki, T. F. George, and K. Morokuma, *J. Chem. Phys.* **65**, 48 (1976).

39. A. Kormornicki, K. Morokuma, and T. F. George, *J. Chem. Phys.* **67**, 5012 (1977).

40. J. R. Stine and J. T. Muckerman, *J. Chem. Phys.* **65**, 3975 (1976).

41. J. R. Stine and J. T. Muckerman, *Chem. Phys. Lett.* **44**, 46 (1976).

42. J. R. Stine and J. T. Muckerman, *J. Chem. Phys.* **84**, 1056 (1976).

43. N. C. Blais, D. G. Truhlar, and C. A. Mead, *J. Chem. Phys.* **89**, 6204 (1988).

44. S. Chapman and R. K. Preston, *J. Chem. Phys.* **60**, 650 (1974).

45. P. J. Kuntz and A. C. Roach, *J. Chem. Soc. Faraday II* **68**, 259 (1972); **69**, 926 (1973).

46. P. J. Kuntz, J. Kendrick, and W. N. Whitton, *Chem. Phys.* **38**, 147 (1979).

47. G. Parlant and E. A. Gislason, *J. Chem. Phys.* **91**, 4416 (1989).

48. M. Desouter-Lecomte and J. C. Lorquet, *J. Chem. Phys.* **66**, 4006 (1977).

49. N. C. Blais and D. G. Truhlar, *J. Chem. Phys.* **79**, 1334 (1983).

50. M. F. Herman, *J. Chem. Phys.* **81**, 764 (1984).

51. J. R. Stine and J. T. Muckerman, *J. Phys. Chem.* **91**, 459 (1987).

52. C. W. Eaker, *J. Chem. Phys.* **87**, 4532 (1987).

53. G. Parlant and M. H. Alexander, *J. Chem. Phys.* **92**, 2287 (1990).

54. M. MacGregor and R. S. Berry, *J. Phys. B* **6**, 181 (1973).

55. (a) M. Sizun, E. A. Gislason, and G. Parlant, *Chem. Phys.* **107**, 311 (1986); (b) M. Sizun, G. Parlant, and E. A. Gislason, *Chem. Phys.* **133**, 251 (1989).

56. R. K. Preston and J. S. Cohen, *J. Chem. Phys.* **65**, 1589 (1976).

57. R. K. Preston and R. J. Cross, *J. Chem. Phys.* **59**, 3616 (1973).

58. J. Krenos, R. K. Preston, R. Wolfgang, and J. C. Tully, *J. Chem. Phys.* **60**, 1634 (1074).

59. C. Schlier, U. Nowotny, and E. Teloy, *Chem. Phys.* **111**, 401 (1987).

60. G. Ochs and E. Teloy, *J. Chem. Phys.* **61**, 4930 (1974).

61. W. Meyer, P. Botschwina, and P. Burton, *J. Chem. Phys.* **84**, 891 (1986).

62. D. Gerlich, U. Nowotny, C. Schlier, and E. Teloy, *Chem. Phys.* **47**, 245 (1980).

63. G. Niedner, M. Noll, J. P. Toennies, and C. G. Schlier, *J. Chem. Phys.* **87**, 2685 (1987).

64. M. Baer, G. Niedner-Scatteburg, and J. P. Toennies, *J. Chem. Phys.* **91**, 4169 (1989).

65. P. M. Hierl, V. Pacak, and Z. Herman, *J. Chem. Phys.* **67**, 2678 (1977).

66. F. A. Houle, S. L. Anderson, D. Gerlich, T. Turner, and Y. T. Lee, *J. Chem. Phys.* **77**, 748 (1982).

67. (a) K. Tanaka, J. Durup, T. Kato, and I. Koyano, *J. Chem. Phys.* **74**, 5561 (1981); (b) K. Tanaka, T. Kato, and I. Koyano, *J. Chem. Phys.* **75**, 4941 (1981).

68. S. Chapman, *J. Chem. Phys.* **82**, 4033 (1985).

69. K. M. Ervin and P. B. Armentrout, *J. Chem. Phys.* **83**, 166 (1985).

70. M. Baer, C. L. Liao, R. Xu, S. Nourbakhsh, G. D. Flesch, G. Y. Ng, and D. Neuhauser, *J. Chem. Phys.* **93**, 4845 (1990).

71. R. K. Preston, D. L. Thompson, and D. R. McLaughlan, *J. Chem. Phys.* **68**, 13 (1978).

72. M. Sizun and E. A. Gislason, *J. Chem. Phys.* **91**, 4603 (1989).

73. M. Sizun, G. Parlant, and E. A. Gislason, *J. Chem. Phys.* **88**, 4294 (1988).

74. E. A. Gislason and P. M. Guyon, *J. Chem. Phys.* **86**, 677 (1987).

75. W. N. Whitton and P. J. Kuntz, *J. Chem. Phys.* **64**, 3264 (1976).

76. P. J. Kuntz and W. N. Whitton, *Chem. Phys.* **16**, 301 (1976).

77. J. D. Kelley and H. H. Harris, *J. Phys. Chem.* **83**, 936 (1979).

78. M. Barat, J. C. Brenot, J. A. Fayeton, J. C. Houver, J. B. Ozenne, R. S. Berry, and M. Durup-Ferguson, *Chem. Phys.* **97**, 165 (1985).

79. E. Herbst, L. G. Payne, R. L. Champion, and L. D. Doverspike, *Chem. Phys.* **42**, 413 (1979).

80. K. B. Sorbie and J. N. Murrell, *Mol. Phys.* **29**, 1387 (1975).

81. J. R. Krenos, K. K. Lehman, J. C. Tully, P. M. Hierl, and G. P. Smith, *Chem. Phys.* **16**, 109 (1976).

82. R. Polak, *Chem. Phys.* **16**, 353 (1976).

83. J. R. Stine and J. T. Muckerman, *J. Chem. Phys.* **68**, 185 (1978).

84. C. W. Eaker and G. C. Schatz, *J. Phys. Chem.* **89**, 2612 (1985).

85. C. W. Eaker and J. L. Muzyka, *Chem. Phys. Lett.* **119**, 169 (1985).

86. I. Koyano and K. Tanaka, *J. Chem. Phys.* **72**, 4858 (1980).

87. C. W. Eaker, G. C. Schatz, N. DeLeon, and E. J. Heller, *J. Chem. Phys.* **81**, 5913 (1984).

88. C. W. Eaker and G. C. Schatz, *J. Chem. Phys.* **81**, 2394 (1984).

89. S. L. Anderson, F. A. Houle, D. Gerlich, and Y. T. Lee, *J. Chem. Phys.* **75**, 2153 (1981).

90. J. T. Muckerman, in *Theoretical Chemistry: Advances and Perspectives*, D. Henderson, Ed., Academic Press, New York, 1981, Vol. 6A, p. 1.

91. C. W. Eaker and G. C. Schatz, *Chem. Phys. Lett.* **127**, 343 (1986).

92. A. Carrington and R. A. Kennedy, *J. Chem. Phys.* **81**, 91 (1984).

93. G. C. Schatz, J. K. Badenhoop, and C. W. Eaker, *Inter. J. Quantum Chem.* **31**, 57 (1987).

94. J. K. Badenhoop, G. C. Schatz, and C. W. Eaker, *J. Chem. Phys.* **87**, 5317 (1987).

95. N. C. Blais, *J. Chem. Phys.* **49**, 9 (1968).

96. M. Godfrey and M. Karplus, *J. Chem. Phys.* **49**, 3602 (1968).

97. P. J. Kuntz, M. H. Mok, and J. C. Polanyi, *J. Chem. Phys.* **50**, 4623 (1969).

98. A. W. Kleyn, J. Los, and E. A. Gislason, *Phys. Reports* **90**, 1 (1982).

99. M. M. Hubers, A. W. Kleyn, and J. Los, *Chem. Phys.* **17**, 303 (1976).

100. J. A. Aten, M. M. Hubers, A. W. Kleyn, and J. Los, *Chem. Phys.* **18**, 311 (1976).

101. J. A. Aten, G. E. H. Lanting, and J. Los, *Chem. Phys.* **19**, 241 (1977).

102. J. A. Aten, G. E. H. Lanting, and J. Los, *Chem. Phys.* **22**, 333 (1977).

103. J. A. Aten and J. Los, *Chem. Phys.* **25**, 47 (1977).

104. R. Düren, *J. Phys. B* **6**, 1801 (1973).

105. C. W. A. Evers and A. E. de Vries, *Chem. Phys.* **15**, 201 (1976).

106. C. Evers, *Chem. Phys.* **21**, 355 (1977).

107. J. A. Aten, C. W. A. Evers, A. E. de Vries, and J. Los, *Chem. Phys.* **23**, 125 (1977); **24**, 287 (1977).

108. C. W. A. Evers, A. E. de Vries, and J. Los, *Chem. Phys.* **29**, 399 (1978).

109. C. W. A. Evers, *Chem. Phys.* **30**, 27 (1978).

110. A. P. M. Baede, D. J. Auerbach, and J. Los, *Physica* **64**, 134 (1973).

111. A. W. Kleyn, M. M. Hubers, and J. Los, *Chem. Phys.* **34**, 55 (1978).

112. A. W. Kleyn, V. N. Khromov, and J. Los, *Chem. Phys.* **52**, 65 (1980).

113. A. W. Kleyn, E. A. Gislason, and J. Los, *Chem. Phys.* **52**, 81 (1980).

114. A. W. Kleyn, V. N. Khromov, and J. Los, *J. Chem. Phys.* **72**, 5282 (1980).

115. G. Parlant, M. Schroeder, and S. Goursand, *Chem. Phys.* **75**, 175 (1983).

116. T. Mochizuki and K. Lachmann, *J. Chem. Phys.* **65**, 3257 (1976).

117. C. E. Young, C. M. Scholeen, A. F. Wagner, A. E. Proctor, L. G. Pobo, and S. Wexler, *J. Chem. Phys.* **74**, 1770 (1981).

118. M. H. Alexander, *J. Chem. Phys.* **69**, 3502 (1978).

119. D. G. Truhlar, J. W. Duff, N. C. Blais, J. C. Tully, and B. C. Garrett, *J. Chem. Phys.* **77**, 764 (1882).

120. N. C. Blais, D. G. Truhlar, and B. C. Garrett, *J. Chem. Phys.* **78**, 2956 (1882).

121. S. R. Kinnessley and J. N. Murrell, *Mol. Phys.* **33**, 1479 (1977).

122. B. A. Garetz, L. L. Poulson, an J. I. Steinfeld, *Chem. Phys.* **9**, 385 (1975).

123. B. Garetz, M. Rubison, and J. I. Steinfeld, *Chem. Phys. Lett.* **28**, 120 (1974).

124. S. Miret-Artes, G. Delgado-Barrio, O. Atabek, and R. Lefebvre, *Nouv. J. Chem.* **6**, 431 (1982).

125. I. Last, M. Baer, I. H. Zimmerman, and T. F. George, *Chem. Phys. Lett.* **101**, 163 (1983).

126. K. Yamashita and K. Morokuma, *J. Chem. Phys.* **91**, 7477 (1989).

127. G. E. Zahr, R. K. Preston, and W. H. Miller, *J. Chem. Phys.* **62**, 1127 (1975).

128. E. E. Nikitin, M. Y. Ovchinnikova, D. V. Shalashilin, *Chem. Phys.* **111**, 313 (1987).

129. W. H. Miller and C. W. McCurdy, *J. Chem. Phys.* **69**, 5163 (1978).

130. C. W. McCurdy, H. D. Meyer, and W. H. Miller, *J. Chem. Phys.* **70**, 3177 (1979).

131. H. D. Meyer and W. H. Miller, *J. Chem. Phys.* **70**, 3214 (1979).

132. H. D. Meyer and W. H. Miller, *J. Chem. Phys.* **71**, 2156 (1979).

133. H. D. Meyer and W. H. Miller, *J. Chem. Phys.* **72**, 2272 (1980).

134. M. F. Herman, *J. Chem. Phys.* **76**, 2949 (1982).

135. M. F. Herman, *J. Chem. Phys.* **81**, 764 (1984).

136. J. R. Laing and K. F. Freed, *Chem. Phys.* **19**, 91 (1977).

137. M. Desouter-Lectomte, D. Dehareng, B. Leyn-Nihant, M. T. Praet, A. J. Lorquet, and J. C. Lorquet, *J. Phys. Chem.* **89**, 214 (1985).

138. V. Staemmler and F. A. Gianturco, *Inter. J. Quantum Chem.* **28**, 553 (1988).

139. F. A. Gianturco and F. Schneider, *Chem. Phys.* **111**, 113 (1987).

STATISTICAL ASPECTS OF ION–MOLECULE REACTIONS

JÜRGEN TROE

Institut für Physikalische Chemie, Universität Göttingen,
Göttingen, Germany

CONTENTS

I. INTRODUCTION

The quantitative treatment of reactive processes continues to be one of the greatest challenges in the field of chemical physics. A rigorous theory would require complete knowledge of the interaction forces between the reactants, that is, of the potential energy surfaces involved, and of the dynamics of the

State-Selected and State-to-State Ion–Molecule Reaction Dynamics, Part 2: Theory, Edited by Michael Baer and Cheuk-Yiu Ng. Advances in Chemical Physics Series, Vol. LXXXII. ISBN 0-471-53263-0 © 1992 John Wiley & Sons, Inc.

process, that is, of the propagation of wave packets on these surfaces. Neither of these problems is solved today in a satisfactory manner: potential energy surfaces are generally known only to a limited extent; rigorous quantum-scattering calculations are often extremely difficult.

In this situation *statistical rate theories* are of great value. They do not attack the problem of the potential energy surfaces, which still have to be implemented in rather crude and fragmentary ways, but they help to bypass the difficulties of dynamical theories such as quantum-scattering or classical trajectory calculations. Instead of considering the complete time evolution of the system, they concentrate on the rate- or flux-determining *bottlenecks* of the reaction. By assuming "equilibrium" or "statistical" populations of the states (or channels) of the system during the approach of the bottleneck, the time-consuming treatment of the evolution of the system before this final event is avoided. The price to pay for this considerable simplification is the loss of rigor: either the system behaves in this way, that is, it is "statistical", or it does not, that is, it reacts "nonstatistically." Only the comparison with rigorous treatments can ultimately justify the statistical approach. However, the gain in simplicity of the method is enormous and the practical success of statistical calculations strengthens the hope that this model is close to reality.

Statistical rate theories have been formulated with different degrees of sophistication since the early days of reaction kinetics: The Arrhenius equation was based on statistical arguments; statistical models followed closely the advances in statistical mechanics and thermodynamics; transition state theory (TST), quasiequilibrium theory (QET), phase-space theory (PST), RRKM theory, SACM, Langevin, and many other theories fall into the category of statistical rate treatments. The question arises whether they all can be identified as special cases of *one general statistical theory*, just employing different types of avoidable or unavoidable simplifications, or whether the differences between theories, or theory and experiment, reflect basically different kinetic behavior. Because of their relatively simple long-range potentials, ion-molecule systems are particularly suitable for the investigation of this question.

Statistical theories can rely on two different basic assumptions: one may postulate *adiabatic* reaction behavior near to the bottlenecks of the process, or one may allow for complete *nonadiabatic* distribution of the populations in this range. In the adiabatic approach, one fixes the distance r between the centers of mass of the reactants (or other coordinates characterizing the progress of reaction) and one calculates the energy eigenvalues $V_i(r)$ of the system at this point. The dependence of $V_i(r)$ on r characterizes *adiabatic channel potential curves*, the maxima of these curves along r define state- or *channel-resolved threshold energies* E_{0i} which are considered to correspond to the bottlenecks of the process. Calculating one-way fluxes through these channel-resolved bottlenecks and assigning equal statistical weight to all open channels, completes a *statistical adiabatic channel model (SACM)* of the reaction. SACM represents the most general statistical treatment of the adiabatic type and thus appears most useful for a comparison with other approaches such as variational transition state, phase space, or RRKM theory which are either formulated in a less sophisticated way, are special cases of SACM, or implicitly contain some degree of nonadiabaticity.

SACM revived the compound nucleus model of nuclear reactions (see Blatt and Weisskopf, 1952) and applied it to chemical reactions (Quack and Troe, 1974, 1975a, 1975b). It may be understood as a state-resolved transition-state theory going beyond microcanonical or canonical TST. Using simplified adiabatic channel potential curves $V_i(r)$ of the bond-energy–bond-order type in the beginning (Jungen and Troe, 1970; Gaedtke and Troe, 1973; Quack and Troe 1974), it recently proceeded to more rigorous calculations of the $V_i(r)$ (Troe, 1988; Troe, 1989; Nikitin and Troe, 1990; Maergoiz et al., 1991; Quack, 1990). Applying the general theory of nonadiabatic processes, the validity of the adiabatic assumption has also been investigated (Maergoiz et al., 1991), indicating at which energies of the relative motion of the reactants nonadiabatic curve-hopping sets in.

Obviously, uni- and bimolecular processes of ion–molecule systems in no way are special such that they could not be analyzed in the framework of general kinetic theories. There are, of course, different "individual chemistries," different types of long-range potentials, and, perhaps, more frequent electronically nonadiabatic processes. However, this does not justify the long-lasting separation of the fields of neutral and ion–molecule reactions and the development of different formalisms. In this sense the present article provides a common view of neutral and ion–molecule reaction aspects. SACM has successfully been applied to ion-dipole capture processes (Troe, 1987a) using the same formalism as applied earlier to complex–forming bimolecular reactions of neutral species (Quack and Troe, 1975a). However, more accurate adiabatic channel potential curves $V_i(r)$ were used in the more recent work. Independent treatments in the field of ion–molecule reactions proved to coincide with the SACM treatment, partly being "reinventions," partly applying particularly accurate calculations of the channel potential curves. Among these approaches are the perturbed rotational state method (Takayanagi, 1978, 1982; Sakimoto and Takayanagi, 1980; Sakimoto, 1980, 1982, 1984, 1985), the adiabatic invariance method (Bates, 1982, 1983; Bates and Mendas, 1985; Morgan and Bates, 1987), the adiabatic rotational state calculations by Dubernet and McCarroll (1989, 1990), and the rotationally adiabatic capture calculations by Markovic and Nordholm (1989a). In the absence of an anisotropy of the potential between the reactants, phase-space theory forms a special case of SACM; Langevin and locked-dipole treatments of ion–molecule capture fall into the same category. Detailed discussions of this limiting case have been given, for example, by Chesnavich and Bowers (1978, 1982). Recent reviews of this field by Clary (1988) and Bates and Herbst (1988) contain additional references to earlier reviews and articles.

It is not surprising that statistical rate calculations for ion–molecule capture processes often coincide with dynamical calculations such as classical trajectory or quantum-scattering calculations. This was emphasized long ago for situations where the different theories lead to identical fluxes through the bottlenecks of the reaction, see, for example, Slater (1959). In this sense, the approach of the analytical representaton of classical trajectory results from Su and Chesnavich (1982) by SACM calculations (Troe, 1987a) is not unexpected. Likewise, the adiabatic capture centrifugal sudden approximate (ACCSA) or more accurate quantum-scattering capture calculations by Clary (1984, 1985, 1987a, 1987b) essentially agree with SACM calculations; some differences at low temperatures (Troe 1987a) most probably are numerical artifacts of the ACCSA treatment. If equal fluxes through the bottlenecks are calculated anyway, of course, the more economic method appears preferable. This often is the statistical theory.

So far we have only considered bimolecular ion–molecule reactions that are governed by capture of the reactants on an attractive potential energy surface. For these processes adiabatic statistical theories are very appropriate. If the reactions by association of the reactants form bound complexes, which can redissociate or dissociate toward the products by simple bond fission or complex elimination, the second stage of the reaction again can be treated by statistical rate theory. In this case, however, specific rate constants $k(E, J)$ for unimolecular decompositions of the complex (with total energy E and total angular momentum J) have to be calculated. Obviously, capture and dissociation processes are reverse and, therefore, linked by microscopic reversibility, that is, describable by the same theoretical approach. As a consequence, unimolecular and complex-forming bimolecular reactions are treated by the same statistical theory. We do not concentrate in this chapter on the unimolecular fragmentation of energy- and angular-momentum-resolved molecular ions and their statistical analysis in terms of QET, TST, RRKM, or SACM calculations (for recent reviews see, for example, Lifshitz, 1987). Unfortunately, in several of these theories the J-dependence of the dissociation rates has been treated inadequately. PST calculations (Chesnavich and Bowers, 1978) and their extension to anisotropic potentials by simplified SACM (Troe, 1983) provide more appropriate procedures. Although we do not go into the details of this problem here, the results are needed for the treatment of complex-forming bimolecular reactions considered later on.

Complex-forming bimolecular ion–molecule reactions can be governed by quite different potential energy surfaces and the literature is full of examples (see, for example, Henchman and Paulson, 1988, 1989), showing multiple intermediate structures with "loose" and/or "rigid" entrance and exit potentials and all types of intermediate rearrangement barriers. It is not the purpose of this chapter to look into the vast number of different possible cases. Instead, simple examples of a statistical treatment of such reactions are considered, which indicate the necessity of taking into account energy E and angular momentum J aspects of the process (Troe, 1987b).

This chapter does not include collisional energy transfer processes contributing to ion–molecule reactions, either in complex-forming bimolecular reactions or in ion–molecule association reactions. However, it should be remembered that the high-pressure limiting bimolecular ion–molecule association is also a capture process for which the present statistical treatment applies.

II. BASIC RELATIONSHIPS OF STATISTICAL RATE THEORIES

A. Biomolecular Processes

Scattering processes in statistical rate models are characterized (Quack and Troe, 1975a, 1981) by a statistical S matrix of the form

$$|S_{fi}|^2 = \begin{cases} W(E, J, \dots)^{-1}, & \text{for strongly coupled channels,} \\ \delta_{fi}, & \text{for weakly coupled channels,} \end{cases} \tag{2.1}$$

where a process between the initial quantum state i and the final quantum state f is considered, E denotes total energy, J denotes total angular momentum, and the total number of strongly coupled channels $W(E, J, \dots)$ depends on E, J, and other good quantum numbers. Equation (2.1) describes matrix elements, which are averaged over a certain energy range such that enough resonances of the collision complex are included, and which are averaged over initial states with random phase distributions. The statistical postulate of *equal weight for strongly coupled channels* in Eq. (2.1) is easily identified. The scattering cross section from Eq. (2.1) follows as

$$\sigma_{fi} = \pi |\delta_{fi} - S_{fi}|^2 / k_i^2 \tag{2.2}$$

with

$$k_i = \sqrt{2\mu E_t / \hbar^2}, \tag{2.3}$$

where μ is the reduced mass of the colliding pair and E_t is its initial relative translational energy. Averaging Eq. (2.2) over the not easily observable coordinates leads to the degeneracy averaged inelastic and reactive total cross section

$$\sigma_{ba} = \frac{\pi}{g_a k_a^2} \sum_{J=0}^{\infty} (2J+1) \frac{W(E,J,a)W(E,J,b)}{W(E,J)}, \qquad (2.4)$$

where a corresponds to the selectable initial state, b to the selectable final state, k_a corresponds to k_i from Eq. (2.3), and g_a is the degeneracy of the initial state. One has, for example, $g_a = (2j_{a1}+1)(2j_{a2}+1)$ for two colliding linear rotors with rotational quantum numbers j_{a1} and j_{a2}. $W(E,J,a)$ and $W(E,J,b)$ are those numbers of strongly coupled channels that emerge from the states a or b of the reactants or products and lead into the collision complex. $W(E,J)$ is given by the sum $W(E,J,a) + W(E,J,b)$.

The ratio $W(E,J,b)/W(E,J)$ denotes the probability that a collision complex with given E and J dissociates into the state b and not into the state a. If this probability approaches unity [i.e., $W_b(E,J) \gg W_a(E,J)$], one has a capture situation such that

$$\sigma_a^{cap} = \frac{\pi}{g_a k_a^2} \sum_{J=0}^{\infty} (2J+1) W(E,J,a). \qquad (2.5)$$

Experimentalists may be interested in state-resolved cross sections as a function of the translational energy E_t such as given by Eqs. (2.4) or (2.5), in orientational effects described by the corresponding relationships omitting the degeneracy averaging, or in E_t-averaged quantities such as thermal rate constants for selected initial quantum states a or completely unselected reactants. The latter, with initial and final state selection, follow from Eq. (2.4) as

$$k_{ba} = \frac{kT}{h} \left(\frac{h^2}{2\pi\mu kT} \right)^{3/2} \sum_{J=0}^{\infty} (2J+1) \int_0^{\infty} \frac{dE}{kT} \exp\left(-\frac{E}{kT} \right) \left[\frac{W(E,J,a)W(E,J,b)}{W(E,J)} \right]$$

$$(2.6)$$

or, in the case of capture, as

$$k_a^{cap} = \frac{kT}{h} \left(\frac{h^2}{2\pi\mu kT} \right)^{3/2} \sum_{J=0}^{\infty} (2J+1) \int_0^{\infty} \frac{dE}{kT} \exp\left(-\frac{E}{kT} \right) W(E,J,a). \qquad (2.7)$$

If there is no resolution of the states of the reactants 1 and 2, and the products, the additional averaging leads to

$$k = \frac{kT}{h}\left(\frac{h^2}{2\pi\mu kT}\right)^{3/2}\frac{1}{Q_{\text{vibrot},1}Q_{\text{vibrot},2}}\sum_{a,b}\sum_{J=0}^{\infty}(2J+1)$$

$$\times\int_0^{\infty}\frac{dE}{kT}\exp\left(-\frac{E}{kT}\right)\left[\frac{W(E,J,a)W(E,J,b)}{W(E,J)}\right] \quad (2.8)$$

or, in the case of capture, to

$$k^{\text{cap}} = \frac{kT}{h}\left(\frac{h^2}{2\pi\mu kT}\right)^{3/2}\frac{1}{Q_{\text{vibrot},1}Q_{\text{vibrot},2}}$$

$$\cdot\sum_{a}\sum_{J=0}^{\infty}(2J+1)\int_0^{\infty}\frac{dE}{kT}\exp\left(-\frac{E}{kT}\right)W(E,J,a). \quad (2.9)$$

Q_{vibrot} denotes rovibrational partition functions. Equations (2.7) and (2.8) only correspond to a single electronic potential surface. If the reactants are in degenerate or nearly degenerate electronic states, multiplication with the factor of $Q_{\text{el}}/Q_{\text{el},1}Q_{\text{el},2}$ has to be performed, where $Q_{\text{el},1}$ and $Q_{\text{el},2}$ are the electronic partition functions of the reactions and Q_{el} is that electronic partition function of the complex that is relevant for the reaction. It may well be that several of the electronic states of the reactants do not contribute to the considered reaction such that the factor $Q_{\text{el}}/Q_{\text{el},1}Q_{\text{el},2}$ falls below unity and, in the case of near degeneracy, depends on the temperature.

B. Unimolecular Processes

A consequence of the symmetry of the scattering matrix $|S_{fi}|^2 = |S_{if}|^2$ are the principles of microscopic reversibility $\sigma_{ba}g_a k_a^2 = \sigma_{ab}g_b k_b^2$ and of detailed balancing $k_{\text{forward}}/k_{\text{reverse}} = K_{eq}$, where K_{eq} denotes the equilibrium constant. Capture or association processes correspondingly are linked to dissociation processes such that numbers of closely coupled, or "open," channels $W(E,J,\ldots)$ are the same for both directions of the reaction and form the key quantities of a statistical theory. Specific rate constants $k(E,J,a)$ of species with given energy E and total angular momentum J, which dissociate unimolecularly into product states a, then are given by

$$k(E,J,a) = \frac{W(E,J,a)}{h\rho(E,J)}. \quad (2.10)$$

Equation (2.10) represents the basic formula of statistical unimolecular rate theory. By considering microscopic reversibility (or detailed balancing) one may show that Eq. (2.10) on all levels of averaging is consistent with the expressions for cross sections (or rate constants) of bimolecular processes that were given in Section II A. Combining Eq. (2.10) with an equilibrium population $f(E, J)$ of the reactant, that is,

$$f(E, J) = \frac{\rho(E, J) \exp(-E/kT)}{Q_{vibrot}} \tag{2.11}$$

the thermally averaged dissociation rate constant

$$k_{diss} = \sum_{J=0}^{\infty} (2J + 1) \int_0^{\infty} dE \, f(E, J) k(E, J) \tag{2.12}$$

follows as

$$k_{diss} = \frac{kT}{h} \frac{Q^*}{Q_{vibrot}} \tag{2.13}$$

with

$$Q^* = \sum_{J=0}^{\infty} (2J + 1) \int_0^{\infty} \frac{dE}{kT} W(E, J) \exp\left(-\frac{E}{kT}\right). \tag{2.14}$$

Verifying the consistency of Eqs. (2.10), (2.12)–(2.14) with the corresponding capture quantities, one reproduces the principle of detailed balancing

$$\frac{k_{diss}}{k^{cap}} = \frac{Q_{elvibrot,1} Q_{elvibrot,2}}{Q_{elvibrot}}, \tag{2.15}$$

where all partition functions have been calculated relative to the same arbitrary zero point of the energy scale.

It should be noted that the fully state-resolved expression (2.10) contains information about product state distributions. As long as Eq. (2.10) is valid, therefore, the fraction $P(E, J, i)$ of dissociation products in state b is given by

$$P(E, J, b) = \frac{k(E, J, b)}{\sum_i k(E, J, i)} = \frac{W(E, J, b)}{W(E, J)}. \tag{2.16}$$

It should also be noted that the ratio $W(E, J, b)/W(E, J)$ in Eq. (2.4) corresponds to the ratio $k(E, J, b)/k(E, J)$ and, hence, represents the probability that a reaction complex dissociates into the channel b.

The various statistical theories calculate the key quantity $W(E, J, i)$ on different levels of sophistication. In the most general case, for a single channel characterized by a complete set of quantum numbers, it should correspond to the transmission coefficient of a wave packet evolving in this channel. It, therefore, should include tunneling through and reflection above the channel energy barrier. Neglecting these effects, it would be set equal to zero at energies below the channel threshold energy E_{0i}, and equal to unity at energies above E_{0i}. The "activated complex partition function" Q^* of Eq. (2.14) then can also be rewritten as

$$Q^* = \sum_i \exp(-E_{0i}/kT), \tag{2.17}$$

where the summation is extended over all channels. Equation (2.17) may be evaluated for fixed activated complexes as a "true" partition function or, for channels with maxima E_{0i} located at varying positions as a "pseudo"-partition function.

The various statistical theories at this stage very easily can be classified: PST represents E_{0i} by the sum of centrifugal barriers and rovibrational energy levels of separated dissociation fragments in a unimolecular reaction or separated reactants in a bimolecular reaction. An anisotropy of the potential is neglected. Q^* then can be factorized into

$$Q^* = Q_{\text{vibrot},1} Q_{\text{vibrot},2} Q^*_{\text{cent}} \tag{2.18}$$

with the centrifugal pseudo-partition function

$$Q^*_{\text{cent}} = \sum_{J=0}^{\infty} (2J + 1) \exp\left(-\frac{E_0(J)}{kT}\right), \tag{2.19}$$

where $E_0(J)$ denotes the centrifugal barriers of the lowest reaction channels. Rigid RRKM theory assumes activated complexes that are localized at a fixed activated complex. In this approach angular momentum conservation would be difficult to obey rigorously if simple bond fissions are considered. RRKM theory with flexible transition states avoids this artifact by implementing more reasonable Hamiltonians and using variational transition state theory, that is, applying the old maximum free-energy criterion. Variational TST calculates Q^* at fixed r and determines its minimum value along the reaction coordinate r. Often simplified eigenvalues are employed in this calaculation. However, one could base a variational TST calculation also on accurate adiabatic channel potential curves $V_i(r)$ and calculate Q^* by

$$Q^* = \min_r \sum_i \exp\left(-\frac{V_i(r)}{kT}\right) \tag{2.20}$$

such as proposed by Quack and Troe (1977). This would eliminate artifacts from variational TST which are due to oversimplified channel eigenvalue models. It is, however, also evident that variational TST and SACM are not identical. Variational TST may correspond to a larger degree of non-adiabaticity than included in SACM. The differences are particularly easy to elaborate for ion–dipole capture rates. They can amount to about $\pm 50\%$ (Markovic and Nordholm, 1989).

III. ADIABATIC CHANNEL EIGENVALUES AND CHANNEL THRESHOLD ENERGIES

According to Sections I and II, the basic problem of statistical theories is the determination of the channel threshold energies E_{0i} and numbers of open channels $W(E, J)$, that is, the counting of channels whose E_{0i} are below a given energy E. Neglecting tunneling and nonadiabatic behavior, these energies indicate when a channel is open $(E \geqslant E_{0i})$ or closed $(E \leqslant E_{0i})$. Transmission coefficients for wave-packet evolution, tunneling, or reflection (which corresponds classically to recrossing trajectories), complete the picture. In adiabatic determinations of the E_{0i} one chooses a "reaction coordinate." For dissociation–association processes, most logically this is the distance r between the centers of mass of the fragments of the dissociation (or the reactants of the association). The choice of reaction coordinate has to be made in a way that minimizes nonadiabatic effects.

At fixed r, the solution of the eigenvalue problem in general will be very time consuming. It requires knowledge of the potential energy surface, and it corresponds to a multidimensional problem. However, in selected cases the problem has been solved to a sufficient accuracy such that the characteristic features of an accurate SACM treatment can be demonstrated.

A. Rigid Activated Complexes

The eigenvalue problem for "rigid activated complexes," that is, potential energy surfaces with pronounced maxima in the forward and the reverse direction of the reaction, is common for many direct bimolecular reactions of the atom transfer type, isomerization, or complex elimination processes. It is of relevance for many ion–molecule reactions involving major rearrangements of the collision complex accompanied by the overcoming of internal energy barriers. A proper choice of the coordinate system and the reaction coordinate here is crucial in order to minimize vibrationally nonadiabatic effects. Curvilinear "natural" coordinates have long been investigated for this purpose (Hofacker, 1963; Marcus, 1964, 1965). There is a great lack of accurate *ab initio* calculations of the potential in barrier ranges of this type. Since the potential governs the eigenvalues problem in a crucial manner, a great deal of ad hoc estimates or parameter fitting is still required at this point. However, for selected cases the problem was solved such that a guideline is provided. Reaction path methods provide an optimum approach (Miller et al., 1980) to the problem. As an example for adiabatic channel eigenvalues in the case of a potential that has a very steep maximum in the forward and reverse direction, the formaldehyde system should be mentioned. *Ab initio* calculations of activated complex frequencies for the elimination process $H_2CO \rightarrow H_2 + CO$ here are available (Waite et al., 1983; Polik et al., 1990; Miller et al., 1990). These calculations show dramatic variations of the frequencies along the reaction coordinate. It would have been hopeless to estimate these frequencies without knowledge of the potential. Even having *ab initio* calculations of the frequencies in hand, some scaling of these frequencies, introducing of anharmonicity corrections, and fitting of the potential barrier was required in order to reproduce the very detailed experimental results of this system. For this reason, guessing of activated complex frequencies $v_i^{\ddagger}$, of activated complex rotational constants $A^{\ddagger}$ and $B^{\ddagger}$ (assuming a symmetrical top) and of the barrier height E_0 today cannot be avoided. Of course, one should limit the number of fit parameters to an absolute minimum and be aware of the ambiguity of the approach.

Placing the maxima of the adiabatic channel potentials on the top of the barrier of the potential energy surface, in this case channel threshold energies have to be estimated by

$$E_{0i} \approx E_0 + B^{\ddagger} J(J+1) + (A^{\ddagger} - B^{\ddagger})K^2 + \sum_{j=1}^{s-1} (n_j + 1/2)h v_j^{\ddagger}. \tag{3.1}$$

Equation (3.1) is used in all statistical theories, RRKM, SACM, or rigid TST, which all coincide for rigid activated complexes. There is no basic difference at this point between neutral or ionic reaction systems.

B. Isotropic Potentials

If there is no anisotropic contribution to the interfragment potential between two reactants 1 and 2, the adiabatic channel potential curves $V_i(r)$ and the corresponding channel threshold energies E_{0i} can very easily be determined. In this case, the eigenvalues of the reactants 1 and 2 do not change during their approach. Only the isotropic, "radial," often attractive potential $V_{rad}(r)$ and the centrifugal potential $E_{cent}(r)$ of the rotating reaction complex vary with r. In this case, the adiabatic channel potential curves are given by

$$V_i(r) \approx V_{rad}(r) + E_{cent}(r) + E_{vibrot,i1}(1) + E_{vibrot,i2}(2) \tag{3.2}$$

where $E_{vibrot,i1}(1)$ denotes the complete set of rovibrational states of the reactant 1 (analogous for 2). The position of the channel maxima now exclusively depends on $V_{rad}(r) + E_{cent}(r)$. In the simplest case of a dominant electrostatic attraction potential

$$V_{rad}(r) \approx -C_n/r^n \tag{3.3}$$

and a simple quasidiatomic centrifugal potential

$$E_{cent}(r) \approx l(l+1)\hbar^2/2\mu r^2 \tag{3.4}$$

with the orbital rotational quantum number l, the channel potential curves have the form

$$V_i(r) \approx -C_n/r^n + l(l+1)\hbar^2/2\mu r^2 + E_{vibrot,i1}(1) + E_{vibrot,2}(2). \tag{3.5}$$

The maxima of these potentials are located at

$$r_{0i} \approx \left(\frac{nC_n\mu}{l(l+1)\hbar^2}\right) \tag{3.6}$$

and have the values

$$E_{0i} \approx \left(\frac{l(l+1)\hbar^2}{2\mu}\right)\left(\frac{n-2}{n}\right)\left(\frac{l(l+1)\hbar^2}{\mu C_n}\right)^{2/(n-2)}$$
$$+ E_{vibrot,i1}(1) + E_{vibrot,i2}(2). \tag{3.7}$$

One should note that, in the case of an ion–dipole potential with $n = 2$, the maxima are located at $r_{0i} = \infty$ as long as

$$l(l+1) \leqslant 2\mu C_2/\hbar^2 \tag{3.8}$$

having the values

$$E_{0i} = E_{vibrot,i1}(1) + E_{vibrot,i2}(2). \tag{3.9}$$

Otherwise, they are located at $r_{oi} = 0$ with $E_{0i} = \infty$. Equations (3.4) and (3.5) and the corresponding results for more complex radial potentials $V_{rad}(r)$ form the basis of loose activated complex theory, PST, or the Gorin model. The Langevin model for ion-induced dipole capture, or the locked-dipole capture model are based on Eqs. (3.6)–(3.9) as well, see below.

C.   Cos θ Anisotropies for Atom + Linear Reactant Systems

In the following we assume that the anisotropy of the potential is strong enough to lift the degeneracy of the rotational states of the reactants 1 and 2, but not sufficiently strong to modify the vibrational states of 1 and 2. For simplicity, we then omit the eigenvalues of these vibrational states $E_{vib}(1) + E_{vib}(2)$ in most of the following equations. At first, we consider a quasiatomic reactant colliding with a linear species or, in the reverse direction, a linear species dissociating into a quasiatomic and a linear fragment. One may formulate the Hamiltonian either in a body-fixed or in a space-fixed frame. At first, we consider a potential energy surface with a simple $\cos\theta$ anisotropy (θ = angle between the axis of the linear reactant and the line connecting the centres of mass of the reactants). The potential then may be represented in the form

$$V(r,\theta) = V_{rad}(r) + \frac{V_0(r)}{2}[1 - \cos\theta]. \tag{3.10}$$

Here, $V_{rad}(r)$ again denotes the radial potential, that is, the potential minimum upon variation of θ, and $V_0(r)$ denotes the barrier height of the angular potential upon variation of θ. In the field of ion–molecule reactions, the simplest potential of the $\cos\theta$ anisotropy is that of the charge-dipole interaction. In this case, the radial potential is directly related to the angular barrier by $V_{rad}(r) = -V_0(r)/2$. For simplicity we consider this case in the following, but we emphasize that different relations between $V_{rad}(r)$ and $V_0(r)$ can also easily be accounted for.

The Hamiltonian of the system is written as

$$\hat{H} = \hat{H}_{kin} + \hat{H}_{rot} + \frac{\hat{L}^2}{2\mu r^2} + V(r,\theta), \tag{3.11}$$

with the radial kinetic energy $\hat{H}_{kin}$ set equal to zero in order to calculate adiabatic channel eigenvalues. $\hat{H}_{rot}$ corresponds to the rotational energy of the linear reactant (rotational constant B), $\hat{L}$ to the relative angular momentum. The solution of the eigenvalue problem for small values of the ratio $V_0(r)/B$ is obtained by second-order perturbation theory (second-order Stark effect). For large values of the ratio $V_0(r)/B$, anharmonic oscillator expansions can be used, the intermediate range is bridged by numerical matrix diagonalizations.

The appropriate correlation diagrams for the transition from free rotor through weakly perturbed rotor to anharmonic oscillator states have been derived by Nikitin and Troe (1990). In the range of perturbed rotor states, the good quantum numbers are the total angular momentum J, the linear rotor angular momentum j, and, if one neglects the Coriolis interaction, the projection Ω of j on the line connecting the centers of mass. $\hat{L}^2$ has also off-diagonal matrix elements in Ω, which connect different Ω by Coriolis coupling. This contribution has been evaluated numerically and, for special cases, analytically (Nikitin and Troe, 1990; Turulski and Niedzielski, 1990). It was found to be negligible as far as rate constants are concerned. However, it influences the distribution of Ω sublevels of j, for example, the orientation of diatomic fragments in triatomic dissociation systems (Dashevskaya et al., 1990).

In the present work we neglect Coriolis coupling, emphasizing, however, the possibility of its simple quantitative estimate. Without going into the finer details of accurate adiabatic channel calculations (Nikitin and Troe, 1990), we only retain the leading contributions. For sufficiently large l (with $l \approx J$), in the perturbed rotor range one obtains (for the charge-linear dipole case)

$$V_i^0(r) \approx Bj(j+1) + \frac{(J+1/2)^2\hbar^2}{2\mu r^2} - \frac{V_0^2(r)}{8B}F(j,\Omega), \qquad (3.12)$$

with

$$F(j,\Omega) = \frac{3\Omega^2 - j(j+1)}{j(j+1)(2j-1)(2j+3)} \qquad (3.13)$$

and $F(0,0) = 1/3$. The coeffcient of the next term of the series, a V_0^4 term, is also known analytically (Wollrab, 1967; Troe, 1989). In the anharmonic oscillator range, the channel potential curves approach

$$V_i^\infty(r) \approx \frac{J(J+1)\hbar^2}{2\mu r^2} + V_{rad}(r) + (2j - |\Omega| + 1)\sqrt{V_0(r)B}$$

$$+ (\Omega^2 + 2|\Omega|j - 2j^2 + |\Omega| - 2j - 2)B/4. \qquad (3.14)$$

The coefficient of the next term (a $V_0^{-1/2}$ term) again is known analytically (Wollrab, 1967; Troe, 1989). The intermediate range is conveniently bridged by numerical matrix inversion (Shirley, 1963). An analytical switching approximation has also been proposed (Troe et al., 1991) that indicates at which r one has to switch from the perturbed rotor expansions $V_i^0(r)$ to the anharmonic oscillator expansions $V_i^\infty(r)$. Replacing the r dependence by the reduced parameter $\lambda = V_0(r)/2B$, the expressions for $V_i^0(r)$ should be used at $\lambda \leqslant \lambda_x$, where the switching value λ_x of λ is given by

$$\lambda_x \approx 1.6j^2 + 1.8. \qquad (3.15)$$

For the charge-dipole potential [with $V_{rad}(r)/B = -\lambda$ and $V_0(r)/2B = \lambda$, omitting the centrifugal energy], for example, for the lowest channel $(j, \Omega) = (0, 0)$ one has the following analytical approximation:

$$V_i^0/B \approx -\lambda^2/6 + 11\,\lambda^4/1080 - 9.03 \times 10^{-4}\lambda^6, \tag{3.16}$$

$$V_i^\infty/B \approx -\lambda + \sqrt{2\lambda} - 1/2 - 1/(16\sqrt{2\lambda}) - 5.08 \times 10^{-2}/\lambda. \tag{3.17}$$

For the $(j, \Omega) = (1, 0)$ channel, one has

$$V_i^0/B \approx 2 + \lambda^2/10 - 73\,\lambda^4/7000 + 5.8 \times 10^{-4}\lambda^6, \tag{3.18}$$

$$V_i^\infty/B \approx -\lambda + 3\sqrt{2\lambda} - 3/2 - 9/(16\sqrt{2\lambda}) - 0.736/\lambda, \tag{3.19}$$

and, for the $(j, \Omega) = (1, \pm 1)$ channels, one has

$$V_i^0/B \approx 2 - \lambda^2/20 + 19\,\lambda^4/56000 - 4.64 \times 10^{-6}\lambda^6, \tag{3.20}$$

$$V_i^\infty/B \approx -\lambda + 2\sqrt{2\lambda} - 1/2 - 1/(16\sqrt{2\lambda}) + 0364/\lambda. \tag{3.21}$$

Equations (3.16)–(3.21) are accurate to better than about 1%. They can easily be transformed to situations where the radial potential V_{rad} differs from $V_{rad}/B = -\lambda$. Similar expressions for higher channels are also available (Troe et al., 1991).

We illustrate the adiabtic channel potential curves for the charge-linear dipole interaction (i.e., $V_{rad} = -V_0/2$) in Figs. 1–4 (neglecting the appearance of avoided crossings because of symmetry restrictions). Figure 1 shows the lowest channel curves in the absence of centrifugal energy. The channels with $|\Omega| = j$ are purely attractive, the channels with $\Omega = 0$ (and small $|\Omega|$ for larger j) have maxima at varying values of λ, that is, varying values of r. Obviously, each channel has its individual activated complex, placed at quite different distances $r^\ddagger$. Variational procedures in TST, therefore, can only try to identify an "average position" of $r^\ddagger$. Because of the wide range of $r^\ddagger$ values, variational TST will in general not lead to the same results as SACM. Depending on the sign of $F(j, \Omega)$ from Eq. (3.13), at large r (or small λ), the channels are either attractive or repulsive. With increasing j, the fraction of attractive channels decreases from 1 to about 1/3, see Fig. 2.

The presence of overall rotation of the system $(J \gg 1)$ adds the centrifugal energy $E_{cent}(r)$ to the channel potential curves $V_i(r)$. Figure 3 shows the result for the $(j, \Omega) = (0, 0)$ channel. Centrifugal barriers arise, which strongly modify the activated complex positions $r^\ddagger$ for the attractive channels, whereas $r^\ddagger$

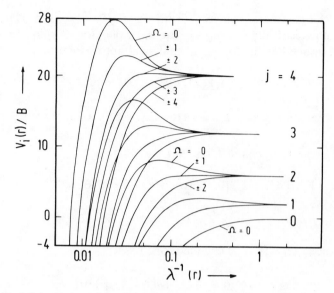

Figure 1. Adiabatic channel potential curves for charge-linear dipole interactions (orbital rotational quantum number $l = 0$, $\lambda = q\mu_D/Br^2$).

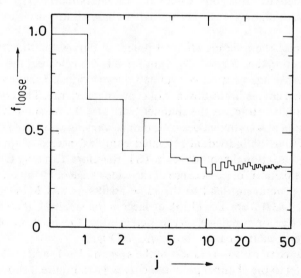

Figure 2. Fraction f_{loose} of attractive adiabatic channel potential curves for charge-linear dipole interactions ($l = 0$).

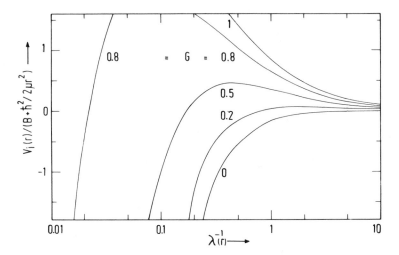

Figure 3. Dependence of the adiabatic channel potential curve for $(j, \Omega) = (0, 0)$ on orbital angular momentum $[G = l(l + 1)\hbar^2/2\mu q\mu_D]$.

varies much less for the repulsive channels. Whereas channel maxima in Figs. 1 and 3 are located in the range $\lambda < 100$, for repulsive channels with higher j values they move into the range $\lambda > 100$. Figure 4 shows channel potential curves with large j values, again calculated by matrix inversion.

The question arises where, with increasing λ at the channel maxima and decreasing $r^{\ddagger}$, the range of the electrostatic multiple potential is left. *Ab initio* calculations of potential energy surfaces show that the multipole long-range potentials dominate at r values larger than about 3–8Å (see, e.g., Troe, 1986, Lester, 1972; Kutzelnigg et al., 1973, Harding, 1989, 1990), whereas valence forces take over at small r. We, therefore, estimate the r values that correspond to a given λ value. As a typical example, we choose the $H_3^+ + HCN$ charge-dipole system characterized by a rotational constant of $B/hc = 1.48 \text{ cm}^{-1}$ and a dipole moment $\mu_D = 2.98$ D for HCN. In the charge-dipole case one has

$$\lambda = q\mu_D/Br^2 \tag{3.22}$$

($q = $ elementary charge), such that $r \approx 70$Å for $\lambda = 100$. At $\lambda = 10,000$ and $r = 7$ Å, one therefore leaves the range of applicability of the charge-dipole potential. Figures 1, 3, and 4 show when this is going to happen. Then, the actual radial potential $V_{rad}(r)$ and the anisotropy barrier $V_0(r)$ may be introduced into the relations for the adiabatic channel potential curves such

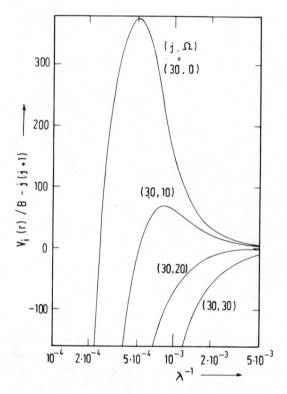

Figure 4. As Fig. 1, for $(j, \Omega) = (30, 0)$, $(30, 10)$, $(30, 20)$, and $(30, 30)$.

as before. Finally, the oscillator range of the complex with its BEBO-type channel pattern and different types of anisotropy of the potential is reached.

The appearance of a charge-induced dipole contribution, in addition to the charge-permanent dipole interaction considered before, to the radial potential is easily accounted for, since only $V_{rad}(r)$ but not $V_0(r)$ has to be modified. Figures 1, 3, and 4 then are simply changed by the addition of a j-independent charge-induced dipole term. Of course, the positions of the channel maxima will change. According to Eq. (3.22), for a given value of λ, smaller r values are obtained for molecules with smaller μ_D and larger B. For instance, HCl instead of HCN has $B/hc = 10.6\,\mathrm{cm}^{-1}$ and $\mu_D = 1.08D$. Figure 5 shows adiabatic channel potential curves which, already for $j = 5$, have channel maxima approaching the range of valence forces.

Figure 6 shows the threshold energies E_{0i} of the lowest attractive channels of the charge-linear dipole system as a function of the overall angular momentum; Fig. 7 gives the corresponding E_{0i} for repulsive channels. Some

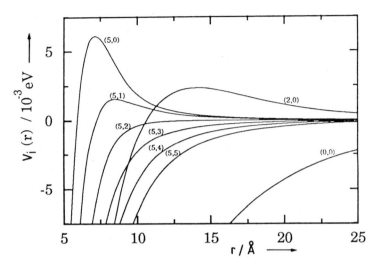

Figure 5. Adiabatic channel potential curves for charge-HCl interaction (from Markovic, 1989).

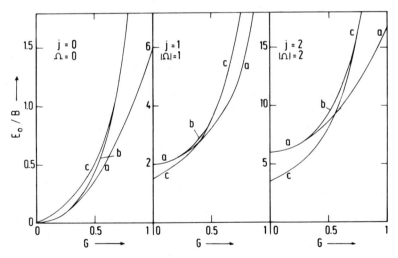

Figure 6. Adiabatic channel threshold energies for charge-linear dipole interactions (a = perturbed rotor limit, b = accurate, c = oscillator limit).

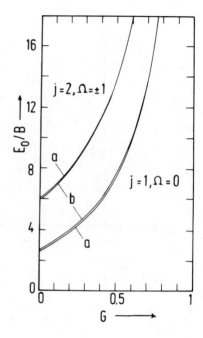

Figure 7. As Fig. 6, for repulsive channels (a = oscillator limit, b = accurate).

higher channels are illustrated in Figs. 8 and 9. The accurate values (curves b) are compared with perturbed rotor and anharmonic oscillator approximations (curves a and c), which can be derived analytically from Eqs. (3.12)–(3.14). One has (Troe, 1987)

$$E_{0i} \approx Bj(j+1) + \frac{BG^2}{2[F(j,\Omega) + \alpha B/\mu_D]^2} \tag{3.23}$$

in the perturbed rotor approximation (charge-linear permanent + induced dipole, polarizability α, dipole moment μ_D, reduced mass μ, $J \approx l \gg j$) v, where G is given by

$$G = J(J+1)\hbar^2/(2\mu g\mu_D) \tag{3.24}$$

and for the anharmonic oscillator approximation

$$E_{0i} \approx B\{(2j - |\Omega| + 1)^2/(2[1 - G]) + (\Omega^2 + 2|\Omega|j - 2j^2 + |\Omega| - 2j - 2)/4\}. \tag{3.25}$$

One notes that Eqs. (3.23)–3.25) are relevant for attractive channels, whereas only Eqs. (3.24) and (3.25) apply to repulsive channels. One realizes

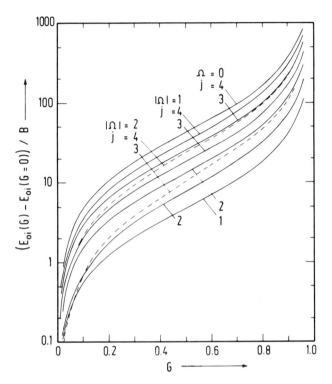

Figure 8.   As Fig. 6, for higher channels.

easily that perturbed rotor and anharmonic oscillator ranges both contribute. Rate constant calculations based purely on perturbation treatments (Turulski and Niedzielski, 1990), therefore, will fail at higher temperatures. One also notes that the channel threshold energies cannot be separated into the sum of a centrifugal barrier of the type of Eq. (3.7) and $a(j, \Omega)$-dependent term. Therefore, averaging over large ranges of j and Ω, such as required in rate constant calculations, can only be done by numerical techniques. The analytical channel eigenvalues approximations of Eqs. (3.12)–(3.21), however, simplify this procedure. The results agree with the accurate calculations to an excellent degree, because switching errors between perturbed rotor and anharmonic oscillator expressions are averaged out.

D.   Cos Θ Anisotropies for Atom + Symmetric Top Reactant Systems

If, instead of an atom + linear reactant system, an atom + symmetric top system is considered, the techniques of Section III.C can be extended. The

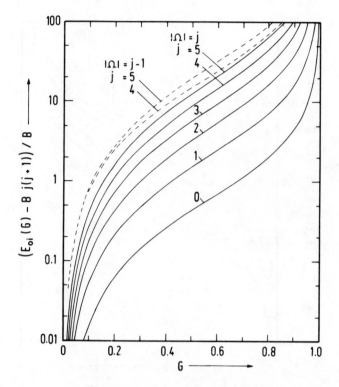

Figure 9. As Fig. 6, for higher channels.

perturbed rotor expansion of Eq. (3.12) now is modified by replacing $Bj(j+1)$ by the free symmetric top energy levels $Bj(j+1) + (A-B)k^2$ (with $k \leq j$). The coefficient $F(j, \Omega)$ of Eq. (3.13) is replaced by

$$F(j, \Omega, k) = \frac{[(j+1)^2 - k^2][(j+1)^2 - \Omega^2]}{(j+1)^3(2j+1)(2j+3)} - \frac{(j^2 - k^2)(j^2 - \Omega^2)}{j^3(2j-1)(2j+1)} \quad (3.26)$$

and a term $-V_0(r)k\Omega/2j(j+1)$ is added, whereas the anharmonic oscillator expansion of Eq. (3.14) takes the form

$$V_i^\infty(r) \approx \frac{J(J+1)\hbar^2}{2\mu r^2} + V_{rad}(r) + (2j - |\Omega + k| + 1)\sqrt{V_0(r)B}; \quad (3.27)$$

see Peter and Strandberg (1957), Shirley (1963), and Troe (1987). Tabulated eigenvalues up to $j = 4$ are available (Shirley, 1963). Possibly, the switching condition of Eq. (3.15) can be extended to this case as well. In any case, the limiting channel threshold energy expressions of Eqs. (3.23)–(3.25) can be directly transferred to this case by replacing $F(j, \Omega)$ in Eq. (2.23) and $|\Omega|$ in Eq. (3.25).

E. Cos$^2 \Theta$ Anisotropies for Atom + Linear Reactant Systems

Charge-quadrupole interactions are characterized by $\cos^2 \Theta$ anisotropies. Adiabatic channel potential curves for this case have been elaborated to much smaller extent (Troe, 1987). However, in principle, the necessary derivations are all available. The eigenvalue problem here corresponds to that of the spherical wave equation. Perturbed rotor and anharmonic oscillator eigenvalue expansions here are available in analytical form (Abramovitz and Stegun, 1965). Also complete tabulations for small j have been published. Numerical matrix inversions obviously can also be performed easily. We do not discuss this case in further detail in the following, partly because the approach would be quite analogous to Section III.C, but also because the detailed relationships have not yet been elaborated (for some results, see Troe (1987).

F. Nonadiabatic Effects

In order to estimate quantitatively the validity of statistical adiabatic channel calculations, an estimate of nonadiabatic "channel-hopping" probabilities is required. This can be obtained by the use of the well-developed general theory of nonadiabatic processes (see, e.g., Nikitin and Umanski, 1984). One may distinguish various types of nonadiabaticity: (1) Coriolis coupling mixing adjacent Ω channels; (2) nonadiabatic coupling at curve crossings or pseudocrossings (Landau–Zener nonadiabaticity); and (3) nonadiabatic transitions between distant channels (global nonadiabaticity). Because Coriolis mixing becomes important only when the channels are very close to each other at large r, channels barriers probably are not avoided by this effect (Dashevskaya et al., 1990). We, therefore, do not consider this type of nonadiabaticity in the following. Global nonadiabaticity sets in when the Massey parameter ξ falls below about 3, where

$$\xi = \Delta V_i (r/hv). \tag{3.28}$$

Here ΔV_i is the energy difference of the channels between which the transition is to occur, r is the corresponding distance between the reactants, and v is the relative velocity. Transitions in the range of the channel maxima are most relevant. Identifying V_i with $2B$ and r with 10–$100 \, \text{Å}$, one estimates that "global" nonadiabaticity in general sets in only at collision energies exceeding several $1000 \, \text{cm}^{-1}$ (see also Maergoiz et al., 1991). Landau–Zener "accidental" nonadiabaticity can be more important, such as elaborated in detail for dipole–dipole collisions (Maergoiz et al., 1991). However, Fig. 1 shows that the corresponding by-passing of the channel threshold energies nevertheless will be a relatively rare event in the lowest channels. We, therefore, assume that adiabatic treatments of ion–molecule capture rate constants in the thermal range are not markedly influenced by nonadiabatic effects such that SACM gives accurate results.

IV. CAPTURE RATE CONSTANTS AND CROSS-SECTIONS

Having discussed adiabatic channel potential curves $V_i(r)$ and the corresponding adiabatic channel threshold energies E_{0i}, one may directly proceed to the calculation of capture rate constants and cross sections. The general treatment of complex-forming bimolecular reactions also requires the determination of numbers of open channels, which will be presented in the next section. At first we consider purely isotropic potentials where the rate constants correspond to those of loose activated complex theory including Langevin and locked dipole expressions. Afterward we discuss "rigidity effects" due to anisotropies of the potential.

A. Isotropic Potentials

In the case of isotropic, "loose" potentials the reactant rovibrational energy levels $E_{\text{vibrot},i1}(1) + E_{\text{vibrot},i2}(2)$ are r independent additive terms in the adiabatic channel potential curves $V_i(r)$ and the threshold energies E_{0i}. Therefore, their contributions cancel in capture rate constant calculations, that is, $Q_{\text{vibrot},1} Q_{\text{vibrot},2}$ in Eq. (2.9) cancels against the corresponding sum and integral over $W(E, J, a) \exp(-E/kT)$ such that

$$k^{\text{cap}} = \frac{kT}{h} \left(\frac{h^2}{2\pi\mu kT} \right)^{3/2} Q^*_{\text{cent}}, \tag{4.1}$$

with the centrifugal partition function

$$Q^*_{\text{cent}} = \sum_{l=0}^{\infty} (2l + 1) \exp\left[-E_0(l)/kT \right]. \tag{4.2}$$

For a charge-permanent + induced dipole potential (omitting the anisotropy)

$$V(r, \theta) = -q\mu_D/r^2 - q\alpha/2r^4 \tag{4.3}$$

the centrifugal maxima $E_0(1)$ then are equal to zero for

$$l(l+1) \leqslant 2\mu q\mu_D/\hbar^2 \tag{4.4}$$

and equal to

$$E_0(l) = \frac{[l(l+1)\hbar^2/2\mu - q\mu_D]^2}{2\alpha q^2} \tag{4.5}$$

otherwise. Because large ranges of l are relevant under all practical conditions, the sum in Eq. (4.2) can be replaced by an integral such that the result

$$Q^*_{\text{cent}} = (2\mu q/h)\sqrt{\pi\alpha kT/2} + 2\mu q\mu_D/\hbar^2 \tag{4.6}$$

is obtained. Combining this with Eq. (4.1) leads to

$$k^{\text{cap}} = 2\pi q\sqrt{\alpha/\mu} + 2\pi q\mu_D\sqrt{2/\pi\mu kT}. \tag{4.7}$$

The first term is identical with the Langevin capture rate constant, the second with the "locked-dipole" rate constant. Equation (4.7) and its derivation, thus, demonstrate the identity between simple ion–molecule capture theories and loose transition state or phase-space theory (Troe, 1985).

Quite analogously, charge-locked quadrupole capture rate constants are calculated with the radial potential

$$V_{\text{rad}}(r) = -qQ/2r^3 \tag{4.8}$$

(Q = quadrupole moment). Centrifugal maxima now are equal to

$$E_0(l) = \frac{2qQ}{27}\left(\frac{l(l+1)\hbar^2}{\mu qQ}\right)^3. \tag{4.9}$$

The corresponding centrifugal partition function follows as

$$Q^*_{\text{cent}} = \int_0^\infty dx \exp\left(-\frac{2qQ}{27kT}\left(\frac{x\hbar^2}{\mu qQ}\right)^3\right)$$
$$\subset \Gamma(4/3)(27kT\mu^3 q^2 Q^2/2\hbar^6)^{1/3} \tag{4.10}$$

such that the charge-locked quadrupole capture rate constant is equal to

$$k^{\text{cap}} = \frac{kT}{h}\left(\frac{2\pi\hbar^2}{\mu kT}\right)^{3/2}\Gamma(4/3)\left(\frac{27kT\mu^3 q^2 Q^2}{2\hbar^6}\right)^{1/3}. \tag{4.11}$$

For radial potentials with sums of several multipole components, in general no analytical expressions for charge-locked multipole capture rate contants are obtained. However, their numerical treatment by analogy to the foregoing procedure is trivial.

The calculation of cross sections for locked-multipole capture is likewise simple. Because the contributions of different a states in Eq. (2.5) cancel, one has

$$\sigma^{cap}(E_t) = \frac{h^2}{8\alpha\mu E_t} \int_0^{l_{max}(E_t)} dl(2l+1), \qquad (4.12)$$

whre l_{max} is that value of l for which $E_t = E_0(l)$, that is, for which E_t is still sufficient to overcome the centrifugal barrier. For the locked-dipole case, this leads to

$$\sigma^{cap}(E_t) = \pi q\mu_D/E_t + \pi q\sqrt{2\alpha/E_t}, \qquad (4.13)$$

for the locked-quadrupole case, to

$$\sigma^{cap}(E_t) = \frac{3\pi}{2^{4/3}}\left(\frac{qQ}{E_t}\right)^{2/3}. \qquad (4.14)$$

B. Anisotropic Charge-Linear Dipole Potentials

The anisotropy of the potential introduces "rigidity" (in the language of TST) and leads to a reduction of the capture rate constants from Section IV B through an increase of the channel threshold energies E_{0i}. This effect can be represented by a "rigidity factor" f_{rigid} (smaller than unity) to be multiplied with the corresponding rate expression from phase-space theory (or loose TST), see Cobos and Troe (1985). Su and Bowers (1973) expressed this phenomenon by a dipole-locking constant C (smaller than unity), which was multiplied with the second term at the r.h.s. of Eq. (4.7). For $\alpha = 0$, f_{rigid} and C coincide. The average dipole orientation (ADO) method for calculating C (Su and Bowers, 1973, 1975; Su et al., 1978) led to capture rate contants that are somewhat lower than SACM or classical trajectory results (Su and Chesnavich, 1982; Troe, 1985). Today the ADO model, therefore, is superseded by the latter methods. There are also a series of other models with simplifying assumptions about the extent of locking, which differ from SACM or trajectory calculations and are, therefore, not considered here.

In the following we determine capture rate constants and cross-sections for the charge-linear dipole case and obtain rigidity factors or dipole-locking constants by comparison with the results of Section IV B. At first, we consider limiting low temperature results. For state-resolved capture rate constants, one has

$$k_n^{\text{cap}}(T \to 0) = \frac{kT}{h} \left(\frac{h^2}{2\pi\mu kT} \right)^{3/2} Q_n^* \tag{4.15}$$

with (at $J \approx l \gg 1$)

$$Q_n^* \approx \int_0^\infty dl(2l + 1) \exp\left(-\frac{E_{0n}(l) - E_{00n}}{kT} \right), \tag{4.16}$$

where E_{00n} denotes the energy of state n for separated reactants. For attractive channels, Eq. (3.23) applies such that, by integration over G,

$$k_{j,\Omega}^{\text{cap}}(T \to 0) = 2\pi q \sqrt{\alpha/\mu} + \mu_D^2 F(j,\Omega)/3\mu B \tag{4.17}$$

is obtained. For repulsive channels [i.e., $F(j,\Omega) < 0$], at $T \to 0$ the capture rate constants approach zero because of nonvanishing barrier heights for these channels (at $l = 0$).

The result of Eq. (4.17) is remarkable in the sense that it removes the divergence of the charge-locked dipole capture rate constant of Eq. (4.7) at $T \to 0$. This is due to the modification of the r^{-2} radial potential by the added adiabatic zero point energy of the anisotropy-induced hindered rotation of the reactants relative to each other. At the same time the corresponding rigidity factor $f_{\text{rigid}}(T \to 0)$ and dipole locking constant $C(T \to 0)$ go to zero.

Next, we consider limiting high-temperature results for the charge-linear permanent dipole case. With increasing collision energies and angular momenta, all channel threshold energies $E_{0i}(G)$ now approach the results given by the harmonic oscillator approximation of Eq. (3.25). In this case, the calculation of the overall capture rate constant

$$k^{\text{cap}}(kT \gg B) = \frac{kT}{h} \left(\frac{h^2}{2\pi\mu kT} \right)^{3/2} \frac{Q^*}{kT/B} \tag{4.18}$$

with an "activated complex" partition function

$$Q^* = \sum_{v,l} g_{v,l} \exp\left[-E_{0i}(v,l)/kT \right] \tag{4.19}$$

again is straightforward, leading to

$$Q^* \approx \frac{2\mu q \mu_D}{\hbar^2} \int_0^\infty v\,dv \int_0^1 dG \exp\left(-\frac{v^2 B}{2(1-G)kT} \right) \tag{4.20}$$

and, hence,

$$Q^* = \frac{2\mu q \mu_D}{\hbar^2}\left(\frac{kT}{2B}\right). \tag{4.21}$$

One should note that Eq. (4.20) contains the degeneracy factor $g_{v,l} = g_v \cdot g_l = v(2l + 1)$ because the bending oscillator of the linear complex is doubly degenerate. k^{cap} follows as

$$k^{cap}(kT \gg B) = \pi q \mu_D \sqrt{2/\pi\mu kT} \tag{4.22}$$

such that a high-temperature limiting rigidity factor

$$f_{rigid} = 1/2 \tag{4.23}$$

arises. Equation (4.23) also follows from more general considerations (Troe, 1989), which are not limited to the charge-dipole potential but to arbitrary potentials for which the anisotropy barrier $V_0(r)$ is proportional to the radial potential $V_{rad}(r)$. Not all of these cases can be treated analytically such as the charge-dipole problem, but the numerical solution is also straightforward. The charge-quadrupole case, for example, has to be treated numerically.

After considering limiting low- and high-temperature rate constants for charge-linear dipole capture, we now proceed to intermediate temperature conditions. Approximate analytical solutions were discussed by Troe (1987a). However, accurate numerical adiabatic channel calculations or numerical calculations, based on the sufficiently accurate approximate eigenvalue expressions of Section III.C, are also feasible relatively easily. We express the result in terms of the rigidity factor $f_{rigid} = k^{cap}(T)/k_{iso}^{cap}(T)$, where the capture rate constant for the isotropic potential k_{iso}^{cap} is given by Eq. (4.7). The numerical result is shown in Fig. 10. The given curve can be approximated empirically by

$$f_{rigid} \approx 0.5[1 - \exp(-(1 + 5\pi kT/3B)\sqrt{4\pi kT/3B})]. \tag{4.24}$$

Our results shown in Fig. 10 agree quantitatively with results by Markovic (1989).

It appears important to note a yet unexplained discrepancy between the SACM results of Fig. 10 and classical trajectory calculations. The analytical representation of the trajectory results by Su and Chesnavich (1982) gave

$$k^{cap} \approx 2\pi q \sqrt{\alpha/\mu}(0.62 + 0.3371 y\sqrt{B/kT}) \quad \text{for } kT/B \leqslant y^2/8,$$

$$k^{cap} \approx 2\pi q \sqrt{\alpha/\mu}(0.9754 + 0.19[y\sqrt{B/kT} + 0.7198]^2) \quad \text{for } kT/B > y^2/8,$$

$$\tag{4.25}$$

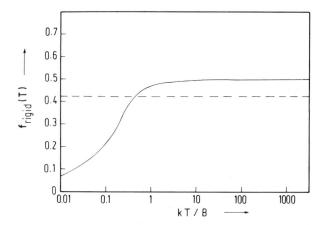

Figure 10. Rigidity factors of thermal charge-linear dipole capture rate constants [example $H_3^+ + HCN$, solid line: SACM, eq. (4.24), dashed line: classical trajectory results by Su and Chesnavich (1982), and Markovic and Nordholm (1989b)].

where $y = \mu_D/\sqrt{\alpha B}$. For $\alpha \to 0$, this corresponds to a temperature-independent rigidity factor $f_{\text{rigid}} = 0.422$. This result has been reproduced in classical trajectory calculations by Markovic and Nordholm (1989b), which seems to rule out numerical artifacts. Therefore, the discrepancy between the trajectory result and the high-temperature SACM limit $f_{\text{rigid}} = 0.5$ from Eq. (4.23) waits for an explanation. The low-temperature drop of f_{rigid} from SACM is clearly a quantum effect not accounted for by classical trajectory calculations.

In reality, at high temperatures the short-range components of the potential become increasingly important. Figure 11 compares SACM, classical trajectory, and ACCSA quantum-scattering calculations with experimental results for the $H_3^+ + HCN$ reaction. In the range of the SIFT measurements, the agreement between all approaches is better than 10%. The mentioned 20% discrepancy between SACM and classical trajectory calculations for the pure charge-permanent dipole case looses its importance because the charge-induced dipole contribution starts to dominate at these relatively high temperatures. Figure 11 also documents the failure of the old ADO approach. Although being not completely accurate in the charge-permanent + induced dipole case (neglected coupling), Eq. (4.7) with the second term multiplied by the rigidity factor (or dipole locking constant) from Eq. (4.24) today probably provides the most reliable (and quickest) prediction of charge-dipole capture rates.

The discrepancy between SACM and classical trajectory calculations at $kT < B$ is not unexpected because the latter method neglects quantum effects.

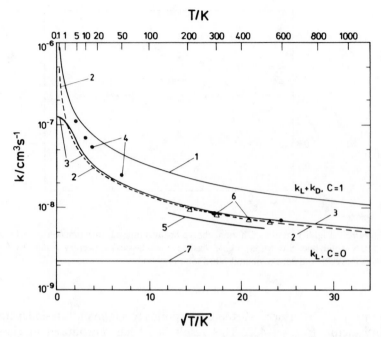

Figure 11. Thermal rate constants for the reaction $H_3^+ + HCN \rightarrow H_2 + H_2CN^+$ [1: phase space theory, $C =$ dipole locking constant $= 1$; 2: classical trajectory calculations, from Su and Chesnavich (1982); 3: SACM, from Troe (1987a); 4: ACCSA calculations by Clary (1984, 1985); 5: ADO results, from Su and Bowers (1973, 1975); 6: SIFT experiments from Clary et al. (1985); 7: Langevin rate constant, $C = 0$).

However, at the lowest temperatures one also notes a discrepancy between SACM and ACCSA quantum scattering calculations. This becomes even more pronounced if state-resolved rate constants are considered (Troe, 1987, Dubernet and McCarroll, 1989). Figure 12 compares adiabatic rotational state rate constants for $j = 0$ and $j = 1$, which are identical with SACM results, for the reaction $H_3^+ + HCl$ with ACCSA results. Since the ACCSA results even exceed the low-temperature SACM result of Eq. (4.17), one has to conclude that there is a low-temperature artifact of the ACCSA approximation. It appears also worth noting the shallow rate constant minimum near $5\,K$ of the $k^{\mathrm{cap}}(j = 1)$ curve. This is due (Troe, 1987a) to the maximum of the $(j, \Omega) = (1, 0)$ adiabatic channel potential curve whereas the $(1, \pm 1)$ curves are purely attractive.

We briefly also look on the effect of anisotropy on the capture cross sections elaborated in Eqs. (4.12)–(4.14) for isotropic potentials. Again, the

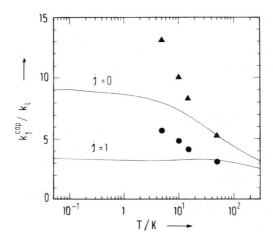

Figure 12. State-resolved thermal capture rate constants for charge + HCl interaction [solid lines: adiabatic rotational state calculations from Dubernet and McCarroll (1989) in agreement with SACM calculations; discrete points: ACCSA results from Clary (1984, 1985)].

anisotropy effects reducing the cross sections are largest at low collision energies where only the attractive channels contribute. Evaluating Eq. (4.12) with $l_{\max}(E_t)$ obtained by inversion of Eq. (3.23) gives

$$\sigma^{\text{cap}}(E_t \to 0) = \frac{\pi q \mu_D}{E_t} \sqrt{\frac{2E_t}{B}\left(\frac{\alpha B}{\mu_D^2} + F(j, \Omega)\right)}, \tag{4.26}$$

which, for the isotropic charge-induced dipole case, agrees with Eq. (4.13) and, for the charge-permanent dipole case, gives

$$\sigma^{\text{cap}}(E_t = 0) = \pi q \mu_D \sqrt{2F(j, \Omega)/(E_t B)}. \tag{4.27}$$

The E_t^{-1} divergence at $E_t \to 0$ is, thus, replaced by an $E_t^{-1/2}$ divergence. Cross sections for high temperatures can be elaborated in a similar fashion. An energy-independent rigidity factor $f_{\text{rigid}}(E_t) \approx 1/2$ arises. A comparison with the classical trajectory results by Bei et al. (1989) is also illustrative, confirming conclusions from the earlier discussion of rate constant results. The availability of reliable experimental ion–molecule reaction rate contants, in particular at low (see Rebrion et al., 1988) and very low (Mazely and Smith 1988; see also rigorous quantum-scattering calculations by Sakimoto, 1989) temperatures, allows for meaningful comparisons with theory. We do not give such a comparison here. However, we note some particularly typical aspects.

Often the experimental data fall below the theoretical capture calculations. This can be explained by redissociation without reaction of the collision complex (see Section V). An extreme example of this type can be encountered with open-shell systems where several different electronic states correlate with the separated reactants. It has been argued (Troe, 1988) for neutral reaction systems, such as $O + OH \rightarrow HO_2^* \rightarrow O_2 + H$, that only the lowest bonding of the potential energy surfaces lead to reaction and that higher surfaces do not contribute at all. Similar behavior might also be observed for reactions of open-shell ions (Clary et al., 1990). A quantitative estimate of the nature (attractive or repulsive at short ranges) of the electronic surfaces involved requires quantum-chemical calculations. For collisions of C^+ ions, 2/3 of the electronic states correlating with C^+ were calculated by Clary et al. (1990) to be attractive whereas for N^+ ions a ratio of 1/3 was obtained. More work in this field needs to be done.

C. Anisotropic Charge-Nonlinear Dipole Potentials

The treatment of charge-nonlinear dipole capture follows on the same lines as that given for the charge-linear dipole case. However, the calculations become more complicated due to the higher dimensionality of the eigenvalue problem involved. The charge + symmetric top dipole case has been elaborated in detail in the past, adiabatic channel potential curves were discussed in Section III D where analytical expressions were given for the limiting perturbed rotor and harmonic oscillator ranges.

In the following, at first, we consider the symmetric top case. With the channel potential curves of Section III D, the channel threshold energies can easily be derived. In the perturbed rotor limit, they are equal to

$$E_{0i} = Bj(j + 1) + (A - B)k^2 \tag{4.28}$$

as long as $k\Omega \geqslant 0$ (with $0 \leqslant k \leqslant j$) and

$$l(l + 1) \leqslant (2\mu q\mu_D/\hbar^2)k\Omega/j(j + 1). \tag{4.29}$$

For $k\Omega \geqslant 0$ and $l(l + 1)$ different from Eq. (4.29), they are

$$E_{0i} \approx Bj(j + 1) + (A - B)k^2 + \frac{B[G - k\Omega/j(j + 1)]^2}{2[F(j, \Omega, k) + \alpha B/\mu_D^2]}. \tag{4.30}$$

Where the harmonic oscillator limit is approached, that is, at large l or for repulsive channels ($k\Omega < 0$), one has

$$E_{0i} \approx B(2j - |\Omega + k| + 1)^2/2(1 - G). \tag{4.31}$$

Obviously, for $k = 0$, all results from the charge-linear dipole case are reproduced. For $k \neq 0$, different relationships are obtained. The repulsive channels $k\Omega < 0$ do not contribute to the capture rate constants in the low temperature limit. For the attractive channels with $k\Omega > 0$, where Eqs. (4.28) and (4.29) dominate, with these equations alone one would obtain

$$k_{j,\Omega;k}^{cap}(T \to 0) = 2\pi q\mu_D \sqrt{2/\pi\mu k T} \, k\Omega/j(j+1). \qquad (4.32)$$

One notes a divergence of k^{cap}, which, however, is removed in further thermal averaging. Since $k\Omega/j(j+1)$ also corresponds to the important top orientation parameter $\langle \cos^2 \theta \rangle$ (Bernstein and Levine, 1988), important state-specific orientation effects are predicted. Specific examples are given by Morgan and Bates (1987), Clary (1987b), Troe (1987a), and others. In high-temperature thermal averaging, the state-specific effects are completely washed out and the limiting high-temperature results of the charge-linear dipole case are approached.

State-specific effects become even more pronounced for the general charge-asymmetric top dipole case. A few sufficiently detailed treatments are now available such as the charge $+ H_2O$ calculations by Sakimoto (1982) and the $N^+ + H_2O$ calculations by Dubernet and McCarroll (1990). The latter work, in particular, discusses under which conditions the asymmetric top can be replaced by an average symmetric top. Figure 13 illustrates the

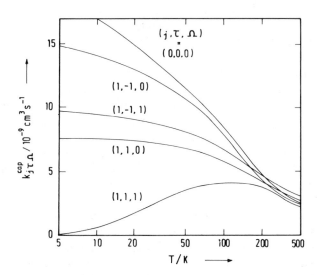

Figure 13. State-resolved thermal capture rate constants for the reaction $N^+ + H_2O$ (from Dubernet and McCarroll, 1990).

strong state-specificity at low temperatures. Because of the repulsive character of some channel potential curves, the corresponding rate constants become very small at $T \rightarrow 0$ K. Near to 300 K, the state specificity is nearly lost.

V. BIMOLECULAR ION–MOLECULE REACTIONS WITH REDISSOCIATING COLLISION COMPLEXES

If two reactants have formed a collision complex, which proceeds without redissociation toward the products, one has the capture situation discussed in Section IV. If there is redissociation, the reduction of the rate below the capture value is governed by the competition between unimolecular processes of the complex, forward-reaction and redissocation, such as discussed in Section II. A detailed description of this dynamics requires the (generally not available) knowledge of the complete potential energy surface. Therefore, in the following, we only can give illustrative examples for possible "scenarios." We first briefly discuss number of open channels $W(E, J)$ and the corresponding specific rate constants. Afterward, the reaction yields $W(E, J, b)/W(E, J)$ in Eqs. (2.4), (2.6), and (2.8) are considered and implemented in the overall rate parameters.

A. Specific Rate Constants for Rigid Activated Complexes

If there are pronounced barriers in the forward and reverse direction for rearrangement of the complex, the channel threshold energies of rigid activated complexes of Section III A are relevant and standard rigid activated complex theory applies. $W(E, J)$ is then determined by accurate state counting following the Beyer–Swinehart routine (Stein and Rabinovitch, 1973; Astholz et al., 1979). The J dependence also has to be accounted for (see Troe, 1984). The possibility of convoluting low-frequency, nearly classical modes with high-frequency harmonic oscillators (Astholz et al., 1979, see appendix of this reference) should particularly be mentioned. A counting procedure analogous to that for $W(E, J)$ leads to the density of states $\rho(E, J)$ and, hence, with Eq. (2.10), to the specific rate constants $k(E, J)$. Because the formalism is completely routine today, we do not go into the details, but we emphasize that the treatment remains as uncertain as the activated complex parameters in reality are.

One interesting, perhaps hypothetical, case deserves particular attention. One may imagine that a weakly bound complex is formed, which either redissociates or rearranges over a mild barrier involving only few low-frequency oscillators, whereas the majority of high-frequency skeleton oscillators of the complex are energetically practically not accessible. In this case, $W(E, J)$, omitting convolution with the high-frequency modes, is given by

$$W(E,J) \approx \sum_{K=-J}^{+J} W^{\ddagger}(E - E_{0b}(J) - (A^{\ddagger} - B^{\ddagger})K^2), \quad \text{for} \quad E - E_0(J) > (A^{\ddagger} - B^{\ddagger})K^2,$$

$$(5.1)$$

and

$$W(E,J) \approx \sum_{K=-K_{max}}^{+K_{max}} W^{\ddagger}(E - E_{0b}(J) - (A^{\ddagger} - B^{\ddagger})K^2) \qquad (52)$$

otherwise, with $E - E_{0b}(J) \approx (A^{\ddagger} - B^{\ddagger})K^2_{max}$, and with the activated complex parameters such as explained in Section III. A, and E_{0b} denoting the threshold energy of the process. Equation (5.1) and (5.2) may be relevant in relation to the corresponding expressions for a redissociation of the loose bond fission type.

B. Specific Rate Constants for Loose Activated Complexes

For simple bond fissions, in general, numbers of open channels for loose rotor-type coordinates have to be convoluted with those for rigid skeleton vibrations. The convolution procedure is as mentioned in the previous section (Astholz et al., 1979). We, therefore, only have to consider expressions for $W(E,J)$ for the loose coordinates with the various possible reactant combinations.

With an isotropic interaction potential, this problem has been solved by phase-space theory. We summarize the results in the following, as far as analytical results have been derived (Chesnavich and Bowers, 1978; Troe, 1983). For an atom + linear reactant pair, one has (partly in classical form)

$$W(E,J) \approx (2J + 1) \approx \sqrt{(E - E_{0a})/B} - J^2 \quad \text{for } 0 \leqslant J \leqslant \sqrt{(E - E_{0a})/B} \quad (5.3)$$

and

$$W(E,J) \approx (E - E_{0a})/B \quad \text{for } J \geqslant \sqrt{(E - E_{0a})/B}. \qquad (5.4)$$

For an atom + spherical top reactant pair, one has

$$W(E,J) \approx (2J + 1)(E - E_{0a})/B - 4J^3/3 \quad \text{for } 0 \leqslant J \leqslant \sqrt{(E - E_{0a})/B} \quad (5.5)$$

and

$$W(E,J) \approx 4[(E - E_{0a})/B]^{3/2}/3 \quad \text{for } J \geqslant \sqrt{(E - E_{0a})/B}. \qquad (5.6)$$

For a linear + linear reactant pair, only the limiting low-J and high-J expressions have been elaborated analytically. One has for small J

$$W(E,J) \approx (2J + 1)2(E - E_{0a})^{3/2}[\sqrt{B_1} + \sqrt{B_2} - \sqrt{B_1 + B_2}]/3B_1 B_2 \quad (5.7)$$

and

$$W(E,J) \approx (E - E_{0a})^2/2B_1B_2 \quad \text{for large } J. \tag{5.8}$$

A plausible extension of Eqs. (5.3)–(5.6) would be the use of Eq. (5.8) at $J \geqslant \sqrt{(E - E_{0a})}/\sqrt{B_1 B_2}$; and the use of Eq. (5.7) for smaller J, subtracting a suitable term from the r.h.s. such that Eq. (5.8) is met at $J \approx \sqrt{(E - E_{0a})}/\sqrt{B_1 B_2}$.

Likewise, for linear (B_1) + spherical top (B_2) reactants, one has for small J

$$W(E,J) \approx \frac{(E - E_{0a})^2}{2B_1^{1/2}B_3^{3/2}} \arcsin \sqrt{\frac{B_2}{B_1 + B_2}} \tag{5.9}$$

and

$$W(E,J) = 8(E - E_{0a})^{5/2}/15B_1B_2^{3/2} \tag{5.10}$$

for large J. Again, the transition between the low-J and high-J expressions can be approximated like for the linear + linear case and in analogy to Eqs. (5.3) and (5.4). The corresponding expressions for spherical top + spherical top reactants are, for small J,

$$W(E,J) \approx (2J + 1)8(E - E_{0a})^{5/2}/[15B_1B_2\sqrt{B_1 + B_2}] \tag{5.11}$$

and, for large J,

$$W(E,J) \approx (E - E_{0a})^3/6B_1^{3/2}B_2^{3/2}. \tag{5.12}$$

The anisotropy of the potential leads to a reduction of the $W(E,J)$ below the PST expressions of Eqs. (5.3)–(5.12). This effect has not yet been elaborated in detail by rigorous SACM calculations. However, the trends of these rigidity effects can already be recognized. At energies close to the threshold energy, due to the large fraction of loose adiabatic channel potential curves such as illustrated in Fig. 2 for the charge + linear reactant case, the $W(E,J)$ curves follow the PST results of Eqs. (5.3)–(5.12). With increasing energy, however, they fall below this result such as observed experimentally in some cases and interpreted in the present way (Troe, 1988, 1989). If the potential energy surface with decreasing r would keep its long-range character, $W(E,J)$ would be reduced by an energy-independent rigidity factor $f_{\text{rigid}}(E,J)$. However, a short-range increase of the anisotropy could result in an increasing drop below the PST result at increasing energy. Since this effect needs a more careful elaboration, we do not go into further details here.

We repeat that the numbers of open channels from Eqs. (5.3)–(5.12), with the mentioned rigidity corrections, in general have to be convoluted with harmonic oscillator contributions corresponding to the vibrations of the separated reactants.

C. Reaction Yields for Vibrational and Rotational Channel Switching

In the following we consider some examples of potential energy surfaces that can lead to strong deviations of the ion–molecule reaction rate constants from the capture rates discussed in Section IV. For instance, we consider the loose entrance/rigid exit potential suggested to apply to the $CH_4 + O_2^+ \rightarrow H + CH_2O_2H^+$ reaction (Barlow et al., 1986); see Fig. 14. $CH_4 + O_2^+$ capture is characterized by $W(E, J, a)$, reorganization of the $CH_4 \cdot O_2^+$ complex to $CH_3O_2H^+$ is characterized by $W(E, J, b)$. The value of the energy gap $\Delta E_g = E_{0b} - E_{0a}$ here is of crucial importance. It sensitively enters the yields ratio $W(E, J, b)/W(E, J)$ and thus governs the experimental rate constant. Depending on the value of ΔE_g, one may have stronger or weaker negative temperature coefficients of the rate constant, or rate constant minima may appear. Figure 15 gives an example for this behavior.

There are essentially two mechanism that can be responsible for the mentioned effects such as the rate constant minima shown in Fig. 15: *vibrational channel switching* and *rotational channel switching* (Troe, 1987b). In rotational channel switching, the threshold energies E_0 for the entrance $[E_{0a}(J)]$, and the exit $[E_{0b}(J)]$ channels may show different J dependences. Quite generally, a rigid activated complex will have a stronger J dependence

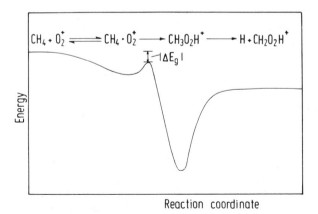

Figure 14. Schematic representation of the potential diagram for the reaction $CH_4 + O_2^+ \rightarrow H + CH_2O_2H^+$.

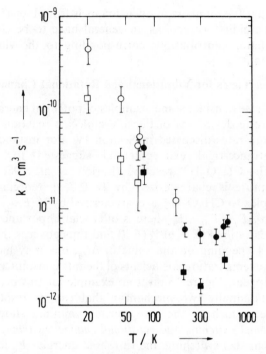

Figure 15. Experimental rate constants for the reaction $CH_4 + O_2^+ \rightarrow H + CH_2O_2H^+$ (upper points) and $CD_4 + O_2 \rightarrow D + CD_2O_2D^+$ (lower points) from Barlow et al. (1986).

of $E_0(J)$ than a loose activated complex. Figure 16 gives an example for the $CH_4 + O_2^+$ reaction, assuming $E_{0a}(J)$ to be dominated by an isotropic charge-induced dipole interaction

$$E_{0a}(J) \approx \hbar^4 [J(J+1)]^2 / (8\mu^2 \alpha q^2) \qquad (5.13)$$

and assuming $E_{0b}(J)$ to be given by

$$E_{0b}(J) \approx \Delta E_g + C_b J(J+1) \qquad (5.14)$$

where C_b corresponds to the rotational constant of a rigid activated complex, which is probably close to that of the $CH_4O_2^+$ collision complex. At a certain J value depending on the magnitude of ΔE_g, there is a "rotational channel switching that is, $E_{0b}(J)$ exceeds $E_{0a}(J)$, although $E_{0b}(J=0) < E_{0a}(J=0)$. As a consequence, there is an increasing amount of redissociation of the complex with increasing J at increasing temperatures. For an illustration of this effect,

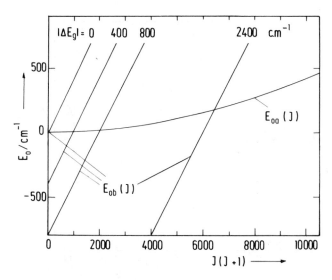

Figure 16. Rotation-dependent loose entrance and rigid exit channel threshold energies for pure rotational channel switching [$\Delta E_g = E_{0b} - E_{0a}$, from Troe (1987b), see text].

we put $W(E, J, b)/W(E, J)$ equal to unity at $E \geqslant E_{0b}(J)$ and equal to zero at $E < E_{0b}(J)$. The corresponding thermal rate constant k can be evaluated analytically (Troe, 1987b) with the results shown in Fig. 17. Owing to redissociation of the collision complex, there arises a change of k from the temperature independent Langevin capture rate constant $k_L = 2\pi q \sqrt{\alpha/\mu}$ to a k with a negative temperature coefficient. Rate constant minima are also observed at relatively low temperatures as long as $|\Delta E_g|$ is sufficiently small. In practice, vibrational effects will be superimposed on this "pure rotational channel switching" behavior; however, the effect of competing J dependences of $E_{0a}(J)$ and $E_{0b}(J)$ will always be there, such as illustrated by the examples of Figs. 16 and 17. A number of possible vibrational effects in rotational channel switching were considered by Troe (1987b).

In the following, we proceed to a discussion of pronounced "vibrational channel switching" situations, neglecting additional rotational effects. Vibrational channel switching can be the result of different energy dependences of specific rate constants for dissociation involving loose activated complexes and for rearrangements involving rigid activated complexes (see Sections V A and V B). For the loose activated complexes, with $W(E, J)$ contributions such as given by Eqs. (5.3)–(5.12), there will be often much larger values of $W(E, J)$ near to threshold than for the rigid activated complexes. This is due to the more rotational character of loose

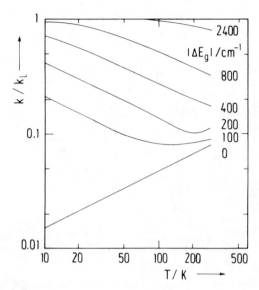

Figure 17. Thermal rate constants for rotational channel switching (example of Fig. 16, k_L = Langevin capture rate constant).

activated complexes compared to the vibrational character of rigid activated complexes. On the other hand, in the classical limit, rotor-type $W(E, J)$ expressions often have a weaker energy dependence than oscillator-type expressions. With $E_{0a}(J = 0) > E_{0b}(J = 0)$, this can lead to double crossings of the corresponding $k(E, J)$ curves. Figure 18 gives a qualitative illustration of this effect. Double crossings of $k(E, J)$ or $W(E, J)$ curves result in yield functions $Y(E, J) = W(E, J, b)/W(E, J)$ that have minima of the type shown qualitatively in Fig. 19. Convolution of such $Y(E, J)$ with $W(E, J, a)$ in thermal rate constants, see Eq. (2.8), then can result in increased negative temperature coefficients of k at low temperatures and in rate constant minima. The phenomenon obviously is complicated because the double crossings of $k(E, J)$ will be strongly influenced by J. In addition, rotational and vibrational channel switching effects may be superimposed. However, the two switching mechanisms often will operate in the same direction and enhance each other.

A quantitative treatment of specific reaction systems today is still not possible as long as detailed knowledge from potential surface calculations, about entrance and exit activated complexes, is not available. In this situation, simple calculations for hypothetical reaction systems appear illustrative. In the following we give such an example. We assume that, for spherical top + spherical top reactants at large J, Eq. (5.12) applies for $W(E, J, a)$, and

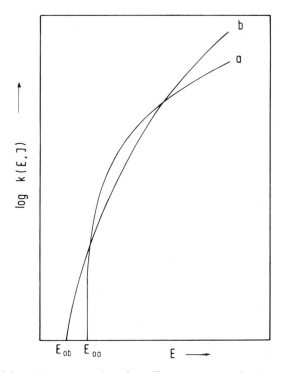

Figure 18. Schematic representation of specific rate constants for loose (*a*) and rigid (*b*) activated complex processes (see text).

that no convolution with reactant oscillators is required in the considered energy range. Likewise, we assume that the six rotor-type modes of relative motion of the two reactants are converted into five oscillator-type and one rotor-type (K rotor) mode at the rigid activated complex exit without that other high-frequency modes would have to be convoluted in. The considered case would correspond to a shallow energy minimum of the collision complex. Using classical expressions for numbers of open channels of rotor-type and low-frequency modes, then $Y(E, J) = W(E, J, b)/[W(E, J, a) + W(E, J, b)]$ can take the form

$$Y(E, J) \approx [1 + C(E - E_{0a}(J) + E_{za})^3/(E - E_{0b}(J) + E_{zb})^{5.5}]^{-1} \quad (5.15)$$

at $E > \max[E_{0a}(J), E_{0b}(J)]$ and $Y(E, J) = 0$ otherwise [E_{za} and E_{zb} denote "semiclassical" zeropoint energy corrections, see Troe (1983)]. Depending on the values of $\Delta E_g = E_{0a} - E_{0b}$, E_{za}, E_{zb}, and C, quite different $Y(E, J)$ can

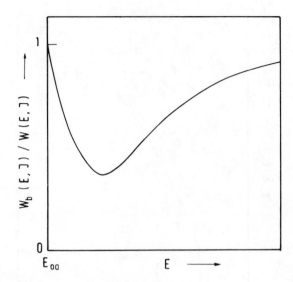

Figure 19. Reaction yield function for complex forming bimolecular reaction with loose entrance/rigid exit activated complexes (see text).

be simulated with different magnitudes of minima. However, in all cases a qualitative behavior similar to that of Fig. 19 is obtained.

If the yield function $Y(E, J)$ is of the form of that sketched in Fig. 19, it can be approximated by a function of the type

$$Y(E, J) \approx \exp(-E/\alpha_1) + [1 - \exp(-E/\alpha_2)], \tag{5.16}$$

with $\alpha_2 \gg \alpha_1$. The parameters α_1 and α_2 are fitted empirically to the result of complete $W(E, J)$ calculations ($E_{0a} = 0$). Neglecting a J dependence and assuming an energy dependence of $W(E, J, a) \propto E^s$ in Eq. (2.8), thermal averaging over $Y(E, J)$ leads to

$$\frac{k(T)}{k^{\mathrm{cap}}(T)} \approx \frac{\displaystyle\int_0^\infty E^s \exp(-E/kT)[\exp(-E/\alpha_1) + 1 - \exp(-E/\alpha_2)]\, dE}{\displaystyle\int_0^\infty E^s \exp(-E(kT))\, dE} \tag{5.17}$$

$$= \frac{\alpha_1\alpha_2 + 2\alpha_1 kT + (kT)^2}{\alpha_1\alpha_2 + (\alpha_1 + \alpha_2)kT + (kT)^2}. \tag{5.18}$$

The minimum of $Y(E, J)$, being given by

$$Y_{\min} \approx \alpha_1/\alpha_2 + [1 - (\alpha_1/\alpha_2)^{\alpha_1/\alpha_2}] \qquad (5.19)$$

at $E(\min Y) = \alpha_1 \ln \alpha_2/\alpha_1$, corresponds to a rate constant minimum of

$$(k/k_{\text{cap}})_{\min} \approx 2\sqrt{\alpha_1/\alpha_2} \qquad (5.20)$$

at $kT(\min k/k_{\text{cap}}) = \sqrt{\alpha_1/\alpha_2}$.

The model calculations of Eqs. (5.16)–(5.20) only correspond to simplified vibrational channel switching. In reality, rotational channel switching will be superimposed. Furthermore, SACM rigidity effects will reduce the PST expressions for $W(E, J, a)$ with increasing energy to an increasing extent. All of these effects result in an enhancement of negative temperature coefficients of k at low temperatures. For example, Eq. (5.18) for vibrational channel switching at most contributes with T^{-1} to k/k_{cap}, rotational channel switching can add another factor T^{-1} (see Fig. 17), energy-dependent rigidity effects may add about another factor of T^{-1}. Therefore, there may be negative T coefficients of up to about $k/k_{\text{cap}} \propto T^{-3}$. The experimental results of Fig. 15 are consistent with this order of magnitude.

It should be emphasized that the given discussion of redissociation effects has to remain of illustrative character. More quantitative treatments would require assistance from quantum-chemical calculations of activated complex structures. As long as these are not available, only qualitative interpretations of experimental data are possible. Before closing this chapter, it should be remembered that the discussed redissociation effects appear in thermal rate constants as well as in kinetic energy dependent reaction cross sections.

In summary, the present review has demonstrated the versatile possibilities of a statistical approach to ion–molecule reactions. The method is very powerful and its results are probably close to reality. However, quantitative predictions have to rely on the quality of the input data, that is, of the properties of the potential energy surfaces involved. It was not possible to include collision-associated processes such as termolecular ion–molecule association, processes involving electronically nonadiabatic transitions or radiative associations. It should be emphasized that these reactions also present considerable statistical aspects, which would deserve a separate analysis.

References

Abramovitz, M. and Stegun, I. A. (1965), *Handbook of Mathematical Functions*, Dover, New York.

Astholz, D. C., Troe, J. and Wieters, W. (1979). *J. Chem. Phys.* **70**, 5107–5116.

Barlow, S. E., Van Doren, J. M., De Puy, C. H., Bierbaum, V. M., Dotem, I., Ferguson, E. E., Adams, N. G., Smith, D., Rowe, B. R., Marquette, J. B., Dupeyrat, G. and Durup-Ferguson, M. (1986), *J. Chem. Phys.* **85**, 3851–3859.

Bates, D. R. (1982), *Proc. Roy. Soc. London, Ser. A.* **384**, 289–300.

Bates, D. R. (1983), *Chem. Phys. Lett.*, **97**, 19–22.

Bates, D. R. and Mendas, I. (1985), *Proc. Roy. Soc. London Ser. A* **402**, 245–255.

Bates, D. R. and Herbst, E. (1988), in *Rate Coefficients in Astrochemistry*, T. J. Millar and D. A. Williams, Eds., Kluwer Academic Publ., Dordrecht, pp. 17–40.

Bernstein, R. B. and Levine, R. A. (1988), *J. Phys. Chem.* **92**, 6954–6958.

Bei, H. C., Bhowmik, P. K., and Su, T. (1989), *J. Chem. Phys.* **90**, 7046–7049.

Blatt, J. and Weisskopf, V. (1952), *Theoretical Nuclear Physics*, Wiley, New York, Chap VIII.

Chesnavich, W. J. and Bowers, M. T. (1978), in *Gas Phase Ion Chemistry*, M. T. Bowers, Ed., Academic Press, New York, Vol. 1, pp. 119–151.

Chesnavich, W. J. and Bowers M. T. (1982), *Progr. in React. Kin.* **11**, 137–268.

Clary, D. C. (1984), *Mol. Phys.* **53**, 3–21.

Clary, D. C. (1985), *Mol. Phys.* **54**, 605–618.

Clary, D. C. (1987a), *J. Phys. Chem.* **91**, 1718–1727.

Clary, D. C. (1987b), *J. Chem. Soc. Far. Trans. 2*, **83**, 139–148.

Clary, D. C. (1988), in *Rate Coefficients in Astrochemistry*, T. J. Millar and D. A. Williams, Eds., Kluwer Academic Publ., Dordrecht, pp. 1–16.

Clary, D. C., Smith, D. and Adams, N. G. (1985), *Chem. Phys. Lett.* **119**, 320–326.

Clary, D. C., Dateo, C. E., and Smith, D. (1990), *Chem. Phys. Lett.* **167**, 1–6.

Cobos, C. J. and Troe, J. (1985), *J. Chem. Phys.* **83**, 1010–1015.

Dashevskaya, E. I., Nikitin, E. E., and Troe, J. (1990), *J. Chem. Phys.* **93**, 7803–7807.

Dubernet, M. L. and McCarroll, R. (1989), *Z. Phys. D* **13**, 255–258.

Dubernet, M. L. and McCarroll, R. (1991), *Z. Phys. D* (to be published).

Gaedtke, H. and Troe, J. (1973), *Ber. Bunsenges. Phys. Chem.* **77**, 24–29.

Harding, L. B. (1989), *J. Phys. Chem.* **93**, 8004–8013.

Harding, L. B. (1991), *J. Phys. Chem.* (to be published).

Henchman, M. and Paulson, J. F. (1988), *Radiat. Phys. Chem.* **32**, 417–423.

Henchman, M. and Paulson J. F. (1989), *Far. Trans. Chem. Soc. 2* **85**, 1673–1684.

Hofacker, L. (1963), *Z. Naturforsch.* **18a**, 607–619.

Jungen, M. and Troe, J. (1970), *Ber. Bunsenges. Phys. Chem.* **74**, 276–282.

Lester, W. A. (1971), *J. Chem. Phys.* **54**, 3171–3179.

Lifshitz, C. (1987), *Int. Rev. Phys. Chem.* **6**, 35–51.

Kutzelnigg, W. Staemmler, V., and Hoheisel, K. (1973), *Chem. Phys.* **1**, 27–44.

Maergoiz, A., Nikitin, E. E., and Troe, J. (1991), *J. Chem. Phys.* (to be published).

Markovic, N. (1989), Ph.D. thesis, University of Göteborg.

Markovic, N. and Nordholm, S. (1989a), *J. Chem. Phys.* **91**, 6813–6821.

Markovic, N. (1989b), *Chem. Phys.*, **135**, 109–122.

Marcus, R. A. (1964), *J. Chem. Phys.* **41**, 2614–2623.

Marcus, R. A. (1965), *J. Chem. Phys.* **43**, 1598–1605.

Mazeley, T. L. and Smith, M. A. (1988), *Chem. Phys. Lett.* **144**, 563–569.

Miller, W. H., Handy, N. C., and Adams, J. E. (1980), *J. Chem. Phys.* **72**, 99–112.

Miller, W. H., Hernandez, R., Moore, C. B., and Polik, W. F. (1990), *J. Chem. Phys.* **93**, 5657–5666.

Morgan, W. L. and Bates, D. R. (1987), *Astrophys. J.* **314**, 817–821.

Nikitin, E. E. and Umanski, S. Ya. (1984), *Theory of Slow Atomic Collisions*, Springer Verlag, Heidelberg.

Nikitin, E. E. and Troe, J. (1990), *J. Chem. Phys.* **92**, 6594–6598.

Peter, M. and Strandberg, M. W. P. (1957), *J. Chem. Phys.* **26**, 1657–1659.

Polik, W. F., Guyer, D. R., and Moore, C. B. (1990), *J. Chem. Phys.* **92**, 3453–3470.

Quack, M. (1990), *22nd Inf. Conf. on Photochem.*, Ann Arbor (to be published).

Quack, M. and Troe, J. (1974), *Ber. Bunsenges. Phys. Chem.* **78**, 240–252.

Quack, and Troe, J. (1975a), *Ber. Bunsenges. Phys. Chem.* **79**, 170–183.

Quack, M. and Troe, J (1975b), *Ber. Bunsenges. Phys. Chem.* **79**, 469–475.

Quack, M. and Troe, J. (1977), *Ber. Bunsenges. Phys. Chem.* **81**, 329–337.

Quack, M. and Troe, J. (1981), in *Theoretical Chemistry*, D. Henderson, Ed., Academic Press, London and New York, Vol. VIb, pp. 199–276.

Rebrion, C., Marquette, J. B., Rowe, B. R., and Clary, D. C. (1988), *Chem. Phys. Lett.* **143**, 130–134.

Sakimoto, J. (1980), *J. Phys. Soc. Japan* **48**, 1683–1690.

Sakimoto, K. (1982), *Chem. Phys.* **68**, 155–170.

Sakimoto, K. (1984), *Chem. Phys.* **85**, 273–278.

Sakimoto, K. (1985), *Chem. Phys. Lett.* **116**, 86–88.

Sakimoto, K. (1989), *Chem. Phys. Lett.* **164**, 294–298.

Sakimoto, K. and Takayanagi, K. (1980), *J. Phys. Soc. Japan* **48**, 2076–2083.

Shirley, J. H. (1963), *J. Chem. Phys.* **38**, 2896–2913.

Slater, N. B. (1959), *Theory of Unimolecular Reactions*, Methuen, London, Chapters 6, 9, and 10.

Stein, B. S. and Rabinovitch, B. S. (1973), *J. Chem. Phys.* **58**, 2438–2445.

Su, T. and Bowers, M. T. (1973), *J. Chem. Phys.* **58**, 3027; *Int. J. Mass Spectron Ion Phys.* **12**, 347–356.

Su, T. and Bowers, M. T. (1975), *Int. J. Mass Spectron. Ion Phys.* **17**, 211–212.

Su, T. and Chesnavich, W. J. (1982), *J. Chem. Phys.* **76**, 5183–5185.

Su, T., Su, E. C. F., and Bowers, M. T. (1978), *J. Chem. Phys.* **69**, 2243–2250.

Takayanagi, K. (1978), *J. Phys. Soc. Japan* **45**, 976–985.

Takayanagi, K. (1982), in *Physics of Electronic and Atomic Collisions*, S. Datz, Ed., North Holland, Amsterdam, p. 343.

Troe, J. (1983), *J. Chem. Phys.* **79**, 6017–6029.

Troe, J. (1984), *J. Chem. Phys.* **88**, 4375–4380.

Troe, J. (1985), *Chem. Phys. Lett.* **122**, 425–430.

Troe, J. (1986), *J. Phys. Chem.* **90**, 3485–3492.

Troe, J. (1987a), *J. Chem. Phys.* **87**, 2773–2780.

Troe, J. (1987b), *Int. J. Mass Spectr. and Ion Proc.* **80**, 17–30.

Troe, J. (1988), *Ber. Bunsenges. Phys. Chem.* **92**, 242–252.

Troe, J. (1989), *Z. Phys. Chem. NF* **161**, 209–232.

Troe, A., Troe, J., and Weiss, C. (1991), *Z. Phys. Chem. NF* (to be published).

Turulski, J. and Niedzielski, J. (1990), *J. Chem. Soc. Faraday Trans.* **86**, 1–3.

Waite, B. A., Gray, S. K., and Miller, W. H. (1983), *J. Chem. Phys.* **78**, 259–265.

Wollrab, J. E. (1967), *Rotational Spectra and Molecular Structure*, Academic Press, New York, Appendix 13.

AUTHOR INDEX

Numbers in parentheses are reference numbers and indicate that the author's work is referred to although his name is not mentioned in the text. Numbers in *italic* show the pages on which the complete references are listed.

SUBJECT INDEX